Valuing Mediterranean Forests

Towards Total Economic Value

————————————————

Valuing Mediterranean Forests

Towards Total Economic Value

_____

Edited by

Maurizio Merlo

and

Lelia Croitoru

CABI Publishing

CABI Publishing is a division of CAB International

CABI Publishing
CAB International
Wallingford
Oxfordshire OX10 8DE
UK

CABI Publishing
875 Massachusetts Avenue
7th Floor
Cambridge, MA 02139
USA

Tel: +44 (0) 1491 832111
Fax: +44 (0)1491 833508
E-mail: cabi@cabi.org
Web site: www.cabi-publishing.org

Tel: +1 617 395 4056
Fax: +1 617 354 6875
E-mail: cabi-nao@cabi.org

A catalogue record for this book is available from the British Library, London, UK.

Library of Congress Cataloging-in-Publication Data

Valuing Mediterranean forests : towards total economic value / edited by Maurizio Merlo and Lelia Croitoru.
       p.   cm.
    Includes bibliographical references and index.
    ISBN 0-85199-997-2 (alk. paper)
    1. Forests and forestry--Economic aspects--Mediterranean Region.   2. Forest policy--Mediterranean Region.   I. Merlo, Maurizio.   II. Croitoru, Lelia.   III. Title.

    SD217.M44V25 2005
    634.9′09182′2--dc22

                                                        2004021953

ISBN 0 85199 997 2

The boundaries, denominations and any other information of all maps in this book do not imply on the part of the publisher and editors any judgement on the legal status of the territory or any endorsement or acceptance of such boundaries.

Typeset by AMA DataSet Ltd, UK.
Printed and bound in the UK by Biddles Ltd, King's Lynn.

Contents

Contributors

Ameur Ben Mansoura is Rangeland Management Expert, Arab Center for Studies of Arid Zones and Drylands (ACSAD), Damascus, Syria.

Yves Birot is Research Director, 5 allée Canto Cigalo, 30400 Villeneuve-Lez-Avignon, France. E-mail: yves.birot@free.fr

Pablo Campos is Senior Researcher, Instituto de Economià y Geografía (CIFOR), Pinar 25, 28006 Madrid, Spain. E-mail: pcampos@ieg.csic.es

Alejandro Caparrós is Senior Researcher, Instituto de Economía y Geografía, Consejo Superior de Investigaciones Científicas, Pinar 25, 28006 Madrid, Spain. E-mail: acaparros@ieg.csic.es

Odile Colnard is Researcher at Laboratoire d'Economie Forestière (UMR ENGREF/INRA), 14 Rue Girardet, CS 4216, F-54042 Nancy Cedex, France.

Lelia Croitoru is Researcher at University of Padova, Centre for Environmental Accounting and Management in Agriculture and Forestry (CONTAGRAF), Via Roma 34, Corte Benedettina, 35020 Legnaro (PD), Italy. E-mail: lelia@contagra.unipd.it

Hamed Daly-Hassen is Researcher in Forest Economics at the National Institute of Research on Rural Engineering, Water and Forestry (INRGREF), B.P. 10, Ariana – 2080, Tunisia. E-mail: Dalyhassen.hamed@iresa.agrinet.tn

Kostandin Dano is Chief of Resources Sector, DGFP – General Directorate of Forests and Pastures, Silviculture and Management of Forest Resources Sector, Rue Sami Frasheri Godina no. 4, Tirana, Albania. E-mail: Danoko02@yahoo.com

Department of Forests, Ministry of Agriculture, Natural Resources and Environment, Louki Akrita 26, PC 1414 Nicosia, Cyprus. Contact: Xenos Hadjikyriacou, E-mail: xenosh@yahoo.com

Mohammed Ellatifi is Senior Forestry Officer at Department of Waters, Forests and Desertification Control, PO Box 50070, 20070 Casablanca Ghandi, Morocco. E-mail: m.ellatifi@altbox.org and m.ellatifi@softhome.net

Avi Gafni is Research Coordinator, Forest Division, Land Development Authority, Jewish National Fund (JNF), Mobile Post Shimshon, KKL Eshtaol, Israel 99775. E-mail: AviGa@kkl.org.il

Paola Gatto is Lecturer of Forest Economics at University of Padova, Department of Land and Agro-Forestry Systems, Agripolis, Via Roma, 35020 Legnaro (PD), Italy. E-mail: paola.gatto@unipd.it

Roubina Ghattas is Research Associate at the Applied Research Institute Jerusalem (ARIJ), P.O.Box 860, Caritas St., Bethlehem, West Bank. E-mail: roubina@arij.org

Joso Gra an is DSc Scientific advisor at Forest Research Institute, Jastrebarsko, Cvjetno naselje 41, 10450 Jastrebarsko, Croatia. E-mail: josog@sumins.hr

Nader Hrimat is Assistant Director General at the Applied Research Institute Jerusalem (ARIJ), PO Box 860, Caritas St, Bethlehem, West Bank. E-mail: nader@arij.org

Jad Isaac is Director of the Applied Research Institute Jerusalem (ARIJ), PO Box 860, Caritas St, Bethlehem, West Bank. E-mail: jad@arij.org

Nader Kabbani is Assistant Professor at the Department of Economics, American University of Beirut, PO Box 11-0236, Beirut, Lebanon. E-mail: nader.kabbani@aub.edu.lb

Angelos Kazaklis is Director, Centre for Integrated Environmental Management (CIEM), 39 Androutsou Str., 55132 Kalamaria, Thessaloniki, Greece.

Vassiliki Kazana is Senior Lecturer, Technological Education Institute of Kavala, Department of Forestry at Drama, 1st km Kalampaki-Drama, 66100 Drama, Greece. E-mail: vkazana@teikav.edu.gr and vkazana@spark.net.gr

Marko Kovac is Researcher at Slovenian Forestry Institute, Vecna pot 2, 1000 Ljubljana, Slovenia.

Lado Kutnar is Researcher at Slovenian Forestry Institute, Vecna pot 2, 1000 Ljubljana, Slovenia.

Robert Mavsar is Researcher at Slovenian Forestry Institute, Vecna pot 2, 1000 Ljubljana, Slovenia. E-mail: robert.mavsar@gozdis.si

Américo M.S. Carvalho Mendes is Professor at the Faculty of Economics and Management of the Portuguese Catholic University, Regional Centre of Porto, Rua Diogo Botelho, 1327, 4169-005 Porto, Portugal. E-mail: amendes@porto.ucp.pt

Maurizio Merlo was Professor of Forest Economics and Policy at The University of Padova and Director of the Centre for Environmental Accounting and Management in Agriculture and Forestry (CONTAGRAF), Via Roma 34, Corte Benedettina, 35020 Legnaro (PD), Italy (died August 2003).

Claire Montagné is Researcher at Laboratoire d'Economie Forestière (UMR ENGREF/INRA), 14 Rue Girardet, CS 4216, F-54042 Nancy Cedex, France. E-mail: montagne@nancy-engref.inra.fr

Ibrahim Nahal is Professor Emeritus at Aleppo University, Department of Forestry and Ecology, PO Box 5008, Aleppo, Syria. E-mail: nahalibr@scs-net.org

Abdellah Nédjahi is Director of National Institute of Forest Research (INRF), Arboretum de Baïnem B.P. 37, Chéraga, Alger, Algeria. E-mail: a_nedjahi@hotmail.com

Alexandra Niedzwiedz is Database Manager at Laboratoire d'Economie Forestière (UMR ENGREF/INRA), 14 Rue Girardet, CS 4216, F-54042 Nancy Cedex, France.

Andreas Ottitsch is Programme Manager at the European Forest Institute, Torikatu 34, 80100 Joensuu, Finland. E-mail: andreas.ottitsch@efi.fi

Atakan Öztürk is Research Assistant of Forestry at Kafkas University, Faculty of Forestry, Division of Forest Economics in the Department of Forest Engineering, Artvin, Turkey.

Paolo Paiero is Professor of Forest Botany, Department of Land and Agro-Forestry Systems, Agripolis, Via Roma, 35020 Legnaro (PD), Italy. E-mail: paolo.paiero@unipd.it

Mehmet Pak is Assistant Profesor of Forestry at Kahramanmaraş Sütçüimam University, Faculty of Forestry, Division of Forest Economics in the Department of Forest Engineering, Kahramanmaraş, Turkey.

Marc Palahi is Programme Manager of the MEDFOREX Project, Centre Tecnològic Forestal de Catalunya, Pujada del Seminari, s/n, Solsona, Spain. E-mail: marc.palahi@ctfc.es

Jean-Luc Peyron is Director of Laboratoire d'Economie Forestière (UMR ENGREF/INRA), 14 Rue Girardet, CS 4216, F-54042 Nancy Cedex, France.

Eduardo Rojas-Briales is Professor at Agriculture and Forest Faculty, Polytechnic University of Valencia, Campus de Vera s/n E-46022 Valencia, Spain. E-mail: edrobr@prv.upv.es

Rudolf Sabadi is Professor (retired) at Faculty of Forestry, University of Zagreb, Croatia.

Enrique Sanjurjo is Researcher, Instituto Nacional de Ecología, Anillo Periférico 5000, 04530 Coyoacán, México, Distrito Federal, Mexico. E-mail: sanjurjo@ine.gob.mx

Elsa Sattout is Associate Researcher at the Department of Plant Sciences, Faculty of Food and Agricultural Sciences, American University of Beirut, PO Box 11-0236-Beirut, Lebanon. E-mail: elsa@intracom.net.lb

Roger A. Sedjo is Senior Fellow, Resources for the Future, Washington DC, USA. E-mail: sedjo@rff.org

José Maria Solano is Head of the Forest Planning area at Dirección General para la conservación de la Naturaleza, Ministerio de Medio Ambiente, Gran via de San Francisco, 4, Madrid, Spain. E-mail: jmsolano@mma.es

Salma Talhouk is Associate Professor at the Department of Plant Sciences, Faculty of Food and Agricultural Sciences, American University of Beirut, PO Box 11-0236-Beirut, Lebanon. E-mail: ntsalma@aub.edu.lb

Mustafa Fehmi Türker is Professor of Forestry at Karadeniz Technical University, Faculty of Forestry, Division of Forest Economics in the Department of Forest Engineering, 61080 Trabzon, Turkey. E-mail: mft@ktu.edu.tr

Dijana Vuletić is DSc Project leader at Forest Research Institute, Jastrebarsko, Cvjetno naselje 41, 10450 Jastrebarsko, Croatia. E-mail: dijanav@sumins.hr

Salim Zahoueh is Assistant Professor at Aleppo University, Department of Forestry and Ecology, seconded since 1996 to the Food and Agriculture Organization of the United Nations. He is presently Assistant FAO Representative in the Syrian Arab Republic, Damascus, Syria. E-mail: salim.zahoueh@fao.org.sy

Mohamed Zamoum is Researcher at National Institute of Forest Research (INRF), Arboretum de Baïnem B.P. 37, Chéraga, Alger, Algeria. E-mail: mzamoum@yahoo.fr

Foreword

There is an old expression – not to be able to see the forest for the trees – meaning that one focuses attention on the details (the trees) and misses the bigger picture (the forest). This has very much been the case when one considers past efforts in economic valuation of forest resources. Most economists have dealt with a limited number of the goods and services produced by a forest ecosystem and have made careful estimates for one or two uses. Few have attempted to value a forest in its entirety and thereby provide valuable information that is needed by resource managers and decision makers to better understand the values of forests and take the steps needed for their conservation.

In this volume the editors, the late Maurizio Merlo and Lelia Croitoru, both of the Centre for Accounting and Management in Agriculture and Forestry of Padua University, Italy, have presented an admirable example of taking a much more holistic approach to valuation of forest resources. With their collaborators, they have applied their valuation template to some 18 countries bordering the Mediterranean Sea. Adopting the Total Economic Value (TEV) approach, the various country studies have attempted to include both direct and indirect use values, as well as various non-use values. Whereas direct use values (e.g. including both consumptive uses such as forestry and collection of non-timber forest products, as well as non-consumptive uses such as recreation and hunting) are easier to value, indirect-use values (such as watershed protection or provision of potable water) are usually harder to value in monetary terms. Non-use values, especially values that often relate to cultural or historical uses associated with healthy forests, are the most difficult to value in monetary terms.

The results are somewhat surprising. Although the relative importance of use versus non-use values varies considerably from country to country, watershed-related values such as reducing the risk of erosion, floods and landslides are important benefits in most countries, and can produce 50% or more of the TEV. Forest products per se (e.g. timber) are usually a fairly small part of the TEV. Other country-by-country variations are not surprising: recreational benefits are very important in western European countries and various extractive uses such as firewood collection or grazing are more important in the southern and eastern Mediterranean countries.

When valued as a whole the economic numbers are large: the average TEV from the 18 countries studied is about €133/ha per year – highest in the northern Mediterranean (about €176/ha) and lower in the eastern (€48/ha) and southern Mediterranean (about €67/ha) countries. *Per capita* values also range from about €70 *per capita* per year in the northern countries to less than €11 *per capita* per year in southern and eastern countries. These figures reflect both the difference in values per ha as well as the major differences in forest area per person in different parts of the region.

Merlo and Croitoru's volume raises as well as answers many valuation questions. It illustrates the importance of applying the TEV approach as well as the data requirements to do so, and the necessity of sometimes 'borrowing' economic value estimates from one location and applying them to another site – a process known as 'benefit transfer'. The book presents the common approaches used for obtaining estimates comparable within and across countries and the methodological difficulties encountered. While highlighting what has been accomplished in the area of forest resource valuation, this volume also identifies those topics that need additional work – often in the areas of indirect use and non-use values.

Once decision makers realize the true economic value of the wide range of goods and services provided by healthy forest ecosystems, they will both demand better analysis of these economic values and be willing to spend more for conservation and management of these important natural (and cultural) resources. This volume is an important step in helping researchers, decision makers, managers, and the public to see BOTH the forests and the trees.

John A. Dixon
Lead Environmental Economist (retired)
The World Bank
Washington, DC

Dedication and Acknowledgements

This book is dedicated to the memory of Maurizio Merlo, a wonderful professor and friend

Maurizio Merlo was born and brought up in the outstanding hilly landscape of Vittorio Veneto. He achieved his honours degree in agriculture at the University of Padova in 1970, obtained a PhD degree at London University in 1978 and became full professor of Forest Economics and Policy at the University of Padova in 1985.

Maurizio Merlo was both a brilliant scientist and a talented teacher. He held numerous responsibilities and initiatives at national and international levels. Among others, he was the Director of the Centre for Environmental Accounting and Management in Agriculture and Forestry in Padova and carried out research work for FAO, OECD, the World Bank, the European Union and many Italian governmental institutions. His scientific work is proved by more than 150 publications, of which more than 100 can be found in peer-reviewed journals and are abstracted by CAB International.

Maurizio Merlo was the coordinator and driving force behind this book. Drafting this book was a difficult task; but when he entered the office, he spontaneously transmitted his spirit and energy to everyone around, and everything seemed easier. He believed in the ideas of this book and wished to see it finished and published. Unfortunately, he did not have this chance: on 24th August 2003, a strong heart attack tore him away at only 58 years of age.

In addition to his scientific merits, Maurizio Merlo's honesty, humanity and sense of humour made him an example as a refined colleague and friend. With his young spirit, he always enjoyed helping and encouraging others. The authors of this book remember him not only as a brilliant professor but, above all, as a good friend.

He left a precious heritage that will live on in those following his path.

Acknowledgements

If this book can be said to have a father, it would be Maurizio Merlo. He was the coordinator and driving force behind this work. His full commitment started five years ago, with the initial idea for such a book. He founded a network of authors from 18 Mediterranean countries and organized the work based on a common framework. Deep knowledge, refined spirit of analysis and creativeness characterized everything he was doing: collaboration with authors, discussions and drafting book chapters. His premature death was a great loss for all his friends, colleagues and the scientific community as a whole.

A book such as this owes much to the generous support, both intellectual and practical, of many people and institutions. The greatest thanks go, of course, to the contributing authors. Without their assistance, its completion would not have been possible. Their input has likewise been supported by many others, who are acknowledged in the individual chapters.

This book is the output of a project regading MEDiterranean FORest public goods and EXternalities (MEDFOREX). Its aim is the identification and valuation of MEDFOREX and other outputs provided by Mediterranean forests in all the countries bordering the Mediterranean Sea. The idea of the project was born in 2000 at a meeting of foresters and forest economists from, at that time, only a few countries: Spain, France, Italy and Portugal. This meeting was organized by the Forest Technology Centre of Catalonia (CTFC) and was the first of a series of four meetings where the ideas, framework and first results of this book were thoroughly discussed.

Since then, the MEDFOREX Centre has become the focal point for the project. The MEDFOREX Centre is an European Forest Institute (EFI) regional project centre formed by a consortium of Mediterranean forestry research and training institutions from 15 countries. The centre is coordinated, managed and represented by the CTFC, whose purpose is: (i) to undertake an inventory of Mediterranean forest externalities, services and non-wood forest products; (ii) to promote research regarding the valuation of Mediterranean forest externalities for sound policy design; (iii) to undertake and promote research on decision support tools for the multipurpose management and planning of Mediterranean forests; and (iv) to disseminate and produce value-added information by effective networking and coordination of activities among the consortium participants.

A common working framework was designed during several meetings held by the authors. Many thanks for the financial and organizational support in preparing these meetings are given to CTFC and especially to Pere Riera and Marc Palahi; to the Department of Environment and Housing of the Government of Catalonia; to the Ministry of Environment of Spain; and to the University of Padova.

This book owes much to the helpful discussions held with several colleagues throughout the world. Many thanks are given to Stefano Pagiola for his valuable support through critical insights, helpful discussions and review of several chapters. I am very grateful to Neil Powe, Guy Garrod, Paola Gatto, Giovanna Toffanin, Richard Panting, Yves Birot, Luca Cesaro for helpful discussions, reviewing chapters and for their continuous support.

Gian Luca Schievano and Fabrizio Bordini helped redraw many maps in the book to make them suitable for publication. Tim Hardwick and his team at CABI Publishing worked hard to ensure that this book is published in a timely fashion and reaches the widest possible audience.

Lelia Croitoru

Acronyms and Abbreviations

AAC	allowable annual cut
ACSAD	Arab Centre for Studies of Arid Zones and Drylands (Tunisia)
AEFCS	Administration des Eaux et Forêts et de la Conservation des Sols (Administration of Waters, Forests and Soil Conservation) (Morocco)
ANN	Agence Nationale de la Conservation de la Nature (National Agency for Nature Conservation) (Algeria)
AOAD	Arab Organization for Agricultural Development (Egypt)
ARIJ	Applied Research Institute Jerusalem (Palestine)
Art.	article
BNEDER	Bureau National des Études pour le Développement Rural (National Office of Studies for Rural Development) (Algeria)
CAP	Common Agricultural Policy
CBA	cost–benefit analysis
CDM	clean development mechanism
CELPA	Associação da Indústria Papeleira (Association of Paper Industry) (Portugal)
CER	carbon emission reduction
CESE	Conselho Para a Cooperação Ensino Superior-Empresa (Council for Higher Education–Business Cooperation) (Portugal)
CFS	Corpo Forestale dello Stato (State Forest Corp.) (Italy)
CFT	Chartes forestières de territoire (Land Forest Charters) (France)
CIEM	Centre for Integrated Environmental Management (Greece)
CNEL	Consiglio Nazionale dell'Economia e del Lavoro (National Council for Economy and Labour) (Italy)
CONTAGRAF	Centro di Contabilità e Gestione Agraria, Forestale e Ambientale (Centre for Accounting and Management in Agriculture and Forestry) (Italy)
COP7	Seventh Conference of Parties
COSE	Confederación de Organizaciones de Selvicultores de España (Spanish Forest Owners Federation) (Spain)
CRPF	Centre Régional de la Propriété Forestière (Regional Forest Ownership Agency) (France)
CSIC	Consejo Superior de Investigaciones Científicas (Superior Council for Scientific Research) (Spain)

CTFC	Centre Tecnològic Forestal de Catalunya (Technological Forest Centre of Catalonia) (Spain)
CVM	contingent valuation method
ÇEKUL	The Foundation for Protection and Promotion of the Environmental and Cultural Heritage (Turkey)
dbh	diameter at breast height
DCES	Direction de la Conservation des Eaux et des Sols (Direction for Conservation of Waters and Soils)
DEFCS	Direction des Eaux et Forêts et de la Conservation des Sols (Direction of Waters, Forests and Soil Conservation) (Morocco)
DFCI	Défense des Forêts Contre l'Incendie (Forest Fire Protection) (France)
DGF	Direction Générale des Forêts (Algeria, Tunisia)/Direcção Geral das Florestas (Portugal)/(General Direction of Forests)
DGFP	General Directorate for Forests and Pastures (Albania)
DGPDIA	Direction Générale de la Planification, du Développement et des Investissements Agricoles (General Direction of Planning, Development and Agricultural Investments) (Tunisia)
DM	Deutschmark
DM	dry matter
DPA	Directorate for Protected Area (Albania)
DREF	Direction Régionale des Eaux et Forêts (Regional Direction of Waters and Forests) (Morocco)
DRS	Défense et Restauration du Sols (Soil Defence and Restoration) (Algeria)
EAA	European Economic Accounts for Agriculture (Spain)
EAF	European Economic Accounts for Forestry (Spain)
EC	European Community
EDF	Électricité de France (Electricity of France) (France)
EEAA	Egyptian Environmental Affairs Agency (Egypt)
EWS	Egyptian Wildlife Service (Egypt)
EGS	environmental goods and services
EVRI	Environmental Valuation Reference Inventory
EU	European Union
EUROSTAT	Statistical Office of the European Communities
FACO	Forest Administration Chief Office (Turkey)
FAO	Food and Agriculture Organization
FAO/MOA	Food and Agriculture Organization/Ministry of Agriculture (Lebanon)
FAOSTAT	Food and Agriculture Organization Statistics
FCC	fixed capital consumption (Spain)
FO	final outputs (Spain)
FOA	Forest Owners' Association (Greece)
FORIS	Forest Information System
FSC	Forest Stewardship Council
FU	forage unit(s)
GDF	General Directorate of Forests (Turkey)
GDNPGW	General Directorate of National Parks, Game and Wildlife (Turkey)
GDP	gross domestic product
GEO.C.G	The Geotechnical Chamber of Greece
GHG	greenhouse gases
GMOs	genetically modified organisms
GSF&NE	General Secretariat of Forests and Natural Environment (Greece)
GVA	gross value added (Spain)
HCA	High Commission for Afforestation (Syria)

HFS	The Hellenic Forestry Society (Greece)
HP	hedonic price method
IC	Intermediate consumption (Spain)
IFEN	Institut Français de l'Environnement (French Institute for Environment) (France)
IFN	Inventaire National Forestier (National Forest Inventory) (France)
ILA	Israel Lands Administration (Israel)
IMF	International Monetary Fund
INAG	Instituto Nacional da Água (National Institute for Water) (Portugal)
INE	Instituto Nacional de Estadística (Spain)
INRF	Institut National de Recherche Forestière (National Institute for Forest Research) (Algeria)
INRGREF	National Institute of Research on Rural Engineering, Water and Forestry (Tunisia)
INS	Institut National de la Statistique (National Institute of Statistics) (Tunisia)
INSEE	L'Institut National de la Statistique et des Études Économiques (National Institute of Statistics and Economic Studies) (France)
INSTAT	Institute of Statistics (Albania)
IPCC	Intergovernmental Panel on Climate Change
IPF/IFF	Intergovernmental Panel on Forests/Intergovernmental Forum on Forests
ISAFA-MAF	Istituto Sperimentale per l'Assestamento Forestale e per l'Apicoltura/Ministero dell'Agricoltura e delle Foreste (Experimental Institute for Forest Planning, Management and Apiculture/Ministry of Agriculture and Forests) (Italy)
ISTAT	Istituto Nazionale di Statistica (National Institute of Statistics) (Italy)
IUCN	formerly known as International Union for Conservation of Nature, at present IUCN is known as The World Conservation Union
JNF	Jewish National Fund (Israel)
LCER	long-term carbon emission reductions
LEF ENGREF/INRA	Laboratoire d'Économie Forestière Ecole Nationale du Génie Rural des Eaux et des Forêts/Institut National de Recherche Agronomique (Laboratory of Forest Economics National School of Rural Engineering, of Waters and Forests/National Institute of Agricultural Research) (France)
MAD	Morrocan dirham
MAF	Ministry of Agriculture and Food (Albania)
MALR	Ministry of Agriculture and Land Reclamation (Egypt)
MAMVA	Ministère de l'Agriculture et de la Mise en Valeur Agricole (Ministry of Agriculture and Agricultural Enhancement) (Morocco)
MAP	Mediterranean Action Plan
MAP	Ministère de l'Agriculture et de la Pêche (Ministry of Agriculture and Fishing) (France)
MAPA	Ministerio de Agricultura, Pesca y Alimentación (Ministry of Agriculture, Fishing and Food) (Spain)
MARA	Ministère de l'Agriculture et de la Réforme Agraire (Ministry of Agriculture and Agrarian Reform) (Morocco)
MATE	Ministère de l'Aménagement du Territoire et de l'Environnement (Ministry of Land Planning and Environment) (Algeria)
MATUHE	Ministère de l'Aménagement du Territoire, de l'Urbanisme, de l'Habitat et de l'Environnement (Ministry of Land Planning, Urbanism, Habitat and Environment) (Morocco)
MCPMF	Ministerial Conferences for the Protection of Mediterranean Forests
MCSD	Mediterranean Commission on Sustainable Development

MCWF	Ministry in Charge of Waters and Forests (Morocco)
MDF	Medium Density Fibreboard
MEAT	Ministère de l'Environnement et de l'Aménagement du Territoire (Ministry of Environment and Land Planning) (Tunisia)
MEDFOREX	Mediterranean forest public goods and externalities
MEPPP	Ministry for Environmental Protection and Physical Planning (Croatia)
METAP	Mediterranean Environmental Technical Assistance Programme
MMA	Ministerio de Medio Ambiente (Ministry of Environment) (Spain)
NFF	National Forest Fund (Morocco)
NFI	National Forest Inventory (Italy)
NFP	National Forest Programme (Cyprus, Italy)
NGOs	non-governmental organizations
NIPF	non-industrial private forest (Portugal)
NMP	National Master Plan (Israel)
NOAA	National Oceanic and Atmospheric Administration
NSSG	National Statistical Service of Greece (Greece)
NWFPs	non-wood forest products
NWMP	National Water Management Plan (Morocco)
OECD	Organization for Economic Cooperation and Development
ONF	Office National des Forêts (National Office of Forests) (France)
ONS	Office National des Statistiques (National Office of Statistics) (Algeria)
ORF	Orientations Régionales Forestières (Regional Forest trends) (France)
ÖBF	Österreichishe Budesforste (Federal Forests) (Austria)
PASEGES	Panhellenic Confederation Union of Agricultural Cooperation (Greece)
PCF	Prototype Carbon Fund
PDR	Plan Directeur de Reboisement (Afforestation Leading Plan) (Morocco)
PDRN	Plan de Développement Rural National (Plan for National Rural Development) (France)
PEFC	Paneuropean Forest Certification
PNA	Palestinian National Authority (Palestine)
PNDA	Plan National pour le Développement de l'Agriculture (National Plan of Agricultural Development) (Algeria)
PNR	Plan National de Reboisement (National Reforestation Plan) (Morocco, Algeria)
R&D	Research and Development
REC	The Regional Environmental Centre for Central and Eastern Europe
RTM	Restauration des Terrains en Montagne (Restoration of Mountain Land) (France)
SCEES	Le Service Central des Enquêtes et Études Statistiques (Central Service of Surveys and Statistical Studies) (France)
SE	starch equivalent
SESSI	Service des Études et des Statistiques Industrielles (Service of Industrial Studies and Statistics) (France)
SFE	State Forest Entreprise (Turkey)
SFM	sustainable forest management
SICOP	Sistema de Informação de Cotações de Produtos Florestais na Produção (Forest Information System on Prices in the Production of Products) (Portugal)
SNB	Serviço Nacional de Bombeiros (National Service of Firefighters) (Portugal)
tCER	temporary Carbon Emission Reductions
TCM	travel cost method

TEMA	Turkish foundation for Reforestation, Protection of natural habitats and Combating soil erosion (Turkey)
TERUTI	Teritoire Utilisation (France)
TEV	total economic value
TKV	Turkish Development Foundation (Turkey)
TO	total outputs (Spain)
TOE	tons of oil equivalent
TSI	total social income (Spain)
UA	Undersecretariat for Afforestation (Egypt)
UK	United Kingdom
UMR ENGREF/INRA	Unité Mixte de Recherche École Nationale du Génie Rural des Eaux et des Forêts/Institut National de Recherche Agronomique (Combined Unit of Research National School of Rural Engineering of Waters and Forests/ National Institute of Agricultural Economics) (France)
UN	United Nations
UNCCD	United Nations Convention to Combat Desertification
UNCED	United Nations Conference on Environment and Development
UNDP	United Nations Development Programme
UN-ECE/FAO	United Nations Economic Commission for Europe/Food and Agriculture Organization
UNEP	United Nations Environment Programme
UNEP/MAP	United Nations Environment Programme/Mediterranean Action Plan
UNESCO	United Nations Educational, Scientific and Cultural Organization
UNFCCC	United Nations Framework Convention on Climate Change
UNFF	United Nations Forum on Forests
US$	US dollars
USDA	United States Department of Agriculture (Egypt)
USLE	universal soil loss equation
VAT	value added tax
vs.	versus
WFPs	wood forest products
WTA	willingness-to-accept
WTP	willingness-to-pay

Symbols

°C	degrees Celsius
C	carbon
cm	centimetre
CO_2	carbon dioxide
€	euro
g	gram(s)
ha	hectare(s)
kg	kilogramme(s)
kg CO_2	kilogramme(s) of carbon dioxide
km	kilometre(s)
km^2	square kilometre(s)
K_2O	potassium oxide
m	metre(s)
m^2	square metre(s)

m^3	cubic metre(s)
mm	millimetre(s)
m^3 o.b.	cubic metre overbark
m^3 u.b.	cubic metre underbark
m.t.	metric tonne(s)
N	nitrogen
NA	not available
NC	not calculated
N° (or no.)	number
%	per cent
P_2O_5	phosphorus pentoxide
q	quintal
t	tonnes
tC	tonnes of carbon
tCO_2	tonnes of carbon dioxide
tdm	tonnes of dry matter
TOE	tonnes of oil equivalent
US$	US dollars

1 Introduction

Maurizio Merlo[†] and Lelia Croitoru

University of Padova, Centre for Environmental Accounting and Management in Agriculture and Forestry (CONTAGRAF), Via Roma 34, Corte Benedettina, 35020 Legnaro (PD), Italy

Mediterranean forests, like all forests, produce a wide array of benefits. Timber and other wood forest products (WFPs) come readily to mind, but often they comprise only a minor part of these benefits. Non-wood forest products (NWFPs), such as pine kernels in Lebanon and cork in Algeria and Tunisia, can be of greater importance than WFPs, and often have a high potential to contribute to local economies. In many cases, the most important benefits provided by the Mediterranean forests are public goods and externalities, such as watershed protection and soil conservation. This multifunctionality has long been recognized. Indeed, since as far back as the 15th and 16th centuries, forest policy and management in some countries on the Mediterranean's shores, such as Catalonia and the Republic of Venice, have been aimed primarily at protecting rural welfare and conserving soil and water and only secondarily at timber production. However, a full realization of many benefits has been hampered by the lack of their recognition and of appropriate mechanisms to internalize them. As a result, forests are often degraded or lost, along with their benefits.

Only a few of the many benefits that Mediterranean forests provide enter formal markets, usually WFPs and some NWFPs. Other forest benefits are either traded only in informal markets, as is typically the case for many NWFPs, or do not enter markets at all. This market failure is due, in part, to the very nature of forest services. For example, scenic beauty is a public good that cannot be kept from people irrespective of whether they pay for it, and watershed protection is an externality that is enjoyed by people far downstream from the actual forest. Market failures can result from the lack of clear and enforceable property rights over forests and their benefits. Often this results in pressure on forest resources, such as deforestation and overgrazing in the southern Mediterranean countries.

These failures provide serious challenges to forests and local welfare. They need to be addressed properly in decision making related to forest policy, management and investments. At present, both public policy and private management decisions are often made based on a very partial and incomplete view of the forest benefits. These decisions usually capture only the market values involved, with consideration of non-market values being rare. In many cases, this makes alternative uses of forestland appear more attractive; in others, it makes the benefits of good forest management appear to be minimal. Under these conditions, forest management decisions are often suboptimal, with forests and the benefits they provide often being lost or degraded.

Inadequate recognition of non-market forest values by decision makers is common in the Mediterranean region, where official statistics

[†] Deceased.

usually reflect only marketed, tangible forest products. Relevant valuations of non-market benefits are not only scarce and site specific, but are often disseminated inadequately. The threat to forests is, of course, not solely due to insufficient valuation and inadequate dissemination (Kengen, 1997). To attribute, for example, deforestation in Morocco to a lack of knowledge of forest values would be taking a narrow view of the problem, as it is influenced by various other factors. In other cases, the estimates, even though properly calculated and disseminated, are not captured within the decision-making process because they do not fit to the local needs, preferences or mentality of rural communities (Grimes et al., 1994).

This book addresses the gap in valuation. It is the first effort to estimate the total economic value (TEV) of forests on a large scale in the Mediterranean region. Previous efforts have focused almost exclusively on estimating the value of individual forest benefits, often at a specific site: hydrological services (see Bruijnzeel and Bremmer, 1989; Aylward et al., 1998; Bruijnzeel, 2004); option uses – pharmaceutical (see Simpson et al., 1994; Barbier et al., 1995; Mendelhson and Balick, 1995; Pearce and Puroshothaman, 1995); and other extractive or non-extractive values of NWFPs (see Ruitenbeek, 1989; Godoy et al., 1993; Grimes et al., 1994; Lampietti and Dixon, 1995; De Beer and McDermott, 1996). Few have attempted to estimate the benefits of a nation's forests as a whole (see Adger et al., 1995; Willis et al., 2003).

This book provides a comprehensive analysis of the economic value of Mediterranean forests, including not just commonly measured benefits such as timber but also, more importantly, the public goods and externalities they provide. Uniquely, it brings together forest valuations at the national level from 18 Mediterranean countries, based on extensive data collection by local experts. It uses a coherent analytical framework for collecting these valuations in a consistent way; it analyses these estimates from a per-country and cross-country perspective; and it uses these results to propose policy recommendations to be undertaken locally, within individual countries and across countries.

The book is structured in three parts. Part I provides an overview of the problem and of the approach followed, and summarizes the results. Chapter 2 begins by presenting a broad overview of forests in the Mediterranean region. The forests in the countries that ring the Mediterranean are broadly described in terms of their geography, environment, institutions and socio-economics. Particular emphasis is given to how Mediterranean forest types differ from other forests. The analysis in this book initially was intended to focus specifically on Mediterranean forests. Because most available data do not distinguish between forest types, however, the analysis had to be broadened to include all forests in Mediterranean countries – the first of many compromises that had to be made in light of data constraints. Examining the many diverse benefits that these forests provide requires a consistent and coherent analytical framework. Chapter 3 describes the framework used in this book – that of TEV – and the methods used for valuing the forest benefits in the country chapters (Part II); it discusses in detail the common approaches and the constraints encountered in arriving at comparable results among the country chapters. Chapter 4 then synthesizes the estimated values of forests in the Mediterranean, drawing from the results of the country chapters; it makes a cross-country comparison among the estimates and the valuation methods used in each country and presents aggregated estimates at subregional and Mediterranean levels.

Part II provides detailed national level case studies of 18 countries and territories bordering the Mediterranean Sea (Chapters 5–22). Each chapter follows a similar structure, allowing for comparison across them. Following the framework outlined in Chapter 3, each country case study classifies forest benefits according to the TEV framework. Each benefit is first discussed qualitatively; efforts are then made to estimate quantitatively the value of each benefit. These estimates are based on a wide range of valuation methods and approaches, drawing on official statistical information supplemented by relevant results of local surveys. Although every effort was made to follow a consistent approach, data scarcity and other constraints often forced deviations. The estimates thus obtained are placed within the context of the institutions and policies affecting forests in each country. The chapters identify the areas where further valuation efforts are needed and the gaps in the present policy and institutional framework of each country.

Part III highlights the institutional and policy implications that result from the valuation efforts of the country chapters. Chapter 23 provides an overview of the forest institutions and policies at the Mediterranean level; examines the social, economic and environmental constraints in the countries analysed; and proposes new policy approaches to be undertaken at regional and country level. Recent years have seen increasing attention paid to participatory processes throughout the region. Many countries have adopted explicit decentralization policies. Chapter 24 discusses these processes as key challenges in the Mediterranean and provides a comparative analysis of forest policy elements across the countries studied. Chapter 25 discusses the current networks of cooperation and emphasizes the need for an international agreement on Mediterranean forest conservation and development.

At the methodological level, approaches such as TEV are often discussed but seldom implemented. This book describes the valuation techniques necessary to estimate TEV, discusses in detail the approaches taken to value a wide range of benefits (often in the context of severe data scarcity) and examines ways to overcome the problems encountered. Thus, the book is meant to provide a unique data source for the region and a methodology that can be applied to other parts of the world.

At the policy level, new approaches for a sustainable forest management are needed in the Mediterranean to internalize the provision of positive public goods and externalities and overcome significant social, economic and environmental constraints (soil erosion, risk of floods and rural outmigration) associated with forest degradation. However, insufficient knowledge on the nature and magnitude of forest benefits in the region creates a severe constraint to addressing these issues. This book captures these forest values in a holistic picture at national and regional level and, based on these estimates, proposes realistic policy recommendations for improving sustainable forest management.

References

Adger, N., Brown, K., Cervigni, R. and Moran, D. (1995) Total economic value of forests in Mexico. *Ambio* 24, 286–296.

Aylward, B., Echeverria, J., Fernandez Gonzalez, A., Porras, I., Allen, K. and Mejias, R. (1998) Economic incentives for watershed protection: a case study of Lake Arenal, Costa Rica. In: *Final Report to the Government of The Netherlands Under the Program of Collaborative Research in the Economics of Environment and Development (CREED)*. IIED, TSC and the International Center for Economic Policy, National University at Heredia (CINPE). London, p. 130.

Barbier, E.B., Brown, G., Dalmazzone, S., Folke, C., Gadgil, M., Hanley, N., Holling, C.S., Mäler, K.-G., Mason, P., Panayotou, T., Perrings, C. and Turner, K. (1995) The economic value of biodiversity. In: Heywood, H. (ed.) *Global Biodiversity Assessment*. Cambridge University Press, Cambridge, pp. 823–914.

Bruijnzeel, L.A. (2004) Hydrological functions of tropical forests: not seeing the soils for the trees? *Agriculture, Ecosystems and Environment* 104, 185–228.

Bruijnzeel, L.A. and Bremmer, C.N. (1989) *Highland–Lowland Interactions in the Ganges–Brahmaputra River Basin: a Review of Published Literature*. ICIMOD Occasional Paper, No. 11.

De Beer, J. and McDermott, M. (1996) *The Economic Value of Non-timber Forest Products in Southeast Asia*, 2nd edn. Netherlands Committee of the International Union for the Conservation of Nature, Amsterdam.

Godoy, R., Lubowski, R. and Markandya, A. (1993) A method for the economic valuation of non-timber tropical forest products. *Economic Botany* 47, 220–233.

Grimes, A., Loomis, S., Jahnige, P., Burnham, M., Onthank, K., Alarcón, R., Cuenca, W., Martinez, C., Neill, D., Balick, M., Bennett, B. and Mendelsohn, R. (1994) Valuing the rain forest: the economic value of nontimber forest products in Ecuador. *Ambio* 23, 405–410.

Kengen, S. (1997) Linking forest valuation and financing. *Unasylva* 48, 44–49.

Lampietti, J.A. and Dixon, J.A. (1995) *To See the Forest for the Trees: a Guide to Non-timber Forest Benefits*. Environmental Economics Paper No 013. Environment Department, World Bank, Washington, DC.

Mendelhson, R. and Balick, M. (1995) The value of undiscovered pharmaceuticals in tropical forests. *Economic Botany* 49, 223–228.

Pearce, D. and Puroshothaman, S. (1995) The economic value of plant-based pharmaceuticals. In: Swanson, T. (ed.) *Intellectual Property Rights and Biodiversity Conservation*. Cambridge University Press, Cambridge, pp. 127–138.

Ruitenbeek, H.J. (1989) *Economic Analysis of Issues and Projects Relating to the Establishment of the Proposed Cross River National Park (Oban Division) and Support Zone*. World Wide Fund for Nature, London.

Simpson, R., Sedjo, R. and Reid, J. (1994) *Valuing Biodiversity: an Application to Genetic Prospecting*. Discussion Paper 94–20, Resources for the Future, Washington, DC.

Willis, K., Garrod, G., Scarpa, R., Powe, N., Lovett, A., Bateman, I., Hanley, N. and Macmillan, D.C. (2003) *The Social and Environmental Benefits of Forests in Great Britain*. Report to Forestry Commission Edinburgh. Centre for Research in Environmental Appraisal and Management (CREAM), University of Newcastle, UK.

2 The State of Mediterranean Forests

Maurizio Merlo[1,†] and Paolo Paiero[2]

[1]University of Padova, Centre for Environmental Accounting and Management in Agriculture and Forestry (CONTAGRAF), Via Roma 34, Corte Benedettina, 35020 Legnaro (PD), Italy; [2]Department of Land and Agro-Forestry Systems, Agripolis, Via Roma, 35020 Legnaro (PD), Italy

---

Forests have always played an important role in the lives of the Mediterranean people. Since ancient times, the relationship between man and forest has been very close. Forests have long been exploited for their different uses, such as wood, food, grazing and clearing for agriculture. However, very often, human actions have resulted in the overuse of forest resources, with negative impact on the environment. In many Mediterranean areas, what remains today is a 'heritage of forest depletion and degradation', the price that the natural environment has had to pay for the cultural development in the region (Thirgood, 1981; Tsoumis, 1996). Today, the common feature of the Mediterranean forests is their fragility, instability and frequent degradation (M'Hirit, 1999).

Forests are now affected by contrasting situations in the different Mediterranean areas. In the northern Mediterranean, and especially in Western European countries, many forests are subject to abandonment; whereas in the southern and eastern Mediterranean, forests are exposed to high levels of human pressure. These situations pose complex challenges in terms of increasing threats to the benefits that forests provide.

This chapter gives a broad overview of the Mediterranean forests. The general features of the forests in the countries that ring the Mediterranean are first examined. Particular attention is paid to the Mediterranean forest ecosystems. The main threats linked to the Mediterranean forests are then discussed. The chapter concludes with the consideration of the importance of the forest benefits and the need for forest conservation.

General Features of Forests in the Mediterranean

Coming from the Latin *medius* and *terra*, the term 'Mediterranean' means 'surrounded by land'. Over the years, different interpretations have been given to the region, based on different criteria. This book considers the Mediterranean region as represented by the countries and territories surrounding the Mediterranean Sea. For convenience, they have been divided into the following subregions:

- South: Morocco, Algeria, Tunisia, Libya and Egypt
- East: Palestine, Israel, Lebanon, Syria, Turkey and Cyprus
- North: Greece, Albania, Serbia-Montenegro, Bosnia-Herzegovina, Croatia, Slovenia, Italy, Malta, France, Portugal and Spain.

---

† Deceased.

©CAB International 2005. *Valuing Mediterranean Forests*
(eds M. Merlo and L. Croitoru)

The climate in much of this region is characterized by dry and hot summers and cool and rainy winters. Mediterranean countries, of course, also include substantial areas that do not have a Mediterranean climate; in particular, northern countries and Turkey have substantial temperate areas, while many southern countries have large desert areas. The analysis in this book covers all forest areas in Mediterranean countries, but emphasizes wherever possible the particular role of Mediterranean forest ecosystems.

Forests in the Mediterranean region cover around 73 Mha, or about 8.5% of the region's area, or 1.8% of the total wooded area in the world (Food and Agriculture Organization, 2003).[1] At 115 m³/ha, the average growing stock is 15% higher than the world mean, whereas the average biomass (59 t/ha) is only 54% of the world's average (Table 2.1).[2]

The distribution of forests across the Mediterranean is quite uneven, with 77% in the north and only 8% in the south and 15% in the east.[3] Around 60% of the forest area in the Mediterranean is concentrated in five northern countries: Italy, France, Spain, Greece and Portugal. The prevalence of forests in the north is due largely to natural conditions (such as climate) allowing

Table 2.1. The forests in the Mediterranean region.

Country	Land area (000 ha)	Forest area in 2000			Plantations		Wood volume[c] (m³/ha)	Wood biomass[d] (t/ha)
		000 ha[a]	% of land area	Annual rate of change[b] (1990–2000) (%)	000 ha	% of forest area		
Southern	574,665	6,110	1.1	NC	1,694	27.7	32.4	51.3
Morocco	44,630	3,025	6.8	NS	534	17.7	27	41
Algeria	238,174	2,145	0.9	1.3	718	33.5	44	75
Tunisia	16,362	510	3.1	0.2	202	39.6	18	27
Libya	175,954	358	0.2	1.4	168	46.9	14	20
Egypt	99,545	72	0.1	3.3	72	100.0	108	106
Eastern	99,969	11,026	11.0	NC	2,176	19.7	128.7	70.2
Palestine	618	NA	NA	NA	NA	NA	NA	NA
Israel	2,062	132	6.4	4.9	91	68.9	49	NA
Lebanon	1,024	36	3.5	−0.4	2	5.6	23	22
Syria	18,377	461	2.5	NS	229	49.7	29	28
Turkey	76,963	10,225	13.3	0.2	1,854	18.1	136	74
Cyprus	925	172	18.6	3.7	0	0.0	43	21
Northern	182,077	56,020	30.8	NC	4,198	7.5	121.3	57.5
Greece	12,890	3,599	27.9	0.9	120	3.3	45	25
Albania	2,740	991	36.2	−0.8	102	10.3	81	58
Serbia-Montenegro	10,200	2,887	28.3	−0.1	39	1.4	111	23
Bosnia-Herzegovina	5,100	2,273	44.6	NS	57	2.5	110	NA
Croatia	5,592	1,783	31.9	0.1	47	2.6	201	107
Slovenia	2,012	1,107	55.0	0.2	1	0.1	283	178
Italy	29,406	10,003	34.0	0.3	133	1.3	145	74
Malta	32	NS	NS	NS	NS	NS	232	NA
France	55,010	15,341	27.9	0.4	961	6.3	191	92
Spain	49,945	14,370	28.8	0.6	1,904	13.2	44	24
Portugal	9,150	3,666	40.1	1.7	834	22.7	82	33
Mediterranean	856,711	73,156	8.5	NC	8,068	11.0	115	59
World	13,063,900	3,869,455	29.6	−0.2	187,086	4.8	100	109

[a]The sum of natural forest and forest plantations.
[b]The net change in forests, including expansion of forest plantations and losses and gains in the area of natural forests.
[c]Total volume over bark of living trees above 10 cm diameter at breast height.
[d]Above-ground mass of woody part (stem, bark, branches and twigs) of trees (alive or dead), shrubs and bushes (Food and Agriculture Organization, 2003).
Source: Food and Agriculture Organization (2003) for country and world data; NS = not significant; NA = not available; NC = not calculated.

faster and better growth. In contrast, forests are scarce in southern and eastern countries, where the climate is very hot and dry. In fact, more than 95% of the forests in the 17 countries in this part of the region are concentrated in only three countries: Turkey in the east and Morocco and Algeria in the south (Fig. 2.1).

Forests in the northern countries occupy around 56 Mha, or about 31% of the subregion's area. About 80% of the total wood volume and 75% of the total wood biomass are found in this subregion. The high levels of growing stock and biomass in this part of the Mediterranean are explained in part by the presence in many of these countries of extensive areas of temperate forests. Forests cover only 6 Mha in southern countries and 11 Mha in eastern countries, representing only 1 and 11% of these subregions' areas. In these countries, a large part of forest area consists of sparse, pre-desert shrub vegetation, with low levels of wood volume and biomass. Wood productivity in these subregions is much lower than in the northern subregion. In relative terms, forests are scarcest in Egypt and Libya, occupying 0.1–0.2% of those countries' areas, while they are most abundant in Slovenia, Bosnia-Herzegovina and Portugal, covering about 40–55% of those countries' areas.

The different Mediterranean subregions have diametrically opposed trends in forest area.

During the period from 1990 to 2000, the forest area increased in most northern countries, at annual rates reaching up to 1.7% (in Italy). In Western European countries, this trend can be explained by the abandonment of forestry and agriculture by private owners, resulting in the expansion of forest area. In contrast, several Eastern European countries have experienced a trend of diminishing forest area as a result of illegal cutting, fires and a fragile system of property rights following the political changes of the 1990s. The forest area declined by 0.8% annually in Albania, for example.

Reported trends in southern and eastern countries show increases in forest area (except for Lebanon); however, the degree to which these reflect real forest trends in these countries is uncertain: illegal practices causing forestland destruction are usually accounted for by the statistics only to a limited extent. This subregion is commonly thought to be experiencing significant decreases in the forest and woodland growing stock, biomass and quality (Boydak and Dogru, 1997).

Plantations cover only 11% of the forest area in the Mediterranean. They represent a significant share of the forest area mainly in the south and east, where natural forests are quite limited. In particular, plantations represent an important share of forest area in Egypt, where

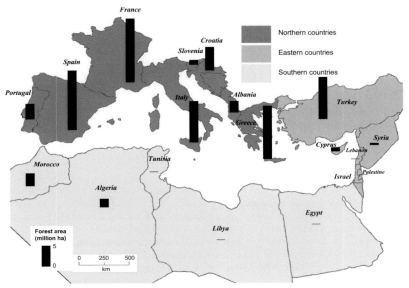

Fig. 2.1. Forests in the Mediterranean countries.

natural forests are practically non-existent, and in Israel and Libya, where intensive afforestation programmes have been implemented. While plantations gained high importance in certain Mediterranean areas, the reasons and objectives for their implementation vary greatly among the different subregions (Box 2.1).

It should be noted that forests (including plantations) in the Mediterranean region do not include only Mediterranean forest types. Countries such as France, Italy and Turkey also include extensive areas of temperate forests. Distinguishing between the Mediterranean forests and other forest types is important for identifying and measuring in economic terms the functions that these forests perform. For example, if temperate forests are known to perform the wood production function well, the Mediterranean forest types are more important for the protective functions they perform.

Mediterranean Forests: Ecological Features and Biodiversity

Mediterranean forest ecosystems consist primarily of forests where broadleaved, evergreen sclerophyllous trees prevail. The area where the Mediterranean climate prevails has probably been best identified[4] with that delimited by the range of olive cultivation (Koppen, 1923). This view is much accepted by bio-geographers and ecologists on the grounds that the olive tree typifies, more than any other, this environment (Lemée, 1967). It is sometimes criticized, however, for failing to consider the effects of elevation and desert encroachment (Thirgood, 1981). Worldwide, Mediterranean forest ecosystems cover around 81 Mha, or nearly 1.5% of the world's total wooded area (M'Hirit, 1999). More than two-thirds of these forests are concentrated in the Mediterranean region

---

Box 2.1. Plantations in the Mediterranean region (Marc Palahí, José Maria Solano and Andreas Ottitsch).

The difficulty in distinguishing between planted and natural forests for some native species and the lack of recent inventories make it difficult to quantify the total area of plantations in many Mediterranean countries. In some countries, planted forests have played an important role in forest trends during the last century. In Tunisia, for instance, planting reversed the loss of forest area, resulting in an increase in forest cover of 0.2% annually between 1990 and 2000 (FAO, 2001b). The annual planting rate in Tunisia is estimated at about 14,000 ha (FAO, 2003). Countries such as Israel and Cyprus have very active national reforestation programmes, with mean annual plantation rates of 500–1000 ha (Scarascia-Mugnozza et al., 2000).

In most Mediterranean countries, the primary objectives of reforestation are soil protection and runoff control; however, in the northern Mediterranean countries, specialized tree plantations of fast-growing species (mostly poplars, eucalyptus and fast-growing pines such as *Pinus radiata* and *Pinus pinaster*) are starting to meet a significant portion of wood needs. For example, plantations now supply about 50% of the national production of industrial wood in Turkey and Italy (Scarascia-Mugnozza et al., 2000) and as much as 81% in Spain (Pandey and Ball, 1998). Vast regions of Portugal and northern Spain as well as south-western France, where suitable edaphic and climatic conditions for cultivation forestry can be found, are undoubtedly the *El Dorado* of intensive forest plantation in the Mediterranean region. A total of 5 Mha of fast-growing species are found in this part of Europe (Arbez, 2001). The transition to plantation forestry has benefited from recently abandoned agricultural lands and a very dynamic wood industry sector in the region.

In most southern and eastern countries, plantation forests are expected to contribute to reducing pressure on fodder and fuel; to provide or diversify rural income, to combat desertification and to reverse deforestation. For example, in Tunisia, forest plantations, in addition to reversing deforestation, have been used to enhance agroforestry practices (by using *Acacia*, *Atriplex* and *Medicago* species) to protect irrigated agricultural land (by planting 5700 ha as windbreaks and shelterbelts between 1990 and 1999) and to combat desertification (17,200 ha were established between 1990 and 1999) (FAO, 2003). Multi-purpose species (such as walnut, pistachio, pecan, hazel and carob) have been widely used, particularly in mountainous areas and in forest clearings.

The trend towards tree planting is expected to continue in the Mediterranean region, driven by different needs and objectives within subregions and individual countries.

itself, and the rest are scattered over the other continents: south-eastern Australia (around Perth), South Africa's Cape region, California and a short strip of Chile in southern America.

Mediterranean forests include a diversity of formations, with different levels of woody vegetation and open areas, for which it is hard to give a strict definition. The difficulty is related to the degraded areas and the sub-Mediterranean forestlands. Throughout history, factors such as forest clearance for wood, agriculture and forest fires provoked changes in Mediterranean vegetation, leaving formations such as *maquis*, the dense shrub formations including plants such as wild olive, myrtle, laurel and juniper; and *garrigue*, composed of aromatic low shrubs and shrublets such as lavender, myrtle, rosemary, cistus, marjoram and thyme, with the occasional higher shrub. Whether these formations are original Mediterranean vegetation or just degraded remnants of better forest types is difficult to say (Thirgood, 1981).

In the Mediterranean region, the Mediterranean forests and *maquis*[5] cover about 56 Mha, i.e. 7.5% of the total countries' land[6] (Table 2.2). Fifty-two per cent of these forests are found in the northern countries, 32% in the east and 16% in the south. Mediterranean forests and *maquis* cover similar shares of land area in northern and eastern countries (~17–18%), but only a negligible share (2%) of land area in southern countries (Fig. 2.2).

In the northern countries, the area covered by the Mediterranean forests and *maquis* is 29 Mha. More than half of this area is found in Spain, where about 90% of the total area covered by *maquis* and *garrigue* is concentrated. The remainder is found mostly in the other western European countries and consists primarily of evergreen oak, beech and deciduous oak.

Mediterranean forests occupy about 18 Mha in the eastern countries, almost all concentrated in Turkey, the country with the largest share of conifers and broadleaves. Because the other countries of this subregion are much smaller, the presence of Mediterranean forests seems almost negligible; however, one should note that they occupy about 19% of Cyprus' land area and 7% of Lebanon's.

In the southern countries, Mediterranean forests occupy about 9 Mha, and consist primarily of evergreen oak and conifers (Aleppo pine,

brutia pine, thuya and junipers). Half of this area is found in Morocco, and most of the rest in Algeria and Tunisia. Compared with the other subregions, other broadleaves such as wild olive, carob and arganier are more prevalent.

According to Quézel (1976), Mediterranean forest ecosystems can be divided into six vegetation categories: (i) thermophilic wild olive and pistachio scrub; (ii) conifer forests of pines (Aleppo, Brutia, stone, etc.), Barbary Thuya and Phoenician juniper; (iii) sclerophyllous evergreen forests of oaks (holm, cork, kermes, etc.); (iv) hilly deciduous forests of oaks (zeen, afares, lebanese, tauzin, etc.), hornbeam, ash and, occasionally, beech; (v) mountain forests of cedars, black pine and beech; and (vi) oro-stage-stands of arborescences juniper and xerophytes. This classification is followed, whenever possible, in the discussion and in the maps in the country case studies, along with an additional 'non-Mediterranean' forest type.

The most important Mediterranean forest type from the economic and environmental viewpoints is the sclerophyllous oak (*Quercus ilex*) (category iii). This forest can be found within *Quercetum ilicis* (*Q. suber*, *Q. coccifera*, etc.) stands located mainly in the central and western Mediterranean and particularly in Spain, Italy, Greece, Algeria and Morocco. The kermes oak (*Q. coccifera*) is most widespread in bush and tree forms, the latter progressively replacing the holm oak between southern Greece and the eastern countries (*Q. pseudococcifera* or *Q. calliprinos*).

The sclerophyllous oaks of *Q. ilicis* are often replaced by their degradation *facies* which in most cases consists of various types of *maquis* (category i): *Q. ilicis*, *Oleo-Ceratonion* (*Olea europaea* var. *oleaster*, *Ceratonia siliqua*, etc.) and *Oleo-Lentiscetum* (*Pistacia lentiscus*, *Mirtus communis*, *Olea europaea*, *Euphorbia dendroides*, *Chamaerops humilis*, etc.). These forest types usually cover large areas between sea level and 500–600 m along the French and Italian coastline and between 500 and 1200 m along the coast of the southern countries.

When the morphological and climate conditions tend to be mesophilic[7], the holm oak (*Q. ilex*) is replaced by the cork oak (*Quercus suber*) in the oceanic sites of the eastern Mediterranean; and by conifer forests of pine (Aleppo, Brutia, stone pine) (category ii) in Italy, Spain,

Table 2.2. Mediterranean forests and *maquis* according to main types and countries.

Country	Conifer (000 ha) Aleppo pine, Brutia pine	Thuyas, junipers	Other conifers	Broad-leaved (000 ha) Evergreen oak	Beech, deciduous oak	Chestnut	Wild olive, carob	Arganier	Maquis, garrigues	Total 000 ha	%
Southern											
Morocco	65	971	144	1,712	24	0	500	700	458	4,574	8
Algeria	843	762	43	1,154	67	0	100	0	0	2,969	5
Tunisia	340	455	2	213	25	0	70	0	0	1,105	2
Libya	5	153	0	1	0	0	50	0	0	206	0
Total	1,253	2,341	189	3,080	116	0	720	700	458	8,854	16
Eastern											
Israel	30	0	0	35	5	0	0	0	0	70	0
Lebanon	5	11	16	33	10	0	0	0	0	75	0
Syria	70	0	25	43	20	0	0	0	0	158	0
Turkey	3,220	458	4,694	590	7,520	183	0	0	823	17,488	31
Cyprus	116	0	5	0	0	0	0	0	52	173	0
Total	3,441	469	4,740	701	7,555	183	0	0	875	17,964	32
Northern											
Greece	528	0	514	743	1,268	28	0	0	0	3,081	5
Albania	—	—	—	302	—	—	0	0	395	1,040	2
former Yugoslavia	50	0	40	35	—	—	0	0	776	960	2
Italy	130	0	114	2,585	1,190	244	2	0	0	4,263	8
France	80	0	105	800	700	135	2	0	300	2,122	4
Spain	1,300	62	1,779	3,155	1,326	163	0	0	10,070	17,855	32
Total	2,088	62	2,552	7,620	4,484	570	2	0	11,541	29,321	52
Total Mediterranean	6,782	2,872	7,481	11,401	12,155	753	722	700	12,874	56,139	100

Adapted from M'Hirit (1999). Calculation of the shares of the different Mediterranean forest areas and the total forest area for each country is not attempted, as not all countries include all Mediterranean vegetation in their definition of forest areas.

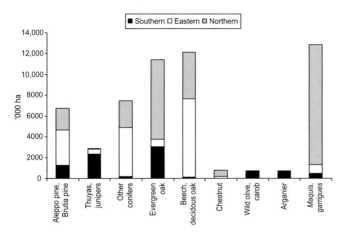

Fig. 2.2. Distribution of the Mediterranean forests and *maquis* among subregions.

Greece, Lebanon and in some mountain sites of Morocco, Tunisia and Algeria. The Mediterranean cypress (*Cupressus sempervirens*) was sometimes found at high elevations.

Towards the eastern Mediterranean, the hilly oak forests are typically submesophilic, and become enriched with deciduous oaks, such as zeen, afares, lebanese tauzin, hornbeam, ash and occasionally beech (category iv). Beech and deciduous oaks are mostly found in Spain, Italy and Turkey, whereas in the mountainous sites (1000–1500 m) of Spain, Italy and France, beech prevails.

At elevations higher than 1000–1200 m, starting from western Europe and moving to Lebanon, broadleaved mesophilic forests can often be found, such as beech. Sometimes they are replaced by coniferous forests of black pine (category v) in the eastern Alps, Corsica, Calabria and the Balkan Peninsula; and by cedar in Lebanon, growing at high altitudes, up to 2000 m.

The Mediterranean junipers (such as *Juniperus oxycedrus*, *J. phoenicea* and *J. excelsa*) (category vi) are prevalent in the western Mediterranean region; for example thuya and other junipers can be found in Morocco, Algeria, Tunisia and Turkey, in rocky and sandy environments where they are often diffused throughout reafforestations. Towards the eastern Mediterranean, they can reach the same altitude as the antique cedar and fir forests.

The multi-purpose nature of the Mediterranean forests has long been recognized. Mediterranean forests are not distinguished by the wood forest products (WFPs) they provide, but rather by the non-wood forest products (NWFPs) and other services. NWFPs, such as cork, game, honey, fruits and mushrooms, besides meeting local needs, contribute to the national economies. For example, cork oak (*Q. suber*) is a very significant source of income for the western Mediterranean and in particular Portugal, Spain, Morocco, Algeria and Tunisia. Other important species for NWFPs in the Mediterranean are, for example, stone pine (*Pinus pinea*) in Lebanon and chestnut (*Castanea sativa*) in Italy, Greece and Spain. Fodder provided by the ground cover and forest trees, such as cork oak, represent vital contributions to the local economies in the region.

Mediterranean forests are notable for the protective and environmental functions, such as protection of watersheds and agricultural soils, water conservation and purification. A large number of Mediterranean species perform these functions well: for example, *Cedrus* spp., *Juniperus* spp. and *Q. ilex* in the southern Mediterranean countries, such as Morocco and Algeria. In addition, plantations with species such as *Pinus halepensis* and stone pine are used, for example, in Israel and with *Casuarina* in Egypt for protection against desertification.

Mediterranean forests offer large recreational opportunities, especially on coastal areas which represent the world's leading tourism destination. One should think also of the cedar reserves in Lebanon, not only as the symbol of their national identity, but one of the most attractive places for international visitors. Mediterranean forests represent a rich reservoir of biodiversity, with their high number of endemic species and plant diversity when compared with other forest ecosystems (Box 2.2).

Threats to Mediterranean Forests

Throughout history, major human-related factors have contributed to forest degradation. From ancient times to the 19th century, overuse of timber for the shipbuilding industry and war needs has strongly affected forests in the whole of the Mediterranean region. The invasions and occupations during the Middle Ages led to forest clearing for agriculture, especially on mountainous steep slopes. Grazing, especially by goats, played an important role in the areas with a dominant pastoral economy: the southern and eastern countries and Spain. Human-made forest fires have always been created in the past in the whole area of the Mediterranean (Thirgood, 1981; Meiggs, 1982). These actions induced a high level of forest destruction in many Mediterranean areas. It is often stressed that destruction was greater in the southern and eastern countries, where the harsh climate and the historical events were especially concentrated. An illustrative example is the disappearance of the famous cedar in Lebanon, now in only patchy remnants; and the dramatic reduction of the forest area in Cyprus (Tsoumis, 1986).

Today, several characteristics and especially differences across the Mediterranean

---

Box 2.2. Biodiversity and Mediterranean forest ecosystems (Yves Birot).

The Mediterranean bears the marks of millennia of interactions between man and nature. Over the centuries, human cultures in the region have emerged and adapted to their environment, leaving a marked impact on local biotic resources, by using, moving or altering them.

The Mediterranean is particularly important for biological diversity due to the huge plant diversity it contains, and because of its high level of endemism. The Mediterranean has the second highest number (13,000) of endemic plant species in the world after the tropical Andes with its 20,000 plant species. Other Mediterranean-like ecosystems also have a very high level of plant species diversity. Mediterranean forests contain about 250 arborescent species, among which 150 are endemic (exclusively or preferably found); moreover, 15 genera are specific to Mediterranean forests (Quézel et al., 1999). Forest areas are generally seen as being the richest ecosystems in terms of biodiversity. Tree species are also known for having the highest level of intraspecific genetic diversity among living organisms.

Unfortunately, significant threats exist to the habitats and species in the region, particularly due to increased human pressure. The Mediterranean Belt is one of the 25 global 'hot spots', characterized by a high level of species diversity and by heavy habitat losses (Myers et al., 2000). The European Union's (EU) 1992 'Habitats' Directive (which implements 'Natura 2000', the network of protected areas of particular importance for the conservation of biodiversity within the EU) identifies 11 biogeographical zones in Europe. Of these, the Mediterranean zone has the highest number of habitats (142) and the highest number of species (386) that need to be protected. Many populations or ecotypes have already disappeared or are endangered.

In the Mediterranean, the major threats to forest biodiversity are: wild fires (especially when their frequency is high), overgrazing, rural decline leading to landscape uniformization, some silvicultural practices, as well as afforestation or reforestation based on 'monocultures'.

Implementing concrete measures for biodiversity conservation is difficult because expectations differ greatly among stakeholders and even within the same group of stakeholders. Some natural scientists are more interested in particular birds, others in some plants, whose ecological requirements are incompatible. In Mediterranean forests, the effect of fire is positive for a bird of the *Oenanthe* genus but negative for turtle species. Moreover, too much focus on protecting a single emblematic species can result in negative effects. In French forests, voluntary undershooting of wild boars by hunters has contributed to a demographic explosion, which has caused problems to crops and led to a dramatic increase in the number of collisions between boars and cars. In the Camargue, overprotection of the flamingo has resulted in heavy predation on paddy fields, and competition with other bird species. Conserving biodiversity is not an impossible ideal, but it does require a balanced and integrated approach. More generally, biodiversity conservation cannot be addressed separately from the economic and social development.

countries have an influence on shaping the development of the Mediterranean forests. Most obviously, northern countries are markedly better off than the rest of the region: *per capita* gross domestic product (GDP) is more than €11,000 in Western Europe, while it is between €1000 and €3000 in most southern and eastern countries. There are also marked demographic differences: annual population growth is around 2% in southern and eastern countries, with peak values in Palestine (3.8%), Syria (2.6%) and Israel (2.4%). In contrast, Western European countries are characterized by an ageing population with negligible growth (United Nations, 2001a). These differences have induced a significant population migration from the southern and eastern to the northern Mediterranean in the last few decades. Although these flows have slowed down recently, this is due to legislative changes rather than economic transformation in the origin countries (Courbage, 2000). Urbanization, which presently stands at about 63% in the region (United Nations, 2001b), is growing rapidly in southern and eastern countries, and particularly in coastal areas and large cities. Urbanization is expected to reach 73% by 2025 (Plan Bleu, 2002). Nevertheless, rural areas still account for a large share of the population in southern and eastern countries.

These characteristics have led to very different threats to forest areas. Demographic growth, relatively low income *per capita* and limited diversification of activities have contributed to a high level of pressure on forest resources in southern and eastern countries. Deforestation for timber and agriculture, excessive grazing and urban encroachment are common phenomena in these countries. These actions, combined with a dry and harsh climate, give rise to important negative effects, such as losses of agricultural fertility, biodiversity losses, increased erosion and desertification.

In northern countries, and particularly in Western European countries, the situation is quite different. During recent decades, high income levels and low demographic growth have resulted in an increasing rate of land abandonment as people sought better economic opportunities in other sectors. Forestry tends to be unprofitable, especially in mountainous areas. This has resulted in an expansion of forest area. While this was considered desirable in the past, it increasingly has come to be seen as having negative consequences, such as an increased risk of forest fires, loss of landscape quality and, to a certain extent, deterioration of traditional rural lifestyles.

Institutional differences accentuate the pressures. Most forest areas in southern and eastern countries are public – indeed, practically all forests are public in Morocco, Syria and Turkey. As a result, forest populations tend to have little incentive to conserve forests, which favours the occurrence of illegal activities. In contrast, the majority of forestland is privately owned in the northern countries, which presently lack a certain level of care.

Coastal areas are particularly critical. The Mediterranean coastal areas are the most important tourist destination, and have a high concentration of population and economic activities. Coastal areas support about 135 million inhabitants, or about 34% of the total population in Mediterranean countries. Population density in coastal areas is about 131 people/km^2, more than 2.5 times higher than the Mediterranean average (United Nations, 2001a; Plan Bleu, 2002). Increasing population density and economic activity in coastal areas give rise to a variety of conflicts, with incompatible activities such as urbanization, agriculture and forestry competing for land, and natural ecosystems being subject to intense human pressure, such as pollution, natural disturbances and destruction.

Conclusions

Mediterranean forests have historically and, in the main, satisfactorily performed numerous functions from watershed protection to provision of WFPs and grazing area. NWFPs, compared with other forest types, have been and still are important for local and national economies. This multi-functionality has been recognized and incorporated into forest management and the policy of several northern Mediterranean countries since a few centuries ago. At the same time, in most southern and eastern countries, this multi-functionality, although theoretically acknowledged, is not represented in guidelines for forest management, which still embrace the sustainable yield principle as a core.

The multi-faceted context of forests growing in the Mediterranean region bears the mark of a space embracing different socio-economic, cultural and environmental conditions. Various threats and, often, conflicting trends affect the Mediterranean forests and their benefits. Therefore, there is a strong need for the value of forestry to be recognized in order that its multi-functionality can be incorporated within forest management guidelines. How these values can be estimated through the lens of economics is considered in subsequent chapters.

Notes

[1]   These estimates are based on Food and Agriculture Organization (FAO) forest data, which include natural forests and plantations. The FAO considers land with a tree canopy of more than 10% and an area of more than 0.5 ha as being forested. Forests are determined by both the presence of trees and the absence of other predominant land uses. The trees should be able to reach a minimum height of 5 m. Young stands that have not yet reached, but are expected to reach, a crown density of 10% or tree height of 5 m are included under forest, as temporarily unstocked areas (Food and Agriculture Organization, 2001a).

[2]   These data do not include 'other wooded land', which often consists of degraded areas where wood volume is much lower. Including these areas would probably have lowered the average wood volume in the Mediterranean and in the subregions to below the world mean.

[3]   If 'other wooded land' is included, the proportion of 'forests and other wooded land' in northern countries falls to around 65% of the Mediterranean total, while it rises to 14% in the south and 21% in the east (see Table 4.13).

[4]   Several analyses have been conducted based on floristic, vegetation structure, climatic and bio-climatic criteria. Many of these delimitations have proved controversial, or at least not fully accepted. For a comprehensive review, see for example Quézel (1985).

[5]   It should be noted that not all this area is included in the 'forest area' definition of the FAO, therefore the data from Figs 2.1 and 2.2 are not always comparable.

[6]   The share is calculated with respect to the areas of the countries for which estimates are available. Countries such as Portugal and Egypt are not included.

[7]   Growing or thriving best in an intermediate environment (as in one of moderate temperature).

References

Arbez, M. (2001) Cultivated forests of South-Atlantic Europe – potentials and research needs. *EFI News* 9, 3–5.

Boydak, M. and Dogru, M. (1997) The exchange of experience and state of art in sustainable forest management (SFM) by ecoregion: Mediterranean forests. *XI World Forestry Congress.* Antalya, Turkey, 13–22 October 1997, Vol. 6, Topic 38.3.

Courbage, Y. (2000) Populations and migrations. In: Vidal-Beneyto J. and Puymège G. (eds) *La Méditerranée: Modernité Plurielle.* Editions UNESCO/Publisud, pp. 169–190.

Food and Agriculture Organization (2001a) *The State of the World's Forests.* FAO, Rome.

Food and Agriculture Organization (2001b) *Global Forest Resources Assessment 2000: Main Report.* FAO Forestry Paper No. 140. FAO, Rome (also available at www.fao.org/forestry/fo/fra/main/index.jsp).

Food and Agriculture Organization (2003) *The State of the World's Forests.* FAO, Rome.

Koppen, W. (1923) *Die Klimate der Erde.* De Gruyter, Berlin.

Lemée, G. (1967) *Précis de Biogéographie.* Masson et Cie, Paris.

Meiggs, R. (1982) *Trees and Timber in the Ancient Mediterranean World.* Oxford University Press, Oxford, UK.

M'Hirit, O. (1999) Mediterranean forests: ecological space and economic and community wealth. *Unasylva* 50, 3–15.

Myers, N., Mittlemeier, R.A., Mittlemeier, C.G., Da Fonseca, G.A.B. and Kent, J. (2000) Biodiversity hotspots for conservation priorities. *Nature* 403, 853–858.

Pandey, D. and Ball, J. (1998) The role of industrial plantations in future global fibre supplies. *Unasylva* 49, 37–43.

Plan Bleu (2002) www.planbleu.org

Quézel, P. (1976) Les chênes sclérophylles en région méditerranéenne. *Option Méditerranéenes* no. 35, CIHEAM, Paris, pp. 25–29

Quézel, P. (1985) Definition of Mediterranean region and the origin of its flora. In: Gomez-Campo, C. (ed.) *Plant Conservation in the Mediterranean Area, Geobotany, 7.* W. Junk, Dodrecht, The Netherlands, pp. 9–24

Quézel, P., Médail, F., Loisel, R. and Barbero, M. (1999) Biodiversity and conservation of forest species in the Mediterranean basin. *Unasylva* 50, 21–28

Scarascia-Mugnozza, G., Oswald, H., Piussi, P. and Radoglou, K. (2000) Forest of the Mediterranean region: gaps in knowledge and research needs. *Forest Ecology and Management* 132, 97–109.

Thirgood, J.V. (1981) *Man and the Mediterranean Forest – A History of Resource Depletion.* Academic Press, New York.

Tsoumis, G. (1986) *The Depletion of Forests in the Mediterranean Region: a Historical Review from Ancient Times to Present.* Scientific Annals of the Department of Forestry and Natural Environment, Aristotelian University of Thessaloniki, Greece, Vol. 28, 265–301.

United Nations (2001a) *World Population Prospects: the 2000 Revision.* Population Division and Social Affairs, United Nations Secretariat.

United Nations (2001b) *World Urbanisation Prospects: the 2001 Revision – Data Tables and Highlights.* Population Division and Social Affairs, United Nations Secretariat.

3 Concepts and Methodology: a First Attempt Towards Quantification

Maurizio Merlo[†] and Lelia Croitoru
University of Padova, Centre for Environmental Accounting and Management in Agriculture and Forestry (CONTAGRAF), Via Roma 34, Corte Benedettina, 35020 Legnaro (PD), Italy

The wide range of benefits that forests provide creates a challenge for analysis. A coherent analytical framework is needed to ensure that these benefits are considered systematically and comprehensively, but without double counting. This chapter describes the conceptual framework used in this book, that of total economic value (TEV). It then discusses the specific approaches used to put this framework into practice.

The Total Economic Value Concept

The framework

Value is a multifaceted concept. The TEV has been used in recent years to identify and, to a certain extent, quantify the full value of the different components of natural resources such as forests. Traditional use and exchange values, as considered by classical economics (Smith, 1776), have been complemented by other value types, including option values (Weisbrod, 1964), quasi-option values (Arrow and Fisher, 1974; Henry, 1974), and vicarious, existence and bequest values (Krutilla, 1967). Figure 3.1 classifies them more straightforwardly[1] into direct and indirect use values, option and non-use (bequest and existence) values.

Use values are benefits derived from actual forest use.

1. *Direct use values* are benefits derived from the direct use of forests. They can be consumptive, such as extraction of timber, firewood or forest fruits, or non-consumptive, such as recreation and landscape quality.

2. *Indirect use values* are benefits derived from the forest ecosystem's functions, such as watershed protection, water purification and carbon sequestration.

3. *Option values* are benefits derived from having the option of using forests in the future, directly or indirectly. Whether they belong to the use or non-use values categories is still a controversial matter in the environmental literature. For simplicity, this book considers them as an individual category.

Non-use values are benefits not linked to the actual use of forests.

1. *Bequest values* arise from placing a value on conservation of a particular forest feature for future generations, for example conserving a forest site for the enjoyment of one's heirs.

[†] Deceased.

©CAB International 2005. *Valuing Mediterranean Forests*
(eds M. Merlo and L. Croitoru)

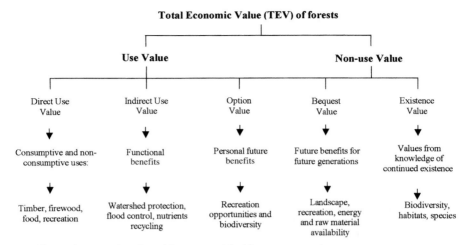

Fig. 3.1. The total economic value of forests. Modified from Pearce and Moran (1994).

2. *Existence values* derive from the knowledge of the existence of a particular forest characteristic, such as habitat conservation.

It should be stressed that the TEV is only a part of the total value of an ecosystem. TEV is an anthropocentric concept that stresses values that bring benefits to human beings, whether directly or indirectly. Many also consider forests and other ecosystems to have *intrinsic value* independent of human preferences and of whether they contribute to human welfare (Goulder and Kennedy, 1997; Millennium Ecosystem Assessment, 2003). In addition, certain forests may also have value for traditional, historical or religious reasons; this may be called *socio-cultural value*. Some natural scientists have articulated a notion of *ecological value* in reference to causal relationships between parts of an ecosystem and the characteristics, structures and processes that ensure its continued functioning (Farber *et al.*, 2002; Gren *et al.* (1994) call this 'glue value', and the Millennium Ecosystem Assessment (2003) describes it as the 'regulating services' of an ecosystem. Some of these other values can, in part, be captured within the TEV; for example, intrinsic and socio-cultural value may, to some extent, form part of existence value and be elicited through contingent valuation surveys. This is not always possible, however, and not always acceptable to the stakeholders (Turner *et al.*, 2000, 2003). A further discussion about the

different value types under the TEV is given in Box 3.1.

The concepts

The theoretical framework of the TEV can be related to a pragmatic view of market, potential market and non-market values of forest resources, depending on whether forest outputs are sold, can be sold or cannot be sold on real markets. The TEV framework can also be adapted to encompass private and public goods and externalities with various gradations, including club and local goods (Fig. 3.2).

When well-functioning markets for goods and services exist, their value can be readily observed. The value of goods such as timber, honey or cork, for example, can usually be measured by observing their *market value*.[2] In contrast, *non-market values* are related to goods or services that have no market price – the so-called 'free' goods: firewood for self-consumption, biodiversity conservation, watershed protection and carbon sequestration. For some of these goods, such as firewood, there are markets for substitute products (charcoal, for example). Other benefits can be realized through the market place, providing a *potential market value*. For example, in many Mediterranean countries, markets for carbon sequestration are missing, but they can be created in the future,

Box 3.1. A cautionary note: towards the TEV of forests or not? (Américo Carvalho Mendes).

This book attempts to quantify, in *monetary* terms, the use values (direct and indirect), the option and non-use values of Mediterranean forests. More or less comprehensively, the country chapters quantify the use values. Some go further and attempt to quantify some of the option, bequest and existence values. However, even had sufficient data been available to quantify all these values, that still would not be a full valuation of forests, and the kind of valuation accomplished would rely on a very strong assumption: the assumption of actual or possible *permutability* of all forest goods and services.

What is value?
Let us start by defining a general concept of value or, more precisely, of a valuation process. A valuation process of things or of other kinds of objects (human actions, actions of other beings) has three dimensions:

- it is a process of separation and objectivation, i.e. a process through which someone is established as a valuation 'subject' and something is established as a valuated 'object' ontologically distinct from the valuator;
- it creates a relationship of signification between the subject and the object, i.e. the object will mean something to the subject;
- this meaning motivates actions from the subject.

This implies that valuation processes rely on social norms regulating the modes of separation and access of persons to the objects of value; the meanings of objects, the communication of those meanings to other people and the modes of action of human beings on the objects they value.

What are the different types of values of things?
Use value is the valuation of things based on *functionality* (What is this thing for? What can we make with it?) and *comparability* (Is this thing as useful as that one for the purpose at stake?). *Exchange value* is the valuation of things based on *functionality*, *comparability* and *permutability* (How much is this thing worth in terms of another thing?). Exchange value requires, in addition to use value, the institutions making possible the permutation of things among human beings. The 'total economic value' of forests is an exchange value concept. So it implicitly relies on the strong assumption that all forest goods and services are or can be made permutable once the appropriate institutions are in place.

Are these all the values of things?
Even if the strong assumption mentioned above holds, and even if there are enough data to value all forest goods and services in monetary terms, would that be a full valuation of forests? I do not think so. Things, including forest goods and services, may have other values besides use and exchange values. One is '*informative value*': things have meanings which human beings communicate among themselves. So, the basis for valuation here is *communicability*. Meaning is a kind of value which is not quantifiable. Nowadays this type of value is being recognized within the 'cultural value' of forests.

Another basis for valuation of things which leads to a non-quantifiable type of value is *community*. This is the valuation involved when things are transferred among human beings as donations. The purpose of these transactions is to establish or reinforce a personal relationship among the human beings involved. I call this the '*symbolic exchange value*' of things because, even though the donor does not negotiate with the beneficiaries, he expects (but cannot be sure) they will respond with the kind of personal relationship he would like to have with them. This type of transaction is very frequent within households and among members of the same local community. It is easy to find relevant examples in private and community forestry.

Another unquantifiable type of value is based on *sensibility*, i.e. on the feelings things generate in human beings who perceive them. These feelings are personal and not fully transmissible. This '*aesthetic value*' is certainly related to all the others, but has its own autonomy: things may have an aesthetic value for someone without necessarily being used to make other things, to be exchanged for other things, to be donated or to be communicated. It is obvious that forests have this kind of value.

The last basis for valuation is *identity*. There are things, also in forestry, which are part of an individual or a community's identity. They cannot be transferred separated from the human beings whose identity they belong to. I call this the '*identity value*' of things. A part of the 'cultural value' of forest fits into this type of valuation.

continued

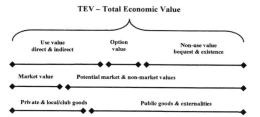

Fig. 3.2.   Possible value types within total economic value. From Croitoru *et al.* (2000).

as they are clearly emerging in other places (Prototype Carbon Fund, 2001, 2002; Pagiola *et al.*, 2002; World Bank, 2003). Finally, for some other goods, there is no closely comparable substitute and little potential for an actual market (water).

Private goods are goods for which consumption by one individual excludes others. They are excludable (to those who pay a price) and rival (one piece of goods can only be consumed by one individual). They are generally 'tangible' goods, such as timber. In contrast, *public goods* are non-excludable and non-rival. As an example of public goods, a beautiful landscape can be enjoyed by all passers-by without an individual's enjoyment of it detracting from that of another – until congestion occurs. They can also be public 'bads', such as smoke generated by a factory, the negative effects of which are collectively 'consumed' by downstream communities.

The notion of *externality* is quite complex (see Baumol and Oates, 1975) and many definitions and classifications are found in the literature.[3] *Externalities* occur when the consumption or production activity of one person or firm

affects another person's utility without being fully or directly reflected by market prices. In other words, we talk about uncompensated costs or benefits arising from economic activity. For example, the benefit of soil conservation due to a good forest management is a positive externality for the agriculturists and rural communities living downstream, who usually do not support the costs of its provision. Soil erosion due to deforestation and overgrazing is a very common negative externality in southern Mediterranean countries.

Between pure private and public goods, regarded as polar cases (Samuelson, 1954, 1955), there is a wide array of *impure* or *mixed* goods, with various degrees of rivalry and potential excludability. Among them, the club goods[4] are characterized by special ownership or membership restrictions, with the purpose of maximizing the utility in their consumption by imposing a certain level of exclusion. Examples include grazing by forest villagers, who often pay a low price or no price at all, especially in the southern Mediterranean. Similarly, timber and firewood collection is often free (or charged at a purely symbolic price) for forest communities.

Figure 3.2 shows that most direct use values (timber, firewood and forest fruits) are market private goods, while moving along to the right, the various use and non-use values become potential market or non-market public goods and externalities. This book aims at estimating all these values, giving particular attention to the non-market public goods and externalities. By their very nature, public goods and externalities are more intangible than private goods and, therefore, more difficult to quantify. Estimating these outputs should be based on a general context defining the purpose and the broad issues related to the valuation.

The context

The most appropriate approach to forest valuation depends on the reasons for the valuation efforts. The most common reasons are to assess the overall contribution of forests to social and economic well-being; to understand how and why the stakeholders use forests as they do; and to assess the relative impact of alternative actions so as to help decision making.

This book is aimed primarily at the first objective. It focuses on measuring the current benefits provided by Mediterranean forests. At any given time, forests provide a specific *flow* of services, depending on the type of forest, its condition (the *stock* of the resource), type of management and socio-economic context. Thus, the book estimates the annual *flows* of outputs provided by forests as they stand now, under current management practices. Many of these outputs are benefits with positive value. Other outputs are social costs, including the externalities and public 'bads' caused by human factors (poor forest management) or natural hazards (dry climate, exceptional temperature or irregular rainfalls). As such, they have negative value.

A very different type of analysis would be required for the second and third objectives. Assessing the relative impacts of alternative actions would require an analysis of the *changes in the flows of services* that would result from each alternative under consideration. Understanding the reasons for stakeholder behaviour would require an examination of the specific subset of the costs and benefits of forests that they receive. Local populations, for example, are likely to receive many of the direct use values of forests but few of the indirect use values. These are all important issues, with important implications for policy. Although the discussion in this book often touches on these issues, particularly in the country chapters, they are not the main objective of the analysis.

The positive and negative forest outputs are shown in Fig. 3.1. The positive outputs can be easily divided among the TEV categories. Also some negative outputs can be easily attributed to the different value categories. For example, erosion, floods and avalanches due to poor forest management are negative externalities affecting the indirect use values of forests. Other negative outputs are, however, more difficult to separate,[5] as they affect two or more categories of value. For example, the damage due to forest fires reduces not only the direct use values (wood forest products (WFPs) and non-wood forest products (NWFPs)) but, above all, the indirect, option and non-use values. For this reason, Table 3.1 presents all the negative outputs without distinguishing them among the different TEV categories.

Identification and quantification of the different forest benefits is undertaken with consideration of the links between the different forest uses. In the Mediterranean context, the multiple uses of forests are interdependent and often competing. For example, erosion caused by excessive timber extraction in a watershed is a common feature in southern Mediterranean countries. Identifying the extent to which a forest use provides a benefit and the threshold above which an increase of this benefit is associated with a social cost should be based on a reference point.[6] Some authors have tried to define what is positive and what is negative making reference to legal boundaries or, more precisely, property rights (Hodge, 1991). This approach, though appealing for its simplicity, seems just a first step. In fact, the definition of acceptable reference points is a more complex issue, requiring investigation in relation to environment, land uses, goods and services provided (Gatto and Merlo, 1999).

Methods and Approaches for Valuing Mediterranean Forests

In all country chapters,[7] estimating forest values was based on the TEV framework. The valuation efforts required not only a common overall framework, but, in particular, similar approaches for valuing forest benefits. This was meant to ensure that the country chapters are as

Table 3.1.   Outputs of Mediterranean forests according to the total economic value.

	TEV categories	Positive outputs	Negative outputs
1. Use values	1.1. Direct use values	Timber, firewood, cork, resin, hunting, grazing, sparto grass, honey, decorative plants, mushrooms, recreation, medicinal plants, berries and truffles	Damage by forest fires Erosion, floods and avalanches due to poor or no forest management Pollen and other allergic factors Loss of recreation opportunities due to intensive plantation forestry and poor management Loss of landscape values to excessive expansion of forest land use Loss of biodiversity and landscape values due to plantation forestry
	1.2. Indirect use values	Protection: watershed management, soil conservation, avalanche prevention and flood prevention Landscape quality Micro-climate regulation Water quality and purification Conservation of the local ecosystem	
2. Option values		Personal future recreation and environmental interests Potential source of energy and raw materials Potential unknown source of biodiversity and medicinal plants, etc. Potential use of unused landscape resources	
3. Non-use values	3.1. Bequest values	Landscape, recreation, energy and raw material availability, biodiversity and environmental conditions, such as related to carbon storage affecting future generations	
	3.2. Existence values	Biodiversity and environmental conditions, e.g. related to carbon storage, affecting other species, Respect for the right or welfare of non-human beings including the forest ecosystem	

Source: adapted from Merlo and Rojas (2000).

homogeneous as possible and that results are comparable across countries. In certain cases, however, the scarcity of data necessitated the use of slightly different methods and approaches. This often poses difficulties in interpreting and comparing the results.

Scope

The original intention of the study was to analyse and value the benefits of Mediterranean forest types. However, with few exceptions, available information does not separate data on Mediterranean forest types from that of other forest types. Only Croatia and France provided data that allowed Mediterranean forests to be distinguished consistently from other forests. Because of this, the scope of the study was adapted to cover all forests in Mediterranean countries. Wherever possible, results for Mediterranean forests[8] are distinguished and compared with those of other forest types, but

the overall analysis covers all forests in each country.

Differences in national definitions also prevented a common definition of forests from being adopted. Comparability would have been enhanced if a consistent definition of forests could have been used. Attempting to adopt a common definition would, however, have further reduced the available data in most countries. It can be noted that among the northern countries, several adopted the Food and Agriculture Organization (FAO) data on forest area and 'other wooded land'; while for others, the data, even though taken from national statistics, are similar to the FAO estimates. For the southern countries, national inventories were preferred in order to account for the typical Mediterranean vegetation (e.g. *Stipa tenacissima* in Morocco) which is representative for these countries.

Within these forest areas, the analysis attempts to estimate *all* of the benefits they are generating annually, no matter to whom these benefits accrue. Some benefits, such as most direct use values, will be enjoyed primarily by local populations. Others, such as many indirect use values, may be received by other groups within the same country but at some distance from the forests themselves. Watershed protection benefits, for example, typically are enjoyed by populations far downstream from the forests that provide them. Indeed, the beneficiary populations need not be in the same country as the forest itself. This is particularly true of global benefits such as carbon sequestration (which helps to mitigate climate change) and biodiversity conservation. It is also true of benefits such as recreation: when the visitors to a forest are foreigners, the increase in well-being that the forests provide is received by citizens of another country. The analysis in a few individual chapters discusses which groups benefit from which forest values. The TEV estimates themselves, however, do not consider these distinctions; they aggregate all benefits generated by a given country's forests. The distinction becomes important in discussing policy issues, however.

As the study focuses on estimating the annual flow of benefits from forests, a base year for the analysis had to be selected. The year chosen was 2001, the most recent for which reasonably complete data were available in each of the countries studied. Inevitably, some data were not available for this year; in this case, the most recent applicable data were used. Data collected on an annual basis also often had to be complemented by data from one-off studies on particular issues. The results of these studies then had to be adjusted to bring them in line with the chosen base year. Comparability of results was aided by the fact that several of the countries use a common currency, the euro. All prices[9] used in the analysis were adjusted for inflation and converted to euros, using average exchange rates for 2001 (International Monetary Fund, 2004).

Data collection

A structured questionnaire was developed, reflecting as far as possible the underlying theory of the TEV. It primarily aimed at identifying and quantifying in physical and monetary terms the public goods and externalities provided by the Mediterranean forests, in addition to other

forest outputs. The questionnaire was structured in ten sections according to the following topics:

1. *Basic forest data*: forest area, types of forests, level of degradation, main forest functions, ownership patterns, average size of forest properties or management units.

2. *Macroeconomic forest-related indicators*: contribution of forestry and related activities to the gross domestic product (GDP), national employment and trade.

3. *Availability of statistics on forest outputs*: the extent to which physical and monetary data on market and non-market forest outputs are available.

4. *Identification of forest public goods and externalities*: adapting a given list of forest outputs (Table 3.1) divided according to the TEV framework, to the forest peculiarities in each country.

5. *Direct use values*: physical and monetary estimates of the identified direct forest uses, a brief outline on the allocation of property rights on forest products and a summary of the forest institutional and administrative framework is required.

6. *Indirect use values*: physical and monetary estimates of indirect forest uses, description of valuation methods used and level of valuation.

7. *Option values*: physical and monetary estimates of option values, followed by a description of valuation methods used.

8. *Bequest and existence values*: physical and monetary estimates of bequest and existence values and description of valuation methods used.

9. *Other externalities affecting forest TEV*: physical and monetary estimates on the externalities not accounted for within the preceding sections.

10. *Final considerations*: official and non-official sources of data used as references.

Methods and approaches for valuing Mediterranean forests

This section introduces the methods and techniques used for estimating forest values in the country chapters. It is not intended to provide a detailed technical description of each valuation technique, as this can be found in numerous other publications (Hufschmidt *et al.*, 1983; Braden and Kolstad, 1991; Winpenny, 1991; Dixon *et al.*, 1994; Organization for Economic Cooperation and Development, 1995). Rather, this section discusses the specific application of these techniques to the problem of valuing the benefits of Mediterranean forests.

A wide array of valuation methods has been developed. They are commonly divided into demand curve approaches and non-demand curve approaches. Demand curve approaches seek to estimate the value of goods and services by explicitly estimating the consumers' demand for them. These approaches include (Bateman, 1994): (i) *revealed preference methods*, based on observations of actual consumer behaviour in markets for the goods and services of interest, or in surrogate markets for related goods and services; and (ii) *expressed (or stated) preference methods*, based on elicitation of consumers' willingness to pay for a benefit or willingness to accept a compensation for a loss. Non-demand curve approaches value environmental benefits and costs via market priced goods and cost-based methods.

Measures based on observed behaviour are usually preferred to measures based on hypothetical behaviour, and more direct measures are preferred to indirect measures. However, the choice of valuation technique in any given instance depends on the characteristics of the case and data availability. Figure 3.3 shows a taxonomy of these methods, while Table 3.2 summarizes the application of these methods and the indicators used in the individual country chapters. The following sections discuss the methods used in more detail.

Estimating direct use values

Estimating direct use values is usually more straightforward than other TEV components. Many outputs are commercialized extensively in the market, and their quantities and average prices are often known. However, some outputs are subject to thin local markets, with considerable intra-country variation in quantities and prices; other outputs, although marketable, are consumed free of charge. Valuation of these outputs is more difficult and involves more advanced techniques than the use of market price. Five major categories of direct use values are considered in most chapters: (i) WFPs, including timber and firewood; (ii) NWFPs such

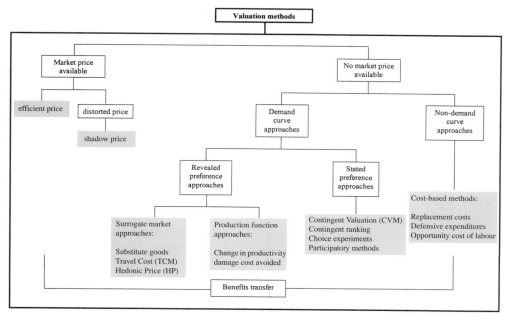

Fig. 3.3.   A taxonomy of valuation methods.

Table 3.2. The use of valuation techniques in the Mediterranean countries.

Value type	Outputs	Valuation techniques	Physical indicators	Monetary indicators used (€)	Countries
Direct use values	WFPs for sale	Market price	Quantity of WFPs (m³)	Roadside price[a]	All countries
	WFPs for subsistence	Market price of similar goods	Quantity of WFPs (m³)	Roadside price of firewood	Morocco, Tunisia, Syria
				Shadow price	Turkey
	Net growth of timber	Market price	Quanity of timber growth (m³)	Half of stumpage price	Syria, Cyprus, Greece, Croatia, Slovenia, Italy, Portugal, France[b]
	NWFPs	Market price	Quantity of NWFPs (t)	Market price	All countries
		Opportunity cost of labour	Quantity of myrtle, rosemary, carob (t)	Cost of harvest and harvest fees	Tunisia
		Opportunity cost of labour and other raw materials	Quantity of carob exports (t)	Net benefit from carob exports	Lebanon
		Substitute goods	Quantity of forage units (FU) corresponding to 1 kg of acorns	Price of barley	Tunisia
	NWFPs for subsistence	Market price of similar goods	Quantity of medicinal plants (t)	Market price of similar goods	Lebanon, Syria
	Grazing	Substitute goods	Quantity of forage grazed (FU)	Price of barley/hay	All countries, except Egypt, France, Slovenia
	Hunting	Fees (permits, licences)	Hunters (no.)	Fees and game value	Morocco, Tunisia, Lebanon, Turkey, Cyprus, Greece, Albania, Slovenia, Portugal
		CVM	Hunters (no.)	Consumer surplus	Croatia[c], Italy
	Recreation	Cost-based method	Visits (no.)	Travel costs	France
		TCM, CVM	Visits (no.)	Consumer surplus/visit	Tunisia, Cyprus[b], Italy, Spain, Portugal
		Cost-based method	Visits (no.)	Travel costs, entrance fees	Israel, France, Turkey
Indirect use values	Watershed protection (water-related issues)	Damage costs avoided	Area protected by forests (ha)	Expenses avoided by the presence of forest cover	Morocco, Algeria, Tunisia, Syria, Greece, Italy, Portugal
		Preventive expenditures	Area under high risk of soil erosion (ha)	Budget expenses for preventing erosion	Albania
	Soil conservation	Change in production function	Incremental agricultural yield (t)	Market price of agricultural production	Tunisia
	Carbon sequestration/ emissions	Shadow price	Net change of carbon sequestered in forest biomass and soils (tC)	Shadow price of carbon	All countries

continued

Table 3.2. *Continued.*

Value type	Outputs	Valuation techniques	Physical indicators	Monetary indicators used (€)	Countries
Option, bequest and existence values	Pharmaceuticals	Rent capture	Plant species (no.)	Market price of pharmaceuticals	Turkey
	Biodiversity conservation	Cost-based approach CVM	Protected area (ha)	Annual expenses for preserving biodiversity	Tunisia, Lebanon, Turkey, Greece, Italy, Spain, Portugal
				WTP for preserving biodiversity	Italy, France
	Potential use of environmental services	CVM	Population number Forest area (ha)	WTP for future recreational opportunities	Croatia
Negative externalities	Erosion, floods and landslides	Replacement cost	Erosion rate (m³/ha)	Cost of restoring erosion	Morocco, Algeria, Syria, Albania, Greece, Italy
		Change in production function (quantitative valuation) and replacement cost (monetary valuation)	Loss of soil nutrients (t)	Cost of fertilizers	Turkey
	Damage caused by forest fires	Restoration cost/or value of damage	Area burnt by fires (ha)	Cost of restoration/or value of wood and other NWFPs and services lost due to forest fires	All countries, except for France and Spain
	Watershed damages caused by deforestation, overgrazing	Cost-based approach	Area affected by deforestation (ha)	Cost of repairing watershed damage	Morocco
	Desertification	Change in production function	Loss of agricultural productivity	Market value of the agricultural loss	Algeria
	Damages due to poaching	Change in production function (quantitative valuation) and replacement cost (monetary valuation)	Forest area affected by poaching (ha)	Cost of reconstituting the game population	Morocco
	Losses of natural and landscape quality due to illegal acts	Cost-based approach	Illegal acts (no.)	Value of fines paid	Tunisia, Albania, Greece, Italy
	Allergy and other negative effects to human health	Cost-based approach (defensive expenditures)	Forest area (ha)	Cost of applying insecticides	Cyprus, France
	Agricultural damage by forest game	Damage cost	Number of farmers	Compensation to farmers	Slovenia, France

[a]Stumpage price for France. [b]The 'net growth' was considered as option value. [c]Used the benefits transfer of the CVM from Italy.
CVM = contingent valuation method; TCM = travel cost method.

as cork, honey and mushrooms; (iii) grazing; (iv) recreation; and (v) hunting.

Wood and non-wood forest products

Market prices are used to value both WFPs and NWFPs based on the quantities traded on the market and their market price. Market prices are acceptable as valuation proxies, provided that markets for these products are efficient. If prices are distorted due to public policy or other types of failures, they need to be adjusted by eliminating distortions (*shadow pricing*) (Gittinger, 1982; Monke and Pearson, 1989).

In this book, these prices are used in combination with estimates of current harvest rather than the sustainable harvest or potential yield. These data are generally obtained from national statistics. These are generally readily available for WFPs, although they may in some cases omit significant amounts of removals – particularly where illegal harvesting is common, as in several countries in the southern Mediterranean.

Ideally, stumpage prices – the prices net of extraction and transport costs – would be used to value WFPs. In many countries, however, stumpage prices of WFPs are not available at the national level, as local stumpage prices vary substantially even within the same country. For this reason, monetary valuation is based on the more readily available roadside prices of WFPs. To a certain degree, this results in overestimation, as it includes the costs of extraction and transport to the roadside. The value of WFPs collected for free by forest users (often for subsistence) is estimated using the market price of similar goods (e.g. firewood sold in other areas in Morocco and Tunisia) or shadow prices after eliminating distortions (in Turkey, for example).

In some northern countries, the annual timber growth is higher than the removals, as a result of either forest protection management or abandonment. The difference between them (the 'net growth') is often considered to have a positive value. As the 'net growth' is, by definition, not harvested, its value is assumed to be less than that of timber that is actually exploited. As an admittedly crude approximation, monetary valuation in all countries is based on half of the stumpage price. In contrast, for other countries (usually southern), illegal practices often result in unsustainable wood harvest. These quantities

represent a private benefit for forest users and are estimated based on the roadside prices of similar goods.

Damage to WFPs due to forest fires is measured as far as possible by using the replacement costs (based on the costs of standing trees, reforestation and extinguishing costs); when these are not available, the value of burnt wood and other forest losses is calculated. NWFP losses are valued based on the burnt forest area and the average value of NWFPs per hectare of forests, as calculated in each country chapter. In some instances, these estimates are weighted according to the estimated degree of damage. As little information is known in this regard, best-guess estimates are sometimes used, resulting in rough estimates.

Data on quantities of NWFPs[10] are usually less readily available than for WFPs. In some cases, the only quantity data available are for a subset of production, such as exported quantities. In these cases, the resulting estimates clearly underestimate the real value of these products. Other quantities consumed for subsistence or traded on informal markets are taken from expert opinion, and monetary estimation is based on domestic prices (as in Lebanon, for example).

Monetary valuation of NWFPs is generally based on farmgate prices. However, in several southern Mediterranean countries, local markets for NWFPs are so thin that both the quantities traded and the associated prices escape statistics. In most of these cases, only a partial valuation can be undertaken. Where market price data are not available, valuation uses the opportunity cost of labour (as in Lebanon) or the costs of harvest (as in Tunisia). However, in a few instances, the gap in the data resulted in the use of potential yield and the average market price, obtaining the gross value of the NWFPs (pine kernels in Lebanon). The estimates obtained in this way run the risk of being highly overestimated and often dubious. When these estimates belong to a country with a small forest area, the calculated values per hectare of forests appear much greater compared with other countries and, therefore, even more contradictory.

The *opportunity cost of labour approach* uses the forgone employment opportunities used to secure a non-marketed benefit as an estimate of its value. This approach can be used for estimating the benefits of collection of minor NWFPs, where labour is the main input and prices are not available, either because the goods are self-consumed or because markets are scarce. This approach is used in this book to estimate, for example, the value of rosemary and myrtle in Tunisia and the value of carob in Lebanon.

Grazing

The case of grazing is similar to that of WFPs and NWFPs in that it results in a tangible, observable output. Unlike most WFPs and NWFPs, however, the actual product that is harvested, fodder, is generally unpriced. Grazing in forests is practised either for free or by a token tax paid by forest users. In this case, valuation uses data on the quantity of fodder grazed in forests annually. As there is no price for fodder *per se*, the substitute goods approach is used. The *substitute goods approach* approximates the value of goods for which no price (or only an unreliable price) is available using the market price of substitute goods. Of course, the extent to which the value of the substitute goods reflects the true value of non-marketed goods depends on the degree of substitutability between the goods in question. Most country chapters use barley as a comparator, while others use hay. In both cases, forage units are converted into units of the comparator goods based on a comparison of nutrient content, and then valued using the price of the comparator (after adjusting for subsidies and other distortions).[11]

Recreation and hunting

Two main approaches are used to value recreation and hunting benefits. These include the *contingent valuation method* (CVM) and the *travel cost method* (TCM). The CVM creates a hypothetical market for an environmental service and then asks consumers in a survey directly how much they would be willing to pay for that service, or be willing to accept to lose it (Mitchell and Carson, 1989; Carson, 1991; Hausman, 1993; National Oceanic and Atmospheric Administration, 1993). The consumer surplus of that benefit can then be derived from the responses. Available CVM studies provide information on the recreational value of forest sites in several northern countries, and for hunting in Italy.

TCM is usually employed for estimating the recreational value of a site (Hufschmidt *et al.*, 1983; Navrud, 1992; Menkhaus and Lober, 1996).[12] It is based on the assumption that the expenses that consumers incur to travel to a particular site, including all transport costs and the opportunity costs of time spent travelling to the site, provide information on the value they place on recreation at that site. It is important to note, however, that the cost of travel itself is not used as the estimate of value: rather, it is used to derive the consumers' demand curve. Like CVM, TCM has been applied extensively in the northern Mediterranean countries (Croatia, Greece, Italy and Spain) to estimate the value of recreational benefits in forests, based on consumer surplus. Except for Tunisia, where results of TCM surveys in reference to nature parks are available, such surveys were not found in the southern and eastern countries.

In the absence of consumer surplus estimates, actual payments – permit or licence fees and the value of game – are used for the other countries. These measures, in addition to being partial, are also qualitatively different from those used in northern Mediterranean countries. Whereas TCM and CVM attempt to measure consumer surplus, i.e. the area between the demand curve and the price curve, entrance fees are actual payments, i.e. the quadrangle under the price curve, and the net revenues of environmental clubs represent a part of producer surplus.

Use of permit prices is likely to underestimate the true value of forest services (such as hunting and recreation), as these are administered prices rather than prices derived from the interaction of supply and demand. In contrast, methods such as CVM, if properly applied, are more likely to lead to estimates that are close to the full economic value of these services.

Estimating indirect use values

By their very nature, indirect use values are much harder to value than direct use values. They are almost always non-market values, demanding the use of more indirect valuation techniques. Moreover, as discussed below, cause and effect relationships are often difficult to establish, making even these techniques difficult to apply.

Watershed protection

The watershed protection function of forests is among the most difficult to value. The impacts of watershed protection are felt at some distance from the forests that provide them and often are not immediate. Moreover, downstream impacts are the cumulative result of all upstream impacts – forests are only one of many factors that contribute to downstream impacts. The downstream impact of forests also tends to vary substantially from place to place, depending in part on the characteristics of the forests and in part on the characteristics of the downstream areas.

Because of the difficulties in establishing a clear cause and effect relationship between forests and downstream water services, many of the country chapters are unable to estimate watershed protection values. Most chapters which manage to estimate this value use some variant of the *damage costs avoided* approach. Use of the *replacement cost technique* generates a value for environmental goods or services by estimating the costs of replacing them, i.e. they attempt to estimate the additional costs that would be incurred were the forests not present, compared with the situation in which the forests are present. The *restoration cost technique* is a variant of this technique. It measures the value of environmental goods by using the costs of recreating (restoring) the original environmental goods or service. In some cases, valuing the magnitudes of these impacts is based on observations of the consequences of loss of forests in specific areas (change in agricultural productivity downstream); in others, it is based on estimates of what it would cost to build and operate an infrastructure that would provide the same service that forests are now providing (hydraulic works or water storage capacity). In two cases (Albania and France), the estimates are based on the defensive expenditures which the countries make in an effort to avoid losing the forest's protective function. The *defensive expenditure approach* places a value on environmental goods or services by estimating the cost of preventing a reduction in the level of benefits derived from a particular area.[13]

Similar approaches are used to estimate the damage that can occur in cases where forests are

poorly managed, except that rather than using the hypothetical costs that would be borne in the absence of forests, the actual costs borne are used.

It should be noted that replacement costs are usually not good indicators of the value of damage incurred. Problems arise when potential rather than actual expenditures are used, as it is not always clear that the environmental benefit in question justifies the costs of replacing the damage (Bishop, 1999). Use of actual expenditures can underestimate the damage, as replacements rarely replace all the services coming from the original ecosystem. It can also overestimate, as the replacement may be undertaken inefficiently. Use of potential expenditures can easily overestimate the loss of an ecological service, as the replacement can be too expensive to make sense. Some argue that in most cases, the replacement costs tend to overestimate the value of damage (Heal, 1999).

The *production function* approach (also known as the *change in productivity* approach) is used in a few cases. This approach estimates non-market forest benefits (or costs) through the induced change of value of marketed goods or service. The production function approach involves two steps. First, the induced change in the quantity of the affected goods or service is estimated. Secondly, the value of the impact is estimated as appropriate, often by using the market prices of the affected goods or service, or by using a replacement cost. For example, the value of soil protection in Turkey is obtained by first estimating the resulting decline in fertility on eroded lands, and then by using the price of fertilizer needed to replace the lost soil fertility. A similar approach is followed in Tunisia, in this case using the market price of the incremental agricultural yield.

The *hedonic price method (HP)* is generally employed to value the influence of a specific environmental amenity feature on the market price of goods or services (Bartik and Smith, 1987; Graves, 1991; Palmquist, 1991). The most common applications of this technique are the *property value approach* and the *wage differential approach*. Only one application of the property value approach is used in this book (Lebanon) for estimating the damage to the land quality caused by a landslide.

Carbon sequestration

Despite being an indirect use value, carbon sequestration is actually relatively straightforward to value, as the quantities involved can be estimated relatively easily. Two sources of data are used to estimate the quantity of carbon stored annually in forest biomass. The FAO (Food and Agriculture Organization, 2000; United Nations Economic Commission for Europe/Food and Agriculture Organization, 2000) calculates the rate of carbon stored in woody biomass by deducting annual fellings and natural losses from the annual increment. National conversion factors are applied for transforming the volume of stemwood into woody biomass and then into carbon. This approach is used in most northern Mediterranean countries and Syria. For the other southern Mediterranean countries, the estimates are drawn from the national communications to the United Nations Framework Convention on Climate Change (UNFCCC) and are based on the 1996 revised version of International Panel on Climate Change (IPCC) methodology. This methodology aims at realizing national inventories of greenhouse gas emissions, based on specific recommendations and coefficients. Wherever possible, estimates of carbon sequestered in forest soils are included. These estimates were only available for a few countries, however.

The primary difficulty in estimating the value of the carbon sequestration function lies in determining the value of the carbon stored. Several studies (Ayres and Walter, 1991; Nordhaus, 1991a,b, 1993a,b; Cline, 1992; Peck and Teisberg, 1992; Titus, 1992; Maddison, 1993) have attempted to estimate the benefits of reducing carbon dioxide emissions. The resulting estimates have varied widely, from as low as US$0.3/tC (€0.3/tC at 2001 prices) (Nordhaus, 1991a,b) to as high as US$221/tC (€234/tC) (Cline, 1992). Fankhauser (1995) reviewed and refined previous research by modelling the impacts of climate change on separate regions of the world. He proposed a benchmark figure of US$20/tC, which has been widely used. This estimate is used in the country chapters. Adjusting for inflation and converting to euros results in a price of about €20/tC.

A market for carbon sequestration is currently emerging, based on the rules of the Kyoto Protocol's Clean Development Mechanism. It is based on project-based emission reduction transactions and trades of emission allowances (Prototype Carbon Fund, 2001, 2002). A World Bank review of the state and trends of carbon markets found quantity-weighted average prices for Kyoto-compliant emission reductions to be between €11.1 and €15.3/tC, depending on who takes the risk of Kyoto not being ratified (World Bank, 2003). These prices were not available at the time that the research reported here was conducted, however. Moreover, they would not necessarily have been the most appropriate prices to use: current transactions are for a range of emission reduction projects which explicitly do not include forest conservation. These prices are therefore not necessarily conclusive, though they are indicative of the range of values that carbon sequestration might take. The estimates reported in the country chapters use Fankhauser's €20/tC estimate. In the synthesis (Chapter 4), results using this estimate are compared with two other scenarios, with carbon prices of €10/tC and €15/tC.

Estimating option, bequest and existence values

Efforts to value the option, bequest and existence values of biodiversity are scarce everywhere, and Mediterranean countries are no exception. Non-use values, bequest and existence values generally can only be measured using stated preference methods such as CVM. CVM surveys have only been conducted in a few northern countries (France, Italy and Spain), however, and only for specific sites such as individual protected areas. They can be used to derive estimates of existence value for reserves and parks (site level) but are difficult to extrapolate to all forests.

Option value can be estimated with a slightly broader range of techniques, as in part it derives from potential future use values. The analysis in the chapters focuses on two particular aspects: the conservation of biodiversity and the potential for finding products of pharmaceutical value. Most countries use cost-based methods to estimate biodiversity conservation value, based on the annual payments made by various organizations to preserve biodiversity in protected forest areas. Again, these estimates apply mostly to the protected areas and not to the forest as a whole. Only one country (Turkey) attempted to estimate the value of potential pharmaceutical products, using a technique which estimates the potential rent which might be generated, based on the number of species at risk, the number of drugs based on plant species and the number of hectares likely to support medicinal plants.

Benefits transfer

Benefits transfer refers to the use of estimates obtained (by whatever method) in one context to estimate the values in a different context. For example, an estimate of the benefit obtained by tourists visiting a given forest might be used to estimate the benefit obtained from visiting a different forest. Benefits transfer has been the subject of considerable controversy in the economics literature, as it has often been used inappropriately. Formal tests have shown that it can be quite inaccurate (Brouwer and Spaninks, 1999; Barton and Mourato, 2003; Ready *et al.*, 2004). A consensus seems to be emerging that benefits transfer can provide valid and reliable estimates under certain conditions. These conditions include the requirement that the commodity or service being valued is identical at the site where the estimates were made and at the site where they are applied, and that the populations affected have identical characteristics (Brookshire and Nell, 1992; Kirchhoff *et al.*, 1997; Brouwer, 2000). Of course, the original estimates being transferred must themselves be reliable for any attempt at transfer to be meaningful. A wide bibliography on the benefits transfer studies is catalogued in the Environmental Valuation Reference Inventory (EVRI) (Environment Canada, 2004).

Benefits transfer is used extensively in this book. The use is related mainly to extrapolating local estimates to a wider area, at regional or national scale. For example, the recreation benefit in a national park is extrapolated to the total area of national parks in Tunisia; and the benefit per hunter obtained from a game reserve is extrapolated to the total number of hunters in Italian forests. In a few cases, benefits transfer is used for estimating a local benefit in one country, based on the estimate valued in a similar site of another country; for example hunting (Croatia) and recreation (Cyprus) are valued based on the average consumer surplus for hunting and recreation estimated in Italy.

It should be recognized that the use of this approach can easily bring inaccuracies, as different contexts, even though similar to a large extent, are not, however, identical. Moreover, extrapolating average estimates at the national level may result in an even higher level of inaccuracy. For this reason, benefits transfer was not even attempted for many benefits: many option, bequest and existence values remain local level estimates, reflecting the particular conditions of those specific sites.

Common Approaches and Methodological Difficulties

The common approaches adopted in this book were meant to arrive at homogeneous information within and across country chapters. However, in many cases, several factors forced the authors to use different approaches and methods. These differences sometimes limit comparability across the country chapters.

These limiting factors included data availability, valuation difficulties and, not least, the differences in the forest contexts. Several problems were encountered in many of the country studies. The aim of the studies was to arrive at national level estimates of all forest benefits. However, the data scarcity and methodological difficulties often prevented calculation of national level estimates for several forest benefits. When reliable data were not sufficient, the valuation was backed by conservative assumptions (or best-guess estimates), resulting in minimum bound values. Moreover, for certain benefits,[14] making such assumptions was not possible or meaningful; instead, local or subnational estimates were provided[15] when available. Thus, many estimates reflect different levels of valuation which cannot be directly compared.[16]

In some countries these difficulties resulted in missing estimates for several important forest benefits.[17] The missing data make the TEV estimates unrepresentative of the true value of forests.[18] They also distort the apparent shares of TEV generated by the different forest benefits. This is unfortunate, as share of TEV would generally be a useful indicator for many cross-country comparisons, as it abstracts from the very different relative sizes of the countries and their forests and from differences in local price levels. Although shares of TEV are used in several comparisons, these comparisons must be treated with some care. For countries with a particularly large number of 'missing' estimates, the share of TEV of those forest values which could be estimated is not computed.

Some discrepancies also resulted from adopting different sources of data for common indicators (e.g. area). Some country authors preferred to rely on national sources (inventories), rather than on international sources (such as the FAO). National inventories were more likely to be attuned to local conditions and to provide disaggregated data on different forest types within a country; the price for this increased local detail was often reduced comparability with other countries, as each national inventory also had distinct definitions (e.g. for what constitutes a forest) and methodologies for gathering data. Wherever these differences occur, definitions of these indicators are reported in the chapters.

Wherever possible, results are compared with estimates in other parts of the world. However, we are aware of only a few previous attempts to estimate forest values at the national level, e.g. see Adger *et al.* (1995). Other comparisons are with individual benefits, such as recreational use or hunting. Even this comparison is difficult, however. Whereas we attempt to estimate the value of benefits for the entire forest area of a given country, other efforts have typically focused on the benefits generated at a specific site (which may not be representative of the country) or on the benefits that would be generated by a particular project, such as a reforestation project (which are often explicitly unrepresentative, as such projects often try to focus on areas thought to be of high benefit). In either case, the resulting estimates are not easily comparable with those in this book. Analysis of benefits at a particular site is highly dependent on the characteristics of that site, and may differ greatly from our estimates without either necessarily being wrong. Estimates made in the context of a project also differ from the estimates

here in a fundamental way: whereas we try to value the total flow of benefits in a given year, a project analysis attempts to value the change in this flow resulting from the project's activities. These estimates are not comparable.

All in all, the experience drawn from the country chapters shows that in using these methods, the difficulty of valuation increases when moving from direct to indirect use values and, finally, to option, bequest and existence values. Direct use values, represented by marketed or potentially marketable products (WFPs and NWFPs), are usually valued using market prices; grazing is estimated using substitute goods prices; recreational services are valued using CVM or TCM surveys and, in their absence, cost-based methods. Indirect use values, particularly soil conservation and watershed protection, are based mainly on the cost of damage avoided and, in a few countries, defensive expenditures; carbon sequestration is estimated according to the shadow price method. The option, bequest and existence values are usually estimated in reference to cost-based methods and, very rarely, in relation to local CVM surveys.

As expected, the valuation difficulties are directly linked to data scarcity. Most direct use values are based on official statistics that usually offer a large amount of information on quantities traded on the market and average prices. Other direct values (such as recreation and hunting) and most indirect use values are estimated by using other references, such as articles, books and websites (including 'grey literature'), often with very little reference to official statistics. Information from experts derived from personal communications was also considered, when thought to be reliable. Information relevant to option, bequest and existence values is particularly scarce – indeed, only a few estimates of these values could be made, and even these tend to be partial and site specific.

Notes

[1]   It should be noted that if the TEV concept is uniquely defined, its component categories have always been subject to debate. Unclear boundaries between the main types of values led to the inclusion of some value categories within others. For example, vicarious values has been included in the option

values category, whereas option value is sometimes regarded as use value (Pearce and Moran, 1994) and sometimes as non-use value (Hodge and Dunn, 1992, cited by Hodge, 1994). Also, different terminologies are sometimes used; for instance, *active/passive use values* instead of *use/non-use values*.

[2] Observed market values may be distorted by taxes or other factors. However, there are well-established methods to correct such distortions (Gittinger, 1982; Monke and Pearson, 1989).

[3] The multiplicity of definitions and classifications attached to the concept of externalities gives insights into their complexity. The concept of externality goes back to neoclassic economics (Sidgwick, 1887; Marshall, 1890) and later developed by Pigou (1920). For more references on the concept's development, see the following: Viner (1931) introduced the pecuniary and technological externalities; Bator (1958) defined externalities in a broad sense, comprising the most major sources of market failure and divided them into ownership externalities, technical externalities and public goods externalities; Buchanan and Stubblebine (1962) restricted the analysis of externalities to the sphere of technological ones and classified them into relevant externalities, potentially relevant externalities, irrelevant externalities and Pareto relevant externalities; Arrow (1970) also regarded externalities as a particular case of market failures, i.e. associated with the absence of the markets for certain goods; and Varian (1978) classified them into consumption and production externalities.

[4] The concept of club goods was introduced by Buchanan (1965). He gave the example of hunting, which acts as a club good when hunting rights are allocated and enforced. Other examples include 'local' public goods (Tiebout, 1956), such as hospitals and other services, restricted to those living in or gaining access to a certain area, as is often the case of various forest outputs.

[5] In practical terms, a separation often emerges from the valuation method used (Chapter 4). For example, the damage due to forest fires was linked to the direct use values, as valuation considered the replacement costs of wood or the value of the WFPs or NWFPs lost. The losses of natural landscape and biodiversity were primarily related to the option, bequest and non-use values. Even there, the distinctions made are not well defined or clear cut.

[6] This is further related to the aggregation issue. Adding up compatible values, for example timber, firewood and NWFPs produced by 1 ha of forest, makes sense. However, in general, simply adding up independently estimated values for the benefit categories may mis-state the total value (Randall, 1991). Not considering the inter-relationships among the different forest values can induce distortions, such as overestimations of some values and underestimations of others. Therefore, the TEV framework is used to make explicit the value of each forest use but not necessarily to put an aggregate value on nature (Michael, 1995).

[7] The exception is Spain, which adopted a green accounting approach, however, ultimately aimed at valuing the green TEV of forests (Chapter 21).

[8] Maps of forest areas were specially prepared for most country chapters. These maps were drawn according to the Quézel classification (1976, see Chapter 2), which, as far as possible, was adapted to the forest types of each country. These maps pay particular attention to the Mediterranean forest types, insofar as data allow.

[9] The exchange rate for 2001 is US$1 = €0.9.

[10] For simplicity, NWFPs are regarded in *senso stretto*: goods of biological origin, other than wood, derived from forests and allied land uses. They also include cork, Christmas trees, sticks and twigs. Grazing, hunting and recreation are considered separately.

[11] Alternatively, valuation could have been based on grazing fees. However, this approach would have underestimated the true value of grazing, as fees are usually nominal fixed amounts and are not determined in well-functioning markets.

[12] TCM is an example of a *surrogate market approach*, as it relies on markets for goods or services that are related to those for the goods and services of interest. In this case, observations of behaviour in markets for travel are used to deduce preferences for the (unpriced) destination.

[13] This approach is quite close to – and sometimes confused with – the damage costs avoided approach. The distinction between the two is that in the preventive expenditure approach expenses are actually incurred (in order to prevent damage), while in the damage costs avoided approach, they are hypothetical.

[14] Mainly option, bequest and existence values.

[15] For example, in many countries, the recreational value was calculated at the local level and extrapolated to protected areas as a whole.

[16] The comparison between estimates is debatable not only because of the different levels of valuation. Some authors consider that these estimates cannot be aggregated on the grounds that they were derived from very different methodological approaches (such as market values, consumer surplus, actual payments). In this light, these estimates could be neither directly compared nor added up to arrive at the TEV. Other authors argue that, irrespective of the valuation methods used, aggregating the estimated values makes sense as long as the estimates reflect compatible forest uses. In the country chapters the estimates obtained are usually aggregated in order to give a

rough order of magnitude of the estimated values, without any intention to put a precise value on a nation's forests.

[17] For example, the watershed protection function, though important in all Mediterranean countries, could be valued only in some countries.

[18] For this reason, in many country chapters, the term 'estimated TEV' is preferred to the term 'TEV'.

References

Adger, N., Brown, K., Cervigni, R. and Moran, D. (1995) Total economic value of forests in Mexico. *Ambio* 24, 286–296.

Arrow, K. (1970) The organisation of economic activity: issues pertinent to the choice of market versus nonmarket allocation. In: Haveman, R. and Margolis, J. (eds) *Public Expenditure and Policy Analysis*, 2nd edn. Rand McNally College Publishing Company, Chicago, Illinois, pp. 59–73.

Arrow, K. and Fisher, G. (1974) Environmental preservation, uncertainty and irreversibility. *Quarterly Journal of Economics* 2, 312–319.

Ayres, R. and Walter, J. (1991) The greenhouse effect: damages, costs and abatement. *Environmental and Resource Economics* 1, 237–270.

Bartik, T.J. and Smith, V.K. (1987) Urban amenities and public policies. In: Mills, E.S. (ed.) *Handbook of Regional and Urban Economics*, Vol. 2. North-Holland, Amsterdam, pp. 1207–1257.

Barton, D.N. and Mourato, S. (2003) Transferring the benefits of avoided health effects from water pollution between Portugal and Costa Rica. *Environment and Development Economics* 8, 351–371.

Bateman, I. (1994) Research methods for valuing environmental benefits. In: Dubgaard, A., Bateman, I. and Merlo, M. (eds) *Economic Valuation of Benefits from Countryside Steward-ship*, Proceedings of a workshop organized by the Commission of the European Communities Directorate General for Agriculture, Brussels, 7–8 June 1993, Wissenschaftsverlag Vauk. Kiel, Germany.

Bator, F. (1958) The anatomy of market failure. *Quarterly Journal of Economics* 8, 351–379.

Baumol, W. and Oates, W. (1975) *The Theory of Environmental Policy*. Prentice Hall, Englewood Cliffs, New Jersey.

Bishop, J.T. (ed.) (1999) *Valuing Forests: a Review of Methods and Applications in Developing Countries*. International Institute for Environment and Development, London.

Braden, J.B. and Kolstad, C.D. (eds) (1991) *Measuring the Demand for Environmental Quality*. Elsevier, Amsterdam, pp. 17–28.

Brookshire, D.S. and Nell, H.R. (1992) Benefit transfers: conceptual and empirical issues. *Water Resources Research* 28, 651–655.

Brouwer, R. (2000) Environmental value transfer: state of the art and future prospects. *Ecological Economics* 32, 137–152.

Brouwer, R. and Spaninks, F.A. (1999) The validity of environmental benefits transfer: further empirical testing. *Environmental and Resource Economics* 14, 95–118.

Buchanan, J. (1965) An economic theory of clubs. *Economica* 32, 1–14.

Buchanan, J. and Stubblebine, W. (1962) Externality. *Economica* 29, 371–384.

Carson, R.T. (1991) Constructed markets. In: Braden, J.B. and Kolstad, C.D. (eds) *Measuring the Demand for Environmental Quality*. Contributions to Economic Analysis No. 198. North-Holland, Amsterdam, pp. 121–162.

Cline, W.R. (1992) *The Economics of Global Warming*. Institute for International Economics, Washington, DC.

Croitoru, L., Gatto, P. and Merlo, M. (2000) Non-wood forest products (NWFP) as a component of the total economic value (TEV) of Mediterranean forests – first results of an ongoing research. In: Joint FAO/ECE/ILO Committee on Forest Technology, Management and Training, Seminar Proceedings *Harvesting of Non-wood Forest Products*, Mnemen-Izmir, Turkey, 2–8 October 2000.

Dixon, J.A., Scura, L.F., Carpenter, R.A. and Sherman, P.B. (1994) *Economic Analysis of Environmental Impacts*. Earthscan, London.

Environment Canada (2004) Environmental Valuation Reference Inventory (EVRI), website: www.evri.ca

Fankhauser, S. (1995) *Valuing Climate Change. The Economics of the Greenhouse*. Earthscan, London.

Farber, S., Costanza, R. and Wilson, M. (2002) Economic and ecological concepts for valuing ecosystem services. *Ecological Economics* 41, 375–392.

Food and Agriculture Organization (2000) *Global Forest Resources Assessment 2000*. FAO Forestry Paper No. 140.

Gatto, P. and Merlo, M. (1999) The economic nature of stewardship: complementarity and trade-offs with food and fibre production. In: van Huylenbroeck, G. and Whitby, M. (eds) *Countryside Stewardship: Farmers, Policies and Markets*. Pergamon, Amsterdam, pp. 21–46.

Gittinger, J.P. (1982) *Economic Analysis of Agricultural Projects*, 2nd edn. Johns Hopkins University Press for the World Bank, Baltimore, Maryland.

Goulder, L. and Kennedy, D. (1997) Valuing ecosystem services: philosophical bases and empirical methods. In: Daily, G. (ed.) *Nature's Services: Societal Dependence on Natural Ecosystems*. Island Press, Washington, DC, pp. 23–48.

Graves, P.E. (1991) Aesthetics. In: Braden, J.B. and Kolstad, C.D. (eds) *Measuring the Demand for Environmental Quality*. Contributions to Economic Analysis No. 198. North-Holland, Amsterdam, pp. 213–226.

Gren, M., Folk, C., Turner, K. and Bateman, I. (1994) Primary and secondary values of wetland ecosystems. *Environmental and Resource Economics* 4, 55–74.

Hausman, J.A. (ed.) (1993) *Contingent Valuation: a Critical Assessment*. Contributions to Economic Analysis No. 220. North-Holland, Amsterdam.

Heal, G. (1999) *Valuing Ecosystem Services*. Working Paper Series in Money Economics and Finance. Columbia Business School, Columbia University, New York.

Henry, C. (1974) Investment decisions under uncertainty: 'the irreversibility effect'. *American Economic Review* 64, 1006–1012.

Hodge, I. (1991) The provision of public goods in the countryside: how should it be arranged? In: Hanley, N. (ed.) *Farming and the Countryside*. CAB International, Wallingford, UK, pp. 179–196.

Hodge, I. (1994) *Rural Amenity: Property Rights and Policy Mechanisms. The Contribution of Amenities to Rural Development*. OECD, Paris.

Hufschmidt, M., James, D.E., Meister, A.D., Bower, B.T. and Dixon, J.A. (1983) *Environment, Natural Systems, and Development. An Economic Valuation Guide*. Johns Hopkins University Press, Baltimore, Maryland.

International Monetary Fund (2004) *World Economic Outlook*. Website: www.imf.org. Access date: February 2004.

Kirchhoff, S., Colby B.G. and LaFrance, J.T. (1997) Evaluating the performance of benefit transfer: an empirical inquiry. *Journal of Environmental Economics and Management* 33, 75–93.

Krutilla, J. (1967) Conservation reconsidered. *American Economic Review* 57, 777–786.

Maddison, D. (1993) *The Shadow Price of Greenhouse Gases and Aerosols*. Mimeo, CSERGE, University College London and University of East Anglia, Norwich, UK.

Marshall, A. (1890) *Principles of Economic*. Book X, Macmillan, 7th edn, 1916, London.

Mendes, A.C. (2004) *Values, Norms, Transactions and Organisations*. Working paper, Faculty of Economics and Management, Portuguese Catholic University, Porto, Portugal.

Menkhaus, S. and Lober, D.J. (1996) International ecotourism and the valuation of tropical rainforests in Costa Rica. *Journal of Environmental Management* 47, 1–10.

Merlo, M. and Rojas, E. (2000) Public goods and externalities linked to Mediterranean forests: economic nature and policy, *Land Use Policy* 17, 197–208.

Michael, S.G. (1995) Economic Valuation of the Multiple Uses of Forests: the Case of Bwindi Impenetrable National Park (BINP), Uganda. MSc Dissertation, University of Edinburgh, UK.

Millennium Ecosystem Assessment (2003) *Ecosystems and Human Well-being: a Framework for Assessment*. Island Press, Washington, DC.

Mitchell, R.C. and Carson, R.T. (1989) *Using Surveys to Value Public Goods: the Contingent Valuation Method*. Resources for the Future, Washington, DC.

Monke, E.A. and Pearson, S.R. (1989) *The Policy Analysis Matrix for Agricultural Development*. Cornell University Press, Ithaca, New York.

National Oceanic and Atmospheric Administration (1993) Report of the NOAA Panel on Contingent Valuation. *Federal Register* 58, 4602–4614.

Navrud, S. (ed.) (1992) *Pricing the European Environment*. Scandinavian University Press, Oslo.

Nordhaus, W.D. (1991a) A sketch of the economics of the greenhouse effect. *American Economic Review, Papers and Proceedings* 81, 146–150.

Nordhaus, W.D. (1991b) To slow or not to slow: the economics of the greenhouse effect. *Economic Journal* 101, 920–937.

Nordhaus, W.D. (1993a) Optimal greenhouse gas reductions and tax policy in the 'DICE' model. *American Economic Review, Papers and Proceedings* 83, 313–317.

Nordhaus, W.D. (1993b) Rolling the 'DICE': an optimal transition path for controlling greenhouse gases. *Resources and Energy Economics* 15, 27–50.

Organization for Economic and Cooperative Development (1995) *The Economic Appraisal of Environmental Projects and Policies – A Practical Guide*. OECD, Paris.

Pagiola, S., Bishop, J. and Lindell-Mills, N. (2002) *Selling Forest Environmental Services. Market-based Mechanisms for Conservation and Development*. Earthscan, London.

Palmquist, R.B. (1991) Hedonic methods. In: Braden, J.B. and Kolstad, C.D. (eds) *Measuring the Demand for Environmental Quality*.

Contributions to Economic Analysis No. 198. North-Holland, Amsterdam, pp. 77–120.

Pearce, D. and Moran, D. (1994) *The Economic Value of Biodiversity*. Earthscan, London.

Peck, S.C. and Teisberg, T.J. (1992) CETA: a model of carbon emissions trajectory assessment. *Energy Journal* 13, 55–77.

Pigou, A. (1920) *The Economics of Welfare*. Macmillan, London.

Prototype Carbon Fund (2001) *Annual Report 2001*. Prototype Carbon Fund, Washington, DC.

Prototype Carbon Fund (2002) *Annual Report 2002*. Prototype Carbon Fund, Washington, DC.

Randall, A. (1991) Total and non-use values. In: Braden, J. and Kolstad, C. (eds) *Measuring Demand for Environmental Quality*. Elsevier Science Publishers BV, Amsterdam, pp. 303–322.

Ready, R., Navrud, S., Day, B., Dubourg, R., Machado, F., Mourato, S., Spanninks, F. and Rodriquez, M.X.V. (2004) Benefit transfer in Europe: how reliable are transfers between countries? *Environmental and Resource Economics* 29, 67–82.

Samuelson, P. (1954) The theory of public expenditure. *Review of Economics and Statistics* 36, 387–389.

Samuelson, P. (1955) *Economics*. McGraw-Hill Inc., New York, reprinted 1970.

Sidgwick, H. (1887) *Principles of Political Economy*. Macmillan, New York.

Smith, A. (1776) *An Inquiry into the Nature and Causes of the Wealth of Nations*. Cannan, E. (ed.) Putnam, 1904, New York.

Tiebout, C. (1956) A pure theory of local expenditures. *Journal of Political Economy* 64, 416–424.

Titus, J. (1992) The cost of climate change to the United States. In: Majumdar, S.K., Kalkstein, L.S., Yarnal, B., Miller, E.W. and Rosenfeld, L.M. (eds) *Global Climate Change: Implications, Challenges and Mitigation Measures*. Pennsylvania Academy of Science, Pennsylvania.

Turner, K., den Bergh, J., Soderquist, T., Baerendregt, A., der Straaten, J., Maltby, E. and van Ierland, E. (2000) Ecological–economic analysis of wetlands. *Ecological Economics* 35, 7–23.

Turner, K., Paavola, J., Cooper, P., Farber, S., Jessamy, V. and Georgiou, S. (2003) Valuing nature: lessons learned and future research directions. *Ecological Economics* 46, 493–510.

United Nations Economic Commission for Europe/Food and Agriculture Organization (2000) *Global Forest Resources Assessment 2000*. Main Report. United Nations Publications, Geneva.

Varian, H.R. (1978) *Microeconomic Analysis*. Norton and Company Inc., New York.

Viner, J. (1931) Cost curves and supply curves. *Zeitschrift für Nationalökonomie* 3, 23–46.

Weisbrod, B. (1964) Collective-consumption services of individual-consumption goods. *Quarterly Journal of Economics* 78, 471–477.

Winpenny, J.T. (1991) *Values for the Environment: a Guide to Economic Appraisal*. Overseas Development Institute. HMSO, London.

World Bank (2003) *State and Trends of the Carbon Market 2003*. World Bank, Washington, DC.

4   Mediterranean Forest Values

Lelia Croitoru and Maurizio Merlo[†]
University of Padova, Centre for Environmental Accounting and Management in Agriculture and Forestry (CONTAGRAF), Via Roma 34, Corte Benedettina, 35020 Legnaro (PD), Italy

Mediterranean forests provide a wide array of benefits, including direct use values such as extraction of wood forest products (WFPs) and non-wood forest products (NWFPs), grazing and recreation; indirect use values such as water cycle regulation and prevention of erosion and floods; and option, bequest and existence values such as conservation of biodiversity. The importance of these values differs from area to area, depending on the diversity and intensity of the natural and human factors affecting forests. The following sections draw from the results of the individual country chapters[1] to give a picture of the estimated values in the Mediterranean region. In particular, they place these values in the Mediterranean forests context and discuss their importance at country, cross-country and Mediterranean level.

As a first attempt to estimate the total economic value (TEV) of forests over a wide area, this effort is unavoidably subject to several constraints. Although every effort has been made to follow a consistent methodological approach in every country studied, data availability and other constraints have often led to individual variations. Data constraints sometimes prevented country chapters from arriving at estimates for particular categories of values.[2] Even when estimates could be made, data constraints were such that the approaches taken were not always fully comparable across countries.

Direct Use Values

Estimating direct use values is usually more straightforward than other TEV components. Many outputs are extensively commercialized in the market, and their quantities and average prices often are known. However, some outputs are subject to thin local markets, with considerable intra-country variation in quantities and prices; other outputs, although marketable, are consumed free of charge. Valuation of these outputs is more difficult and involves more advanced techniques than the use of market price.

Wood forest products: timber and firewood

Timber and other WFPs such as firewood have always been important to the region, but they are not the main output of most Mediterranean forests. Timber is scarce in the southern and eastern countries, which have small forest areas with low timber productivity and quality. Timber production is more abundant in the northern countries, thanks to larger forest areas and to favourable climate conditions which promote faster timber growth.

[†] Deceased.

Wood removals

In most northern Mediterranean countries, timber accounts for the majority of total removals: 93% in Portugal, 75% in Slovenia and 64% in Croatia. Italy and Greece are exceptions to this pattern: due to their greater area of coppices compared with high forests, firewood extraction exceeds that of timber. In contrast, firewood is more important than timber in southern and eastern countries: it reaches 81% of overall removals in Tunisia, 94% in Morocco and around 100% in Lebanon (where almost all timber is imported).

Although statistics are not always reliable, it is believed that in most southern and eastern countries, a large share of the firewood consumed passes through informal markets or does not enter markets at all. Illegal firewood collection is common; 69% of firewood collection in Morocco is illegal and 42% in Turkey. Illegal firewood collection is most prevalent where institutions are weak, forests are publicly owned, enforcement is limited and rural communities have little or no incentive to preserve forests. Valuation of wood removals considers the quantities of timber and firewood traded on markets and also, wherever data are available, the quantities collected free of charge. The latter can include both local communities' rights of free collection, as in Tunisia, and illegal timber cuts, as in Morocco and Turkey. It should be noted that in certain cases, extraction, whether legal or illegal, can be unsustainable and cause negative externalities.[3]

As discussed in Chapter 3, monetary valuation uses the roadside prices[4] of WFPs traded in the market. For the quantities collected without payment, valuation uses the roadside prices of similar goods (in Morocco and Tunisia), or the shadow price, after eliminating public policy distortions (in Turkey).

Average per hectare timber values are high in Western European countries: €130/ha in Portugal and €72–76/ha in France and Slovenia, for example (Table 4.1). In these countries, forests have never been scarce and national timber production is the main input for timber-based industries.[5] In contrast, in southern and eastern countries, timber value is less than €5/ha of forests. This is due to the low timber productivity and quality and, therefore, prices. Here,

timber-based industries usually depend on imports. The value of firewood is also higher in northern countries, and particularly Western European countries, ranging between €10 and 50/ha, primarily due to higher prices compared with southern and eastern countries.

WFPs represent between 20 and 40% of the TEV in most northern countries, but less than 15% in most southern and eastern countries. In absolute terms, the highest values range from €760 million to €1300 million and are found in the northern Mediterranean countries endowed with extensive forest areas: Spain and France. In contrast, in most southern and eastern countries, forest scarcity leads to a substantially lower value of WFPs, which often are less significant than the NWFPs. Morocco and Turkey are exceptions, with the largest forest areas and the highest value of WFPs among countries in the southern and eastern Mediterranean, respectively.

Net growth of standing timber

Many northern Mediterranean countries have experienced timber removals below the annual increment, resulting in net growth of standing timber. This is often a result of the adoption of forest management practices focused on protection rather than on intensive timber production. In many Western European countries, net growth in recent decades has tended to result from the abandonment of primary activities, such as forestry and agriculture, rather than from deliberate conservation.

This net growth was considered desirable in the past, as it allowed spontaneous regeneration and 'renaturalization' of forests. Today it is sometimes argued that in a fragile ecosystem such as that of Mediterranean forests, poor or lack of forest management can contribute to negative effects, such as increased risk of forest fires and reduced biodiversity and landscape quality (Di Castri, 1996; Fabbio et al., 2003). The extent to which net growth can cause positive and negative externalities is unclear, however. In the country chapters, 'net growth' is generally considered as a source of benefits linked to forest biodiversity and ecological functions, and to increases in potential sources of energy and raw materials.[6]

As discussed in Chapter 3, the net growth of standing timber is estimated based on half of

Table 4.1. Value of timber and firewood (2001).

Country	WFPs removals		Prices		Total value		Value/ha	
	Timber (000 m³)	Firewood[a] (000 m³)	Timber (€/m³)	Firewood (€/m³)	Million €	% of TEV	Timber (€/ha)	Firewood (€/ha)
Southern								
Morocco	615.0	9,600.0	77.6	14.5	187.3	31	5.3	15.5
Algeria	123.7	77.8	5.4	3.2	0.9	0	0.2	0.1
Tunisia	147.7	621.9	13.6	6.6	4.1	3	2.2	2.3
Egypt	0.3	16.2	45.0	27.0	0.4	NC	0.2	6.0
Eastern								
Palestine	1.5	1.5	80.8	48.5	0.2	NC	5.3	3.2
Israel[c]	21.0	NA	20.0	NA	0.4	NC	4.5	NA
Lebanon	0.0	82.3; 11.4[b]	0.0	22.9; 165[e]	3.8	NC	0.0	28.0
Syria[c]	3.6	2.3	70.0	58.3	0.4	1	0.6	0.3
Turkey	10,316.0	23,620.0	40.9	1–3.9	475.1	50	20.4	2.6
Cyprus	11.8	6.6	24.7	8.5	0.3	NC	0.8	0.1
Northern								
Greece	700.0	1,580.0	83.6	20	87.8	19	8.4	5.1
Albania	60.0	200.0	78.0	21	5.7	3	2.6	2.9
Croatia	2,300.0	2,400.0	81.9	13.1	219.7	30	75.8	12.7
Slovenia	1,603.0	539.0	55.4	22.3	100.6	40	76.0	10.3
Italy	3,276.0	5,076.0	64.0	44.8	436.5	20	24.3	26.4
France[f]	3,780.0	2,200.0	28.8	9.4	1,339.0	NC	72.8	16.4
Spain	15,362.0	1,878.5[d]	47.2	13.1	763.0	NC	130.5	1.5
Portugal	14,011.0	774.0	31.3	48.2	467.9	41	130.5	11.3

[a]Includes illegal firewood in Morocco (69% of total firewood), Turkey (42%) and firewood self-consumed in Tunisia (88%).
[b]Production of charcoal (t), in addition to firewood.
[c]Transformation of t to m³ is undertaken by applying an average conversion rate of 1 t = 0.7 m³.
[d]The original estimate expressed in stere (st) has been converted to m³ based on the average conversion rate of 1 st = 0.65 m³ (Sistema de Informação de Cotações de Produtos Flosestais na Produção, 2003).
[e]Per Mt of charcoal.
[f]As the data regarding timber are not clearly separated from those regarding firewood, in this table, timber was considered 'hardwood logs, softwood logs, plywood and other marketed roundwood'; firewood was considered the 'non-marketed wood' (based on Chapter 20). This distinction is inaccurate to a minimal degree.
NC, not calculated (due to insufficient information for other categories of value).
Sources: based on Chapters 5–22, this volume.

the wood stumpage price. Table 4.2 presents the estimates for the countries providing this valuation – mainly northern countries and Syria.[7] The highest net growth occurs in Slovenia (~3.3 m³ /ha), followed by Croatia, Italy and France (~1.9 m³/ha). Cyprus, Greece and Portugal have much lower rates, of about 0.2–0.4 m³/ha. The highest estimated benefit per hectare (~€75/ha) and the greatest share of TEV (35%) is found in Slovenia. For the other countries, the importance of this benefit is much lower.

Wood losses due to forest fires

Forest fires result from a combination of factors, varying markedly between countries. In many northern countries, and particularly in Western Europe, forest fires are mostly caused by poor management of private forests, rural depopulation and negligence. In southern and eastern countries, forest fires are due mainly to deforestation and severe climate conditions. Fires lead to a loss of benefits from all TEV components. It should be noted, however, that in certain

Table 4.2.   Net growth of standing timber (2001).

Country	Net growth of standing timber		Monetary indicators		Total value	
	Million m³	m³/ha	€/m³	€/ha	Million €	% of TEV
Eastern						
Syria	0.1	0.2	35.8	7.1	3.3	8
Cyprus	0.1	0.2	10.0	2.3	0.9	NC
Northern						
Greece	1.5	0.2	10.0	2.3	14.7	3
Croatia	4.8	1.9	20.2	39.0	97.0	13
Slovenia	3.9	3.3	31.3	75.4	87.9	35
Italy	16.0	1.9	20.0	37.2	320.0	15
France	29.2	1.9	10.5	20.3	305.0	NC
Portugal	3.9	1.2	20.0	22.9	75.7	7

NC, not calculated (due to insufficient information for other categories of value).
Sources: based on Chapters 5–22, this volume.

Table 4.3.   Wood losses due to forest fires (2001).

Country	Burnt area[a]		Cost of fires		Total value of wood damages by fires	
	ha	% of forest area	€/ha of burnt area	€/ha of forests	Million €	% of TEV
Southern						
Morocco	2,700	0.03	926	0.3	2.5	−1.2
Algeria	25,000	0.61	880	5.4	22.0	−3.6
Tunisia	2,215	0.25	1,354	3.3	3.0	−2.3
Eastern						
Lebanon	1,200	0.89	4,000	35.2	4.8	NC
Syria	1,500	3.00	1,700	5.3	2.4	−5.9
Turkey	5,807	0.03	1,429	0.4	8.3	−0.9
Cyprus	190	0.05	1,578	0.8	0.3	−0.2
Northern						
Greece	33,700	0.52	884	5.0	32.8	−7.1
Albania	9,000	0.88	1,000	8.8	9.0	−5.0
Croatia	4,600	0.19	583	1.1	2.7	−0.4
Italy	41,019	0.48	1,480	7.1	60.7	−3.0
Portugal	40,000	1.21	3,420	41.5	136.9	−12.0

[a]Average for the last 10 years.
NC, not calculated (due to insufficient information for other categories of value).
Sources: based on Chapters 5–22, this volume.

situations, fires can also produce positive effects, such as increasing the quantity of soil nutrients in burned areas. The separation between positive and negative effects is difficult if not impossible at the country level.

The value of losses of WFPs resulting from forest fires is usually estimated by using either the replacement costs or the value of burnt wood. Table 4.3 shows the value of forest fires in relation to the annual average burnt area of each country. Forest fires affect extensive areas in Western European countries. Portugal and Italy have the largest burnt areas and the highest restoration costs – about €1500–3400/ha of burnt area. The comparative loss in value due to forest fires is worst in Portugal, reducing the TEV by 12%. Among southern countries, Algeria is particularly affected by fires.[8] In these countries, the costs of restoring forest areas damaged by fire vary between €580 and €4000/ha of burnt area. The range may be partly due to the different costs used in the individual country chapters.

Grazing

In the past, grazing in forests was a significant activity in all Mediterranean countries. Its importance has declined considerably in many northern countries, and especially in Western Europe. In southern and eastern countries, intensive grazing practices are still a common source of livelihood. They often result in decreased fodder productivity and soil degradation. In many Mediterranean countries, grazing is a free public right of forest communities; sometimes it is granted explicitly by forest legislation, as in Tunisia. In other countries, grazing access to forests is charged a nominal fee, as in Greece.

Average fodder productivity is low in many countries: about 70–130 forage units (FU)/ha of forests in Lebanon and Turkey and as little as 30 FU/ha in Syria. This is often a result of excessive grazing in forests in the past, which is notable in countries where a small share of agricultural area is devoted to fodder production (e.g. 2.9% in Turkey; Bann and Clemens, 2001). Unsustainable grazing is an important issue in southern countries, as it causes high rates of environmental degradation. Fodder production and consumption can vary considerably within countries, depending on grazing patterns, types of forests and climate conditions. In Portugal, production of fodder varies from 360 FU/ha of *Pinus pinea* forests, to 412 FU/ha in *Quercus suber* stands, to as much as 450 FU/ha in *P. pinaster*, *Eucalyptus* sp. and *Castanea* sp. stands. In Tunisia, fodder production ranges between 50 FU/ha in southern semi-desertic areas to around 600 FU/ha in northern humid zones. In north-eastern Algeria, the average fodder productivity in forests is about 240 FU/ha; in steppe areas, it is much lower, ranging between

Table 4.4. Grazing (2001).

Country	Grazing forest area[a] (000 ha)	Consumption of forest fodder			Monetary indicators		Total value of fodder	
		FU/ha of grazed area	FU/ha of forests	Total (million FU)	€/ha of forests	€/ha of grazed area	Million €	% of TEV
Southern								
Morocco	7200.0	208	167	1500.0	28.3	35.4	255.0	42
Algeria	NA	244; 30[b]; 60–100[c]	373	1530.0	32.9	NA	206.5	33
Tunisia	NA	510; 50[d]–600[e]	510	481.0	73.9	NA	69.7	54
Eastern								
Palestine	NA	NA	219	5.0	21.9	21.9	0.5	NC
Lebanon	NA	NA	71	9.6	6.4	NA	0.9	NC
Syria	50.0	30	3	1.5	0.4	3.6	0.2	0
Turkey	5800.0	131	37	759.0	10.9	38.8	225.0	24
Northern								
Greece	5300.0	330	292	1749.0	34.9	42.9	227.4	50
Albania	600.0	300	175	180.0	17.5	30	18.0	10
Croatia	NA	NA	358	858.0	5.4	NA	13.0	2
Italy	2000.0	200	47	400.0	7.0	30	60.0	3
Spain	NA	NA	120	3112.0	7.8	NA	203.0	NC
Portugal	3657.5	770[f]; 420[g]	635	2097.2	34.0	8.9[f]; 26.6[g]	112.3	10

[a]For many countries where grazing forest areas are not indicated, grazing is practised in the entire forest areas.
[b]In degraded steppe.
[c]In steppe areas covered by palatable vegetation.
[d]Semi-desert areas in southern Tunisia.
[e]Areas in northern Tunisia.
[f]In scrubland grazed area.
[g]In forest grazed area.
Source: based on Chapters 5–22, this volume.

30 FU/ha of degraded steppe and 60–100 FU/ha of steppe covered by palatable vegetation.

As discussed in Chapter 3, grazing is valued using the substitute goods approach. Most country chapters use barley as a comparator, assuming that the nutrient content of 1 FU is similar to that of 1 kg of barley. Fodder is then valued using a shadow price of €0.1/FU, based on the local market prices of barley, after deducting subsidies. Some country chapters use the quantity of hay equivalent (or dry matter), where 1 quintal (q) of hay corresponds to 330 FU. Table 4.4 summarizes the main results of the valuation. High fodder consumption in forests occurs in Algeria and Morocco, countries with extensive grazing areas. In Algeria, intensive grazing occurs in forests because it is excluded from agricultural lands most of the year. Among northern countries, grazing remains common in the Portuguese *montados* and Spanish *dehesas*, where it represents the most important commercial production, and in Greece, because of the large rural population. Overall, the importance of grazing is high in most southern and eastern countries and Turkey, accounting for around 24–55% of the TEV, and low in most northern countries, where grazing accounts for less than 10% of the TEV.

Non-wood forest products

Mediterranean forests supply a diversity of tangible NWFPs with potential for contributing to local economies. Cork, mushrooms and honey are the main individual NWFPs encountered in Mediterranean forests; a variety of other NWFPs are also found in smaller amounts.

A considerable literature has developed in recent years on the value of NWFPs, spurred by a study (Peters *et al.*, 1989) estimating that products harvested from a tropical forest in Peru could be worth as much as US$422/ha/year (€493/ha/year at 2001 prices). Although this particular estimate is almost certainly an overestimate, other studies indicate that collection of non-timber products can yield non-negligible returns. A review of studies (Godoy *et al.*, 1993) reports a per hectare value varying between US$0.75 and US$422 (~€0.8–493/ha at 2001 prices). Godoy *et al.* (2000) find aggregate

estimates of about US$18–24/ha (~€19–25/ha at 2001 prices) of rainforests in Central America.

Cork

Cork is a significant product in the Mediterranean region. Cork oak (*Q. suber*) covers more than 2 Mha and provides a source of livelihood for many thousands of people (Varela, 1999). However, during the last decades, cork production has diminished significantly throughout the region. In many Maghreb countries, deforestation and overgrazing have contributed to reducing the cork oak area, causing a steep decline in employment in cork harvesting. In many Western European countries, cork production has decreased mainly as a result of forest abandonment. In addition, the development of potential substitutes (e.g. plastic) has affected the cork market.

Cork is valued based on the quantities commercialized and their average prices. Cork production is mostly concentrated in the western Mediterranean countries and Italy. Portugal, as the world's leading cork producer, stands out with the highest benefit in both absolute and relative terms: it obtains €550/ha of cork oak area, accounting for 35% of the TEV. Cork oak forests in the Portuguese *montados* have received much attention since the 19th century, with cork production concentrated in the hands of individual producers or family associations. Cork is also significant in Tunisia, where it generates about 7% of the TEV, a share greater than that of WFPs and of all the other NWFPs, but smaller than that for grazing. It provides a benefit of about €200/ha of cork oak area, much of which is received by local communities. The role of cork in the other countries is much lower.

Mushrooms

In the Mediterranean region, mushrooms are considered public property. Collection of mushrooms is usually regulated by local authorities through permits. In most countries, collection is free up to a limit. In many cases, particularly in the southern Mediterranean, regulation is either not established or not enforced. In Western European countries, a large part of mushroom production is a by-product of recreational activities, enjoyed by visitors and society as a whole.

Valuation is based on quantities traded in the markets and, whenever statistics are available, on quantities collected for free. These estimates are often partial: in Slovenia, Portugal and Turkey, they fail to include the quantities collected for free; in others, even the marketed quantity is partial, for example by only considering exports (as in Tunisia). In eastern Mediterranean countries, no estimates for mushroom production are available. The scarcity of data in many southern and eastern countries does not imply the lack of mushrooms. Rather, it reflects the fact that mushrooms are collected for free and sold on thin local markets, for which quantities and prices are not known. The average market price is used to value mushrooms. Average prices are highest in Western European countries, at about €5–6/kg, thanks to developed markets and high demand for the product. Mushrooms account for less than 4% of the TEV (Table 4.5). Mushroom benefits are concentrated mostly in the northern countries.

Honey

Valuation is based on the estimated quantities produced by beehives placed in forests and the average market price. The highest value per hectare, about €90/ha, is found in Lebanon, as a result of the high market price; here, the demand for honey is met mostly by imports. In Egypt, honey generates a similar value. It should be noted that Egypt is the world's leading honey producer; however, only a small part of total honey production can be attributed to forests. Honey is also relatively important in Cyprus, Lebanon and Greece. The role of honey in other countries is almost negligible, accounting for less than 3% of the TEV.

Other NWFPs

Other NWFPs include various fruits or plants growing in the forests. Some, such as chestnuts, berries, acorns and medicinal plants, can be widely found in all Mediterranean countries. Others are characteristic to individual countries and areas, such as tan bark from *Acacia*

mollissima in Morocco and alfa (*Stipa tenacissima*) in Algeria. In most countries, they are less important than the NWFPs discussed previously. Valuation is mostly based on the quantities[9] traded on markets and their average prices. Additional quantities collected for free are sometimes estimated by using the opportunity cost of labour (medicinal plants in Lebanon) and the costs of harvest and harvest fees (rosemary, myrtle and Aleppo pine acorns in Tunisia).

The case of pine kernels in Lebanon illustrates some of the potential problems that can be encountered in this kind of valuation. Pine kernels are a high-yield, high-value NWFP. Multiplying out available data on yield (480 kg/ha), price (€20.3/kg) and area (5400 ha) results in a total value of €52.5 million. This is equivalent to as much as €9700/ha of stone pine area, or €392/ha over the entire forest area. If correct, this result would be truly exceptional: pine nuts in Lebanon by themselves would have a greater value, on a per hectare basis, than the entire estimated TEV in any other Mediterranean country. There is good reason, therefore, to suspect that something is amiss. Perhaps the available yield data reflect optimal conditions, and are not representative; perhaps the available price data include costs other than raw materials, or are for high-quality products that only represent a small share of total output; or perhaps the quantity measure used in the yield data does not correspond to that used in the price data, because of processing or wastage along the chain.[10] Any of these problems would result in overestimates. Unfortunately, although *every* effort was made to verify data used in the calculations to avoid such problems, this was not always possible. The problem is magnified in this case by the relatively small area of forest in Lebanon, which gives such results a very high weight in the average value.

Most NWFP values (Table 4.5) are broadly in the range of those found by Lampietti and Dixon (1995). Estimated benefits tend to be higher in Western European countries than in most southern and eastern countries, but this may reflect a wider data availability on NWFPs. Table 4.5 shows that the importance of NWFPs varies considerably, from as little as 1% to as much as 42% of the TEV.

Table 4.5. Non-wood forest products in the Mediterranean countries (2001).

	Cork			Mushrooms			Honey			Other NWFPs		Total NWFPs		
Country	000 t	€/ha[a]	Total (million €)	000 t	€/ha	Total (million €)	000 t	€/ha	Total (million €)	€/ha	Total (million €)	€/ha	Million €	% TEV
Southern														
Morocco	151.0[b]	24.1; 1.2	6.8	1.0	1.1	6.1	4.0	3.4	19.4	0.6	3.5	6.3	35.8	6
Algeria	12.0	23.6; 1.3	5.4	NA	NA	NA	1.6	0.1	0.5	0.4	1.7	1.8	7.6	2
Tunisia	11.6	198.2; 9.6	9.0	0.0	0.0	0.0	0.2	1.9	1.7	11.4	10.8	22.8	21.5	17
Egypt	NA	NA	NA	NA	NA	NA	1.6	88.9	6.4	NA	NA	88.9	6.4	NC
Eastern														
Palestine	NA	NA	NA	NA	NA	NA	NA	NA	NA	24.9	0.6	24.9	0.6	NC
Israel[c]	NA	NA	NA	NA	NA	NA	NA	NA	NA	NA	NA	NA	NA	NC
Lebanon	NA	NA	NA	NA	NA	NA	1.0	89.6	12.1	516.3	69.7	605.9[f]	81.8	NC
Syria[c]	NA	NA	NA	NA	NA	NA	0.0	0.4	0.2	6.7	3.1	7.1	3.3	8
Turkey	NA	NA	NA	11.0	0.5	11.1	NA	NA	NA	3.5	72.3	4.0	83.5	9
Cyprus	NA	NA	NA	NA	NA	NA	0.9	8.8	3.4	5.1	2.0	13.8	5.3	NC
Northern														
Greece	NA	NA	NA	NA	NA	NA	10.0	6.9	45.0	0.3	2.2	7.3	47.2	10
Albania	NA	NA	NA	NA	NA	NA	NA	NA	NA	0.4	0.4	0.5	0.5	NC
Croatia	NA	NA	NA	1.0[c]	0.8	2.0	0.5	0.8	2.0	1.7	4.3	3.3	8.3	1
Slovenia	NA	NA	NA	0.8	3.4	3.9	1.9	4.9	5.8	17.8	20.8	26.2	30.5	12
Italy	97.9	42.2; 0.5	4.2	14.6[d]	9.9	85.5	3.7	2.6	22.5	10.1	87.2	23.2	199.4	9
France	10.0	0.2	3.0	NA	NA	NA	NA	NA	NA	7.9	118.2[e]	8.0	121.2	NC
Spain	62.3	NA; 1.6	41.0	17.8	2.5	64.0	NA	NA	NA	1.8	46.0	5.8	151.0	NC
Portugal	193.0	548.1; 118.4	390.7	6.5	4.9	16.2	4.5	2.3	7.6	17.5	57.8	143.2	472.4	42

[a] In this column, the first indicator refers to €/ha of cork oak area; the second refers to €/ha of forests.
[b] Expressed in stere.
[c] Of which 0.3 for market and 0.7 for self-consumption.
[d] Of which 2.6 for market and 12 for self-consumption.
[e] Includes truffles, miscellaneous, mushrooms and other fruits.
[f] The estimate is suspect; see text.
Source: based on Chapters 5–22, this volume.

Losses of NWFPs due to forest fires

Intensive fires may provoke a complete loss of NWFPs, while fires which do not completely burn the vegetation cover may cause only partial damage. The time lag for the NWFPs to regenerate depends on the intensity of fires and the type of NWFP. Certain NWFPs can even regenerate better in open areas created by forest fires than in the habitat of shaded high forests (Pagiola, 1996).

As explained in Chapter 3, NWFP losses are valued based on the burnt forest area and the average value of NWFPs per hectare of forests, as calculated in the previous sections. These valuations are undertaken in only a few country chapters (Morocco, Tunisia, Algeria, Syria and Greece), and the results show a negligible importance, of less than 1% of the TEV.

Recreation, landscape and tourism

Recreation and landscape quality have always been significant benefits of Mediterranean forests. Their importance has increased tremendously over the last decades, as population and income growth have prompted an increased demand for tourism, especially in coastal areas. Today, the Mediterranean region is the world's leading tourism destination, with more than 140 million visitors in coastal regions (Plan Bleu, 2004). The demand for recreation in many developed countries is reflected by a willingness to pay around €2.5–5/day-visit in various forest sites with recreational and aesthetic value. A study of Croatian coastal forests found that, in most areas, landscape benefits were greater than all other benefits combined (Pagiola, 1996).

In Croatia, Greece, Italy and Spain, the estimated values of recreational benefits are based on consumer surplus estimates derived from travel cost method (TCM) and contingent valuation method (CVM) studies. There have been a large number of such studies in these countries.[11] In Croatia and Greece, distinct estimates of consumer surplus per visit are calculated for both protected areas and other forest sites with recreational value. Combining these indicators with the total number of visitors to protected and other recreational forests gave an estimate of total recreational benefits and benefits per hectare for each type of area. These estimates were then aggregated or extrapolated to arrive at national level estimates. In France, the costs of travel are used and considered the lower bound estimate for the true value of recreational services (around five times less than the consumer surplus).

CVM and TCM studies are extremely scarce in the southern and eastern Mediterranean. TCM results are available only for a protected area in Tunisia, which is used to arrive at an estimate for recreational benefits in that country's national parks. In the absence of estimates of consumer surplus, other approaches had to be followed. In Turkey, revenues from entrance fees are used to obtain an estimate of the recreational value of forest sites. In Israel, the total costs of travel by car to forests is applied. In Lebanon, the net revenues of environmental clubs[12] are used to obtain a partial value of recreational benefits. The qualitative difference in these results (consumer surplus versus actual payments) makes cross-country comparisons dubious (see Chapter 3).

Table 4.6 shows the estimated recreational benefits of forests. Some apply to protected areas (national parks and reserves), some to other forest sites with recreational value, and some to all forests in a given country. These differences, together with the methodological differences in the way they are derived, do not allow for homogeneous cross-country comparisons. However, France stands out with the highest estimate (€115/ha), exceeding by far any other forest benefit; followed by Italy (€20–50/ha). These high values of recreational benefits reflect both a high social demand for outdoor activities (which often is associated with NWFPs, such as mushrooms and berry picking, birdwatching and hunting) and the large number of visitors[13] to French and Italian forests.

The highest estimate for recreational value at a site (€450/ha of national park) is found in Croatia; it reflects a high willingness to pay for visiting national parks such as Plitvice Lakes, Paklenica and Krka waterfalls, which receive many visitors in a relatively small area (2.5% of total forest area). Likewise, the relatively high estimated value in Turkey and Israel (€120–170/ha) results from the small area allocated for recreational purposes. Indeed, as a

Table 4.6. Value of forest recreation (2001).

Country	Type of recreational area	Recreational area (000 ha)	No. of day visits (000)	Value/ha of recreational area (€)	Value/ha of forests (€)	Value/ day-visit (€)	Total value (000 €)	% TEV	Valuation method
Southern									
Tunisia	National parks	69.0	93.4	7.2	NA	5.7	500	0.4	TCM
Eastern									
Israel	All forests	NA	3,000.0[e]	NA	167.7	5.5	16,500	NC	Cost of travel
Lebanon	Reserves and other recreational sites	NA	NA	NA	NA	7.5[a]	300[b]	NC	Net revenues[c]
Turkey	Recreational sites	16.0	NA	122	NA	0.3	1,900	0.3	Entrance fees
Cyprus	All forests	385.6	73.4	NA	4.6	2.5	1,800	NC	Benefit transfer (CVM)
Northern									
Greece	National parks	337.0	271.0	12.0	NA	15.0	4,100	1.7	CVM
	Recreational sites	6,166.0	1,500.0	0.6	NA	2	3,800	3.8	CVM
Croatia	Reserves	29.0	10.0	2.9	NA	8.3	27,800	7.9	TCM
	National parks	61.7	2,000.0	450.0	NA	13.9	83	NC	TCM, CVM
Italy	All forests	8,600.0	68,000.0– 168,000.0	NA	19.7–48.8	2.5	170,000– 420,000	NC	TCM, CVM
France	All forests	NA	394,000.0	NA	114.5	4.4	1,718,000	NC	Cost of travel
Spain	National parks	622.0	4,920.0[d]	95.0	NA	5.2	25,000	NC	TCM, CVM
	Other protected areas	2,316.0	14,445.0[d]	32.4	NA	5.2	75,000	NC	TCM, CVM
Portugal	All forests	3,349.0	6,000.0	NA	4.9	2.8	16,500	1.5	CVM

[a]Result of a CVM study referred to only one reserve (Bsharre cedar grove).
[b]This represents the net revenues from three forest reserves (donations).
[c]Net revenues of environmental clubs organizing recreational activities in Lebanese forests, including the donations collected at entrance in the three forest reserves of Al-shouf, Bsharre and Ehden. Some part of this value also includes forest-based activities.
[d]Only the number of day visits of the people willing to pay.
[e]The minimum bound of the interval 3–10 million day visits.
Source: based on Chapters 5–22, this volume.

result of visitor congestion, many of these areas are subject to overuse and degradation. The other recreational estimates for national parks vary between €7/ha of national park in Tunisia and €95/ha of national park in Spain; while for other sites with recreational value, they lie between €5/ha of forests in Portugal and €20/ha of forests in Italy.

Large variations in recreational benefits according to the site are also found in other non-Mediterranean countries. For example, the recreational benefits of forest sites have been found to be US$8/ha (~€8/ha at 2001 prices) in Mexico (Pearce *et al.*, 1993), US$12/ha (€14/ha) in Kenya (Brown and Henry, 1989) and US$52/ha (€56/ha) in Costa Rica (Tobias and Mendelsohn, 1991). Some of these estimates are not exclusively for forest-based recreation, however, while others are for areas that may not be representative. The US$52/ha in Costa Rica, for example, is based on the country's most popular tourist destination, the Monteverde Cloud Forest.

Across the Mediterranean basin, there are large differences in consumer surplus between protected areas (~€5.7/day-visit in Tunisia, €7.5/day-visit in Lebanon and €14–15/day-visit in Croatia and Greece) and for other forest sites with recreational value (~€2–2.5/day-visit for Italy, Cyprus and Greece). These differences reflect the special features of protected areas in terms of landscape quality.

The highest values of forest recreation are found mostly in the developed countries of the Mediterranean. Here, recreation is increasingly becoming a remunerated service, through payments for entrance and complementary products (Merlo *et al.*, 2000; Mantau *et al.*, 2001). However, the benefits are mainly captured by visitors and the tourism industry, while the costs of forest management are supported by the limited number of forest owners and managers. This income misallocation needs to be addressed.

In southern and eastern countries, recreation is enjoyed commonly as a free public good. Efforts to capture these values have been implemented only rarely and mostly in protected areas. A few key issues can be noted, with important policy implications:

1. In many protected areas, most recreational benefits are enjoyed by foreign rather than national visitors. In other words, the values are enjoyed outside the national boundaries.

2. For other sites with recreational value, there are usually no or low entrance fees. Recreational benefits at these sites are often threatened by degradation resulting from overcrowding and overuse.

3. In many recreational sites, the local population derive little income from tourism activities.

Hunting

Hunting is regulated in all Mediterranean countries. Game and shooting rights are state property, and the states sell shooting and hunting permits and specify the areas where hunting is permitted, as well as the game breeding reserves where it is forbidden.[14] The exception is Syria, where hunting is forbidden in all forests. Illegal hunting (poaching) is common in all southern Mediterranean countries. The number of illegal hunters in Morocco and Lebanon is at least double that of the legal hunters. Illegal hunting brings private benefits to hunters, and so is included among direct use values with a positive sign. On the other hand, it may have a negative impact on the sustainability of game populations. These environmental costs reduce the social value of hunting.

For many hunters, particularly in northern countries, enjoying nature is often an inherent part of the hunting experience. The benefits of hunting, therefore, include an element of recreation. In southern countries, hunting benefits are primarily a source of livelihood. Valuation is undertaken at the national level, using a variety of techniques. The results of CVM surveys and other measures suitable for benefit transfer are available only for Italy and Croatia. In the absence of such estimates, actual payments – permit or licence fees and the value of game, and cost of travel – are used for the other countries. As discussed in Chapter 3, the benefits received by illegal hunters are estimated on half the permit price. The damages due to illegal hunting are valued only for Morocco, using the restoration cost method. As in the case of recreation, the use of dissimilar valuation methods limits the comparability of results.

CVM surveys show benefits of about €215–250/hunter in Croatia and Italy. Use of fees or permit prices suggests much lower values,

Table 4.7. Benefits and costs of hunting (2001).

Country	Hunting area (ha)	Number of hunters	Value/hunter (€)	Value/ha of hunting forest area (€)	Value/ha of forests (€)	Total value (000 €)	% TEV	Valuation method
Southern								
Morocco	NA	NA	NA	NA	−4.0	−35,000	−5.7	Taxes, licence fees
Benefits	NA	30,000–50,000	130–210	NA	0.9	8,900	1.4	
Costs of damages due to poaching	9,050,000	NA	NA	−4.9	−4.9	−43,900	−7.1	Restoration costs
Tunisia	NA	13,200	147.0	NA	2.1	1,900	1.5	Permit price, licence fees
Eastern								
Lebanon	NA	600,000	20.0	NA	88.9	12,000	NC	Licence fees
Turkey	NA	350,000	43.8	NA	0.7	15,300	1.6	Licence fees, permit price
Cyprus	NA	19,330	61.0	NA	3.1	1,200	NC	Permit price
Northern								
Greece	900,000	272,000	98.0	29.6	4.1	26,700	5.8	Permit price
Albania	NA	10,550	30.3	NA	0.3	300	0.2	Permit price
Croatia	300,000	47,000	215.0	33.3	4.0	10,100	1.4	Benefit transfer[b]
Slovenia	NA	13,200	431.8	NA	4.9	5,700	2.3	Fees, volunteer work
Italy	1,153,000[a]	285,000	250.0	61.7	8.3	71,200	3.3	CVM
France	NA	10,000,000[c]	10.0	NA	6.4	96,000	NC	Cost of travel
Portugal	NA	219,000	97.3	NA	6.4	21,300	1.9	Permit price and other expenditures

[a]ISTAT (1997).
[b]Based on CVM results in Italy.
[c]Hunting visits.
Source: based on Chapters 5–22, this volume.

ranging from €20 to €60/hunter in Lebanon, Albania, Turkey and Cyprus to €100–150/hunter in the other countries. The greatest benefits are found in Slovenia, as a result of the specific hunting context: in addition to paying fees, hunters are required to perform yearly volunteer work. Aggregating the two values results in about €430/hunter. Data on the distribution of hunting benefits between foreign and local hunters are available only in Morocco, Tunisia and Albania, where hunting benefits for foreigners are larger than those for the locals.[15]

Table 4.7 shows the estimates of hunting and the valuation methods used. Probably the most striking result is that for Morocco, where the value of hunting activities is negative (–€35 million). This is due to the magnitude of the damage to wildlife being much greater than the private benefits obtained by hunters. If this damage had also been valued in the other southern Mediterranean countries, the value of hunting in those countries would probably also have been estimated to be much lower. As with recreation, the highest absolute values of hunting benefits are found in the northern Mediterranean countries and, in particular, in Western Europe, varying within €21–71 million.

In most northern Mediterranean countries, hunting benefits are about €4–8/ha, with Portugal, France and Italy having higher values. The highest value is found in Lebanon, mainly due to the large number of hunters and the small forest area. The lower value of hunting in the few southern Mediterranean countries for which estimates were made may only be apparent due to the application of qualitatively different valuation methods (fees versus CVM). The lack of comprehensive studies in the southern and eastern Mediterranean on the number of hunters (especially illegal ones) may also lead to an underestimation. It is interesting to note that in terms of share of the TEV, hunting benefits appear larger than recreational benefits in all countries except for Italy and Morocco. This is a result of the different methods and levels of valuation: on the one hand, the widespread use of actual payments (hunting fees) tends to result in higher estimates compared with the consumer surplus used for recreation valuation in these specific contexts; on the other, in many countries, valuation of recreation refers only to

protected areas, while the estimates of hunting are results at the national level.

Converting total values into values per hectare of designated hunting areas was possible only in a few countries, with results within €30–62/ha of hunting forest area. For other countries, the size of the hunting area is not specified. In southern countries, illegal hunting is believed to take place throughout the forest area.

Valuation efforts undertaken in other parts of the world show relatively high values for hunting in developed countries. Hunting benefits in the American state of South Dakota were found to be about US$64/ha (€82/ha) of hunting area (Johnson and Linder, 1986), for example. In contrast, estimates of hunting value in several developing countries[16] were much lower, about US$1–16/ha of hunting area (~€1–18/ha of hunting area in 2001 prices) (see Lampietti and Dixon, 1995). The large difference between these estimates has been explained by the use of different valuation methods: consumer surplus in the developed countries and producer surplus (value of game caught) in most developing countries.

Indirect Use Values

By their very nature, indirect use values are much harder to quantify than direct use values. They are almost always non-market values, demanding the use of more indirect valuation techniques. Moreover, as discussed below, cause and effect relationships are often difficult to establish, making even these techniques difficult to apply.

Water-related services

Water-related issues have always been of central importance to the Mediterranean region, much of which is characterized by water stress,[17] severe climate conditions and soils susceptible to erosion, floods and landslides. Water stress is particularly severe in countries in the south and east Mediterranean (North Africa and Middle East), who have only 1250 m³/*capita* on average (World Bank, 1994). In some of these countries, water is at high risk of scarcity, while in some others, water is already scarce.[18]

Growing population and inefficient water use put additional pressure on water resources. In contrast, countries in the northern Mediterranean benefit from a greater water endowment (> 5000 m³/*capita*) and usage patterns tend to put less pressure on water resources. As the economic value of water is not reflected in markets, however, growing demand may result in water shortages developing in these countries as well. It should be noted that the level of water scarcity and stress varies considerably even within each country.

Water and watershed protection

In such a fragile environment, the role of forests in protecting water supplies, purifying water, regulating water flows and conserving soil is particularly important. Forests have generally been thought to provide a wide range of water services. Recent research has stressed a more nuanced view of the role of forests in providing water services, however (Bruijnzeel and Bremmer, 1989; Aylward *et al.*, 1998; Calder, 2000; Bruijnzeel, 2004; Tognetti *et al.*, 2004). While forests do tend to improve water quality, their impact on dry season water flows varies depending on local conditions, and their impact on total annual flow is generally negative (because of high evapotranspiration from trees). Forests also help to reduce downstream flood risk, but this effect is on a more limited scale than is commonly assumed. Indeed, it is increasingly argued that the presence of forests does not help to reduce the most damaging floods which occur once in a lifetime; rather, at a large scale, dams, drainage channels and the extent of water use are more significant (Kaimowitz, 2004). Research on forest–hydrology links in the Mediterranean has been limited to a few site-specific studies, almost all in the northern Mediterranean. Although the knowledge base is limited, its conclusions are broadly similar (Bellot *et al.*, 1999, 2001; Lavabre *et al.*, 2000).

Table 4.8 summarizes the main approaches used to value the watershed protection benefits of forests in the country chapters providing these valuations. Data constraints drive most country chapters to focus on particular aspects of the potential downstream damage. In Tunisia, for example, estimates focus on the cost of reservoir sedimentation (using the cost of building new reservoir capacity to replace that lost to siltation) and on the potential loss of production in irrigated agriculture (using a production function approach). In Greece, estimates focus on the cost of replacing the water supply (using the cost of drilling new wells). These differences are partly a reflection of local conditions (such as the prevalence of irrigated agriculture, or the number of reservoirs vulnerable to siltation) but also a reflection of the availability of data. As a result, each country estimate is only able to capture a part of the potential benefit of watershed protection. Some estimates are particularly imperfect: defensive expenditures, for example, may be quite inadequate for the task (and thus tend to underestimate the benefit of preserving forests) or used quite inefficiently (and thus tend to overestimate).

Even within a country, there can be considerable variation in the level of watershed protection benefits that forests provide. Some countries were able to undertake a disaggregated analysis which considered the benefits of different forests (or forests in different areas) separately, and then added them up to arrive at an aggregate benefit. Thus, the estimated value of watershed protection in Morocco (€31/ha) reflects an average value of different levels of protection against erosion in different forest areas. They range from €20/ha of area covered by Saharan acacias (*Acacia gummifera, A. horrida* and *A. cyclops*) to €38/ha of area covered by other forest types (*Cedrus atlantica, Pinus* spp., *Juniperus* spp., *Abies pinsapo, Quercus ilex, Q. suber* and *Q. faginea*) (weighted area mean). In most cases, however, data constraints prevent countries from undertaking a disaggregated valuation and force them to use country-wide averages.

The results of these estimates for the country chapters providing these valuations are summarized in Table 4.8 and Fig. 4.1. These results show that watershed protection benefits tend to be high, both in absolute and in relative terms. In several countries, watershed protection appears to be the single most valuable forest benefit. In Syria, Greece and Italy, watershed protection accounts for more than 50% of the TEV (although Greece's figure falls sharply if the costs of damage in degraded watersheds are subtracted, see Table 4.9 below). In Tunisia,

Table 4.8. Valuation of water-related issues (2001).

Country	Water-related issue	Area of protective forests (Mha)	Value/ha of protective forests (€/ha)	Value/ha of total forests (€/ha)	Total value (M€)	% TEV	Method of valuation/type of costs involved
Southern							
Morocco	Watershed protection function:			30.8	278.7	47	Damage cost avoided
	Acacia gummifera, A. horrida, A. cyclops	1.0	20.1				
	Stipa tenacissima	3.3	26.1				
	Argania spinosa, other coniferous and broad-leaved	0.9	30.2				
	Tetraclinis articulata	0.6	34.2				
	Cedrus atlantica, Pinus spp., *Juniperus* spp., *Abies pinsapo, Quercus ilex, Q. suber, Q. faginea*, mattoral/*maquis*, plantations	3.2	38.2				
Algeria	Protection of water supply and soil conservation	NA	NA	25	478.0[a]	77	Not indicated
Tunisia	1. Reduced reservoir sedimentation	NA	NA	26.3	24.8	19	1. Damage cost avoided: cost of replacing water storage capacity
	2. Agricultural soil conservation						2. Production function approach: increased value of agricultural yield
Eastern							
Syria	Protection of lowland areas against erosion, floods, landslides	NA	157.2	87.6	40.4	99	Damage cost avoided: annual public expenses for soil conservation and hydraulic work maintenance
Turkey	1. Protection against erosion, avalanches	NA	130.0[b]	NA	NA	NC	1. Damage cost avoided: costs of dam, bench, sleeper, dry walls (against erosion) and other technical precautions (against avalanches)
	2. Water purification						2. Damage cost avoided: costs of afforestation and other technical precautions

continued

Table 4.8. *Continued.*

Country	Water-related issue	Area of protective forests (Mha)	Value/ha of protective forests (€/ha)	Value/ha of total forests (€/ha)	Total value (M€)	% TEV	Method of valuation/type of costs involved
Northern							
Greece	1. Agricultural soil protection 2. Protection of water supply	NA	NA	45.2	293.6	64	1. Production function approach: increased value of agricultural yield 2. Damage cost avoided: costs of drillings
Albania	Protection against erosion	0.5	6.0	2.9	3.0	NC	Defensive expenditures: public expenditures for erosion control
Croatia	Protection of water supply	1.15	18.0	10.0	20.7	2	Damage cost avoided: costs of water protection
Italy	Protection against erosion, floods, landslides	NA	NA	154	1321.0	62	Damage cost avoided: annual public expenses for soil conservation and hydraulic works
France	Protection of landscape and soils against forest fires, avalanches, landslides and floods	NA	NA	7.8	118.0	—	Defensive expenditures: annual public expenses for fire prevention and combatting and other defence costs
Portugal	1. Protection of agricultural soil 2. Protection of water supply	NA	NA	23.3	78.1	7	1. Production function approach: increased value of agricultural yield 2. Damage cost avoided: public expenditures on watershed management

[a]Of which 102.5 M€ is attributed to forests and €375.5 M€ to steppe vegetation.
[b]Result of a local level survey in Altindere National Park, updated to 2001 prices.
Source: based on Chapters 5–22, this volume.

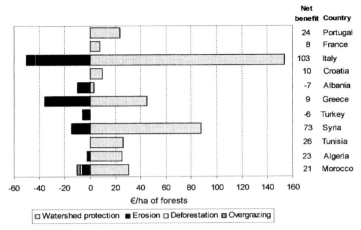

Fig. 4.1. Average benefits of watershed protection and costs of degradation (2001).

watershed protection is second in value only to grazing. Watershed protection also accounts for a large share of TEV in Morocco and Algeria. This is particularly notable in light of the fact that watershed protection benefits are likely to have been underestimated, as they only capture part of the upstream–downstream linkages in each country. It should also be noted that in some cases where the estimated watershed protection benefits are particularly low, the estimates are particularly weak. Thus, Albania's estimate of about €3/ha may well reflect the inadequacy of government funding for watershed protection.

These estimates show that water and soil protection functions are an important forest benefit in most Mediterranean countries. The reasons for the high level of watershed benefits differ somewhat across the study area. In the southern Mediterranean, they often reflect the vulnerability of hydraulic infrastructure and irrigated areas and the high value of water in arid climates subject to considerable rainfall variation. In the northern Mediterranean, where water is much more abundant and hydraulic infrastructure much more developed, the high value of watershed protection results primarily from the higher value of infrastructure at risk and the higher costs of repairing damage.

Table 4.8 also illustrates the variability of watershed protection benefits. Where data were sufficient to do so, the watershed protection benefit was estimated separately for that portion of the forest that actually provides protective benefits. As can be seen, this can vary quite

substantially from the average level of benefits provided by the forest as a whole. In Syria, for example, the value of watershed benefits in forests that protect lowland agricultural areas is more than twice the average value of watershed benefits provided by the country's forests as a whole. In Tunisia, most protective benefits are generated in only 9% of the country's area, where the most important storage reservoirs are located. In Italy, the role of forests is especially important in areas with steep slopes and erodible soils in the Alps, the Apennines and Sardinia, which would be exposed to a high risk of floods and landslides in the absence of forest cover. This has important implications for the need to target interventions.

Negative externalities linked to water-related issues

Forests do not always generate benefits. Some forests, due to their characteristics or context, might cause damage. Other forests cause damage because they have been mismanaged, through excessive collection of firewood and other NWFPs, overgrazing or poor forest management. Such forests can create damage, such as increased erosion and dam siltation and increased risk of floods, landslides and avalanches. Likewise, deforestation reduces the area of forest and leads to a loss of the services that had been provided. In the southern Mediterranean, this can lead to desertification.[19]

Table 4.9. Costs of watershed degradation (2001).

Country	Degradation aspect	Physical indicator	Area of affected forests (ha)	Value/ha of affected forests (€/ha)	Value/ha of total forests (€/ha)	Total value (million €)	% of TEV
Southern							
Morocco	Watershed degradation due to:	NA	NA	NA	16.6	94.6	−15
	Erosion due to poor forest management (m³/ha)	(2.5–5)[a]; 10[b]		8.7	6.6	57.2	−9
	Deforestation (ha)		31,000	372.0	2.0	11.5	−2
	Overgrazing (ha)		7,240,000	3.1	3.9	22.5	−4
	Firewood overcollection (ha)		110,000	30.8	0.6	3.4	−1
Algeria	Land degradation due to:	NA	NA	NA	NA	44.4	−7
	Reservoir sedimentation		NA	NA	NA	41.4	−7
	Desertification (ha)		301,716	9.9	NA	3.0	−0
Tunisia	Watershed degradation due to forest fires (ha)		2,215	27.2	NA	0.1	−0.5
Eastern							
Syria	Watershed erosion	3.5[c]–20[d]	200,000	34.0	14.7	6.8	−17
Turkey	Watershed erosion due to deforestation	1.24[e]–4.5[f]	NA	NA	5.9	121.3	−13
Northern							
Greece	Watershed erosion due to poor forest management	6.47	2,000,000	116.5	35.8	232.9	−51
Albania	Watershed erosion due to poor forest management	30[g]	200,000	50.0	9.7	10.0	NC
Italy	Watershed degradation due to:		NA	NA	50.6	432.4	−20
	Erosion due to poor or no forest management (m³/ha)	2.18[h]–13[i]	2,000,000	32.7	7.6	65.4	−3
	Floods, landslides, avalanches	NA	NA	NA	42.7	367.0	−17

[a]Middle or higher Atlas.
[b]Rif mountain.
[c]Average rate in degraded forest area.
[d]Maximum erosion rate, resulting from fires on coastal slopes.
[e]Average rate for the Mediterranean part of Turkey (15,000,000 ha) (United Nations Environment Programme/Mediterranean Action Plan, 2000).
[f]Average rate for the whole country (United Nations Environment Programme/Mediterranean Action Plan, 2000).
[g]Maximum rate of erosion on degraded forest area.
[h]Alpine environments (southern slopes).
[i]South Italy.
Source: based on Chapters 5–22, this volume.

Similar techniques are used to value these negative externalities as are used to value positive externalities. There is an important difference in how the data are used, however. As discussed in the previous section, positive externalities are generally valued by estimating the damage that the presence of forests avoids. The damage, in that case, is hypothetical. In this case, however, estimates are based on efforts to measure the *actual* damage that results from poorly managed forests.

EROSION CAUSED BY POOR FOREST MANAGEMENT. For some countries, estimates of erosion are obtained by using the universal soil loss equation (USLE);[20] for others, the valuation method applied is not specified. It should be noted that erosion rates vary substantially from place to place; therefore, extrapolating erosion rates can result in inaccuracies. The costs caused by this erosion are then estimated using a replacement cost method based on a variety of approaches: the cost of purifying water from silted reservoirs (Algeria); the costs of dredging accumulated silt downstream of the watershed (Morocco, Syria, Greece and Italy); and the cost of fertilizers for replacing soil nutrients (Turkey).

These cost estimates have important limitations. Dredging costs can overestimate damage, especially if they are applied to the entire quantity of eroded soil. In general, only part of the soil lost is actually deposited downstream. The costs of water treatment can be a reliable estimate, as far as no other less costly options exist for restoring the damage caused by siltation. The prices of fertilizers can capture the value of on-site damage, but not that of other off-site effects.

As discussed in Chapter 3, replacement costs are usually not very good indicators of the value of damage incurred. Use of actual expenditures can underestimate the damage, as replacements rarely replace all the services coming from the original ecosystem; it can also overestimate, as the replacement may be undertaken inefficiently. Use of potential expenditures can easily overestimate the loss of an ecological service, as the replacement can be too expensive to make sense.

Table 4.9 shows the estimated negative externalities from poorly managed forests in the Mediterranean. High levels of erosion occur in the southern and eastern countries, where they reach on average 10 m³/ha/year in the Rif mountains of Morocco, 20 m³/ha/year on the coast of Syria and 30 m³/ha/year in the most degraded forest areas in Albania. In these countries, erosion is caused mainly by poor forest management, overgrazing, excessive fuelwood collection and deforestation for conversion to agriculture. The absence of estimates from other countries in this part of the basin reflects the lack of data and does not imply that they are unaffected by this phenomenon. The extent of the area affected varies. In Tunisia, soil erosion affects 56% of the country's area, the highest proportion in North Africa, while another 17%, located in the central and southern parts of the country, is highly sensitive to desertification (Ministère de l'Environnement et de l'Aménagement du Territoire, 1998; Plan Bleu, 2000a). In Lebanon, erosion is thought to affect around 65% of the country. However, no comprehensive quantitative studies of soil loss are available, as no central authority is charged with monitoring and preventing soil erosion (Plan Bleu, 2000b).

In the northern Mediterranean countries, soil erosion is due mainly to forest fires resulting from land abandonment by private owners and lack of on-site protection. Site-specific estimates show erosion rates of 2.1 m³/ha in Italian alpine forests and 13 m³/ha of forests in southern Italy; and an average of 6.47 m³/ha of degraded forests in Greece. Even though erosion rates tend to be lower than in the southern Mediterranean countries, the damage they cause can be substantial.

The resulting monetary estimates of damage are highest in Greece (€117/ha of affected forests), followed by Albania (€50/ha of affected forests) and Italy (€33/ha of affected forests). To a large extent, this is due to the high estimates of costs of dredging per hectare. Overall, erosion in poorly managed forests is sufficiently important to reduce the TEV of forests by 51% in Greece and 9% in Morocco.

DAMAGE CAUSED BY FLOODS, LANDSLIDES AND AVALANCHES. These types of damage occur in areas where forest cover is not able to perform water and soil protection functions, mainly due to poor forest management or insufficient vegetation cover. These types of damage are estimated only for Italy, based on the expenses undertaken for restoring the affected areas. Valuation shows an average cost of €43/ha of forests, representing some 17% of the TEV of Italian forests.

WATERSHED DEGRADATION DUE TO DEFORESTATION. Deforestation for conversion to agricultural uses is common in most southern and eastern Mediterranean countries. It causes negative externalities in terms of reduced soil protection in watersheds. In Morocco, where deforestation affects 31,000 ha yearly, this type of damage is

estimated based on restoration costs, resulting in an average estimate of €2/ha for forests or €370/ha for deforested land. This type of damage accounts for some 2% of the TEV.

DESERTIFICATION. The risk of desertification affects the southern Mediterranean countries and is mainly linked to the harsh climate, combined with deforestation, overgrazing, poor forest management and the absence of vegetation cover. The negative externalities linked to desertification are valued only for Algeria, based on the change in productivity on 300,000 ha

---

Box 4.1.   The value of forest carbon (Roger A. Sedjo[a]).

The build-up of greenhouse gases in the atmosphere is believed to be a cause of global warming. Concerns over the consequences of warming on human society have been reflected in the host of scientific research, policy discussions and international negotiations that has been undertaken on sources and sinks for greenhouse gases, and particularly carbon dioxide. The role of forests as a carbon source, which releases carbon into the atmosphere, and the forest's potential to reduce atmospheric carbon by sequestrating carbon into forest biomass and soils are important in both research and policy discussion. Plants capture carbon in the process of biological growth and store that carbon in their cells during their lifetime and sometimes beyond. Trees, by accumulating large volumes of biomass over their long lives, are particularly effective in capturing and storing carbon. Forest soils also capture and store substantial volumes of carbon over time via their interaction with tree and plant roots. Both old and young forests provide carbon sequestration services. Old steady-state forests serve as carbon-holding facilities, while young growing forests sequester new carbon by virtue of their biological growth. Forest destruction and decomposition releases carbon back into the atmosphere. However, carbon continues to be held captive in dead trees and long-lived wood products until they are burned, decay or otherwise release carbon back into the atmosphere.

   While it is widely recognized that most of today's increases in atmospheric carbon are due to fossil fuel burning, it is estimated that a substantial amount of the historic and current build-up in atmospheric carbon is the result of land use changes that release carbon into the atmosphere. Since land use practices have contributed to the problem, they may also be able to contribute to the mitigation of atmospheric carbon build-up. Early work in this area showed that forest expansion cannot be the total solution; the task is too large. However, the Intergovernmental Panel on Climate Change (2001) estimated that up to 20–30% of the anticipated atmospheric carbon build-up over the next 50 years could be offset by appropriate forest and land use activities. Furthermore, in moderating the build-up of forest carbon, sequestration could 'buy time' for more sophisticated carbon-free technologies to be developed.

   The potential role of forests in capturing carbon is straightforward. If areas that previously had been deforested, and thereby had contributed to carbon build-up in the atmosphere, were reforested, the process would be reversed and these lands would return to permanent forest, thereby sequestering positive net amounts of carbon. Similarly, the same could occur for afforested lands that had previously never been forested. However, such an approach may not be as inexpensive as it seems. Land is often taken from forestry because there are alternative higher value uses. To reforest these areas would not only involve the costs of reforestation but also the loss (opportunity costs) of the use of the land for the other purposes. Nevertheless, if the benefits of mitigating atmospheric carbon build-up are large, the reforestation of the land may be justified. Furthermore, a forest need not sit idly serving only to sequester carbon. By limiting timber harvest to net growth, a working forest can serve as a carbon sink even as it continues to provide timber and other services. Thus, the provision of carbon sequestration services need not severely limit the forest's production of other outputs.

[a]In remembrance of my good friend Maurizio Merlo.

affected annually. It results in an estimate of €10/ha of degraded land, accounting for less than 1% of the TEV.

Carbon sequestration

Mediterranean forests can play an important role in carbon sequestration (Box 4.1). On the one hand, this relates to the forests' high ability to adapt to local conditions; on the other, to the soils' absorption capacity, which is thought to be greater than that of other forest ecosystems of northern Europe (Scarascia-Mugnozza *et al.*, 2000). Notwithstanding the considerable potential of Mediterranean forests for sequestering carbon, at present their role in doing so is limited. It also varies greatly from one area to another, depending on the natural conditions (climate, forest growth) and human actions (deforestation, abandonment) affecting forests.

Valuation of carbon sequestration begins by estimating the quantity of carbon stored annually in forest biomass. As noted in Chapter 3, most northern Mediterranean countries[21] and Syria use estimates based on Food and Agriculture Organization (FAO) methodologies (Food and Agriculture Organization, 2000; United Nations Economic Commission for Europe/Food and Agriculture Organization, 2000), while other southern Mediterranean countries use estimates drawn from the national communications to the United Nations Framework Convention on Climate Change (UNFCCC). The country chapters then use Fankhauser's €20/tC estimate for the value of carbon sequestration (Fankhauser, 1995). The discussion here is based on this estimate, as well as two other scenarios, with carbon prices of €10C and €15/tC. Table 4.10 presents the estimates of carbon sequestered in forest biomass. As might be expected, countries with substantial forest areas, such as France, Spain and Turkey, have the highest estimated carbon sequestration in woody biomass. The largest estimate is found in France, of about 11 MtC. Turkey's estimate of 7.9 MtC should be considered with caution, however, as it does not take into account the carbon released through illegal felling. Should that amount be included in the calculations, the estimates would also be lower in Turkey, Albania and many southern countries.

In contrast, forests in Algeria, Morocco and Lebanon are net carbon sources, as a result of slow forest growth and strong human pressure. The highest carbon losses, of about 0.6–0.9 MtC, are found in Morocco; here forests are estimated to contribute around 11% of the country's CO_2 emissions, mainly due to high rates of firewood collection (Ministère de l'Aménagement du Territoire, de l'Urbanisme, de l'Habitat et de l'Environnement, 2001). The estimates in Algeria of about 0.3 MtC are due to the conversion of forests and prairies to other land uses, which is responsible for around 12% of the total country's emissions of CO_2 (Ministère de l'Aménagement du Territoire et de l'Environnement, 2001).

Estimates of carbon stored in forest soils are available for Cyprus, Croatia, Slovenia and France. In the first two countries, they are 3–4 times greater than those in forest biomass (see footnotes to Table 4.10). The highest value, of 6.2 MtC, is found in Croatia, the country with the most extended forest area among the first three; a lower value of about 5.2 MtC is attributed to France, referring only to the annually reforested area of 74,200 ha.

Mediterranean forests sequester around 0.01–1.08 tC/ha of forests annually: the lower bound is found in Albania, as a result of the large extent of timber cut; the upper bound occurs in Croatia and Slovenia, though this may be partially due to the different method of calculation compared with other countries. Losses of carbon in the southern and eastern Mediterranean countries vary within 0.08–0.53 tC/ha, with the highest in Lebanon, which is characterized by a small forest area and intensive timber cut.

The differences in the physical estimates lead to corresponding variations among countries in monetary valuations. Thus, the highest value of carbon sequestration is about €21–22/ha in Croatia and Slovenia. Another group of countries, primarily in the northern and eastern Mediterranean, have values of about €3–9/ha. In Cyprus and Albania, forests provide almost no carbon sequestration benefits. On the other hand, soils sequester carbon worth about €50/ha (Cyprus and Croatia). As noted, forests in Morocco, Algeria and Lebanon have a net reduction in value as a result of carbon emissions.

With few exceptions, the importance of carbon sequestration is much lower than that of

Table 4.10. Carbon sequestration in Mediterranean forest biomass (2001).

Countries	Net quantity of carbon sequestered/emitted annually		Total value of carbon			Monetary indicators			% of TEV		
	tC/ha	000 tC	Valued at €10/tC	Valued at €15/tC	Valued at €20/tC	Valued at €10/tC	Valued at €15/tC	Valued at €20/tC	Valued at €10/tC	Valued at €15/tC	Valued at €20/tC
			Million €	Million €	Million €	€/ha	€/ha	€/ha	%	%	%
Southern											
Morocco	−0.10	−583.3	−5.8	−8.7	−11.7	−1.0	−1.5	−2.0	−0.9	−1.4	−1.9
Algeria	−0.08	−341.5	−3.4	−5.1	−6.8	−0.8	−1.2	−1.7	−0.6	−0.8	−1.1
Tunisia	0.25	220.9	2.2	3.3	4.4	2.5	3.7	4.9	1.7	2.6	3.4
Eastern											
Israel[f]	0.27	NA	NA	NA	NA	NA	NA	NA	NC	NC	NC
Lebanon	−0.53	−72.1	−0.7	−1.1	−1.4	−5.3	−8.0	−10.7	−0.8	−1.2	−1.5
Syria	0.28	130.5	1.3	2.0	2.6	2.8	4.2	5.7	3.3	4.9	6.4
Turkey	0.38	7,920.0	79.2	118.8	158.4	3.8	5.7	7.7	9.1	13.1	16.7
Cyprus[a]	0.07	27.0	0.3	0.4	0.5	0.7	1.1	1.4	2.7	3.8	4.8
Northern											
Greece	0.07	463.0	4.6	6.9	9.3	0.7	1.1	1.4	1.0	1.5	2.0
Albania	0.01	12.0	0.1	0.2	0.2	0.1	0.2	0.2	0.1	0.1	0.1
Croatia[b]	1.08	2,700.0	27.0	40.5	54.0	10.9	16.3	21.7	3.8	5.7	7.4
Slovenia[c]	1.05	1,227.0	12.3	18.5	26.6	10.5	15.8	21.0	5.1	7.5	9.8
Italy[d]	0.39	3,100.0	31.0	46.5	62.0	3.9	5.8	7.8	1.5	2.3	3.0
France[e]	0.73	11,000.0	110.0	165.0	220.0	7.3	11.0	14.7	NC	NC	NC
Spain	0.17	4,400.0	44.0	66.0	88.0	1.7	2.5	3.4	NC	NC	NC
Portugal	0.44	1,450.0	14.5	21.8	29.0	4.4	6.6	8.8	1.3	1.9	2.6

In addition, the quantity of carbon stored in forest soils is:

[a]93,000 tC or 0.24 tC/ha of forests.
[b]6.2 MtC or 2.48 tC/ha of forests.
[c]78,300 tC or 0.07 tC/ha of forests.
[d]5.2 MtC or 0.35 tC/ha of forests (70 tC/ha of reforested land).
[e]The minimum value of total quantity of carbon. It ranges between 3.1 MtC (Cesaro and Pettenella, 1994) and 6.9 MtC (United Nations Economic Commission for Europe/Food and Agriculture Organization, 2000).
[f]A local level estimate in Yatir Forest (2400 ha).
Source: based on Chapters 5–22, this volume.

other forest benefits. It represents a significant portion of the forests' TEV only in Turkey (17%), followed by Slovenia (10%) and Croatia (7%). In all other countries, carbon sequestration only accounts for a few per cent of forest TEV, even when valued at €20/tC. The share of carbon sequestration in total value is even lower if lower carbon prices are assumed.

The carbon sequestration in forest soils is more significant than that in forest biomass. The estimate for Croatia forms 17% of the TEV. Had that also been estimated for other countries, the importance of the carbon sequestration function would have been significantly greater.

Option, Bequest and Existence Values

To date, few efforts have been made to value these benefits in the Mediterranean. The scarcity of data is to a large extent related to the difficulty of valuation itself. Reliable estimates usually result from indirect methods (stated preference approaches), whose application is difficult and costly. In addition, the results are often site specific and cannot be extrapolated meaningfully to a wider level.

Biodiversity conservation

With a remarkable number of 13,000 endemic plants, the Mediterranean region is considered a 'hot spot' for biodiversity conservation (Myers et al., 2000). The rich biodiversity of the Mediterranean forests is reflected in a wide range of indicators. Mediterranean forests have around twice as many woody species as European forests (247 versus 135) and one and a half times as many as Californian Mediterranean forests (170) (Quézel et al., 1999). In particular, Mediterranean forests host a larger number of tree species (100 versus 30) and a higher animal diversity[22] (11.3 versus 4.2) than European forests (Scarascia-Mugnozza et al., 2000).

Efforts to value the option, bequest and existence values of biodiversity are scarce everywhere, and Mediterranean countries are no exception. CVM surveys were conducted in only a few northern countries (Croatia, France, Italy and Spain), resulting in estimates of existence value for reserves and parks (site level). For example, an annual willingness to pay for the existence values of parks and other protected areas of about €5–15/person was found in Italy (based on Signorello, 1990).

In the absence of such surveys, cost-based methods were used in the other countries, based on the annual payments of various organizations for preserving biodiversity in protected forest areas (protected area level). These estimates do not distinguish among the option, bequest and existence value of biodiversity in these areas. In addition, as these payments are usually investments with long-term benefits, they do not distinguish between the value of biodiversity flows and stocks. It should be noted that the methodological differences in the results obtained by different methods (consumer surplus and actual payments) applied to different levels of valuation (site and protected area levels) do not allow a homogeneous cross-country comparison among the estimates.

Table 4.11 shows the estimates obtained. They capture only a very small part of the option, bequest and existence values of forests; therefore, in most countries, their share of less than 2% of the estimated TEV is unlikely to be a fair representation of their true importance. The estimates derived from the cost-based method allow calculation of the biodiversity value per hectare of national parks only in Turkey (€0.5/ha of protected area) and Tunisia (€90/ha of protected area), with large differences in the levels of expenses for biodiversity conservation. For the countries where the specific area to which the biodiversity expenses are attributed is not reported, average estimates per hectare of forests were calculated. Despite the underestimation problem, the estimate for Tunisia is higher than that of WFPs (€6.6/ha versus €4.5/ha, see Table 4.1). The weakness of these estimates does not allow any strong conclusions to be drawn about the value of biodiversity conservation.

The value of biodiversity in Mediterranean forests is reduced by a variety of human actions. Forest fires and deforestation, illegal hunting and harvesting of plants, and poor forest management all cause harm to the health of the forests and the biodiversity they contain. Most country chapters use the value of fines paid to the local and regional authorities for illegal

Table 4.11. Option, bequest and existence values of forest biodiversity (2001).

Country	Area (ha)	Value/ha of affected area (€/ha)	Benefits			Type of area	Valuation method	Costs (cost-based method)		
			Value/ha of forests (€/ha)	Total value (million €)	% TEV			Value/ha (€/ha)	Total value (€million)	% TEV
Southern										
Tunisia	69,000	90.4	6.6	6.2	4.8	Parks and reserves	Cost-based method	0.3	0.2	−0.2
Lebanon	NA	NA	6.4	0.9	0.9	Not specified	Cost-based method	0.5	0.1	−0.1
Eastern										
Turkey	2,590,455	0.5	0.1	1.3	0.1	Protected areas	Cost-based method	NA	NA	NC
Cyprus	NA	NA	NA	NA	NA	NA	NA	0.7	0.3	−2.2
Northern										
Greece	NA	NA	2.0	12.7	2.8	Protected areas	Cost-based method	0.5	3.0	−0.7
Albania	NA	NA	NA	NA	NA	NA	NA	1.4	1.4	−0.8
Croatia	2,500,000	NA	59.7	149.4	NC	All forests	CVM	NA	NA	
Italy	NA	NA	2.9	25.0	1.2	Protected areas (*in situ*)	Cost-based method	0.2	1.8	−0.1
France	NA	NA	24.1	362.0	NC	All forests	CVM	NA	NA	
Spain	NA	NA	39.9	1,039.0	NC	All forests	Cost-based method	NA	NA	NC

Source: based on Chapters 5–22, this volume.

actions in forests as a place-holder for the resulting damage.[23] The corresponding estimates are all quite small, of less than €1.5/ha of forests. Only in Cyprus and Lebanon is a different approach used, based on the annual costs of controlling pest infestation resulting from poor forest management.

Pharmaceutical value

The option value of forests as habitats for species likely to produce pharmaceutical substances and drugs is widely recognized. The widespread availability in Mediterranean forests of species with high potential to provide positive effects on human health is notable. Despite local knowledge of the forests' potential to provide these benefits, little interest has been shown in estimating these values, and even less in capturing them within decision making of pharmaceutical production.

One method to estimate the option value of pharmaceuticals derived from genetic materials uses a model developed by Pearce and Puroshothaman (1992). It is a 'rent capture' approach which estimates the option value as a function of the number of species at risk, the number of drugs based on plant species and the number of hectares likely to support medicinal plants. In the Mediterranean region, estimates based on this method were found only for Turkey. This model was adapted, taking into account: the number of forest species yielding medicinal products; the royalty rates that would be payable to the host country; a coefficient of rent capture; the likely value of internationally traded pharmaceutical products; and the forest area. It gives an annual value of €109.1 million, or €5.3/ha (11.5% of the TEV). The value is likely to be higher in areas of high biodiversity, such as protected areas, where genetic resources are most likely to be found. Like most such estimates, it focuses exclusively on pharmaceuticals, ignoring wild relatives of agricultural plants (Bann and Clemens, 2001).

Worldwide, efforts to value undiscovered pharmaceutical drugs have focused mainly on tropical forest ecosystems. For example, Adger *et al.* (1995) report a value of around US$5.6/ha (€5.5/ha at 2001 prices) per year for Mexican tropical forests. Other estimates related to

tropical forests show much lower values, of around US$0.01–21/ha per year (€0.01–21/ha at 2001 prices) (Pearce and Puroshothaman, 1995) and about US$0.9–1.3/ha per year (€0.9–1.3 at 2001 prices) (Mendelhson and Balick, 1995).[24] Among these results, the estimate for Mexico is similar to that for Turkey.

Conclusions

Table 4.12 summarizes the estimates of forest values in the Mediterranean. According to these estimates, the overall average total economic value of forests in the 18 countries studied is about €133/ha (national averages are weighted by forest area to arrive at an overall average[25]). The average value is highest in the countries of the northern Mediterranean, and substantially lower in the eastern and southern Mediterranean. This value, as well as the subregional averages, are probably substantial underestimates, as important forest benefits are not estimated for many countries.[26] The degree of underestimation is probably greatest in eastern and southern countries, as data constraints tend to be much greater; the gap between the TEV in these countries and that in northern countries is thus probably smaller than it appears here.

Table 4.13 shows these estimates in *per capita* rather than per hectare terms. Overall, forests in the Mediterranean provide annually about €50 *per capita* of benefits to people living in the region. Average benefits are higher in northern countries (about €70 *per capita*) and lower in southern and eastern countries (< €11 *per capita*), although again the underestimation of benefits in the latter group needs to be borne in mind. *Per capita* benefits clearly depend not only on the per hectare benefits of forests, but also on forest area and population size. This relationship is explored further in Fig. 4.2. Southern and eastern countries with very small forest areas cluster at bottom left. In countries with at least 0.1 ha of forest *per capita*, there seem to be two distinct patterns: one group of northern countries tends to have benefits *per capita* that increase more or less in direct proportion to forest area *per capita*, and another group composed primarily of southern and eastern countries has benefits *per capita* which increase very little even as forest area *per capita* increases. Once again,

Table 4.12. Total economic value of Mediterranean forests (€/ha).

Country	Direct use values						Indirect use values			Option, bequest and existence values	Estimated TEV[a]
	WFP	Grazing	NWFPs	Recreation	Hunting	Total	Watershed protection	Carbon sequestration	Total		
Southern											
Morocco	21	28	4	NA	−4	49	21	−2	19	NA	68
Algeria	−5	33	1	NA	NA	30	23	−2	21	NA	51
Tunisia	1	74	23	NA	2	101	26	5	31	6	138
Egypt	6	NA	NA	NA	NA	6	NA	NA	NA	NA	6
Average	12	32	4	NC	−2	46	22	−2	20	NC	67
Eastern											
Palestine	8	22	25	NA	NA	55	NA	NA	NA	NA	55
Israel	5	NA	30	168	NA	203	NA	NA	NA	NA	203
Lebanon	−7	7	129	2	89	220	NA	−11	−11	6	215
Syria	3	NA	7	NA	NA	10	73	6	79	NA	89
Turkey	23	11	4	NA	1	39	−6	8	2	5	46
Cyprus	2	NA	14	5	3	24	NA	6	6	NA	30
Average	22	10	5	1	1	40	−4	8	4	5	48
Northern											
Greece	11	35	7	1	4	58	9	1	10	2	70
Albania	−3	18	NA	NA	NA	15	−7	NA	−7	NA	8
Croatia	126	5	3	11	4	149	10	NA	82	60	291
Slovenia	161	NA	26	NA	NA	187	NA	22	22	NA	210
Italy	81	7	23	20	8	139	104	8	112	3	254
France	109	NA	8	115	6	238	8	22	30	24	292
Spain	29	8	6	4	NA	47	NA	3	3	40	90
Portugal	124	34	143	5	6	312	24	9	33	NA	344
Average	67	10	16	32	3	125	18	8	27	25	176
Mediterranean	47	13	12	21	2	95	14	7	21	17	133

Averages shown for each country group (Southern, Eastern, Northern) are weighted by the forest area for each country; for countries for which estimates are available, the average estimates considered forest benefits '0', thus arriving at underestimations. NC = not calculated due to insufficient information.

[a]The estimate for Algeria includes only the values produced by forests and not those produced by steppe.

The average for Israel is calculated in reference to JNF forest area; NWFP estimates for Lebanon do not include pine kernels and honey due to the overestimation; the estimate for Spain does not include the owners' self-consumption of the recreational services.

Table 4.13. Forest benefits *per capita* in the Mediterranean region.

Country	Forest area Total (million ha)	Forest area Per capita (ha)	Estimated TEV Total[a] (million €)	Estimated TEV Per capita (€/year)	TEV *per capita*/ GDP *per capita* (%)
Southern					
Morocco	9.0	0.3	616	20.8	1.9
Algeria	4.1	0.13	206	6.6	0.6
Tunisia	0.9	0.09	129	13.2	0.7
Egypt	0.1	0	NC	NC	NC
Total[b]/Average[c]	14.1	0.1	951	6.9	0.5
Eastern					
Palestine	0	0.01	1	0.3	0.0
Israel	0.1	0.01	18	2.7	0.0
Lebanon	0.1	0.03	29	6.6	0.2
Syria	0.5	0.03	41	2.4	0.2
Turkey	20.7	0.3	948	13.6	0.6
Cyprus	0.4	0.48	12	14.6	0.1
Total/Average	21.8	0.22	1049	10.7	0.4
Northern					
Greece	6.5	0.61	458	43.2	0.4
Albania	1	0.32	9	2.8	0.2
Croatia	2.5	0.55	728	161.8	3.1
Slovenia	1.2	0.58	251	125.5	1.3
Italy	8.6	0.15	2,393	41.5	0.2
France	15	0.25	4,363	73.3	0.3
Spain	26	0.64	2,345	57.3	0.4
Portugal	3.3	0.32	1,193	117.0	1.2
Total/Average	64.1	0.38	11,740	69.8	0.4
Mediterranean	100	0.36	13,740	49.3	0.4

Sources: Population and GDP data from World Bank, 2004. NC = not calculated due to insufficient information.
[a]Most of these estimates are those reported by the country chapters and are not calculated based on the average estimates obtained in Table 4.12. In addition, they do not include the estimates considered overvalued in the previous sections (for example pine kernels and honey in Lebanon) and the values obtained outside the forest areas; in these cases, the total estimated TEV might therefore be lower than that obtained in the country chapters.
[b]Refers to the total forest area (million ha) and total estimated TEV (million €).
[c]Refers to *per capita* forest area (ha) and *per capita* estimated TEV (€).

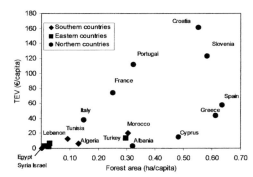

Fig. 4.2. Relationship between *per capita* forest area and *per capita* forest benefits.

part of the gap is likely to be due to a greater underestimation of benefits in the latter group. This is almost certainly an important part of the reason for Spain not being grouped with the other northern countries, for example. However, harsher climatic conditions and greater pressure on forest areas also probably play an important part in this relationship.

Direct use values contribute more than 70% and indirect use values around 15% of the TEV. The dominance of direct use values in total forest benefits is found in all country groups, the magnitude of which differs considerably from area to area. It should be borne in mind, however, that direct use values are much easier to measure. Indirect use values are much harder to estimate, and option, bequest and existence values harder still. As a result, both of the latter categories are likely to be substantially underestimated. Indeed, given the valuation constraints, it is surprising that indirect use values nevertheless appear to contribute a substantial portion of the TEV.

WFPs generally account for only a small part of total benefits – less than a third, on average. In southern countries, their importance is dwarfed by that of grazing. Their high share of total benefits among eastern countries is driven primarily by the result for Turkey, which is in turn affected by the failure to estimate watershed protection benefits in this country. Even with their imperfections, these estimates clearly demonstrate how unwise it is to base forest policy solely on timber values.

Grazing and NWFPs tend to provide a much greater share of forest benefits in southern and eastern countries than WFPs (even discounting Lebanon's result). These benefits tend to be much lower in northern countries, with the notable exception of Portugal, with its cork production. Recreation and hunting benefits have been measured imperfectly in southern and eastern countries; much work remains to be done to improve the estimates for these areas. In northern countries, these benefits rival or even exceed, in some cases, those of WFPs.

Watershed protection is an important benefit in several countries, notably Italy, Syria and the three *maghreb* countries. It would probably have played an important role in several other countries as well, if it had been possible to better estimate its value. This is perhaps the area in which it would be most important to improve our understanding, both in terms of understanding the magnitude of the benefits and in terms of understanding the conditions under which specific forest areas are most likely to generate watershed protection benefits. In the case of recreation benefits, for example, many market-based mechanisms are evolving to internalize externalities (Merlo *et al.*, 2000). This requires much more purposive action in the case of indirect benefits such as watershed protection (Pagiola *et al.*, 2002; Pagiola and Platais, 2004).

Carbon sequestration in trees provides relatively low benefits. These are highest in the northern countries (€8/ha), where net growth of forests tends to result in high net sequestration. In many southern countries, ongoing deforestation means that forests can be net sources of carbon emissions. Carbon sequestration in soils, for which only a few estimates are available, is thought to be much higher than sequestration in above-ground biomass, but quantifying this impact remains difficult (averaging ~€12/ha based on the few estimates obtained).

The estimates for the option, bequest and existence values are scarce and partial, showing the biodiversity value ranging between €2 and €60/ha accross the countries. The option value of pharmaceutical products found only in Turkey (€5/ha) suggests positive (not yet accounted) benefits also for other Mediterranean countries.

The discussion in this chapter and Chapter 3 has made it clear that these estimates suffer from a variety of methodological and empirical shortcomings. As such, they should not be taken as giving a precise value for these areas; rather, they have an indicative value and are meant to provoke further refinements. As a first effort to value a wide range of benefits at such a large scale, it is often constrained by scarcity or lack of estimates of important benefits at country and regional level. However, the available estimates show, at least in part, the situations of the different country groups and, in particular, of individual countries. In the south and east, a large share of the rural population is poor and dependent on forests; the unsustainable practices (wood exploitation, excessive grazing and other illegal actions) cause a high level of degradation which affects forests and the communities nearby. In the northern countries and especially in the Western European countries, the greater economic development in the past decades resulted in a certain lack of care and poor forest management, often responsible for high risks of forest fires and other hazards. Addressing these issues through policy measures is important for enhancing the benefits and reducing the associated costs (Chapter 23).

Notes

[1] All the estimates reported in the tables and discussed in the text are drawn from Chapters 5–22, this volume.

[2] Because some countries were only able to obtain estimates for a subset of the categories of forest value, it would be misleading to analyse the share of the TEV that these values represent. These countries include Egypt, Palestine, Israel, Lebanon, Turkey, Cyprus, Albania, Slovenia and France.

[3] These externalities are discussed separately in 'Negative externalities linked to water-related issues'.

[4] As noted in Chapter 3, this results in a certain overvaluation of timber, as the roadside price includes the costs of extraction and transport to the roadside. Only in France are the average stumpage prices available and used.

[5] This is not true in all northern Mediterranean countries: in countries such as Italy and Cyprus, timber-based industries are largely dependent on imports.

[6] In this sense, net growth can be regarded as creating option value. It was so considered in Chapter 20 (France), for example.

[7] The estimate for Syria, however, should be regarded with caution, as it does not consider the real extent of illegal timber cuts. Had that been accounted for, it might have resulted in a much lower, if not negative, value.

[8] However, some authors argue that the total area reported (25,000 ha) as affected by fires is in fact affected by a variety of other causes as well, such as deforestation (Ministère de l'Aménagement du Territoire et de l'Environnement, 2002).

[9] These quantities do not include substantial amounts coming from other wooded lands, such as orchards.

[10] For example, 120 kg of coffee in dried cherry form typically results in 50.4 kg of roasted coffee beans. Using yield data for dried cherry with price data for roasted beans would, therefore, result in a substantial overestimation of the value of production.

[11] For Spain, see Campos *et al.*, Chapter 21, this volume; for Greece, see Kazana and Kazaklis, Chapter 15, this volume; for Italy, see for example Merlo and della Pupa (1994), Bellù and Cistulli (1997), Tempesta (2000), Romano (2002) and Scarpa *et al.* (2003).

[12] Environmental clubs are private societies or non-governmental organizations (NGOs) that organize recreational and other nature-related activities in Lebanon.

[13] The number of visitors are national level estimates derived from either national-level studies (France) or official statistics in Italy (Istituto Nazionale di Statistica, 1998). Other countries do not have corresponding national level estimates, so visitor numbers are estimated by extrapolating local level studies. This is likely to result in a degree of underestimation.

[14] Control of hunting is usually delegated to regional or local authorities, such as Regions and Provinces in Italy or Forest Service districts in Albania. In several countries, exercising hunting rights requires prior possession of permits for carrying firearms which are renewed annually (Morocco) or just issued once (Lebanon).

[15] This is a result of larger payments imposed on visitors from overseas (in Albania €100/permit for foreigners against €10/permit for locals) and often to a higher number of foreign hunters.

[16] Venezuela (Thobjarnarson, 1991), Cameroon (Korup National Park) (Infield, 1988; Ruitenbeek, 1988), Zaire (Ituri Forest) (Wilkie, 1989); Malaysia (Sarawak) (Caldecott, 1987) and Nigeria (Cross River Park) (Ruitenbeek, 1989).

[17] Water stress is defined as an average water availability of less than 1700 m^3/*capita*/year; water scarcity is defined as an availability of less than 1000 m^3/*capita*/year (Falkenmark and Widstrand, 1992).

[18] According to World Bank (1995) data, water is already scarce in Palestine (105 m^3/*capita*), Israel (375 m^3/*capita*), Syria (385 m^3/*capita*), Tunisia (489 m^3/*capita*) and Algeria (655 m^3/*capita*). Morocco is close to reaching water scarcity, having 1083 m^3/*capita*/year. It receives an average rainfall of 150 billion m^3, of which around 120 billion m^3 is lost by evaporation. The remaining 30 billion m^3 enters the water cycle annually with 10 billion m^3 infiltrating to aquifers and 20 billion m^3 contributing to surface flow. Approximately 2.5 billion m^3 from groundwater flows ends up in rivers, resulting in a total surface water supply of about 22.5 billion m^3/year, or about 75% of Morocco's total water resources (World Bank, 1995).

[19] The United Nations Convention to Combat Desertification and its Causes (1994) defines desertification as 'land degradation in arid, semi-arid and dry sub-humid areas, resulting from various factors, including climatic variations and human activities'.

[20] The universal soil loss equation (USLE) predicts soil loss from a plot from the plot's characteristics and management practices used on it. It is expressed as $A = R \times K \times L \times S \times C \times P$, where: A is the soil loss/ unit area; R is the rainfall erodibility factor; K is the soil erodibility factor; L is the normalized field length factor; S is the slope factor; C is the normalized cropping management factor; and P is the normalized conservation practice factor (Wischmeier and Smith, 1978). Application of this method at watershed level is problematic, however, as the USLE is designed to be used at the plot level. As much of the soil eroded from one plot is deposited elsewhere within the landscape, the USLE tends to overestimate erosion at the landscape level.

[21] Except for Croatia, whose estimate is calculated by applying the average growth rate (2% of the growing stock) to the total estimated carbon stock.

[22] Calculated as the ratio between species richness and area.
[23] This approach is deeply unsatisfactory on many levels – it is based on administratively set fees; and it only reflects illegal actions that were detected. Moreover, the assumption that all illegal actions are harmful to biodiversity (and, conversely, that legal actions are not) is itself dubious. Nevertheless, it was thought preferable to include an imperfect measure rather than to omit this effect altogether.
[24] For other estimates of option values of pharmaceuticals in tropical forests, see Simpson et al. (1994), Ruitenbeek (1989) and a review by Pearce (1996).
[25] Given that many negative externalities have not been estimated for many countries, it could be argued that the estimate is not a lower bound.

References

Adger, N., Brown, K., Cervigni, R. and Moran, D. (1995) Total economic value of forests in Mexico. *Ambio* 24, 286–296.

Aylward, B., Echeverria, J., Fernandez Gonzalez, A., Porras, I., Allen, K. and Mejias, R. (1998) Economic incentives for watershed protection: a case study of Lake Arenal, Costa Rica, In: Final Report to the Government of The Netherlands under the Program of Collaborative Research in the Economics of Environment and Development (CREED). IIED, TSC and the International Center for Economic Policy, National University at Heredia (CINPE), London, p. 130.

Bann, C. and Clemens, M. (2001) *Turkey Forest Sector Review – Global Environmental Overlays Program Final Report*. World Bank, Washington, DC.

Bellot, J., Sanchez, J.R., Chirino, E., Hernandez, N., Abdelli, F. and Martinez, J.M. (1999) Effect of different vegetation type cover on the soil water balance in semi-arid areas of south eastern Spain, *Physics and Chemistry of the Earth (B)* 24, 353–357.

Bellot, J., Bonet, A., Sanchez, J.R. and Chirino, E. (2001) Likely effects of land use changes on the runoff and aquifer recharge in a semiarid landscape using a hydrological model. *Landscape and Urban Planning* 778, 1–13.

Bellù, L. and Cistulli, V. (1997) *Economic Valuation of Forest Recreation Facilities in the Liguria Region (Italy)*. CSERGE working paper GEC 97–08.

Brown, G. and Henry, W. (1989) *The Economic Value of Elephants*. London Environment Economics Centre Paper 89–12. International Institute for Environment and Development, London.

Bruijnzeel, L.A. (2004) Hydrological functions of tropical forests: not seeing the soils for the trees? *Agriculture, Ecosystems and Environment* 104, 185–228.

Bruijnzeel, L.A. and Bremmer, C.N. (1989) *Highland–Lowland Interactions in the Ganges–Brahmaputra River Basin: a Review of Published Literature*. ICIMOD Occasional Paper, No. 11.

Caldecott, J. (1987) *Hunting and Wildlife Management in Sarawak, Malaysia*. World Wildlife Fund, Washington, DC.

Calder, I. (2000) Land use impacts on water resources. In: *Land–Water Linkages in Rural Watersheds* Electronic Workshop, 18 September–27 October 2000, FAO, Background Paper No. 1.

Cesaro, L. and Pettenella, D. (1994) Un'analisi degli effetti delle politiche forestali nella prevenzione dei cambiamenti climatici. *Rivista di Economia Agraria* 11.

Di Castri, F. (1996) Mediterranean diversity in a global economy. *International Symposium on Mediterranean Diversity*, ENEA, Rome, pp. 21–30.

Fabbio, G., Merlo, M. and Tosi, T. (2003) Silvicultural management in maintaining biodiversity and resistance of forests in Europe, *Journal of Environmental Management* 67, 67–76.

Falkenmark, M. and Widstrand, C. (1992) Population and water resources: a delicate balance. *Population Bulletin* 47, 2–36.

Fankhauser, S. (1995) *Valuing Climate Change. The Economics of the Greenhouse*. Earthscan, London.

Food and Agriculture Organization (2000) *Global Forest Resources Assessment 2000*. FAO Forestry Paper No. 140.

Godoy, R., Lubowski, R. and Markandya, A (1993) A method for the economic valuation of non-timber tropical forest products. *Economic Botany* 47, 220–420.

Godoy, R., Wilkie, D., Overman, H., Cubas, A., Cubas, G., Demmer, J., McSweeney, K. and Brokaw, N. (2000) Valuation of consumption and sale of forest goods from Central American rain forests. *Nature* 406.

Infield, M. (1988) *Hunting, Trapping, and Fishing in Villages Within and on the Periphery of the Korup National Park*. World Wide Fund for Nature, London.

Intergovernmental Panel on Climate Change (2001) Technological and economic potential of options to enhance, maintain, and manage biological carbon reservoirs and geo-engineering. In: *Greenhouse Gas Emission Climate Change 2001: Mitigation*. Cambridge University Press, Cambridge, UK.

Istituto Nazionale di Statistica (1997) *Statistiche della Pesca e Caccia*. ISTAT, Rome.

Istituto Nazionale di Statistica (1998) *Annuario delle Statistiche del Turismo*, ISTAT, Rome.

Johnson, C. and Linder, R. (1986) An economic valuation of South Dakota wetlands as a recreation resource for resident hunters. *Landscape Journal* 5, 33–38.

Kaimowitz, D. (2004) The great flood myth. *New Scientist*, 19 June.

Lampietti, J. and Dixon, J. (1995) *To See the Forest for the Trees: a Guide to Non-timber Forest Benefits.* Environment Department Papers, Paper No. 13, World Bank, Washington, DC.

Lavabre, J., Andréassian, V. and Laroussine, O. (2000) *Eaux et Forêts. La Forêt un Outil de Gestion des Eaux.* ECOFOR, CEMAGREF Editions.

Mantau, U., Merlo, M., Sekot, W. and Walker, B. (2001) *Recreational and Environmental Markets for Forest Enterprises.* CAB International, Wallingford, UK.

Mendelhson, R. and Balick, M. (1995) The value of undiscovered pharmaceuticals in tropical forests. *Economic Botany* 49, 223–228.

Merlo, M. and della Pupa, F. (1994) Public benefits valuation in Italy: a review of forestry and farming applications. In: *Economic Valuation of Benefits from Countryside Stewardship.* Wissenshaftsverlag Vauk, Kiel, Germany, pp. 117–131.

Merlo, M., Milocco, E., Panting, R. and Virgilietti, P. (2000) Transformation of environmental recreational goods and services provided by forestry into recreational environmental products. *Forest Policy and Economics* 1, 127–138.

Ministère de l'Aménagement du Territoire et de l'Environnement (2001) *Projet National ALG/98/G31: Elaboration de la Stratégie et du Plan d'Action National des Changements Climatiques.* Communication Nationale Initiale, MATE.

Ministère de l'Aménagement du Territoire et de l'Environnement (2002) *Plan National d'Actions pour l'Environnement et le Développement Durable PNAE-DD.* MATE.

Ministère de l'Aménagement du Territoire, de l'Urbanisme, de l'Habitat et de l'Environnement (2001) *Morocco – First National Communication – United Framework Convention on Climate Change.* MATUHE, Rabat.

Ministère de l'Environnement et de l'Aménagement du Territoire (1998) *Programme d'Action National de Lutte Contre la Désertification.* Note de synthèse, MEAT, Paris.

Myers, N., Mittelmeier, R.A., Mittelmeier, C.G., Da Fonseca, G.A.B. and Kent, J. (2000) Biodiversity hotspots for conservation priorities. *Nature* 403, 853–858.

Pagiola, S. (1996) *Republic of Croatia – Coastal Forest Reconstruction and Protection Project – Annex J.* Economic Analysis. Report no. 15518-HR. World Bank, Environment Department, Washington, DC.

Pagiola, S. and Platais, G. (2004) *Payments for Environmental Services: From Theory to Rractice.* World Bank, Washington, DC.

Pagiola, S., Landell-Mills, N. and Bishop, J. (2002). Making market-based mechanisms work for forests and people. In: Pagiola, S. Bishop, J. and Landell-Mills, N. (eds) *Selling Forest Environmental Services: Market-based Mechanisms for Conservation and Development.* Earthscan, London, pp. 261–290.

Pearce, D. (1996) *Can non-market values save the world's forests?* Paper presented at the International Symposium of Non-market Benefits of Forestry organized by the Forestry Commission, Edinburgh, UK, June 1996.

Pearce, D.W. and Puroshotaman, S. (1992) *Protecting Biological Diversity: the Economic Value of Pharmaceutical Plants.* Global Environmental Change Working Paper 92–27, CSERGE, University of East Anglia and University College, London.

Pearce, D. and Puroshothaman, S. (1995) The economic value of plant-based pharmaceuticals. In: Swanson, T. (ed) *Intellectual Property Rights and Biodiversity Conservation.* Cambridge University Press, Cambridge, pp. 127–138.

Pearce, D., Adger, N., Brown, K., Cervigni, R. and Moran, D. (1993) *Mexico Forestry and Conservation Sector Review: Substudy of Economic Valuation of Forests.* Draft Report. World Bank, Washington, DC.

Peters, C.M., Gentry, A.H. and Mendelssohn, R.O. (1989) Valuation of an Amazonian Rainforest. *Nature* 339, 655–656.

Plan Bleu (2000a) *Profiles de Pays Méditerranéens: Tunisie – Enjeux de Politiques d'Environnement et de Développement Durable.* Plan Bleu.

Plan Bleu (2000b) *Mediterranean Country Profiles: Lebanon – Environment and Sustainable Development Issues and Policies.* Plan Bleu.

Plan Bleu (2004) Issues and Concerns: Tourism and Sustainable Development, Plan Bleu, www.planbleu.org Access date: May 2004.

Quézel, P., Médail, R., Loisel, R. and Barbero, M. (1999) Biodiversity and conservation of forest species in the Mediterranean basin. *Unasylva* 50, 21–28.

Romano, D. (2002) *An Assessment of Italian Environment Valuation Studies, with Emphasis on CVM.* Paper of the Dipartimento di Economia Agraria e delle Risorse Territoriali, Università degli Studi di Firenze.

Ruitenbeek, H.J. (1988) *Social Cost–Benefit Analysis of Korup Project Cameroon*. World Wide Fund for Nature, London.

Ruitenbeek, H.J. (1989) *Economic Analysis of Issues and Projects Relating to the Establishment of the Proposed Cross River National Park (Oban Division) and Support Zone*. World Wide Fund, London.

Scarascia-Mugnozza, Oswald, H., Piussi, P. and Radoglou, K. (2000) Forests of the Mediterranean region: gaps in knowledge and research needs. *Forest Ecology and Management* 132, 97–109.

Scarpa, R., Tempesta, T. and Thiene, M. (2003) La domanda escursionistica della montagna veneta: un'analisi tramite modelli di conteggio con varianza flessibile. *Rivista di Economia Agraria* 1, 47–78.

Signorello, G (1990) La stima dei benefici di tutela di un'area naturale: un'applicazione della contingent valuation/Estimate of protection benefits of an natural area: application of a contingent valuation. *Genio Rurale* 9, 59–66.

Simpson, R., Sedjo, R. and Reid, J. (1994) *Valuing Biodiversity: an Application to Genetic Prospecting*. Discussion Paper 94–20, Resources for the Future, Washington, DC.

Sistema de Informação de Cotações de Produtos Florestais na Produção (2003) *Forest Information System on Products Prices in the Production*. On-line data, SICOP.

Tempesta, T. (2000) Il contributo delle aree protette allo sviluppo economico della montagna: uno studio nel Parco Naturale delle Dolomiti Ampezzane. XXXVII Convegno SIDEA *Innovazione e Ricerca nell'Agricoltura Italiana'*, Bologna 14–16 Settembre 2000.

Thorbjarnarson, J. (1991) An analysis of the spectacled ciman (*Caiman crocodilus*) harvest program in Venezuela. In: Redford, K.H. and Robinson, J.G. (eds) *Neotropical Wildlife Use and Conservation*. University of Chicago Press, Chicago, Illinois, pp. 217–235.

Tobias, D. and Mendelsohn, R. (1991) Valuing ecotourism in a tropical rainforest reserve. *Ambio* 20, 91–93.

Tognetti, S., Mendoza, G., Aylward, B., Southgate, D. and Garcia, L. (2004) *A Knowledge and Assessment Guide to Support the Development of Payment Arrangements for Watershed Ecosystem Services (PWES)*. World Bank, Washington, DC.

United Nations Convention to Combat Desertification and its Causes (1994) *Test of the United Nations Convention to Combat Desertification and its Causes*. http://www.unccd.int/convention/text/convention.php

United Nations Economic Commission for Europe/Food and Agriculture Organization (2000) *Global Forest Resources Assessment 2000. Main Report*. United Nations Publications, Geneva.

United Nations Environment Programme/Mediterranean Action Plan (2000) *Problèmes et Pratiques de Lutte Anti-érosive en Turquie*. Rapport rédigé par Orhan Dogan, UNEP/MAP.

Varela, M.C. (1999) Cork and the cork oak system. *Unasylva* 50, 42–44.

Wilkie, D.S. (1989) Impact of Roadside Agriculture on Subsistence in the Ituri Forest of Northeastern Zaire. *American Journal of Physical Anthropology* 78, 485–494.

Wischmeier, W.H. and Smith, D. (1978) Predicting rainfall erosion losses: a guide to conservation planning. *USDA-ARS Agriculture Handbook No. 537*. Washington, DC.

World Bank (1995) *From Scarcity to Security – Averting a Water Crisis in the Middle East and North Africa*. Working Paper. World Bank, Washington, DC.

World Bank (2004) *World Development Indicators 2004*. World Bank, Washington, DC.

5 Morocco

Mohammed Ellatifi

Department of Waters, Forests and Desertification Control, PO Box 50070, 20070 Casablanca-Ghandi, Morocco

Introduction

Morocco (Al Maghrib), officially the Kingdom of Morocco (Al Mamlaka al Maghribiya), is situated at the extreme west of North Africa (Fig. 5.1). It covers 710,850 km² and is home for around 30 million inhabitants, with an annual growth rate of 1.7%. Urban dwellers make up 55.2% of the total (Ministère de la Prévision Economique et du Plan/Directorate of Statistics, 2000).

Beside the numerous plains which lie along the coasts of the Atlantic Ocean and the Mediterranean Sea, four major mountain ranges break the monotony of the flat country: the Rif in the north, the Middle Atlas in the centre-north, the High Atlas in the centre and the Anti Atlas in the centre-south. They are bordered by large areas of plateaus and inter-montane valleys. The lowest point is Sebkha Tah (55 m below sea level) and the highest is Jabal Toubkal (4165 m). To the extreme south of the country lie the sandy wastes of the Moroccan Sahara desert.

The climate is of the Mediterranean type, very sunny and bright throughout the year, with an irregular distribution of rainfall, concentrated during winter, whereas the summer is hot and dry. In fact, all Emberger's climatological stages can be found in Morocco: desertic, arid, semi-arid, subhumid, humid and high mountain, the latter being very cold and snowy.

Forests in Morocco, as in other Mediterranean countries, grow in fragile biotopes where even a single shrub is important for its multiple uses as fuel, fodder, fruit, medicine and for its conservation influences on soil, water and the landscape (Ellatifi, 2002). Some forest outputs, such as timber, cork and fuelwood, pass through traditional markets, reflecting their value given by market prices. Other various environmental services do not register a market price, though they present a high value for people. Therefore, the search for a full value, the total economic value (TEV), of Moroccan forests is crucial for policy making and management in order to increase people's welfare. This study is the first attempt towards the quantification of the TEV of Moroccan forests.

Forest Resources

Area and people

Forests cover around 5.7 Mha, or 8.1% of the country's area. If esparto grass formations are also included, the forest area reaches 9.05 Mha, i.e. 12.7% of the total land area. Natural forests cover 5.2 Mha (53% of the total), of which around 20% are coniferous and 80% are broadleaved. They also include the mattorals/*maquis* Mediterranean formations, covering some 4.5% of the total forest area. Plantations and esparto grass form 5.5 and 36.7% of the total forest area, respectively (Table 5.1).

It is argued that centuries ago, the forest area was much greater than the present one. Some biogeographers (Emberger, 1939; Boudy,

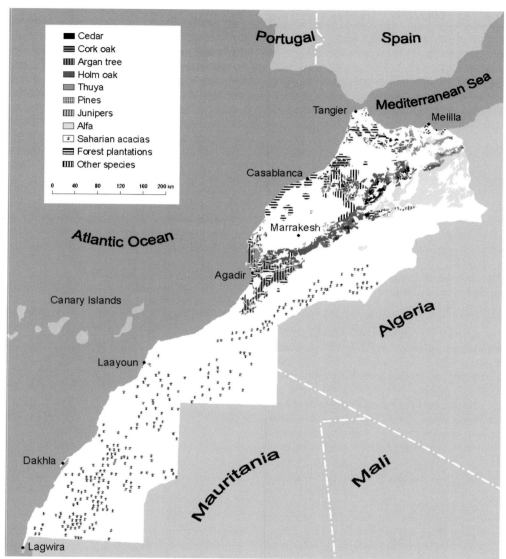

Fig. 5.1. A map of Morocco with forestry species. Source: Ministry in Charge of Waters and Forests (1999) (redrawn).

1958) advocate that around the 4th century AD, Moroccan forests were at their peak, covering some 13.5 Mha or 2.4 times more than the present forest cover. This 'climactic' forest resulted in a cover of 19% if referred to the whole country area, and 30% if the Moroccan southern desertic area is not included.

Until the 9th century, forests were in a very good state of conservation. It is from the 10th

century onwards that some deforestation commenced with the development of large cities, without, however, jeopardizing the national forest cover. Alarming deforestation rates occurred during the 20th century, at the end of which the forest area fell below 6 Mha, including 0.5 Mha of plantation. Under the French protectorate, the Moroccan forests were 'bled' of their capital timber and firewood by the occupants to provide

Table 5.1. Types of forests, growing stocks and annual increment.

Forest type	Tree species	Area[a]		Average[b] growing stock (m³/ha)	Average[b] annual increment (m³/ha)
		000 ha	%		
High forests	Total	1218.0	13.5	100	3.0
	Cedrus atlantica	133.6			
	Abies pinsapo	3.1			
	Pinus spp.	82.1			
	Quercus faginea	9.1			
	Quercus suber	282.5			
	Quercus ilex	707.2			
Coppices	Total	3603.5	39.8	35	1.0
	Tetraclinis articulata	565.7			
	Juniperus spp.	243.2			
	Other coniferous	7.5			
	Quercus ilex	707.2			
	Q. suber	94.1			
	Argania spinosa	871.2			
	Saharian acacias[c]	1011.0			
	Other broadleaved	103.6			
Maquis	Mattoral, shrubs	407.4	4.5	15	0.5
Plantations	Total	502.3	5.5	70	5.0
	Coniferous	234.4			
	Broadleaved	267.9			
Esparto grass (*Stipa tenacissima*)		3318.2	36.7	12 q (green)	1 q (green)
Total		9049.5	100		

[a]Ministry in Charge of Waters and Forests (2002).
[b]Personal estimation.
[c]*Acacia gummifera, A. horrida, A. cyclops*, etc.
q (green) refers to quintals of fresh matter.

trains and other vehicles during the First World War. Since the 1950s, with the rural population growth and the development of cities, deforestation in Morocco has increased drastically. Clear-cutting for conversion to agriculture, overgrazing and overcollection of firewood, in addition to forest mismanagement and misuse, resulted in a large amount of negative externalities such as soil erosion, loss in agricultural production, dam siltation and biodiversity resources depletion. Nowadays, deforestation continues to take place at an average annual rate of 31,000 ha (Ministry in Charge of Waters and Forests, 1999).

Typologies

Table 5.1 presents the main forest typologies, along with the growing stock and average annual increment.

Man-made plantations are established either by afforesting new areas or by reforesting areas where original forests had been destroyed by fire, deforestation and other factors. In both cases, the major objectives are production (timber and wood energy), watershed protection, desertification control (sand dune fixation and wind breaks), recreation, landscape improvement, urban forestry or a mixture of such objectives as in multi-purpose plantations.

Afforestation commenced in 1939. During the last half century, the afforested area increased[1] from less than 5000 ha (1950) to 502,400 ha (1999). In 1999, the planted area was composed of conifers (47%), eucalyptus (40%) and other broadleaved species (13%). The main eucalyptus species used in afforestation are *Eucalyptus camaldulensis* (48%), *E. gonphocephala* (32%), *E. sideroxylon* (4%) and *E. grandis* (4%), and the main conifer species are *Pinus halepensis* (64%) and *P. pinaster* (21%).

Table 5.2. Plan National de Raboisement foreseen plantations (000 ha) according to the main objectives.

Major objectives	Coniferous	Broadleaved	Other	Total
Timber production	300	55	0	355
Timber production and protection	100	15	0	115
Soil and water protection	140	45	0	185
Recreation	0	0	7	7
Total	540	115	7	662

The afforested area can be divided into State-owned (74%), communal (18%) and private land (8%) (Ministry in Charge of Waters and Forests, 1999). Plantations are enabled by means of some 30–40 million seedlings produced annually in nurseries throughout the country.

The first Afforestation National Plan (Plan National de Reboisement, PNR) was adopted in 1971 and foresaw for 2000 a total planted area of 662,000 ha, as shown in Table 5.2 (Food and Agriculture Organization/United Nations Development Programme, 1970). To date only 38% of the plantations has been achieved.

This prompted the adoption of a second Afforestation Leading Plan (Plan Directeur de Reboisement, PDR) in 1996. It recommended planting 500,000 ha by 2006, divided according to the major objectives: 42% for water, soil and biodiversity protection; 9% for rationalization of silvo-pastoral activities; 46% for timber, industrial wood and wood energy production; and 3% for landscape improvement. This area is intended to cover State-owned (75%), communal (12%), and private and other land (13%). It should consist of eucalyptus (21%), other broadleaved species (28%), pine (19%) and other conifer species (32%).

PDR aims also to involve local communities and private promoters in the afforestation activities for ecosystem protection and development. The cost of implementing these activities is estimated at €66 million, of which 78% is for seedling plantation, 20% for connected activities and 2% for infrastructures (Ministry in Charge of Waters and Forests, 1999).

Nearly 75% of the Moroccan forests are degraded natural forests. Most of the degraded forest area is due to overgrazing and fuelwood collection, while forest fires annually affect just 2700 ha (Ministry in Charge of Waters and Forests, 1998a,b, 2002).

The major forest species in Morocco according to the classification of Quézel (Quézel,

1976; Quézel and Barbero, 1982) are shown in Fig. 5.2.

Forest vegetation in Morocco can also be considered according to Emberger's climatological classification, as shown in Box 5.1 and Fig. 5.3.

Functions

In a biotope so fragile as that of the Mediterranean, and particularly in Morocco's pre-desertic region, but even in humid and subhumid areas, forest preservation is essential to safeguard the various protection functions performed by forests. Rainfalls are irregular and often torrential, therefore erosion, floods and other disservices are frequent and can be prevented by forests particularly when they are well managed. Wood and non-wood forest products (WFPs and NWFPs) are also important. Nevertheless, it is difficult strictly to distinguish productive from protective forest functions and stands, as well as from the recreational landscape functions. Multi-functionality is the key to understanding Moroccan and, indeed, Mediterranean forests (Ellatifi, 2002). However, it can be estimated that for 30–35% of Moroccan forest area, production, particularly of WFPs, is the dominant function, while the remaining 65–70% is mainly protective, including a large share of NWFPs.

Institutional Aspects

Ownership and size of properties

In Morocco, the 'Forestry Domain' encompasses the 'domanial' State-owned forests, land covered by esparto grass, terrestrial and maritime sand dunes, man-made plantations,

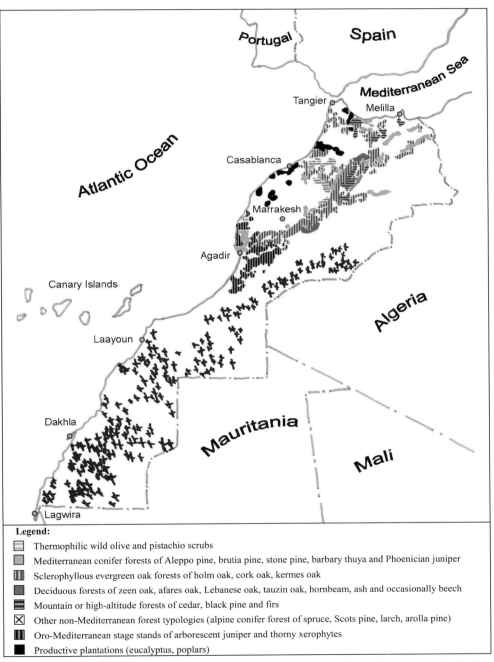

Legend:

▢ Thermophilic wild olive and pistachio scrubs

▨ Mediterranean conifer forests of Aleppo pine, brutia pine, stone pine, barbary thuya and Phoenician juniper

▥ Sclerophyllous evergreen oak forests of holm oak, cork oak, kermes oak

▦ Deciduous forests of zeen oak, afares oak, Lebanese oak, tauzin oak, hornbeam, ash and occasionally beech

▤ Mountain or high-altitude forests of cedar, black pine and firs

☒ Other non-Mediterranean forest typologies (alpine conifer forest of spruce, Scots pine, larch, arolla pine)

▥ Oro-Mediterranean stage stands of arborescent juniper and thorny xerophytes

■ Productive plantations (eucalyptus, poplars)

Fig. 5.2. Map of Moroccan forest species according to the classification of Quézel. Drawn by M. Ellatifi, based on the species geographic distribution shown in Fig. 5.1.

Box 5.1. Bioclimate and forest species distribution according to Emberger's classification.

Wherever on earth in general, in the Mediterranean region and in Morocco in particular, when there is no exaggerated anthropogenic interaction with the ecosystem, vegetation species are distributed in the landscape according to four key factors: light, precipitation (humidity), temperature and chemical composition of the soil (substratum), as detailed below:

- Light: some species, known as 'sciaphyllic' do not thrive in excessive light and prefer to grow under the shade of other dominant species, e.g. *Taxus baccata*.
- Precipitation (humidity): many Mediterranean species, such as *Pinus radiata*, *Quercus suber*, *Q. faginea*, *Cedrus atlantica* and *Abies pinsapo*, cannot grow, or have a 'normal' growth rate when the average annual rainfall is too low, e.g. less than 100 mm.
- Temperature: most Mediterranean species except, for example, *Cedrus atlantica* Manetti and *Juniperus thurifera*, die with a hard frost.
- Soil (substratum): some species, called 'calcifuge', do not thrive in a soil with a high lime content, e.g. *Quercus suber*, *Eucalyptus camaldulensis*, *E. cladocalyx*, *Pinus radiata* and *P. pinaster* var. *atlantica*, while others do not thrive on very salty soils.

Based on these factors, Emberger (1936, 1955) adopted a pluvio-thermic quotient Q_2 to define, for Mediterranean-like ecosystems, six vegetation bioclimatic stages: desertic (Saharian), arid, semi-arid, subhumid, humid and high mountain. Emberger's quotient Q_2 is given by

$$Q_2 = (1000 \times P)/\{[(M + m)/2] \times (M-m)\}$$

where: P is the average annual rainfall (mm); M is the average of temperature maxima for the hottest month of the year and m is the average of temperature minima of the coldest month of the year.
Based on Emberger's pluvio-thermic Q_2, Sauvage and Brignon (1963) drew a provisional map of the different bioclimatic stages in Morocco. Having the temperature 'm' as the abscissa and Q_2 as the ordinate, the map shows the limits of Emberger's different bioclimatic stages and, accordingly, indicates the areas of the major forest species in Morocco (Fig. 5.3).

nurseries located in the 'domanial' State-owned forests and on land that has been, or is to be, afforested, and forestry buildings.

Around 99% of the forests[2] (8.9 Mha) are state owned or 'domanial', while the remaining 1% (90,000 ha) are mainly private – with an average size of around 0.5 ha. This can be explained by the Dahir[3] of 1917, which constitutes the Forest Law in Morocco. According to this Dahir, any wooded area in Morocco is presumed to be state owned. This presumption is to be confirmed and officialized by a delimitation process which is carried out by a commission formed by representatives of the forest service, local authorities (Ministry of the Interior) and local communities. This commission is entitled to receive any complaint or opposition regarding the forest delimitation process from the forest service. If no amicable arrangement is found, the tribunal is the competent body that takes the final decision. Presently, the delimitation process of some 300,000 ha of forests and 102,000 ha of esparto grass is in progress.

Administration and policies

Before 1998, the Moroccan forest service was part of the Ministry of Agriculture and Agrarian Reform (Ministère de l'Agriculture et de la Réforme Agraire, MARA). It included a central administration called the Administration of Waters, Forests and Soil Conservation (Administration des Eaux et Forêts et de la Conservation des Sols, AEFCS) and the provincial services throughout the country. In 1998, the forest service was upgraded to a Delegate Ministry, in charge of Waters and Forests (MCWF), represented at the regional level by Regional Directorates of Waters and Forests (Direction Régionale des Eaux et Forêts, DREF).

MCWF is a governmental department, responsible for: conservation and management of forests, wildlife and fish farming in terrestrial waters; forest, hunting and fishing police; afforestation in the 'Forestry Domain'; erosion control, both hydric and aeolian; and forestry research and experimentation. Besides the MCWF, two important institutions exist: (i) The

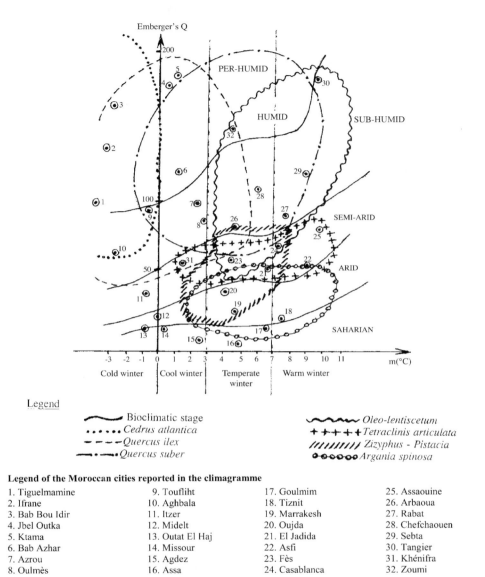

Legend

～ Bioclimatic stage	～ *Oleo-lentiscetum*
••••• *Cedrus atlantica*	++++ *Tetraclinis articulata*
– – – *Quercus ilex*	///////// *Zizyphus - Pistacia*
– • – • *Quercus suber*	●○●○○○ *Argania spinosa*

Legend of the Moroccan cities reported in the climagramme

1. Tiguelmamine	9. Toufliht	17. Goulmim	25. Assaouine
2. Ifrane	10. Aghbala	18. Tiznit	26. Arbaoua
3. Bab Bou Idir	11. Itzer	19. Marrakesh	27. Rabat
4. Jbel Outka	12. Midelt	20. Oujda	28. Chefchaouen
5. Ktama	13. Outat El Haj	21. El Jadida	29. Sebta
6. Bab Azhar	14. Missour	22. Asfi	30. Tangier
7. Azrou	15. Agdez	23. Fès	31. Khénifra
8. Oulmès	16. Assa	24. Casablanca	32. Zoumi

Fig. 5.3. Emberger's climagramme, with the different bioclimatic vegetation stages and the repartition areas of the major forest species in Morocco. Drawn by M. Ellatifi, based on Emberger's pluvio-thermic Q_2 (Sauvage and Brignon, 1963).

Forest National Council includes members of other departments and representatives of local communities and is chaired by the Minister of the MCWF. It holds an annual meeting to analyse the national forest situation, and at province level is represented by Forest Provincial Councils. (ii) The Moroccan Royal Hunting Federation groups together all hunters throughout the country. It holds an annual meeting to examine the situation of wildlife management and hunting activities in the country.

Regarding the legislative aspects, the entire area of the Forest Domain is under the supervision of the governmental Department of

Waters, Forests and Desertification Control (Haut Commissariat aux Eaux et Forêts et à la Lutte contre la Désertification), according to the Dahir of 17 October 1917 which forms the main Moroccan Forest Law. According to this law, the forest engineers are also judiciary police officers who assist the court and the public attorney in any trial of forest proceedings.

All forest goods originating from the forest domain, afforested land or land to be afforested, belonging to the state, to private individuals or to communities are submitted to the forestry regime.

The Forest Department also supervises and exercises police control upon hunting activities (Dahir of 21 July 1923) and fishing in terrestrial waters (Dahir of 1 April 1922). It creates, supervises and manages the national parks (Dahir of 11 September 1934), undertakes water and wind erosion control, restores eroded soils and establishes areas for pastoral improvement (Dahir of 25 July 1969).

The Forest Department also manages the National Forest Fund (NFF) to carry out afforestation and forestry research activities (Dahir of 12 September 1949), protects and manages the dwarf palm tree (*Chamaerops humilis*) formations, and supervises the production and commercialization of the vegetable horse-hair[4] (Dahir of 24 October 1962).

Local rights, customs and practices

The Moroccan forest law grants the 'riverains' (people living in the forest proximity) two major rights: free collection of dead fuelwood for domestic use; and domestic livestock grazing for a token monetary tax. Nevertheless, grazing is not permitted in regeneration areas, which are less than 20% of the total forest area.

In addition, two other special rights of the riverains are recognized: branch cutting for domestic livestock feeding during snowy periods; and agroforestry practices in the Argan forest (*Argania spinosa* L.). In this context, the riverains have the right to collect Argan fruits and cultivate crops in the Argan forest.[5]

Another Dahir was enacted on 20 September 1976 to involve local communities in the development of the forest economy. This Dahir gave deliberative power to local communal councils, in order to settle the communal forestry questions, support the local forest service in its role and assist the Forest Department in elaborating a policy and a strategy for a sustainable forest development.

According to this Dahir, the forestry income is deposited in the commune's budget, subject to the use of at least 20% of this income to finance afforestation of communal land, improve silvo-pastoral activities, fruit tree plantations, water catchments, management of communal shelters or roads, and creation of greenery, or protection of natural sites.

Contribution of the Forest to the National Economy

Moroccan forests contribute 0.4% to the gross domestic product (GDP) and 10% of the agricultural GDP (Ministry in Charge of Waters and Forests, 1998b, 2002). If only the marketable forest products are considered, the forests' contribution to the national economy can be described as follows (Ministry in Charge of Waters and Forests, 2002):

- Covers 30% of the annual national consumption of timber and industrial roundwood, by providing around 600,000 m^3.
- Covers 30% of the annual national energy consumption, by providing around 10 Mm^3 of fuelwood or some 4 million TOE (tonnes of oil equivalent).
- Covers 17% of the national livestock forage consumption, i.e. nearly 1.5 billion forage units (FU) or the equivalent of €288 million per year.
- Provides €550 million as annual income to local forest users communities.

Employment in forestry and forest-related activities represents around 5.2% of the total national employment (in 1999) and around 11% of the employment of the 'Agriculture, forestry and fisheries' sector. It amounts to nearly 100 million work-days per year, i.e. around 370,000 jobs, most of which are involved in silviculture, forest management, afforestation, forestry tracks, fire prevention and combating, and phytopathological treatments. Of the remaining jobs, 28,000 are provided by forestry enterprises, 14,000 jobs by forestry processing, 26,000 jobs by fuelwood collection, 40,000 jobs by grazing activities and another 4544 jobs by the public sector.

The Values of Moroccan Forests

Forest values in Morocco are estimated by using a large variety of methods and approaches, and the overall picture, though incomplete, is presented in Table 5.4.

Direct use values

Many direct use values of Moroccan forests are estimated according to the quantities traded on the market and their average market price. Timber and firewood are valued by using the harvested quantities and their roadside prices. However, firewood collection is worthy of particular attention. The sustainable firewood production from Moroccan forests is 3 Mm^3 (Ministry in Charge of Waters and Forests, 1999), but the total firewood collected annually from the forest domain amounts[6] to 9.6 Mm^3, i.e. a further 6.6 Mm^3. The quantity of firewood within sustainable limits is harvested totally and is composed of 0.6 Mm^3 that are formally sold by the forest department and 2.4 Mm^3 that are informally collected by local communities. The extra 6.6 Mm^3 is collected illegally. Even though it is an illicit overcollection, it is still a direct use value, i.e. a benefit for the people. Nevertheless, as it is greatly above the stipulated annual amount, it damages the forest ecosystem, thus provoking negative effects that are valued in the 'Negative externalities' section.

Several NWFPs are also estimated by means of their market price, among which, various plants or parts of plants – trees, shrubs and herbs – are collected annually from forests for medicinal use. For some of them, such as *Artemisia herba-alba*, *Ceratonia siliqua*, *Cistus* spp., *Myrthus communis*, *Lichen* and *Rosmarinus officinalis*, monetary values can be assessed and are reported in Table 5.4. Nevertheless, there are some other plants of major importance for medicinal use in Morocco for which values could not be estimated: *Ajuga iva* ssp. *pseudoiva*, *Allium sativum*, *Amni visnaga*, *Anthum graveolens*, *Argania spinosa*, *Asparagus acutifolius*, *Atractylis gummifera*, *Calamintha officinalis*, *Capparis spinosa*, *Carum carvi*, *Cedrus atlantica*, *Citrullus colocynthis*, *Colocynthis vulgaris*, *Crocus sativus*, *Cuminus cyminum*, *Daphne gnidium*, *Datura stramonium*, *Euphorbia resinifera*, *Juniperus oxycedrus*, *J. thurifera*, *Laurus nobilis*, *Lavendula stoechas*, *Lawsonia inermis*, *Linum usitatissimum*, *Mandragora automnalis*, *Nigella sativa*, *Origanum vulgare*, *Ormenix mixta*, *O. praecox*, *Peganum harmala*, *Ruta montana*, *Sesamum indicum*, *Tetraclinis articulata*, *Thymus ciliatus*, *T. serpyllum*, *Trigonella foenum graecum*, *Zizyphus lotus* and *Zigophyllum gaetulum* (Ellatifi, 2000b).

The grazing value is estimated by using the substitute goods approach, assuming that the value of 1 FU is equivalent to the market price of 1 kg of barley. It is estimated that around 1500 million FU are consumed by forest grazing, of which 4.8 million FU are overgrazed, thus causing negative externalities such as land degradation. Overall, at a market price of €0.17/kg of barley, the total value of forest grazing in Morocco is estimated at €255 million.

The valuation of hunting is based on the value of licences and taxes paid by foreign and national hunters. There are some 30,000–50,000 hunters in Moroccan forests. The benefit from foreign hunting is €4.9 million, which is high if compared with the €1.3 million derived from national hunting. It should be borne in mind that illegal hunting (poaching) is also commonly practised across the whole forest domain and its value is assumed to be at least twice the value of national hunting (Table 5.3). Therefore, the total economic value of hunting increases to €8.9 million. Some institutional aspects directly linked to hunting regulation in Morocco are reported in Box 5.2.

Indirect use values

Indirect forest uses, services or influences on the environment are numerous, and this is particularly true for Mediterranean Moroccan forests (Ellatifi, 2002). However, it is difficult to express these values in monetary units. One method used is the substitution cost necessary to realize

Table 5.3.   National parks and biosphere reserves in Morocco.

Name	Area (ha)
Toubkal	38,000
Tazekka	12,000
Eastern Higher Atlas	49,000
Al Hoceima	47,000
Ifrane	53,000
Talassentane (Rif)	60,000
Souss-Massa	34,000
Dakhla	1,900,000
Lower Dr'aa	n.i.
Iriqui (Higher Dr'aa)	n.i.
Argan forest Biosphere reserve	2,000,000

Source: Ministry in Charge of Waters and Forests (1999).
n.i. = not indicated.

the same services undertaken by forests. However, when trying to estimate one forest benefit *separately* from the others, there is the risk of neglecting these other benefits (François, 1961). The intrinsic risk is actually of overestimating the values of certain services taken alone and underestimating others, due to the poor knowledge of the various services and the difficulty in finding suitable alternative solutions which would provide the same functions.

The value of the watershed protection function is assessed on the basis of the replacement costs or costs avoided approach. It is assumed that the cost of repairing the multiple types of damage that would occur in the absence of forest cover is a good estimate of the forest protective benefit. It results in a benefit of €278.7 million, which corresponds to the value of water protection performed by the Moroccan forests and other vegetal formations of the Forest Domain (annual average: €30.8/ha of forest).

Option, bequest and existence values

Mediterranean forests, with their trees, shrub and *maquis* composition, possess tremendous option, existence and bequest values. However, no estimate of these values is available, as no survey or study has been yet undertaken in Morocco. To date, ten national parks and biosphere reserves officially established in Morocco can be assumed to have high option, bequest and existence values. In addition, there are 160 established sites of biological and ecological interest,[7] on a total area of 1 Mha.

Negative externalities linked to forests

Moroccan forests are affected by several anthropogenic actions that generate negative TEV components: erosion due to poor forest

---

Box 5.2.   Game shooting in Morocco.

Game and hunting/shooting rights are State owned in Morocco. They are regulated by the 1923 Dahir on hunting police and are controlled by the Forest Department which is also in charge of the country-wide forest and fishing (in terrestrial waters) police. To enable the breeding of game, the Forest Department manages *permanent* hunting reserves, as well as *biannual* and *annual* reserves. In all categories of hunting reserves, game shooting is strictly prohibited throughout the year. Outside reserves, shooting is possible during the open season. Every year, a decree enacted by the MCWF determines the shooting season's dates, and the number of game 'pieces' to be shot per hunter and per day, by game category. In recent years, the authorized hunting quota was on average: five partridges, one hare, five rabbits, ten ducks, 20 quail, 20 snipes, five woodcocks and 40 turtledoves.

Individuals as well as associations can, for a determined period, rent shooting rights from the Forest Department, on certain state-owned forest areas, i.e. *amodiations*. To shoot legally, hunters must:

- Possess a permit for carrying firearms, which is issued by local authorities (Department of Interior) and renewed annually.
- Possess a hunting permit for the game category(ies) to be shot, which is issued by the local forest service.
- Strictly observe the dates of the hunting season, as well as the quota allowed per each category of game.
- Strictly observe shooting prohibition inside hunting reserves.

management; damage caused by forest fires and deforestation; losses of agricultural crop production; and forest degradation due to overgrazing and firewood overcollection.

Valuation of erosion damage due to poor or no forest management is based on the replacement cost of the average amount of soil loss. The erosion rate is around 20 t/ha/year in the Rif mountain, and from 5 to 10 t/ha/year (2.5–5 m³/ha) in the Middle and Higher Atlas (Ministère de l'Agriculture et de la Mise en Valeur Agricole, 1995). Overall, an average amount of 50 Mm³ of soil is lost annually and deposited in dam reservoirs. This corresponds, on average, to a fertility loss of over 7500 ha of agricultural soil per year, evaluated at €14.6 million (Ministry in Charge of Waters and Forests, 1999). In addition, at different off-site locations (downstream of the dams and on flat rain-fed agricultural soils), erosion causes a serious loss in agricultural yearly production of 22,000 ha of arable soil, estimated at around €42.7 million (Ministry in Charge of Waters and Forests, 1998b; and personal estimation). This gives a total value of €57.2 million per year.

The valuation of damages due to forest fires concerns only the anthropogenic forest fires that annually destroy an average forest area[8] of 2700 ha (Ministry in Charge of Waters and Forests, 2002). The total value of this damage amounts to €2.5 million/year and is given by the total replacement costs of standing trees, including the restoration cost of standing trees on the burned area (€0.9 million), the cost of combating/extinguishing the fire (€1.2 million) and the extraordinary costs, exceeding normal operations, to restore and replant the burned area (€0.4 million).

The damage caused by deforestation refers to the standing trees which are illegally cut and the loss of watershed protection services. Deforestation occurs on an annual average forest area of 31,000 ha (Ministry in Charge of Waters and Forests, 1998b). The restoration cost of standing trees illegally cut is €10.6 million. The cost of repairing watershed damage caused by deforestation is valued at an average of €30.8/ha (Box 5.3), or €0.9 million. This results in a total value of €11.5 million.

The losses of carbon due to deforestation and firewood overcollection amount to 583,300 tC. Valued at a shadow price of €19.4/tC, the monetary value of these losses is €11.3 million (Box 5.4).

Overgrazing occurs annually on at least 80% of the forest domain area, i.e. 7.24 Mha. Flocks are forbidden to access the remaining 20% in order to allow stand regeneration and to protect young seedlings and young trees. The valuation of the damage due to overgrazing in the forest domain includes: the value of the annual overharvest of about 4.8 million FU[9] per year (Ministry in Charge of Waters and Forests, 1998b), valued at €0.167/FU; and the damage incurred to the forest soil by the trampling of the flocks, valued at an average of €3.08/ha (i.e. 10% of the watershed forest protective value). Hence, the value of the total forest damage provoked by overgrazing is €22.5 million.

The valuation of the forest damage produced by the illicit overcollection of firewood took into account the degradation that this activity causes to the forest ecosystem. The total firewood volume overcollected by local communities, beyond the forest annual sustainable limit, amounts to 6.6 Mm³ per year (Ministry in Charge of Waters and Forests, 1998a). Even if it is an illicit and forest-degrading activity, it is nevertheless a use value for the rural people who use the overcollected fuelwood, partly for their domestic needs and partly to sell it for income improvement. As such, this volume has been included in Table 5.4 as a use value, separately from the legal firewood harvest (3 Mm³/year). Being a negative activity, it generates a negative externality equivalent to the degradation of a forest area of 110,000 ha/year (Ministry in Charge of Waters and Forests, 1998a). Valued at €30.8 per degraded hectare, the total value of the illicit firewood overcollection is €3.4 million.

Illicit hunting (poaching) is common in Morocco, being practised both by 'authorized' hunters and by poachers – particularly country people. It has very serious consequences on the regression of the game population. Valuation of the damage generated by illicit hunting is based on the regeneration costs of game populations, remaining aware that such an estimation should be considered as partial. Within the framework of its orientation plan for the period 1988–1992, the Forest Department (Direction des Eaux et Forêts et de la Conservation des Sols) in Morocco provided in its budget a total amount of MAD7.5 million (~€727,500) for the protection and the

Box 5.3.   The value of watershed protection services provided by the Moroccan forests.

There are 22 watersheds with a total area of 15 Mha above the major dams in Morocco. In some of them, forest cover is relatively dense and plays its protective roles fully. In others, the forest has been more or less degraded, or has completely disappeared, due to anthropological causes. In 1996, a National Water Management Plan (NWMP) was adopted by the Government of Morocco to tackle this situation. The NWMP estimated a cost of about €291 million for repairing the damage in a watershed area of 1.5 Mha where forests have been totally eradicated (Ministry in Charge of Waters and Forests, 1999). Based on this information, the forests protective services in watersheds amount to about €194/ha of forests.

This indicator is applied to the different forest formations, with a correction factor k, ranging from 0.50 to 0.95, according to the condition of the 'forest' stand and the density of its canopy. Therefore, the practical cost unit to apply is:

$$W = 194 \times k$$

Based on personal experience of the forest situation in Morocco, the following k values are adopted:

- $k = 0.95$ for the stands of *Cedrus atlantica*, *Pinus* spp., *Juniperus* spp., *Abies pinsapo*, *Quercus ilex*, *Q. suber*, *Q. faginea*, mattorals/*maquis* and plantations, covering 3,172,199 ha
- $k = 0.85$ for the stands of *Tetraclinis articulata* (565,720 ha)
- $k = 0.75$ for the stands of *Argania spinosa*, other coniferous and other broadleaved trees (982,375 ha)
- $k = 0.65$ for the steppe of esparto (*Stipa tenacissima*) (3,318,259 ha)
- $k = 0.50$ for the scattered stands of saharian acacias (1,011,000 ha).

Computing $W = 194 \times k$, and applying it to the corresponding forest area, a total value of the water protection services provided by all vegetal stands of the Moroccan Forest Domain is about €1,337.3 million.

If this protection state of the forest is considered to last, on average, 40 years before it is replaced by a new forest stand to sustain the forest protection function, the average annual value of the forest protection function can be estimated as follows:

$$V_F = (1{,}337.4 \text{ million}/9.05 \text{ million})/[(1 + 0.04)^{40})] = €30.8/\text{ha of forest}$$

Box 5.4.   Carbon storage/emission assessment.

For the Moroccan forest species, an average wood volume mass of 0.6 g/cm^3 or 0.6 t/m^3 is adopted. Since the conversion coefficient of dry biomass into carbon equals 0.5, to assess the net annual mass of carbon stored in Moroccan forests, the net annual wood volume increment of the forest (in m^3/ha/year) is multiplied by the approximative carbon conversion factor

$$f = 0.6 \times 0.5 = 0.3$$

Hence,
Moroccan total forest (woody) area[a]: 9,049,553 − 3,318,259 = 5,731,294 ha
Weighted mean annual forest volume increment (see Table 5.1): 1.74 m^3/ha/year
Annual forest volume increment: 5,731,294 × 1.74 = 9,972,500 m^3/year
Annual felling:
Timber: 615,000 m^3
Firewood (legal): 548,700 m^3
Firewood (illegal)[b]: 9,600,000 − 548,700 = 9,051,300 m^3
Weighted average of the forest growing stock (esparto not included): 50.5 m^3/ha
Annual average deforested volume: 31,000 × 50.5 = 1,565,500 m^3
Annual average burned volume: 2,700 × 50.5 = 136,350 m^3

Net annual loss of wood volume:

9,972,500 − (615,000 + 548,700 + 9,051,300 + 1,565,500 + 136,350) = −1,944,350 m³

Therefore:

Net annual loss of carbon = 1,944,350 × 0.3 = 583,300 tC

The value of this annual loss of carbon quantity, using the opportunity cost[c] of US$20 (€19.4) per tC, as adopted by Fankhauser (1995), is: 583,300 × €19.4 = €11,316,000[d]

[a]Esparto steppe (*Stipa tenacissima*), being a non-woody formation, is not taken into account, here, for carbon storage assessment.

[b]The total volume of biomass annually consumed in Morocco reaches 18,300,000 (Ellatifi, 1998; Ministry in Charge of Waters and Forests, 1998b). Of this total volume, 9.6 Mm³ (52.4%) come from *forest*, and the rest (47.6%) from outside forests, i.e. 3,400,000 m³ (18.6%) from agricultural fruit trees and 5,300,000 m³ (29.0%) from agricultural and industrial wastes, and from other non-woody biomass (dung, straw and others).

[c]A shadow price of US$10–20/t of carbon emission is generally accepted as a reasonable estimate of potential damage from climate change (Food and Agriculture Organization, 1997, p. 92).

[d]In its report to the UN Framework for Climate Change (Ministère de l'Aménagement du Territoire, de l'Urbanisme, de l'Habitat et de l'Environnement, 2001), the Moroccan Ministère de l' Aménagement du Territoire, de l'Urbanisme, de l'Habitat et de l' Environnement (MATUHE) gave a total volume of carbon annually emitted by the Moroccan forests, equal to 1.07 MtC, which corresponds to a value of net emission of €20.7 million, applying a price of €19.4/tC.

The average area of 31,000 ha, which is deforested yearly in Morocco, is the official figure given by the Forest Department (Ministry in Charge of Waters and Forests, 1999). Other sources give higher rates of deforestation: the rate of deforestation in Morocco is not constant, but rather exponential (Ellatifi, 2004). To reflect the reality on the ground, this rate should be taken as being equal to 93,000 ha/year (Vallée, 1999).

With a deforested area of 62,000 ha/year, which is twice the figure given by the Forest Department, the net annual loss of carbon would be of 1.05 MtC, with a value of €20.4 million.

With a deforested area of 93,000 ha/year, which is three times the figure given by the Forest Department, the net annual loss of carbon would be of 1.52 MtC, with a value of €29.5 million.

The figure given by the Moroccan MATUHE (Ministère de l'Aménagement du Territoire, de l'Urbanisme, de l'Habitat et de l'Environnement, 2001) corresponds to a deforested area of 62,000 ha/year, approximately. However, since the MATUHE's report does not give any explanation regarding the estimation method for the volume of carbon loss, the official deforestation figure given by the Forest Department (Ministry in Charge of Waters and Forests, 1999) is used along with the author's own estimation, which is considered to be more reliable.

reconstitution of the game population,[10] on a total area of 15,000 ha (Direction des Eaux et Forêts et de la Conservation des Sols, 1988). This gives an approximative amount of €48.5/ha. If the damage caused by poaching is considered not to be a recent phenomenon, as it was already evident some 30 years ago, and that negative impact on game increased annually from bad to worse (exponentially), and if an average compound interest rate of 8% is adopted for the investment of the value of illicit hunting, then the average annual value of poaching is €4.85/ha. Applied to the whole forest domain area, which can, more or less, be considered as subject to poaching, the total value of the damage provoked by illicit hunting is €43.9 million.

Towards the Total Economic Value of Moroccan Forests

For thousands of years, well before the Pharaonic period, Greek, Phoenician and Sabaean civilizations, people have always enjoyed various goods and services from forests, not only timber. Arabic gum, an NWFP extracted from *Acacia senegal*, for example, was used some 5000 years ago (Davison, 1980) for wrapping mummies (Nair, 2000). Pharaonic hieroglyphs mention it under the appellation *kami*. Local communities living in the forest vicinity have always recognized the forest ecosystem as a reservoir of valuable biological resources, indispensable for their food security/ subsistence and their general welfare (Ellatifi, 2002).

If, nowadays, some foresters and other scientists have often overlooked the importance of these forest externalities, other than timber, the gap is now being, slowly but surely, closed. Non-governmental organizations (NGOs), ethnobotanists, conservationists, forest and environment economists and many others are, ever increasingly, focusing their attention on NWFPs or, in a more general term, FORest EXternalities (FOREXs) (Merlo *et al.*, 2000).

Table 5.4.   Values of Moroccan forests (2001 prices).

Valuation method/output (unit)	Physical indicators	Value (000 €)
Direct use values		
Market price valuation		
Timber (m³)	615,000	47,724
Firewood within sustainable limits (m³)	3,000,000	43,650
Firewood illegal overharvest (beyond sustainable limits) (m³)	6,600,000	96,030
Honey (t)	4,000	19,400
Mushrooms (t)	1,000	6,111
Cistus spp. (t)	50	2,425
Virgin cork (st)	33,000	2,240
Miscellaneous cork (including reproduction) (st)	118,000	4,577
Acorns (t)	500	485
Tan bark (of *Acacia mollissima*) (t)	3,550	206
Artemisia herba alba (t)	1,500	131
Myrtle (*Myrthus* spp.) (t)	300	97
Esparto grass (t)	50,000	49
Carob beans (*Ceratonia siliqua*) (t)	1,150	47
Lichen (t)	640	31
Rosemary (*Rosmarinus officinalis*) (t)	23	12
Dry fern (t)	500	10
Briar stumps (t)	85	4
Gum sandarac (t)	1.5	1
Substitute goods		
Grazing (million FU)	1,500	255,000
Permit price, licences, taxes:		
Hunting (no. of hunters)	30,000 ÷ 50,000	8,924
Foreign hunting	—	4,850
National hunting	—	1,358
Poaching	—	2,716
Total direct use values		487,154
Indirect use values		
Cost-avoided method		
Watershed protection		278,717
Total indirect use values		278,717
Negative externalities		
Cost-based method		
Erosion due to poor or no forest management (t of soil loss)	5 ÷ 10[a]; 20[b]	−57,230
Damages caused by deforestation (ha of deforested area)	31,000	−11,536
Restoration cost of standing trees illegally cut	31,000 ha	−10,582
Cost of repairing watershed damage	31,000 ha (valued at €30.8/ha)	−954
Net loss (emissions) of carbon (tC)	583,300	−11,316
Damage due to overgrazing (ha)	7,240,000	−22,479
Damage caused by firewood overcollection (ha)	110,000	−3,388
Damage caused by poaching (illegal hunting) (ha)	9,050,000	−43,892
Losses due to forest fires (ha)	2,700	−2,488
Total negative externalities		−152,329
TEV		613,542

Source: Ministry in Charge of Waters and Forests (1998a, 1999, 2002).
[a]Middle or higher Atlas.
[b]Rif mountain.

It was in 1995, just after the 1994 meeting held in Yogyakarta, Indonesia (Chandrasekharan, 1994), that NWFPs were defined as consisting of 'goods of biological origin other than wood, as well as services derived from forests and allied land uses' (Food and Agriculture Organization, 1995). However, this definition is incomplete, because it does not include other important non-wood forest aspects such as the social, cultural, religious, ornamental, environmental and protection functions of the forest. The definition should, then, be refined to fill this gap. Nevertheless, it allows the inclusion of the various forest services in addition to goods, leaving room for the majority of components that make up, at least theoretically, the TEV of forests.

Certainly the various values considered in this study, to a certain extent also estimated in monetary terms, have been used in an effort to calculate an approximate TEV of Moroccan forests. It is an attempt aimed, above all, at reaching a more comprehensive view of the forest ecosystems, and reveals its real importance in the light of an improved sustainable management of Moroccan forests.

Due to the lack of data, the total values of Moroccan forests could not be estimated, but only a partial TEV. Many TEV components – mainly existence and bequest values – could not be taken into account, because of the unavailability of information. Hence, in this study, the calculated TEV on the one hand neglects Moroccan forest values as far as option and non-use values (bequest and existence values) are concerned, and on the other hand undervalues some direct and indirect use values that have been scarcely or poorly computed and incorporated within the TEV.

The estimated TEV equals €612 million. This value is a minimum bound to be considered as a first approximation towards the true TEV, which is likely to be significantly higher. The breakdown of this value according to its various components is reported in Fig. 5.4.

Table 5.4 shows the amounts of the estimated direct use values (€487.0 million) and indirect use values (€278.7 million), as well the negative externalities (€152.3 million). Forage for grazing alone accounts for over 52% of the direct use values and nearly 42% of the TEV. The two components – forage for grazing and

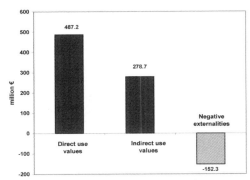

Fig. 5.4. Components of the TEV.

firewood (licit and illicit) – account for 81% of the estimated direct use values and for over 64% of the TEV. It is important to note that timber value occupies only third place, far behind fodder and firewood, with less than 10% of the direct use values; and fifth place in the composition of the estimated TEV, with less than 8%.

Regarding indirect use values, due to a lack of reliable data, only one component, watershed protection, is estimated. It reflects the ecological functions of the Moroccan forests. As expected in a Mediterranean forest ecosystem, the watershed protective value is very high (€278.7 million) and it ranks above all other estimated values of the TEV, with a share of more than 45% (Table 5.5).

Estimated externalities that negatively affect the TEV account for €152.3 million. They amount to approximately 25% of the estimated TEV. It should be emphasized that over 37% of the negative externalities are imputable, and assessed, in terms of damage caused by erosion due to poor or no forest management, at around 29% due to poaching (illicit hunting) and 15% to overgrazing. Deforestation – based on the officially reported data – and carbon emissions come in fourth and fifth positions, with around 7.5% of the negative externalities each (Table 5.5 and Fig. 5.5).

Table 5.5 and Fig. 5.5 give an overview of the estimated TEV components. The table shows that the most important forest function in Morocco is related to watershed protection. This protective function accounts for €278.7 million, or +45% of the TEV. This is followed by forage for grazing with around +42%, firewood (illicit)

Table 5.5.   Overall ranking of the 15 major estimated components of the TEV (2001).

Rank	Estimated component of the TEV	Value (000 €)	% of the TEV
1	Watershed protection	+278,717	+45.4
2	Grazing	+255,000	+41.6
3	Firewood (illegal)	+96,030	+15.7
4	Erosion damage due to poor or no forest management	−57,230	−9.3
5	Timber	+47,724	+7.8
6	Damages caused by poaching	−43,892	−7.2
7	Firewood (legal harvest)	+43,650	+7.1
8	Damage due to overgrazing	−22,479	−3.7
9	Honey	+19,400	+3.2
10	Damage due to deforestation	−11,536	−1.9
11	Net loss of carbon/emission	−11,316	−1.8
12	Hunting	+8,924	+1.5
13	Cork (all categories)	+6,817	+1.1
14	Mushrooms	+6,111	+1.0
15	Damage due to firewood overcollection	−3,388	−0.5
16	Damage due to forest fires	−2,488	−0.4
Total estimated direct use values		+487,154	+79.4
Total estimated indirect use values		+278,717	+45.4
Total estimated negative externalities		−152,329	−24.8
Total economic value (TEV)		613,542	100.0

with +16%, erosion with −9%, timber with nearly +8% and firewood (licit) with +7%. It is important to point out that not all the TEV components are independent. For example, overcollection of firewood and overgrazing also trigger forest degradation, soil erosion and water depletion. This is the case in the Mediterranean basin in general, and in Morocco in particular.

Conclusions and Perspectives

From this pioneering investigation towards the evaluation of the TEV of the Moroccan forests, a total figure of approximately €613.5 million can be given. However, this figure should be considered as a *minimum* estimate,[11] since other important components of this TEV have not been appraised at all, or have been only partially appraised, due to crucial shortages of information, ignorance or poor knowledge of all forest functions. This has been the case of the NWFPs, hunting, fishing, recreation-tourism, air and water quality, relaxation/therapy in the forest environment, landscape improvement, option, bequest and existence values, as well as other negative externalities such as losses of

natural and landscape quality due to illegal actions (Ellatifi, 2002).

Notwithstanding the approximate character (by default) of the estimate of the still unknown real value of the Moroccan forests' TEV, this study has, for the first time in Morocco, presented a start to resolving this uneasy issue. These findings show a comprehensive view of the different aspects of the TEV that, so far, have been either obvious, partly known or completely obscure. Further investigation is required to obtain a more accurate estimate of the TEV of Moroccan forests.

As a share in the estimated TEV, the direct use values represent +79.4%, the indirect use values form +45.4% and the negative externalities are about −25% (Table 5.5). The most important function is not timber production, contrary to what individuals and groups may believe. Timber ranks fifth, with a rather modest +8% share of the TEV. The most important component, with a +45% share, is the protective functions of water, landscape and biodiversity habitats (watershed protection) that forests perform. Conversely, the largest negative forest externality, with 9% of the TEV, is the damage provoked by erosion due to poor or no forest management.

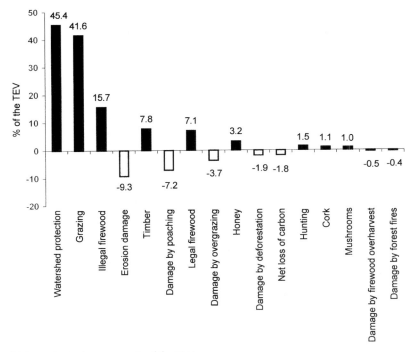

Fig. 5.5. Major estimated components of the TEV.

On the basis of the results and findings outlined in this study, the following can be concluded:

1. Many public goods and services of paramount importance provided by the Moroccan forests are not officially accounted for in the national economy. This, so far, has given underestimated ratios regarding the share of forests in national and sectorial balance sheets. With the TEV estimate of €613.5 million outlined by this study compared with the national GDP of €33,385 million (Ministère de la Prévision Economique et du Plan/Directorate of Statistics, 1999, 2000), the share of Moroccan forests would jump[12] from 0.4%, officially accounted for so far, to 1.84%.

2. Within the framework of sustainable forest management (SFM) as an objective, more of the budget should be allocated to the Forest Department and more incentives allowed to foresters in order to further and better preserve and develop the national forests. In particular, larger and more successful plantations should be produced. This should be undertaken in close cooperation

with grass-roots people, local communities and other concerned departments and organizations in order to reduce negative forest externalities and, therefore, improve the forest TEV. This would also improve the agricultural production downstream and reduce dam siltation (Ellatifi, 1999, 2000a).

3. Further research should be developed towards a refinement of the estimation methods of the forest TEV and a better evaluation of the Moroccan forest TEV, particularly through the appraisal of non-use values. This has been omitted in this study, due to the lack of reliable information.

Notes

[1] Statistics show that planted area reached 82,000 ha in 1960 and 280,000 ha in 1970, comprised of 59% eucalyptus, 7% other broadleaves and 34% conifers.

[2] Personal estimation.

[3] In Morocco, a Dahir is a text law enacted by the King.

[4]   Vegetable horse-hair is a plant fibre which is extracted, in the present case, from the dwarf palm tree (*Chamaerops humilis*).

[5]   These and other rights are regulated by the following legislation: Dahir of 10 October 1917 on forest conservation and harvesting; Vizirial Decree of 15 April 1946, on the silvo-pastoral regime; Dahir of 4 March 1925, on the special statute of the Argan forest; Decision of 1 May 1938, regulating users' rights in the Argan forest; Decision of 22 June 1936, detailing the exercise of user's rights texts (Ellatifi, 2004).

[6]   Of this, 55.3% comes from arborescent forest stands and 44.7% from mattorals/*maquis* (Ellatifi, 1998; Ministry in Charge of Waters and Forests, 1998a).

[7]   They include 79 terrestrial sites (840,000 ha), 43 continental damp sites (35,000 ha) and 38 littoral terrestrial and damp sites (205,000 ha) (Ministry in Charge of Waters and Forests, 1999).

[8]   Average computed for the period 1970–1979.

[9]   This excessive grazing has also been accounted as a direct use value.

[10]   This amount encompasses: (i) the re-introduction of species which have disappeared or those in danger of disappearing: wild guinea-fowl on 5000 ha; Dorca Mohor gazelle on 5000 ha; Berberian Deer on 5000 ha; and (ii) the management, equipment and restocking of seven existing national hunting reserves, situated in the regions of Marrakesh, Fes, Ifrane and Ben Slimane.

[11]   At least to our knowledge of the present state of the different components of the Moroccan forest TEV. We think that, in front of the weight of other non-estimated forest externalities (direct and indirect use values), no other non-estimated negative forest externality(ies) could make the Moroccan forest TEV lower that the value we have estimated in this figure.

[12]   (Estimated TEV)/(GDP) = (613,372/33,384,976) = 0.0184 = 1.84%.

If Moroccan forests were managed sustainably and all present unsustainable (excessive) actions were eliminated, the actual amount of negative externalities would be minimized and, at least some of them, would be, in contrast, transformed into direct and indirect use values, and therefore incorporated into the forest TEV, and also into the country GDP. Within this framework, the new TEV would be superior to its presently estimated value, and inferior to a maximum value which is equal to the sum of all absolute values, i.e. €918.4 million.

The new Moroccan GDP would be superior to €33,384,976,000 and inferior to a maximum which is equal to €33,384,976,000 + 2 × €152,499,000 = €33,689,974,000. Therefore,

    €613,372,000 < New Moroccan forest TEV < €978,370,000

    €33,384,976,000 < New Moroccan GDP < €33,689,974,000

and the new share of Moroccan forests in the country GDP would be:

    1.84% < (new TEV)/(new GDP) < 2.73%.

References

Boudy, P. (1958) *Economie Forestière Nord-africaine*. I,II,III. Larose, Paris.

Chandrasekharan, C. (1994) Non-wood forest products: a global view of potentials and challenges. Paper for the *International Seminar on Management of Minor Forest Products*, Dehra-Dun, India, 13–15 November, 1994, FAO, Rome.

Davison, R.L. (1980) *Handbook of Water-soluble Gums and Resins*. McGraw Hill, New York.

Direction des Eaux et Forêts et de la Conservation des Sols (1988) *Conseil National des Forêts. Cinquième Session*. DEFCS, Rabat.

Ellatifi, M. (1998) Fuelwood consumption: a major constraint for the sustainable management of the forest ecosystem in developing countries. The example of Morocco. In: *Proceedings of Foresea Miyazaki 1998 International Symposium on Global Concerns for Forest Resources Utilization – Sustainable Use and Management*, Miyazaki, Japan.

Ellatifi, M. (1999) Key forestry issues for a strategic forest policy in Morocco. In: *Proceedings of the International Conference on Contributions to Science to the Development of Forest Policies*, IUFRO Division 6, II Divisions Meeting, January 7–15, 1999, Pretoria, South Africa.

Ellatifi, M. (2000a) Rural communities as a cornerstone of sustainable forest management. Paper presented at the *XXI IUFRO World Congress*, Kuala Lumpur, Malaysia, 7–12 August 2000.

Ellatifi, M. (2000b) The situation of non-wood products in Morocco. In: *Joint FAO/ECE/ILO Committee on Forest Technology, Management and Training Seminar Proceedings on Harvesting of Non-wood Forest Products*, Menemen-Izmir, Turkey, 2–8 October 2000.

Ellatifi, M. (2002) Forests in the Biosphere. In: *Encyclopedia of Life Support Systems (EOLSS)*. Oxford, UK, Website: http://www.eolss.net/

Ellatifi, M. (2004) Economy of Forest and Forest Products in Morocco: Balance and Perspectives. PhD Thesis, University of Bordeaux, France.

Emberger, L. (1936) Présentation de la carte phyto-géographique du Maroc au 1/1500000. *Comptes Rendus Séances Mensuelles Société Sciences Naturelles Physiques Maroc* 4, 28–29.

Emberger, L. (1939) *Aperçu Général sur la Végétation du Maroc*. Commentaire de la carte phytogéo-graphique du Maroc 1:1500000. Veröff Geobot Institute, Eidgen Techn Hochsch Rübel Zürich 14, 40–157.

Emberger, L. (1955) Une classification biogéo-graphique des climats. *Revue des Travaux Laboratoire Botanique Faculte de Sciences* 7, 3–43.

Fankhauser, S. (1995) *Valuing Climate Change. The Economics of the Greenhouse*. Earthscan, London.

Food and Agriculture Organization/United Nations Development Programme (1970) *Rapport au Gouvernement du Maroc sur un Plan National de Reboisement (PNR)*. Elaboré sur la base des travaux de A. Metro. FAO/UNDP N0. AT 2803. Rome.

Food and Agriculture Organization (1995) Non-wood forest products for rural income and sustainable forestry. *Non-wood Forest Products* 7. FAO, Rome.

Food and Agriculture Organization (1997) *Forest Valuation for Decision-making. Lessons of Experience and Proposals for Improvement*. Andre Mayer Research Fellowship 1994–1995. FAO, Rome.

François, T. (1961) Evaluation of the utility of forest influences. In: *FAO Forest Influences: an Introduction to Ecological Forestry*. FAO, Rome.

Merlo, M., Milocco, E., Panting, R. and Virgilietti, P. (2000) Transformation of environmental recreational goods and services provided by forestry into recreational environmental products. *Forest Policy and Economics* 1, 127–138.

Ministère de l'Agriculture et de la Mise en Valeur Agricole (1995) *La Désertification au Maroc: Causes, Ampleur et Réalisations*. Terre et Vie, Nos 15–16, Février, Mars et no. 17, Avril 1995, Rabat, Morocco.

Ministère de l'Agriculture et de la Réforme Agraire/ Direction des Eaux et Forêts et de la Conservation des Sols (1978) *Guide Pratique du Reboiseur au Maroc*. DEFCS, Rabat, Morocco.

Ministère de l'Aménagement du Territoire, de l'Urbanisme, de l'Habitat et de l'Environment (2001) *Morocco – First National Communication – United Framework Convention on Climate Change*. MATHUE, Rabat, Morocco.

Ministère de la Prévision Economique et du Plan/ Directorate of Statistics (1999) *Le Maroc en Chiffres*. BMCE Bank, Casablanca.

Ministère de la Prévision Economique et du Plan/ /Directorate of Statistics (2000) *Chiffres Clés 2000*. BMCE Bank, Casablanca.

Ministry in Charge of Waters and Forests (1998a) *Etude de la Consommation Nationale de Bois de Feu au Maroc (1994)*. Synthèse. Cellule d'Études et de Prospection. Casablanca.

Ministry in Charge of Waters and Forests (1998b) *Five-year Period Plan 1999–2003*. Specialized Commission No. 39. Waters and Forests. Preliminary Report, December 1998, Rabat, Morocco (in French).

Ministry in Charge of Waters and Forests (1999) *Le Grand Livre de la Forêt Marocaine*. Mardaga, Belgium.

Ministry in Charge of Waters and Forests (2002) Website: http://www.eauxetforets.gov.ma Access date: October 2002.

Nair, B. (2000) Sustainable utilization of gum and resin by improved tapping technique in some species. In: *Joint FAO/ECE/ILO Committee on Forest Technology, Management and Training Seminar on Harvesting of Non-wood forest products*, Menemen-Izmir, Turkey, 2–8 October 2000.

Quézel, P. (1976) Les forêts du pourtour méditerranéen. In: *Forêts et Maquis Méditerranéens: Ecologie, Conservation et Aménagements*. Note technique MAB, UNESCO, Paris, 2, 9–33.

Quézel, P. and Barbero, M. (1982) Definition and characterization of Mediterranean-type ecosystems. *Ecologia Mediterranea* 8, 15–29.

Sauvage, C. and Brignon, C. (1963) Etages bio-climatiques. In: Atlas du Maroc, section II, carte 6b. Notice explicative par C. Sauvage, Comité National de Géographie du Maroc, Rabat, Morocco, pp. 1–44.

Vallée, S. (1999) *Forêt Marocaine: l' Heure des Bilans*. SYFIA-Maroc, Casablanca.

6 Algeria

Abdellah Nédjahi and Mohamed Zamoum
National Institute of Forest Research (INRF), Arboretum de Baïnem BP 37, Chéraga, Alger, Algeria

Introduction

Algeria covers an area of 2.4 Mkm^2, of which 84% is represented by the Sahara, one of the largest and hottest deserts of the world. The northern regions, where forest formations are located, due to suitable edaphic–climatic conditions, cover slightly more than 10% of the country's area (Direction Générale des Forêts, 2000). Notwithstanding its 1200 km of coastline, giving the reputation of a Mediterranean country, Algeria is strongly marked by its aridity. Climatic conditions are improved in the north by the presence of two mountain ranges running parallel to the Mediterranean coast, stretching from the west to the east: the Atlas Tell and the Sahara Atlas, with the High Plateaux in the middle.

The Atlas Tell extends over 4% of the country's area and concentrates the majority of the natural forests, which are the most economically valuable. However, they are jeopardized by fires, deforestation and strong water erosion, due to intense rains, which have given rise to small coastal and inner plains. The High Plateaux cover 9% of the country's area and consist of barren plains, with tracts of steppe vegetation, containing alfa and brushwood. This fringe marks the limit between the influence of the Mediterranean climate and the steppe. The Sahara Atlas region is characterized by limestone peaks, folded and much eroded, and by a little forest cover. In the south of the region, the Sahara desert begins with the Hoggar and Tassili mountains, considered as the 'cradle of civilization' and characterized by a rather peculiar and specific biodiversity.

The pattern of rainfall, with extreme events and storms, is rather irregular in terms of both quantity and spatial distribution. The eastern coast is characterized by humid and subhumid climate, with rainfall exceeding 1000 mm/year. The western coast is less humid, with rainfall reaching less than 300 mm/year, allowing growth of certain drought-tolerant crops. At the south of the Tell, the climate is semi-arid, with rainfall rarely exceeding 300 mm/year. This area is covered by steppe vegetation, made up of *Graminae* on finely textured clay soils, sparto and drinn on sandy soils, white artemisia (*Artemisia herba alba*) on finely textured soils and halophyte species around saline depressions. At the south of the Sahara Atlas, where rainfall is less than 100 mm/year, the typical arid climate of the Sahara desert begins.

The dramatic climatic differences among regions are responsible for a high demographic concentration along the coasts. Of the total population of 29.1 million people, 64% live in the Tell area, 26.5% in the High Plateaux and only 9.5% in southern Algeria. The uneven distribution could also be related to a rapid urbanization, mainly towards small and medium-size towns, affecting 60% of the total population in 1998 (Ministère de l'Aménagement du Territoire et de l'Environnement, 2000). However, the positive demographic growth caused an increase in the absolute number of the rural population, reaching 12.3 million in 1998 (Office National des Statistiques, 1998).

Forest Resources

Area and people

Algerian forests cover 4.1 Mha, of which 1.5 Mha are natural forests, 0.8 Mha are *maquis* and 0.7 Ma are plantations realized after Algeria gained independence (in 1962). Deforestation during the last 150 years reduced the forest area by 37%, if only natural forests are considered, and by less than 10% if both forests and *maquis* are included.

The current forest cover, including *maquis* and shrub vegetation, occupies 16.5% of the northern regions' area, where they are located, and only 1.7% of the total country's land. This share is clearly insufficient for ensuring a physical and biological equilibrium. It is estimated that at least 7 Mha, or 28% of the northern regions' area, are needed to provide an adequate level of protection. The actual forest cover reaches only 57% of what is advisable.

Despite urbanization, forest areas are densely populated and, as a result of government investment in socio-cultural infrastructure, access to the electricity network and roads in the less favoured areas, this is likely to continue in the future. Indeed, between 1966 and 1998, the population of these areas rose from 5.2 to 5.5 million – or 19.2% of the total in 1998 (Office National des Statistiques, 1998). Poverty is an important issue all over the country, with rates[1] growing from 12.2 to 22.6% during 1988–1995 (World Bank, 1999 cited by Direction Générale des Forêts, 2000). The issue is closely linked to unemployment, which is much diffused in rural areas and, in particular, forests. This contributes to a greater dependence and pressure of local communities on forest resources.

The exploitation of forest and natural vegetation is becoming an increasingly serious issue. Illegal cutting of trees is spreading, giving way to erosion and progressive desertification, particularly in the pine, cedar and oak forests (Zeraia, 1981) and steppe formations (Makhlouf, 1992). Another factor contributing to desertification is urban expansion, which is largely occurring without planning controls and encroaching upon forest areas.

The orientation towards a market economy and the lack of import-related subsidies for many products has favoured an increased exploitation of firewood by a rural population increasingly affected by poverty. Moreover, cork, cedar forests and steppe vegetation are strongly degraded as a result of overgrazing and forest conversion to arable lands by local communities. This is due partly to the economic crisis of the 1990s, rendering rural families landless, or not sufficiently endowed with arable land and pastures for extensive grazing, but also due to illegal grazing by non-residents, resulting in a dramatic increase of livestock over only a few decades.[2] In addition, further degradation is occurring due to natural hazards such as forest fires, which annually damage 0.9% of the forest area (Direction Générale des Forêts, 2000), and biotic factors, such as insects and fungi, affecting an additional 4.3% (Zamoum, 2002; Direction Générale des Forêts, 2002b).

Forest pressure is especially high in places where forest and steppe products are fundamental for meeting the needs of energy, building, grazing activities and traditional handicrafts. Demographic growth and the economic development are key factors in shaping the future state of forests and their conservation or, alternatively, degradation. Alleviating the human pressure on forest resources is an important issue, which can be met by adopting policies based on traditional grazing practices and promoting technological progress linked to the use of raw materials and energy.

Typologies

Total forest area presently is divided between high standing natural forests (36.8%), *maquis* (45.5%) and plantations (17.7%). A comparison between the last two forest inventories of 1955 and 1984 shows the stability of Aleppo pine (*Pinus halepensis*) due to the hundreds of thousands of hectares planted (Table 6.1). The area of cork (*Quercus suber*) woodland was almost halved, while the area covered by holm oak (*Quercus ilex*) and juniper (*Juniperus* sp.pl.) was even more reduced. The regression

Table 6.1. Area cover trends of the main forest species.

Species	Area (ha) 1955[a]	Area (ha) 1984[b]
Aleppo pine	852,000	881,302
Cork oak	426,000	228,935
Zeen and afarès oak	66,000	48,034
Cedar	30,400	16,359
Maritime pine	12,000	31,513
Thuya, juniper, holm oak	1,115,000	218,148
Eucalyptus	—	43,000[c]
Total	2,501,400	1,467,291

[a]Boudy (1955).
[b]Bureau National des Etudes pour le Développement Rural (1984) except for eucalyptus.
[c]Khemici (1993).

of dense forests essentially is due to overgrazing, which limits the possibility of forest regeneration.

The distribution of the main forest species in Algeria is shown in Fig. 6.1. The main species is Aleppo pine (*P. halepensis*), covering 880,000 ha and growing particularly on semi-arid areas with a low wood increment (Kadik, 1983). Cork oak (*Q. suber*) extends over 230,000 ha and can be found mainly in the north-eastern and central parts of Algeria, and only rarely in the west. Zeen and afarès oaks

(*Quercus* sp.pl.) cover around 48,000 ha, and grow in a cool environment within cork stands or alone in Kabylie (Messaoudene, 1989). Cedar (*Cedrus atlantica*) formations are rather sparse and cover some 16,000 ha, particularly on the Central Tell and the Aurès (Nédjahi, 1988). The maritime pine (*Pinus pinaster*) forms natural stands in the north-east and covers 32,000 ha. Eucalyptus (*Eucaliptus* sp.pl.), introduced in northern Algeria and, above all, in the eastern part of the country, extends over 43,000 ha (Khemici, 1993). These species constitute the first group of productive forest types and cover an area of 1.2 Mha, of which 424,000 ha are planted forests.

The second group is composed of holm oak (*Q. ilex*), thuya (*Tetraclinis articulata*) and juniper (*Juniperus* sp.pl.), which covers only 218,000 ha, but plays an essential soil protection role in the semi-arid area. The remainder of the forest area can be divided between protection plantations (0.7 Mha), and *maquis* and bush (1.9 Mha). In addition, alfa vegetation (*Stipa tenacissima*) should be included, covering an area of 2.7 Mha.

Algerian flora diversity includes 3139 species, or 5402 taxa, if the subspecies, varieties and forms are considered. They are scattered over different phyto-geographical sectors and subsectors evolving in various ecosystems

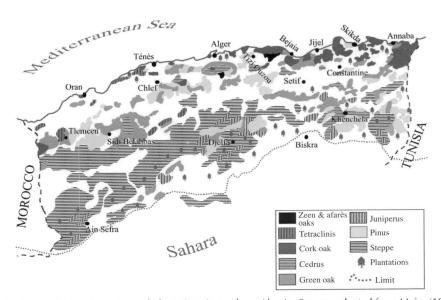

Fig. 6.1. The main forest species and plantations in northern Algeria. Source: adapted from Maire (1926).

and vegetation structures (Quézel and Santa, 1962, 1963; Zeraia, 1983). The Sahara region includes 2800 taxa, distributed as follows: 1100 in Northern Sahara, 850 in Southern Sahara, 500 in Central Sahara, 830 in Sahara high altitudes, 870 in Western Sahara and 1700 in Eastern Sahara.

The country has around 70 tree species, of which 13 are conifers: yew tree (one), thuya (one), cypress (one), juniper (five), fir (one), cedar (one) and pine (three). Deciduous species include poplar (four), willow (five), alder (one), chestnut (one), oak (seven), European hackberry (one), elm (one), fig (three), mulberry (two) and palm (two). Certain species are endemic and quite rare, such as *Taxus baccata*, *Juniperus thurifera*, *Pinus pinaster*, *Alnus glutinosa*, *J. communis*, *Populus nigra*, *Sorbus torminalis*, *Glycyrrhiza foetida*, *Acer campestre*, *A. obtusatum*, *Quercus afares*, *Ficus tiloukat*, *Pistacia atlantica*, *Argania spinosa*, *Olea laperrin*, *Sambucus nigra*, *Cupressus dupreziana* (cypress of Tassili), *Abies numudica*, *Salix triandra*, *Castanea sativa*, *Populus tremula*, *Acer monspessulanum* and *A. opalus*. Extremely rare species are *Cistus rerhayensis*, *Veronica scutellaria* and *Campanula aurasiaca*. The map of Algerian forest vegetation according to the Quézel classification is given in Fig. 6.2.

Functions

Algerian forests play a minor role in production of wood forest products (WFPs), compared with the provision of non-wood forest products (NWFPs) and protection functions. If only productive forests are considered, the potential wood production is around 1.2 Mm^3 (Table 6.2). Of this, 67% is provided by Aleppo pine stands (0.8 Mm^3), with the remainder made up of eucalyptus, zeen and afarès oaks. The forest area under active management is around

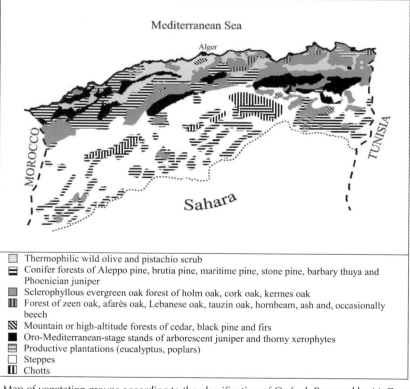

☐ Thermophilic wild olive and pistachio scrub
▤ Conifer forests of Aleppo pine, brutia pine, maritime pine, stone pine, barbary thuya and Phoenician juniper
▨ Sclerophyllous evergreen oak forest of holm oak, cork oak, kermes oak
▥ Forest of zeen oak, afarès oak, Lebanese oak, tauzin oak, hornbeam, ash and, occasionally beech
▧ Mountain or high-altitude forests of cedar, black pine and firs
■ Oro-Mediterranean-stage stands of arborescent juniper and thorny xerophytes
▤ Productive plantations (eucalyptus, poplars)
☐ Steppes
▥ Chotts

Fig. 6.2. Map of vegetation groups according to the classification of Quézel. Prepared by M. Zamoum.

Table 6.2. Growing stock, increment and potential wood production by species.

Species	Growing stock: Industrial roundwood and timber 000 m³	m³/ha	Firewood 000 m³	m³/ha	Total 000 m³	m³/ha	Increment 000 m³	m³/ha	Potential production 000 m³	m³/ha
Aleppo pine	20,787	30	9,640	14	30,427	45	1,217	1.8	856	1.3
Cedar	5,784	328	2,009	114	7,793	442	38.2	2.2	67.2	3.8
Zeen and afarès oak	7,418	129	3,262	57	10,680	186	142.6	2.5	127.3	2.2
Holm oak	1,337	19	1,511	22	2,848	41	58.4	0.8	52.4	0.8
Eucalyptus	1,344	41	1,171	36	2,515	78	122.9	3.8	144.8	4.5
Maritime pine	408	36	264	23	672	59	43.5	3.8	28.5	2.5
Total	37,078	43	17,857	21	54,935	63	1,622.6	1.9	1,276.2	1.5

Source: Direction Générale des Forêts (2000).

1 Mha, all species included, with a potential production of around 0.5 Mm³.

The actual wood production is around 15–20% of that potentially feasible. Removals from natural forests are essentially taken from the Aleppo pine (70%), of which only 20% is industrial timber. Production from oaks, cedars and maritime pines comprises around 30% of the total and they are used mainly as sawnwood.

Two important NWFPs for the Algerian economy are cork and alfa. Cork is a strategic resource due to its multiple uses, having a potential production of more than 20,000 t/year. An improved management and a more rational exploitation could increase the production even in the short term. Alfa production decreased from 30,000 t in 1990 to 10,000 t in 1994, mainly due to the difficult operations involved and the scarcity of the workforce employed in harvesting. In addition, charcoal, medicinal and aromatic plants, forest fruits, mushrooms, resin and honey could greatly contribute to the development of local economies, if properly exploited.

Grazing is an important use of forest, *maquis* and steppe areas, especially because this activity is excluded from agricultural lands for most of the year. Grazing pressure has increased almost fourfold during the last 30 years. Impacts are strong along the north-eastern coastline (Jijel, Tizi Ouzou and Skikda) and less evident in the south (Biskra, Tiaret and Saida). At the same time, forest fodder production is important and well distributed over time and space.

Forests also ensure important services in terms of water regulation and soil protection.

Tourism-related benefits should also be remembered, particularly linked to protected areas according to the international regulations of the IUCN and the United Nations Educational, Scientific and Cultural Organization (UNESCO). If the southern parks of Ahgar and Tassili are not included, the areas of high biodiversity cover 250,700 ha, of which 172,200 ha are due to forest formations. Tourism activities are encouraged by the Forests Services within regional and national parks; however, the development of infrastructures and the access for vehicles and people are somewhat limited.

Fighting desertification is a serious issue in Algeria, calling for the development of integrated programmes aimed at agro-silvo-pastoral activities. Algeria ratified the International Convention against Desertification in 1996. A specific programme, 'Green barrier' (*Barrage vert*) was developed with the aim of rehabilitating the degraded forest areas in the Sahara Atlas with species more suitable for this environment. In addition, dune fixation techniques have also been applied for more than 20 years where the desert is advancing (Makhlouf, 1992).

In 2000, the National Plan for Agricultural Development (Plan National pour le Développement de l'Agriculture, PNDA) was initiated, with the aims of protecting and improving the management of agricultural and forestlands and landscape diversification. Support is given to the local population for improving biodiversity of the cultivated and rural ecosystem. In addition, fruit seedlings are provided to mountain-dwelling people for the creation of family orchards, thus allowing the improvement of food security and protection of natural resources (Direction Générale des Forêts, 2002a).

Institutional Aspects

Ownership and size of properties

Of the total forest area, 74.5% is public (State) property, 4.9% is communal (Communes), 2.3% is collective, 8.0% is private and the remainder is classified as having an undefined status (Direction Générale des Forêts, 2000). Future developments aim to consider the assignment of property rights to local people in exchange for a type of rent agreement with specific engagements for maintenance and conservation. Moreover, most of the land in the steppe area, which covers 20 Mha, is classified as being under tribal property status (*Arouch*) (Ministry of Agriculture, 1977), the majority of which is subject to nomadic grazing.

From the legal–juridical viewpoint, the management of natural spaces, including public areas, is undertaken according to Law 84.12 concerning the general forest regimes; Decree No. 83.458 of July 23, 1983 on national parks; Decree No. 83.143 of 6 March 1987 on definition of national parks; and Law 83.03 concerning environmental protection.

Administration and policies

Within the Ministry of Agriculture, the General Directorate for Forestry (Direction Générale des Forêts, DGF) is in charge of the administration and protection of national forests, enhancing the value of land suitable for forestry, fighting desertification of steppes and pre-desertic areas, as well as drafting and proposing laws and regulations related to forests and nature protection.

All DGF functions which come within the scope of national forest policy aim to integrate environmental and socio-economic concerns, also with respect to international agreements ratified by Algeria. At an operational level, decentralized regional branches are responsible for forest conservation of *wilaya*.[3] They benefit from administrative, legal and technical rights, which enable them to define means and local strategies for forestry development.

Forest Administration Services are organized into 48 conservation branches – one for each *wilaya* – which are divided into 173 forest units at *daïra* level, 501 districts and 1221 forest stations (*triages*), the smallest unit being present in each commune. This network allows a permanent presence of a forest authority which is able to mobilize the different actors at different scales according to real world needs under the State umbrella (Direction Générale des Forêts, 2000).

Besides the DGF, the National Institute for Forestry Research (Institut National de Recherche Forestière, INRF) is responsible for forest research; the National Agency for Nature Conservation (Agence Nationale de la Conservation de la Nature, ANN) undertakes inventories and monitors flora and fauna; and other public authorities administrate national parks and natural reserves.

Forest authorities in charge of forest policy conception are assisted by consultant bodies, in particular the National Council for Forests and Nature Protection (Le Conseil National de Forêts et de la Protection de la Nature) which is formed by representatives of different ministries and institutions interested in forest problems and nature conservation. In adddition, the Superior Council for Hunting (Conseil Supérieur de la Chasse) plays an important role. All this means that decisions are taken after large-scale participation and consultation of the involved parties and interests has taken place.

The Forest Law No. 84.12/1984 regarding forest regime gives the Forest Administration the role of the management and administration of the national forest area. However, this law does not include a provision for public participation regarding the forest management process. A revised version has been proposed for discussion in parliament in order to fill the gaps and improve forest policies.

As far as forest professionals are concerned, they have been given the status of a specific Corps (forest police), which is in a position to implement public measures of general interest in a compulsory manner. Forest policy, translated through a forest development plan, is implemented in practice through various programmes. They aim at improving the administration of the forest area, increasing forest revenues, enhancing the welfare and ensuring the stabilization of the rural people living in and around forest areas.

Contribution of the Forest to the National Economy

Gross domestic product and employment

The role of Algerian forests in financial terms is negligible, representing just 0.02% of the gross domestic product (GDP). However, forests are fundamental for soil protection and conservation, production of WFPs and NWFPs, employment and rural welfare.

The forestry sector plays an important role in less developed regions, where other economic alternatives are hardly developed. The Direction Générale des Forêts (2000) reports a forest employment of 8.5 million work-days, corresponding to 35,500 jobs in 1999. In addition, the forest development programme employs some 60,000–80,000 workers. The exploitation of alfa grass involves around 8000–14,000 harvesters and 600–700 staff for other related jobs (packaging). The cork sector employs 3000 personnel for harvesting and around 1000 for processing. In addition, there are some 5000 permanent staff in wood-processing enterprises represented by both large and small units. Overall, forestry and forest-related activities employ some 110,000–140,000 people, representing around 1.2–1.5% of the country's labour force.

International trade: wood exports–imports

During 1963–1990, national wood removals registered a continuous drop, followed by a slight increase, reaching a peak of 240,000 m^3 in 1993. This was essentially a response to the increased national demand for wood, facilitated by the adoption of more updated tools for woodcutting. Presently, total wood removals are around 200,000 m^3, of which 60% is processed as panels, 20–30% as sawnwood and 10–20% for fuelwood. They supply only 15% of the country's estimated total needs. To increase national production, productive plantations of fast-growing species have already been established. Overall, the demand for timber for panels is met by national production, while 80–90% of the needs of large diameter roundwood are covered by imports. Algerian forests cover all domestic needs for firewood. After the 1970s, firewood consumption has decreased due to the use of gas, which is an easy, clean and cheap energy source. However, after the adoption of market policies, rural populations, becoming more and more afflicted by poverty, have turned again to fuelwood.

Other forest-based industries: tourism and green belts

Recreation-tourism is favoured especially in national parks, in which the management tends to draw public awareness to the needs of biodiversity and landscape conservation. There are eight national parks,[4] covering 165,360 ha, and they are managed by the DGF. The management objectives consist of the conservation of fauna and flora, the promotion of local participation in the ecodevelopment, and the encouragement of an ecotourism compatible with conservation and sustainable development. The creation of four new national parks stretching over 620,000 ha and five new nature reserves covering 500,000 ha are planned. In addition, to meet the target of 10 m^2 of green space per inhabitant, initiatives include the creation of botanical gardens on 30,000 ha, establishment of plant nurseries in each commune (or for agglomerations of > 10,000 inhabitants) and horticultural developments. Other proposals include the development of hunting resources, conservation of wetlands, nature promotion and welfare through rehabilitation of natural habitats, restoration of recreative forests, and the conservation and utilization of genetic pools.

Issues related to knowledge, conservation and enhancement of biological diversity and sustainable management of natural resources are under the joint responsibility of the forest administration and INRF. Numerous interventions are undertaken in forests, steppes, mountains, wetlands, farming fields, botanical gardens and animal zoos. A programme is also in place for the improvement of the genetic base of animals and plants of social and economic interest. The protection and enhancement of genetic pools is administered by the INRF and involves the conservation of genetic banks, and the genetic improvements of trees dedicated to wood and cork production as well as plants for fibre production, such as alfa (Harfouche, 2001).

The Values of Algerian Forests

The estimated quantities and the corresponding monetary values of Algerian forests are presented in Table 6.3, according to the main valuation methods used.

Direct use values

The direct use values of Algerian forests are represented mainly by NWFPs such as cork, honey, alfa and grazing, while WFPs are less significant. Valuation of timber and firewood is based on the quantities traded on the market and their average roadside prices, resulting in €0.7 million and €0.3 million, respectively (Table 6.3). The prices of WFPs are highly

Table 6.3. Values of Algerian forests.

Valuation method/output	Quantity	Value[a] (000 €)
Direct use values		
Market pricing		
Timber[b] (m³)	123,747	671.4
Firewood[b] (m³)	77,773	252.8
Virgin cork[b] (t)	12,000	5,397.3
Honey[c] (t)	1,600	516.2
Alfa (*Stipa tenacissima*) (t)	10,000	81.9
Miscellaneous	—	247.5
Substitute goods pricing		
Grazing[d] (million FU)	1,530	206,550.0
In forests	1,000	135,000.0
In steppic areas	530	71,550.0
Shadow pricing (rent for forest spaces)		
Other direct uses of forest area[b]	—	1,328.7
Total direct use values		215,045.8
Indirect use values		
Cost avoided method		
Soil and water protection (Mha)	19.1	477,973.5
Total indirect use values		477,973.5
Negative externalities[e]		
Cost-based methods		
Erosion due to poor or no forest management (Mm³)	50	−41,481.0
Desertification effects (ha)	301,716	−3,022.5
Carbon losses due to forest land use changes (tC)	341,500	−6,830.0
Losses of WFPs due to forest fires (ha)	25,000	−22,000.0
Losses of NWFPs due to forest fires	25,000	−2,300.0
Total negative externalities		−75,633.5
TEV		617,385.8

[a]Updated to 2001 prices.
[b]Direction Générale des Forêts (2000).
[c]Ministry of Agriculture (2001).
[d]Ministry of Agriculture (2001).
[e]Ministère de l'Aménagement da Territoire et de l'Environnement (2002).

variable across local markets, depending on the species and wood quality.

For several NWFPs, valuation is based on the estimated quantities and the average market prices. Cork is the major NWFP, whose production potential has diminished from 15,000 t in 1980 to an average of 12,000 t due to the regression of cork-exploited areas. At a market price of €450/t, the total value of cork is about €5.4 million.

Valuation of honey is based on a quantity of 1600 t, which is 46% higher than that of the previous year. This is justified by a 30% increase in the number of beehives, which presently reaches 469,330. The monetary value of honey is €0.5 million, which is expected to increase in the future.[5]

Valuation of grazing is based on the the number of forage units (FU) consumed by animals grazing in forests and steppe and the market price of barley. During 1991–2000, the average livestock number was 22 million head, mostly represented by sheep (80%), goats (13%) and cattle (7%) (Ministry of Agriculture, 2002). Forests are used permanently for livestock grazing during the winter by farmers living in the north. They are also used for transhumance grazing by steppe livestock. On average, there are 0.96 million cattle, 0.6 million goats and 4.2 million sheep grazing in the forests (Ministère de l'Aménagement du Territoire et de l'Environnement, 2002). Grazing pressure is high particularly in cork and cedar forests, as available grazing land has become exhausted. The estimated fodder consumption in forests is about 1000 million FU.

The annual grazing in steppe ecosystems is about 530 million FU, which is one-third of the total forage production of 1978. The situation reflects the productivity losses of the degraded steppe soils, mainly due to the intensive grazing over time. Even if half of the steppe area is not degraded, the overall fodder consumption exceeds the land carrying capacity. This is due to the increased number of livestock from one sheep-equivalent/4 ha in 1968 to one sheep-equivalent/0.78 ha in 2001 (Ministère de l'Aménagement du Territoire et de l'Environnement, 2002). The total forage consumption in forests and steppe areas is around 1530 million FU, which gives an economic value of €206.5 million.

Other forest direct uses – cultivation or exploitation of vegetation cover – are enjoyed by local people to which open forest areas are leased in order to avoid forest fires. The rent paid for this leasing (€1.3 million) is assumed to be a minimum proxy for the value of forest products collected from these areas (Direction Générale des Forêts, 2000).

Indirect use values

The edaphic, climatic and economic conditions are among the factors that negatively affect forest ecosystems of southern Mediterranean countries and regions. The consequences can be seen in terms of accentuated degradation of forest cover due particularly to hydraulic erosion and desertification. In this context, forest and steppe cover plays a significant role in soil protection against hydraulic and wind erosion. Valuation of the water and soil conservation function is based on the total forest area (4.1 Mha), the surface covered by steppe (15 Mha) (Ministère de l'Aménagement du Territoire et de l'Environnement, 2002) and a monetary benefit of around €25/ha. This generates a total value of €478 million.

Option, bequest and existence values

Algerian flora is rich in medicinal and aromatic species, such as myrtle, hawthorn, rosemary, lavender and eucalyptus. Their widespread availability in diversified bioclimatic conditions – humid, semi-arid and desertic – is significant for a notable genetic resource potential, albeit little exploited. Presently they are collected only for traditional local uses; meanwhile there is certainly a higher economic interest, but not sufficiently utilized, from pharmaceutical, cosmetics and agri-food industries. The potential to deliver a large quantity of chemical substances such as oil essences, characteristic for each taxonomic group, is largely available in local species (Rahili, 2002). At present, 3300 spontaneous species have been identified, of which 640 are rare and endangered and 256 are endemic. Specific decrees establish the lists of non-cultivated plant species, and other decrees are being prepared for the sustainable management of genetic resources.

Negative externalities linked to forests

Algerian forests are natural formations continuously subject to climatic and human pressures. Considering the historical events that have affected forests and the established constraints, it is easily understandable that they are being degraded progressively. The main species are substituted by *maquis* formations and bush, the role of which is nevertheless important in terms of soil conservation and fixation. The forest sector, through its various initiatives, is providing the essential inputs as a main actor and a crucial partner for any policy aimed at protection against environmental degradation in terms of erosion, desertification and biodiversity losses. On the negative side, the lack of cooperative approaches and agreements for common management between the administration, local authorities and communities should be mentioned. The difficulty in implementing/enforcing already established laws and regulations for preventing forest degradation raises concern, mainly due to the joint pressure of forest conversion into agricultural land, overgrazing, insects and other diseases, fires and urbanization.

Erosion is a big problem on Algerian soils, aggravated by overgrazing and harsh climate (Box 6.1). It gives rise to soil sedimentation which reduces reservoir capacity and water reserves by an annual 50 Mm^3 (Ministère de l'Aménagement du Territoire et de l'Environnement, 2002). If estimated at an annual cost of €0.83/m^3, an overall value of €41 million is obtained.

Desertification affects agricultural and steppe land uses, and the valuation is based on the area annually affected and the corresponding losses of soil productivity. Annual changes in productivity are registered on 301,716 ha/year due to various causes: 24,395 ha are affected by wind erosion; 263,821 ha are very sensitive to desertification, causing an annual loss of 0.6 q equivalent cereals/ha; and 13,500 ha of steppe vegetation are overgrazed, resulting in an annual loss of 0.9 q equivalent cereals/ha. The

costs of these losses are estimated at around €15.3–18.0/q equivalent cereals and are variable from one region to another. Overall, the annual monetary value is €3 million.

Valuation of carbon losses is based on the methodology proposed by the UNFCCC (United Nations Framework Convention on Climate Change) for valuing the emissions and absorptions of greenhouse gases. In 2001, Algerian forests and other vegetation sequestered 4.33 Mt equivalent of carbon dioxide (CO_2). At the same time, conversion of forests and prairies to other land uses led to emissions of 5.59 Mt equivalent of CO_2 (Ministère de l'Aménagement du Territoire et de l'Environnement, 2001). Transformation into carbon quantity[6] yields a net carbon loss of 340,000 tC. Valued at €20/tC, the value of carbon losses due to land use changes is €6.8 million.

The damage caused by forest fires is estimated according to the cost-based method. On average, 25,000 ha were affected by forest fires during 1984–1999 and caused a loss of WFPs of about €22 million (€880/ha, value of burnt wood). In addition, fires provoked losses of NWFPs estimated at €2.3 million. The latter value should be regarded with caution, being a result of benefit transfer (€92/ha) from areas with conditions slightly different from those of Algeria (Benefits *et al.*, 1985, cited by Ministère de l'Aménagement du Territoire et de l'Environnement, 2002).

Towards the Total Economic Value of Algerian Forests

The estimated total economic value (TEV) of Algerian forests is calculated as €617.3 million. Figure 6.3 shows a minimum estimate of the direct use values of about €215.0 million, which is around 35% of the TEV. The indirect use values have the most significant weight, reaching €478 million or 77% of the TEV. The TEV is reduced by the negative externalities components, estimated at an annual minimum of €76 million, corresponding to −12% of the TEV.

Figure 6.4 shows that the most important forest values are soil conservation performed by forest and steppe vegetation (77.0%), followed

Box 6.1. Erosion and desertification in Algeria.

In Algeria, water erosion affects around 12 Mha and contributes to a high level of degradation of the vegetation cover and soils. It greatly affects the Tell area due to three contemporary constraints: mountains with steep slopes; rocks poorly resistant to erosion; and sparse, or even non-existent, vegetation coverage. A strategy against soil erosion and reservoir siltation has been adopted within the project for Soil Defence and Restoration (*Défense et Restauration du Sols*, DRS). Its main objective is the undertaking of works to protect the most badly affected regions and reservoirs, such as the reafforestation of some 800,000 ha.

Table 6.4 shows the proportion of the total land area of each subregion of northern Algeria affected by hydraulic soil erosion. It shows that the hydraulic erosion affects 28% of the total area of northern Algeria. The High Plateaus and Atlas mountains are less eroded than the lowlands and the Tell mountains, with 9 and 15% of their land areas, respectively. In fact, the western regions are the most damaged by apparent hydraulic erosion, affecting 55% of the lowland area and 60% of the Tell mountains area.

Table 6.4. Proportion of land area affected by hydraulic erosion in northern Algeria (%).

	Lowlands and inner basins	High and inner plateaus	Tell mountains	Atlas mountains	Total
West	55	41	60	22	44
Centre	20	5	30	–	23
East	18	3	39	7	20
Total (northern Algeria)	40	9	39	15	28

Source: Hadjiat (1997).

Table 6.5 shows the proportion of each subregion in the total area of northern Algeria affected by hydraulic erosion. It can be seen that hydraulic erosion affecting the lowlands and the Tell mountains together accounts for 83% of the eroded soils. The other 17% of the eroded soils in northern Algeria are located in the High Plateaus and the Atlas mountains. The western region is the most affected, with the hydraulic activity being responsible for the damage of 47% of the total area of eroded soils.

Table 6.5. Proportion of total eroded area per subregion in northern Algeria (%).

	Lowlands and internal basins	High and internal plateaus	Tell mountains	Atlas mountains	Total
West	18.5	4.5	16	8	47
Centre	0.4	1.5	21.5	–	27
East	0.2	1	21	2	26
Total (northern Algeria)	24.5	7	58.5	10	100

Erosion rates can vary from one area to another. From 1986 to 1995, a study carried out in the south of the Blideen Atlas showed that erosion was moderate in both cultivated and not cultivated lands, ranging from 0.1 to 3 t/ha/year, reaching 20 t/ha on the fertilized red soils on bare slopes with a gradient of 35% (Arabi, 1991).

Concerning the Algerian steppe, the tendency towards desertification is certain: the steppe ecosystems show a regression of vegetation cover by more than 50% and a reduction of forage production from 120–150 FU/ha/year in 1978 to 30 FU/ha/year in degraded areas and to 60–100 FU/ha/year in areas covered by palatable vegetation. This trend is shown on a map of sensitivity to desertification produced by remote sensing by the National Centre of Spatial Techniques (Centre National des Techniques Spatiales). This map allows classification of the steppe areas affected by the desertification into zones: already desertified (487,902 ha), highly sensitive (2,215,035 ha), sensitive (5,061,388 ha), moderately sensitive (3,677,035 ha), hardly or not sensitive (2,379,170 ha).

by grazing (33.3%). Other NWFPs (1.2%) have less importance, among which cork, honey and alfa are the main components. The small contribution of NWFPs can be explained by two main factors; the first is due to the limited statistics concerning the use/exploitation of the numerous forest products, and the second is due to the fact that Algerian forest policy is essentially oriented towards its primary purpose, i.e. protection combined with a rational resource exploitation. Timber and firewood are of least importance, contributing to less than 1% of the TEV. Negative externalities such as erosion and desertification, forest fires and carbon losses due to changes in forestland use reduce the TEV by 7.2, 3.9 and 1.1%, respectively.

Finally, it would be interesting to undertake more studies for valuing precisely, from the quantitative and monetary viewpoint, the potential of the NWFPs in Algeria with the purpose of increasing their value. These studies would allow the development of a sound management strategy for the

sustainable development of forests and steppe, with clear positive impacts on the rural economy.

Perspectives and Crucial Issues

The causes of continuous degradation of Algerian forests are similar to those experienced in other Mediterranean forests; the evident climate changes, and the magnitude of erosion and desertification. Together with the new concepts of participatory forestry, common property forestry and communal forestry, agroforestry, biodiversity conservation, sustainable forestry and development, fair governance, acting on the terrain, a challenge is undoubtedly presented to the forest profession

Under present climatic changes, Algerian forests are needed to ensure wood biomass production. In order to address this requirement, mechanisms able to improve the forest's contribution to the GDP should be put in place. Another significant goal is to enhance production of NWFPs, rural employment and income, the ecological and protection functions such as water and soil conservation and biodiversity, and to increase the fight against desertification.

In fact, it should be borne in mind that the government is oriented towards the support of agriculture as the key priority sector for a developing economy and society. It acts towards the sustainable development of rural areas, of which forests are just a component. In 2002, the government created a ministry devoted to rural

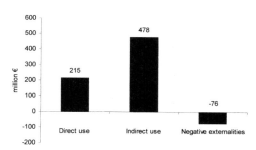

Fig. 6.3. Estimated TEV of Algerian forests.

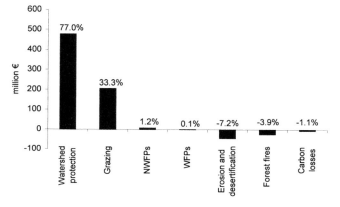

Fig. 6.4. Main components of the TEV of Algerian forests.

development. In the context of the management and conservation of rural heritage, the following objectives concerning forests have been established:

- To contain the regression of forest resources and ensure their exploitation respecting the principle of sustainable management.
- To preserve, regenerate and develop the existing resources through forest practices, exploitation and management plans well integrated within the rural space as a whole.
- To plant 3 Mha of land suitable for forest use, as a measure against erosion and desertification; to accomplish this task, a national plan of afforestation for 1999–2018 has been developed and implemented, with the objective of planting 1.2 Mha.

Management studies have shown that a mobilization (removal) of 500,000 m³ of wood and the harvest of 15,000 t of cork would reduce the pressure on imports. Meanwhile the National Reforestation Plan (Plan National de Reboisement, PNR) should result in a higher self-sufficiency of forest products concerning both wood and secondary NWFPs.

More means are made available for research projects, and for communication to the general public aimed at increasing awareness of forests and the natural heritage. The INRF can count on 12 regional stations distributed amongst the various ecological areas. The research involves many different aspects (genetic, forest zoology, silviculture and management) and the results are made available to forestry and farming sectors. Several research projects and related tasks are given to universities and local research centres within specific agreements. Links are also available with international research institutions in the context of cooperation efforts, particularly with northern Mediterranean (France, Italy and Spain) and southern Mediterranean neighbouring countries (Morocco and Tunisia). The INRF publish two journals a year; one is scientific (*Annale de Recherche Forestière d'Algérie*) and the other technical (*Revue Forestière d'Algérie*).

The main source of financing and incentives is from the State and other potential international sources and is aimed at ensuring the conservation of Algerian forests. Financing is encouraged, particularly for afforestation projects and for production/exploitation activities of wood, cork, alfa and secondary products. Institutional investors as well as private parties are requested to contribute in order to alleviate and assist the State efforts. Cooperation efforts are made in order to favour scientific and technical exchanges, sharing of experiences and competencies at national and international levels, particularly amongst northern–southern Mediterranean countries. Private intervention in forestry is now possible, in addition to public involvement. Contracts can be implemented in order to carry out investments and forest development programmes. Concessions are also available for private parties under specific agreements according to *cahiers de charge*. Non-governmental organizations (NGOs) and society in general, though not presently important players, are gaining importance, weight and scope in forest policies. It is a matter of providing stimuli, encouragement and tasks, and monitoring the ongoing situation regarding forest policy and the rural world.

In an unfavourable context, agriculture and forestry play an important role in the field of renewable resources and for ensuring food security. It is true that the role of forests as a contributor to the GDP is negligible, but they are essential regarding the other functions such as protection, production of WFPs and NWFPs, rural employment, welfare and stability. Therefore, a new forest strategy or, better still, a readjustment, has taken place. It takes responsibility for an integrated rural development, management of natural resources and increased welfare of rural people.

Notes

[1] The rates reflect the population share with food consumption expenses insufficient for covering the basic consumption needs.

[2] For example, it rose from 15,000 to 100,000 cattle in El Kala National Park in only a few years.

[3] *Wilaya* corresponds to a large city in Algeria. In Algeria, there are 48 *wilayas*, each composed of several *daïras* (cantons); each *daïra* is composed of several communes.

[4]   The national parks are Chréa (26,590 ha), Djudjura (18,550 ha), Theniet El had (3420 ha), Belezma (26,250 ha), Tlemcen (8,230 ha), Taza (3800 ha), Gouraya (2080 ha) and El Kala (76,440 ha). In addition, the nature reserves occupy 42,850 ha, divided among Béni-Salah (2000 ha), Babors (2370 ha), Mergueb (16,480 ha) and Macta (22,000 ha). Hunting reserves extend on 42,460 ha, divided into: Djelfa (31,870 ha), Tlemcen (2200 ha), Mascara (7300 ha) and Zéralda (1090 ha) (Direction Générale des Forêts, 2000).

[5]   The honey bee industry is expected to develop through an investment plan of about €52 million within the National Agricultural Development Plan (Plan National de Développement Agricole, PNDA). It involves the creation of 87 nurseries through 25 wilayas with a potential of around 78,000 additional hives.

[6]   Using the conversion coefficient of 0.27.

References

Arabi, M. (1991) Influence de Quatre Systèmes de Production sur le Ruissellement et l'Érosion en Milieu Montagnard Méditerranéen (Médéa, Algérie). PhD Thesis, University of Grenoble, France.

Boudy, P. (1955) Economie Forestière Nord Africaine. Description Forestière de l'Algérie et de la Tunisie. Tome V, Larose, Paris Vème.

Bureau National des Etudes pour le Développement Rural (1984) Inventaire des Terres et Forêts de l'Algérie du Nord. BNEDER, Algeria.

Direction Générale des Forêts (2000) Etude Prospective du Secteur Forestier en Algérie. Direction Général des Forêts.

Direction Générale des Forêts (2002a) Rapport National sur la Mise en Œuvre de la Convention des Nations Unies sur la Lutte Contre la la Désertification. Direction Général des Forêts.

Direction Générale des Forêts (2002b) Bilan de Campagne de Traitement Contre la Processionnaire du Pin, Direction de la Protection de la Nature. Direction Général des Forêts.

Hadjiat, K. (1997) Etat de la Dégradation des Sols en Algérie. Rapport d'expert PNAE, World Bank.

Harfouche, A. (2001) Préservation et Développement des Ressources Génétiques Forestières: Situation et Perspectives. INRF.

Kadik, B. (1983) Contribution à l'Étude du Pin d'Alep (Pinus halepensis Mill.) en Algérie. Ecologie, Ddendrométrie, Morphologie. PhD Thesis, Université de Pierre et Marie Curie, Paris, France.

Khemici, M. (1993) Contribution à l'Étude Éco-éthologique de Phoracantha semi punctatata F. (Coleoptera, Cerambycidae) Ravageur Xylophage de l'Eucalyptus en Forêt de Baïnem. Masters Thesis, University of Science and Technology, Houari Boumedienne, Alger, Algeria.

Maire, R. (1926) Notice Phytogéographique de la Carte Phytogéographique de l'Algérie et de la Tunisie. Gouvern Général de l'Algérie.

Makhlouf, L. (1992) Etude Sédimentologique des Sables du Cordon Dunaire dans le Bassin du Zahrez Gharbi. PhD Thesis, Paris, France.

Messaoudene, M. (1989) Dendroécologie et Productivité de Q. afares Pomel et de Q. canariensis Wild. dans les Massifs Forestiers de l'Akfadou et Béni-Ghobri en Algérie. PhD Thesis, University Aix Marseille III, France.

Ministère de l'Aménagement du Territoire et de l'Environnement (2000) Rapport sur l'État et l'Avenir de l'Environnement. GTZ.

Ministère de l'Aménagement du Territoire et de l'Environnement (2001) Projet National ALG/98/G31: Elaboration de la Stratégie et du Plan d'Action National des Changements Climatiques. Communication Nationale Initiale, MATE.

Ministère de l'Aménagement du Territoire et de l'Environnement (2002) Plan National d'Actions pour l'Environnement et le Développement Durable PNAE-DD. MATE.

Ministry of Agriculture (1977) Rapport Steppe. Ministry of Agriculture, Algeria.

Ministry of Agriculture (2001) Rapport sur la Situation du Secteur Agricole en 2002. Direct. des statistiques agricoles et des systèmes d'information. Ministry of Agriculture, Algeria.

Nédjahi, A. (1988) La Cédraie de Chréa (Atlas Blidéen): Phénologie, Productivité, Régénération. PhD Thesis, University of Nancy, France.

Office National des Statistiques (1998) Recensement Général de la Population Humaine. ONS, Algeria.

Quézel, P. and Santa, S. (1962) Nouvelle Flore de l'Algérie et des Régions Désertiques Méridionales, Vol. 1. CNRS, France.

Quézel, P. and Santa, S. (1963) Nouvelle Flore de l'Algérie et des Régions Désertiques Méridionales, Vol. 2. CNRS, France.

Rahili, G. (2002) Les huiles essentielles ed leurs intérêts. Revue des Forêts Algérienne 3, 28–31.

Zamoum, M. (2002) Quelques eléments pour la préservation de la santé des forêts en Algérie. Revue de Forêt Algérienne 3, 4–7.

Zeraia, L. (1981) Essai d'Interprétation Comparative des Données Ecologiques, Phénologiques et de Production Subero-ligneux dans les Forêts de Chêne Liège de Provenance Cristalline (France Méridionale) et d'Algérie. PhD Thesis, Faculty of Science and Technology, Saint Jérôme, Marseille, France.

Zeraia, L. (1983) *Liste et Localisation des Espèces Assez Rares, Tares, Très Rares et Rarissimes en Algérie*. INRF, Algeria.

7 Tunisia

Hamed Daly-Hassen[1] and Ameur Ben Mansoura[2]

[1]National Institute of Research on Rural Engineering, Water and Forestry (INRGREF),
B.P. 10, Ariana −2080, Tunisia; [2]Arab Center for Studies of Arid Zones and Drylands
(ACSAD), Damascus, Syria

Introduction

Historically, Tunisian forests suffered a great deal of misuse, degradation and depletion due to overgrazing, cropping expansion, fires and inappropriate policies. As a result, the country's wooded area has decreased by more than half since early Roman times, reaching about 1.25 Mha at the start of French colonization, in 1881 (Direction Générale des Forêts, 1992). Deforestation continued, with forests reaching their lowest point at the beginning of the 1950s at some 386,000 ha. It was only after independence, in 1956, that trends of forest regression were reversed with ambitious plantation programmes, resulting in an increase of the forest cover from 2.3 to 5.2% of the land area by 1994 (Box 7.1). In the year of reference, 1998, the total forest cover reached 942,800 ha, representing 5.7% of the country's total area (Direction Générale des Forêts, 2002).

Most of the Tunisian forest area (94%) is located within and along the three mountain chains of the Khroumiries, the Mogods and the Dorsale (Fig. 7.1). These elevations, which dissect the agricultural plains of northern and central Tunisia, are essentially oriented from south-west to north-east. They are characterized further by a rapid north to south aridity gradient, as the mean annual rainfall drops quickly from 1200 mm in the north, to about 300 mm in the south of the area (Ben Mansoura et al., 2001). Among the forest species present, Aleppo pine (*Pinus*

halepensis Mill.) is the most widely distributed tree in the country, due to its well-known adaptation to the prevailing arid and semi-arid conditions (Direction Générale des Forêts, 1995a).

Wood production in Tunisia includes a high proportion of fuelwood (61%) and a lower share of industrial roundwood (30%). The remaining 9% consists of other types of roundwood targeted for various agricultural and domestic uses, such as poles. Despite their low productivity, Tunisian forests play a prominent socio-economic role, as they provide the basic needs and ensure employment for nearly 1 million people, representing about 10% of the country's total population (Direction Générale des Forêts, 2001b). In addition, forests constitute a major land use, offering grazing opportunities to more than half of the country's livestock resources (Ben Mansoura and Garchi, 2001).

Unfortunately, the excessive anthropogenic pressure due to forest users and domestic livestock causes a great deal of environmental degradation. A recent study found significantly rising trends in all types of forest-degrading activities, especially clearing and logging (Ben Mansoura et al., 2001). This is a result of the open access to public forests due to improper enforcement of forest policy. However, the recent orientation towards increased participation of local users in forest management, as well as greater urbanization and its influence on the diversification of economic activities may contribute substantially to forest preservation.

Box 7.1.   History of the Tunisian forest heritage.

Historically, Tunisia used to be a resourceful country in terms of forest cover and timber production. The writings of early historians such as Pliny and Appian, documented by Boudy (1948), pointed out that Tunisia's ancient cities such as Carthage, Utique and Dougga were located in the vicinity of major forest areas. An abundant local wood supply for building construction was the backbone of urban expansion in historic cities. Unfortunately, the country's rich forest heritage suffered a dwindling trend throughout history, as forests shrunk from 3 Mha during early Roman times to 1.25 Mha at the start of French occupation in 1881. This sharp decline in forest cover was due mainly to the expansion of cultivation, especially in the second century BC. Substantial forest losses were also suffered as a result of local uprisings in mountain areas and the resulting political turmoil which prevailed in the country throughout the third century AD. Another period of marked forest clearance occurred in the 11th century AD when severe recurrent droughts in the Arabian peninsula led thousands of desperate migrants from the Bani Hilal tribe to invade Tunisia along with their flocks of grazing livestock.

   Further deforestation continued to occur under French occupation as Tunisia's forest cover declined from 1.25 Mha in 1881 to only 386,000 ha in 1956 when independence was gained. Expansion of colonial farming and excessive extraction of cork and tannins, particularly in the early years of occupation, constituted the major causes of forest regression. Under French rule, deforestation peaked between 1920 and 1930 when more than 100,000 ha of forests were converted into cropping lands. By 1956, cultivation caused the loss of about 864,000 ha of forests, in addition to 2 Mha of steppes that were also cleared and ploughed. As a result, cultivated lands rose dramatically from 1.2 Mha in 1920 to 4 Mha in 1956. Other factors contributing to the reduction of forest cover were forest fires, mismanagement, inappropriate policies and demographic pressure.

   After 1956, despite the rising anthropogenic pressure, awareness of the necessity for forest conservation has been gained. Deforestation trends could be reversed due to extensive forest plantations that have become a strategic priority in agricultural development plans. Forest plantations implemented between 1956 and 1994 covered a total area of about 457,000 ha. They raised the country's total forest cover to nearly 843,000 ha, i.e. 5.2% of all Tunisian land. The illegal conversion of land use from forest to agriculture is estimated to amount to several hundred hectares per year. However, in the 1990s, rangelands continued to rank first among all land use types with a total cover of 30.1%, whereas cultivated lands were second with 29.4%.

Forest Resources

Area and people

Tunisian forests have been exposed to increasing anthropogenic pressure, as the human population grew from 2 million people in 1915 to about 10 million at present. During the last five decades, the number of local forest users increased at an average annual rate of about 2.8%, reaching nearly 1 million people in 1998 (Direction Générale des Forêts, 2001b). However, the growth rate of rural populations is expected to drop from 1.7% in 1994 to only 1% by 2005 (Institut National de la Statistique, 1994).

   The historical presence of forest users with a heavy reliance on forest resources for income and survival has perpetuated a social dilemma and a conflicting relationship with the forest administration. In 1890, the administration was entrusted to own the forest resources on behalf of the State (Snane, 1993). However, depriving the small farmers of their own lands forced them into a greater dependence on forests, with consequences such as clearing forest areas and charcoal making (Hamzaoui, 1993). The first regulation in 1934 gave some usage rights to the communities living within a 5 km radius of a forest area. They consisted of the collection of dead wood and brushwood, grazing in permitted areas, non-commercial use of other forest products and agriculture in non-covered parcels. The Forest Code of 1959 and its revision in 1988 (République Tunisienne, 1993) aimed at halting further forest depletion by banning clearing and cultivation and restricting the usage rights to the communities living inside the forest areas (Snane, 1993).

   Despite the limitations imposed, the socio-demographic pressure and the administration's

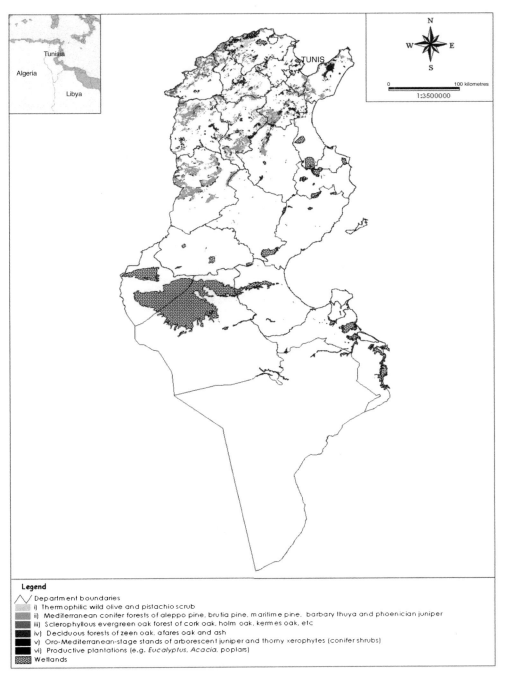

Fig. 7.1. Map of Tunisian forest typologies according to the classification of Quézel. Source: Direction Générale des Forêts (1995a). Adapted by G. Gader (Direction Générale des Forêts).

lack of means to carry out proper surveillance and repression of illegal acts perpetuated the open access to forest resources. This encouraged an excessive use and depletion of forests, especially through clearing for wood and forest conversion for agriculture. This, in turn, increased the conflict between forest users and administration, which has been likened recently by Ben Mansoura *et al.* (2001) to the 'tragedy of the commons' (Hardin, 1968).

Early awareness of the need to involve local populations in the preservation of forest resources led authorities to carry out an ambitious reforestation programme soon after independence. Plantations provided jobs for local inhabitants in remote forest areas, where the unemployment rate remains high. However, work opportunities and rural development actions aimed at the improvement of living conditions failed to curb forest degradation.[1]

As a result, many activities meeting local inhabitants' basic needs were classified as judicial offences and transgressions of the law (Hamzaoui, 1993). In 2000, the major types of forest abuses were illicit grazing (31% of the total), logging (22%) and illegal hunting (poaching) (11%), while the minor ones were clearing (5%), handling of forest products (4%) and ploughing (3%). Significant rising trends of all types of forest abuses were registered, in particular those related to poaching and forest clearing. The ratio of forest violations *per capita* is positively correlated with both rural poverty and extent of forest cover in the various districts of northern Tunisia. There is also a positive relationship between the frequency of occurrence of forest abuses and the extent of soil erosion in the region. Overgrazing provides the best illustration of this issue in northern Tunisia,[2] covering around one-quarter of the total land area, but supporting nearly half of the country's domestic livestock. This is due mainly to the region's more favourable climate compared with central and southern Tunisia where aridity prevails.

Typologies

In 1994, of the total forest cover of 843,000 ha, 72% was natural forest and 28% was plantation (Direction Générale des Forêts, 1995a). High

forests covered 503,000 ha, accounting for 60% of the total forest area. The remainder consisted of degraded forest (Direction Générale des Forêts, 1995a). These were mostly shrub lands where tree presence was either minor, for example *maquis*, or completely lacking, such as *garrigues*.

Tunisian forests are characterized by a high presence of coniferous species (53.7%) compared with pure broadleaved ones (20.3%). Table 7.1 shows that Aleppo pine is the most common dominant species, mainly concentrated in the arid and semi-arid areas. In these regions, other forest species are also encountered, such as stone pine (*Pinus pinea* L.), sandarac tree (*Tetraclinis articulata* Vahl.), eucalyptus (*Eucalyptus* spp.) and acacia (*Acacia* spp.). Cork oak (*Quercus suber* L.) is the most important broadleaved species and grows only in the humid and subhumid areas of northern Tunisia. On better soils, cork oak is replaced by other native oak species, such as pubescent oak (*Quercus faginea* Lam.), but they extend to less than 1% of all forests. Similarly, maritime pine (*Pinus pinaster* Sol.) is concentrated in higher fertility areas in north-western parts of Tunisia. Stands with a mixed presence of both conifers and broadleaved species are relatively rare in

Table 7.1. Distribution of the forest area according to species dominance (1994).

Dominant species	Area (000 ha)	Proportion (%)
Aleppo pine	296.6	35.1
Maritime pine	3.8	0.5
Thuya	21.8	2.5
Other coniferous species	35.7	4.2
Cork oak	45.4	5.3
Other oak species (such as Pubescent oak)	7.9	0.9
Eucalyptus	28.5	3.3
Acacia	12.6	1.4
Other broadleaved species	29.6	3.5
Mixed woodlands	20.9	2.4
Shrubs within forests	132.9	15.7
Scrubs without trees	194.8	23.1
Other (forest roads, clearing areas)	12.3	2.1
Total forest area	842.8	100.0

Source: Direction Générale des Forêts (1995a).

high forests, but more frequent in degraded woodlands.

Due to aridity, Tunisian forests are characterized by a low standing volume. Their average growing stock amounted to only 32.6 m³/ha in 1994 (Direction Générale des Forêts, 1995a). The mean values range between 25 m³/ha for most conifers and 90 m³/ha for broadleaved species and for maritime pine stands (Direction Générale des Forêts, 1995a). Greater variability is recorded in the mean annual increment, which ranges from 0.45 m³/ha/year for Aleppo pine stands to a peak of 3.1 m³/ha/year for pubescent oak on the most fertile sites (Direction Générale des Forêts, 1995a). Nationwide, the overall annual increment amounted to a low mean value of 0.72 m³/ha (Direction Générale des Forêts, 1995a). Present stand age structure shows that the majority of forest trees are either middle-aged (59%) or mature (30%). Forest stands where juvenile trees predominate constitute a minor proportion of 9%, whereas 2% of stands have an irregular age structure.

Functions

The most important function of Tunisian forests lies in the vital role they play in environmental protection and biological conservation. Due to the permanent vegetation cover provided by forests, soil erosion and dam sedimentation are prevented. Forests also reduce the risk of floods and provide habitat for many endangered animal and plant species. The forest role in maintaining biodiversity and species richness was strengthened in recent years with the expansion of natural reserves and parks, which today cover about 69,000 ha³ (Direction Générale des Forêts, 2002).

In national parks and reserves, the most prevalent uses of forest resources are strictly limited to ecotourism and visitor education. Also, crop cultivation for subsistence in cleared areas and extraction of a variety of edible food, medicinal and aromatic plants make a substantial contribution to fulfil the needs of local forest users. In contrast to their environmental and socio-economic importance, Tunisian forests play a minor role as a source of industrial wood supply, mainly supplemented by imports. Similarly, forests offer a minor contribution towards the total fuelwood consumption of the country, the balance being obtained from orchards.

Institutional Aspects

Ownership and size of properties

Nearly 84% of the Tunisian forests presently are classified as public property. Private properties account for 15% of the forest area and common properties for 1% (Table 7.2). About two-thirds of privately owned forests are marginal lands that were afforested with technical and material assistance from the forest administration in order to halt further soil erosion and protect dams from silting. Degradation affects only 18% of State forests, but as much as 64% of privately owned ones. This may be explained through the efficient role played by the administration in forest protection and the lack of awareness of conservation issues among private forest owners.

No precise data concerning the size of individual properties are available. However, it is

Table 7.2. Ownership and land tenure status of Tunisian forests (1994).

| Tenure status | Area according to forest type | | | | | |
| | High forests | | Degraded woodlands | | Total forest area | |
	000 ha	%	000 ha	%	000 ha	%
Public	581.0	91.3	124.0	59.9	705.0	83.6
Common	8.0	1.2	0.0	0.0	8.0	0.9
Private	47.0	7.5	83.0	40.1	130.0	15.5
Total	636.0	100.0	207.0	100.0	843.0	100.0

Source: Direction Générale de la Planification, du Développement et des Investissements Agricoles (1997).

believed that their average size is quite similar to that of agricultural properties, with most private owners holding less than 5 ha. Larger private holdings may range between 5 and 20 ha. The average size of common forests is far greater than that in private ownership.

From a legal point of view, the land tenure status of public forests is rather complex. Only 58% of all State-owned lands – forests and rangelands included – are actually titled properties with clear boundaries. For the remainder, around half are awaiting a ruling from real estate courts, and for the other half requests made for registration to the forest administration have been rejected due to opposing claims for private and tribal ownership rights (Direction Générale des Forêts, 2001b).

Regardless of land tenure status, regional administrations at district level are in charge of protecting forests from user damage. In addition to surveillance and policing, decentralized forest administrations are also responsible for the implementation of long-term forest management plans, usually prepared by private consultants. Forests are planned according to managerial units divided into 'series' and parcels. The average size of a 'series' is between 2000 and 6000 ha, while parcels usually cover between 50 and 120 ha.

Administration and policies

Since its creation about a century ago, the Tunisian Forest Service has undergone a gradual structural expansion and has evolved into a General Directorate comprising four major departments. This evolution was dictated largely by the substantial expansion of forested and afforested areas between 1956 and 1995. The first and largest department – or directorate – deals with Sylvo-Pastoral Development; the second is devoted to Forest Conservation; the third is concerned with Control and Regulations, among which enforcement and implementation of the Forest Code laws are included; the fourth, introduced by a decree in 2001, is devoted to the Socio-economic Development of Forest Populations. Its creation came about as a result of the amendment of the Forest Code in 1988, particularly Art. 43 and

44, which stressed the necessity to improve the living conditions of local forest populations (Daly-Hassen and Gader, 2001).

Tunisian forest policy is based on the 1959 Forest Code. The protective role of the forest was underlined, banning clearing and cultivation, and reforestation programmes were developed for job creation. Further amendments of the Forest Code were made in 1988 aiming at the participation of local populations in the management of forest resources. According to the Tunisian Forest Code, the management and use of private forests must abide by all forest laws and regulations. The orientation towards greater involvement of forest users in forest management was promoted in 1996 and 2001 by the legal creation of 'Common Interest Forest Groups', members of which are consulted and directly involved as full partners in all forest development plans and operations (Direction Générale des Forêts, 2001a).

The Tunisian Forest Service is also responsible for managing and exploiting nearly 350,000 ha of esparto grasslands (*Stipa tenacissima* L.) and other types of rangelands that belong to the State (Daly-Hassen and Gader, 2001). Foresters also exercise control over about 1.5 Mha of collectively owned esparto grass prairies and grazing lands which are mostly located in central and southern Tunisia. Esparto grasslands are especially important for the country's paper industry, while rangelands offer extensive grazing opportunities for ruminants raised by local populations throughout the country (Ben Mansoura and Garchi, 2001).

Contribution of the Forest to the National Economy

Gross domestic product and employment

Tunisian forests have a negligible economic weight, accounting for only 0.06% of the gross domestic product (GDP) in 1998. Timber-based industries, largely dependent on imported wood, represent around 1.1% of the GDP. The Tunisian forestry sector plays a greater role in employment, with 2.5% of the total labour force. The 1996 payrolls of the forest administration show that 62,000 employees obtained wages from various remunerated forest jobs, with an average work period of 112 days/year. Thus, forest employment contributed to the income of nearly 6.2% of all forest inhabitants. Timber-based industries employed 2.2% of the national active population, and other forest-related activities another 1%.

International trade

Tunisian forests and farmlands cover all the country's domestic needs for fuelwood and charcoal consumption which, together, amounted to approximately 3.7 Mm³ in 1997 (Direction Générale des Forêts, 1998a). In contrast, there was a heavy reliance on imports in order to satisfy the country's consumption of sawnwood, paper pulp, paper and paperboard (Daly-Hassen, 1998) (Table 7.3). Lower rates of dependence on imported wood were recorded for industrial roundwood and wood-based panels. Overall, the rate of dependence on imports was remarkable, with 91.3% in 2001. Exports were generated mostly by paper and high quality pulp manufactured from esparto grass.

Other forest-based industries: tourism and the green belts

Presently, tourism is a major source of foreign currency and represents 6% of the country's GDP. Ecotourism is an emerging economic activity, and the foundations for its further development have already been laid with the creation of seven national (forest) parks with a surface area of 47,000 ha, and 26 recreative forests (Direction Générale des Forêts, 2002).

The fenced parks of Boukornine, Châambi, El-Feija, Ichkeul and Nahli provide spectacular examples of well-managed wildlife habitats offering a variety of natural resources and sightseeing attractions to potential visitors. In addition to their infrastructure and resource diversity, national parks are also equipped with eco-museums providing learning opportunities. Some of these parks are located close to major cities where they offer a valuable recreational space needed by urban dwellers, such as the Nahli park in the outskirts of Tunis, and Boukornine in the vicinity of Hammam-Lif. Others require some travel, as they are located in remote but accessible mountain chains, such as the El-Feija, Châambi and Ichkeul parks. Parks shelter a variety of wildlife species, including many which are endangered such as deer, buffaloes and migrant birds.

Tunisia has been aiming to achieve a minimum ratio of 10 m² *per capita* of green space for all its citizens, particularly urban dwellers. This objective, which should be attained by the year 2005, requires a substantial development of urban forestry. Urban green belts, in which the forest administration is fully involved, along with the active participation of other local partners, represent a complement of the North African Countries Green Belt Project which was initiated a few decades ago. This project was co-founded

Table 7.3. Apparent consumption of timber and reliance on imports (2001).

Production and trade	Timber category (000 m³ equivalent roundwood)						
	Roundwood	Sawnwood	Particle board fibre board	Veneer sheets plywood	Pulp (wood, other fibre)[a]	Paper and paperboard	Timber
Production	92.9	40.2	78.3	12.8	54.4	309.4	147.3[b]
Import	38.4	806.6	41.3	34.7	300.4	523.1	1744.5
Export	0.0	1.5	13.2	12.1	45.4	120.0	192.2
Apparent consumption	131.3	845.3	106.4	35.4	309.4	712.5	1699.6
Reliance on net imports (%)	29.2	95.2	26.4	63.8	82.4	56.6	91.3

[a]The origin of pulp production is esparto grasslands, not forests.
[b]The total production of timber corresponds to the sum of roundwood and other fibre pulp production.
Sources: Food and Agriculture Organization (2002) for pulp production; Régie d'Exploitation Forestière (2002) for production of other timber categories, Institut National de la Statistique (2002) for trade.

by five Arab countries: Algeria, Egypt, Libya, Morocco and Tunisia, under the auspices of the League of Arab States, in a common effort to combat desertification.

The Values of Tunisian Forests

Table 7.4 presents the estimates of Tunisian forest values, grouped into direct and indirect use values, option, bequest and existence values and complemented by estimates of some negative externalities linked to forests. The results of the monetary valuations are shown in 1998 US$ prices (third column) and updated to

2001 € prices (fourth column). Valuation efforts are based on a diversity of methods, such as market price, substitute good price, travel cost method (TCM) for direct use values, shadow price and cost avoided method for indirect use values, and cost-based approaches for bequest–existence values and other externalities. Application of these methods is described in detail for each value category in the subsequent sections.

Direct use values

The direct use values generated by Tunisian forests are represented by timber, firewood,

Table 7.4.   Values of Tunisian forests.

Valuation method/output	Quantity	Value (000 US$)	Value (000 €) 2001 prices[a]
Direct use values			
Market price valuation			
Timber[b] (m³)	146,700	2,071	2,009
Sawlogs	1,300	42	41
Pulpwood	48,400	617	598
Other industrial roundwood	8,200	191	185
Wood sold at the stumpage price[c]	88,800	1,221	1,184
Firewood for sale[b] (m³)	75,700	508	493
Cork[b] (t)	11,618	9,297	9,018
Reproductive cork	8,330	8,719	8,457
Virgin cork	93	17	16
Miscellaneous cork	3,195	561	544
Honey[d] (t)	200	1,759	1,706
Substitute goods pricing			
Grazing (million FU)	481	71,917	69,759
Acorns of cork oak[e] (million FU)	34	5,098	4,945
Firewood for self-consumption (m³)	546,200	1,639	1,590
Cost of harvest and harvest fees			
Rosemary (t)	20,400	704	683
Myrtle[f] (t)	1,900	86	83
Aleppo pine acorns[g] (t)	44,500	3,915	3,798
Stone pine acorns[h] (t)	1,500	132	128
Mushrooms[i] (t)	16	21	20
Capers[j] (t)	151	398	386
Carobs[k] (t)	63	10	10
Snails[l] (t)	113	444	431
Miscellaneous[m]	—	311	302
Permit price			
Hunting[n] (no. hunters)	13,200	2,002	1,942
Value of licences and taxes	—	935	907
Game value	—	719	697
Wages paid for herding out wild boars	—	349	339
TCM			
Recreation (no. visits)	93,400	532	516
Total direct use values		100,845	97,819

Table 7.4. *Continued.*

Valuation method/output	Quantity	Value (000 US$)	Value (000 €) 2001 prices[a]
Indirect use values			
Cost avoided method			
Watershed protection:		25,623	24,854
Reduced dam sedimentation (m³/ha)	1.9–16.2	16,223	15,736
Agricultural soil conservation (Mha)	1	9,400	9,118
Shadow pricing			
Carbon sequestration (tC)	220,890	4,418	4,285
Total indirect use values		30,041	29,140
Bequest and existence values			
Cost-based method			
Biodiversity in parks and reserves (ha)	69,000	2,768	2,865
Forest conservation (ha)	942,800	3,662	3,552
Total bequest and existence values		6,430	6,237
Negative externalities			
Cost-based method			
Losses due to forest fires (ha)	2,215	−3,393	−3,290
Wood, other NWFPs and services	—	−974	−944
Fire prevention and fighting	—	−2,419	−2,346
Damage caused by illegal acts (no.)	3,900	−243	−236
Total negative externalities		−3,636	−3,526
TEV		133,680	129,670

[a]Adjustment to 2001 prices based on US$ inflation rate between 1998 and 2001 (1.086) and exchange rate in 2001 (€1 = US$0.9).
[b]Source: Régie d'Exploitation Forestière (1999).
[c]Roadside price includes stumpage price (US$10.2/m³) and exploitation costs (US$3.5/m³).
[d]Source: Direction Générale des Forêts (1999).
[e]The area of cork oak covers 45,400 ha (Direction Générale des Forêts, 1995a). Acorn production varies greatly between 250 and 2500 kg/ha, depending on changing climatic factors and fruiting cycles. For 1998, harvested acorns were supposed to attain 30% of the maximum annual yield.
[f]Based on the quantity of essential oils of myrtle (3 t) (Institut National de la Statistique,1999) and the extraction rate of 1.56 kg of oils/t of plants. Harvest costs are US$ 24.6/t of plants (Daly-Hassen *et al*., 2003). Harvest fees equal US$38,500 (Régie d'Exploitation Forestière, 1999).
[g]Based on an area of 296,600 ha with a cone production of 150 kg/ha and a harvesting cost of US$8.8/100 kg. Sghaier *et al*. (1997) reported a greater level of grain production topping 350 kg/ha in the Ouergha forest.
[h]Based on an area of 15,000 ha (Direction Générale des Forêts, 1999), a cone production of 100 kg/ha and a harvesting cost of US$8.8/100 kg.
[i]Based on exported quantity and the average harvesting cost which ranges between US$0.87 and US$1.75/kg.
[j]The area opened for harvesting capers is 6350 ha (1995–1998), with an average production rate of 151 t/year (Ministère de l'Environnement et de l'Aménagement du Territoire, 1997) and a harvesting cost of US$2.64/kg.
[k]Based on exported quantity and the average harvesting cost of US$0.15/kg.
[l]Snails are not only collected in forests (943,000 ha), but also in pasturelands (4,680,000 ha), fallow lands (972,000 ha) and arboricultural lands (1,983,000 ha) (Direction Générale de la Planification, du Développement et des Investissements Agricoles, 1997). Valuation was based on the share of forest area in total productive surface. The harvest cost is estimated at US$3.87/kg (Direction Générale des Forêts, 1999).
[m]Corresponds to the harvest fees and the cost of harvest of mushrooms, capers, acorns and other forest products (Régie d'Exploitation Forestière, 1999).
[n]Value of licenses and taxes include: local licence fees (US$310,000), foreign licence fees (US$530,500), hunting concession tax (US$14,100) and game slaughter tax (US$80,400) (Régie d'Exploitation Forestière, 1999). Game value comprises the value of small game (US$561,740) and large game (US$157,160) (Direction Générale des Forêts, 1999). The premium is given to local populations for herding out wild boars (Direction Générale des Forêts, 1999).

grazing, cork and other non-wood forest products (NWFPs), hunting and recreation (Table 7.4). Many of these products are traded on the market and for some, i.e. timber, firewood for sale, cork and honey, valuation is based on the amounts sold on the market and their average price. However, for timber, which is sold by the State at both roadside prices (sawlogs, pulpwood and other industrial roundwood) and stumpage price (other wood), the roadside prices are applied.

Due to the non-availability of market prices, valuation of some outputs is calculated by using the market value of equivalent products. For example, grazing and acorns from cork oak used as fodder are valued on the basis of the forage units (FU) consumed and the market price of barley, of which grazing deserves particular attention. The imputed prices are used, due to the fact that grazing usage rights, granted by the Forest Code to local users, are not sold as such. In 1994, forest cover produced a total forage quantity of 430 million FU (Direction Générale des Forêts, 1995b), corresponding to an average annual productivity of 510 FU/ha. This same productivity is used with the 1998 forest cover to arrive at an estimate[4] of 481 million FU. The market price of 1 kg of barley grain is used, the energy content of which is equivalent to 1 FU (Kayouli and Buldgen, 2001).

The value of firewood harvested for self-consumption, free of charge in compliance with current forest legislation, is based on the harvested quantity and the administration price tags. In local markets, the price of firewood varies between US$2.5 and US$7.5/m³, depending on diameter (Direction Générale des Forêts, 1998a). Due to the prevalence of small categories, an average unit price of only US$3/m³ is used.

For goods for which market prices and prices of substitute products are not available, the costs of harvest and harvest fees[5] are used. For example, the estimate of rosemary plants, before transformation into essential oils, is based on the quantity of extracted oils of 76.4 t and the extraction rate of 3.75 kg oils/t plants (Institut National de la Statistique, 1999). Economic valuation considers both harvest costs of US$26.4/t plants (Daly-Hassen *et al.*, 2003) and harvest fees paid by the enterprises of essential oils of US$165,900 (Régie d'Exploitation Forestière, 1999). The value of myrtle is assessed in a similar

way. Other estimates are based on either harvest costs (Aleppo pine acorns, Stone pine acorns and mushrooms) or harvest fees (miscellaneous) and details are reported in the footnotes of Table 7.4. Of course, these are partial values that underestimate the true value of these forest products.

The economic value of hunting was assessed by summing the value of licences and taxes paid to forest administration by foreign tourists (US$530,500) and local hunters (US$310,000), the value of captured wild game (US$719,000) and the wages paid to the local population for herding out wild boars (US$348,000). Hunting in the forest area is regulated by the Forest Code, which specifies the period, the organization of domestic and foreign hunting and the reserves where hunting is prohibited. The best-known areas for hunting are Ain Draham for wild boars and Cap Bon for birds. All hunters need a permit for carrying firearms. About 12,000 Tunisian hunters are registered with local hunting organizations to which they pay registration and insurance fees of about US$8/year. About 6000 licences for hunting in the forest area are issued annually to Tunisians at US$7/year. Hunters also pay a game tax (e.g. US$17.5/bear in 1998) for animals hunted in the forest area. About 1200 hunting licences are issued to foreigners annually, at prices that vary according to the type of animal hunted; about US$88 for boars and US$880 for thrushes and starlings. Foreign hunters also pay a game tax equivalent to US$88/wild boar.

The recreation value of visiting forests and parks is based on estimates derived from a TCM investigation of visitors to the national park of Ichkeul. It estimated the consumer surplus as US$211,100 of public recreation in 1994 (Aouididi, 1996). Dividing by the number of visits, the visit value was $4.52 in the same year. Assuming a similar consumer surplus per visit to the country's other parks, such as Ichkeul, and adjusting for inflation, the 93,400 visits in 1998 (Direction Générale des Forêts, 2002) generated a benefit of US$532,200. In addition, tourist hunters generated revenues of US$697,900 to the tourism industry. Part of this is net profit and could be used to supplement the estimated benefits from forest recreation if the profit margin was known (Direction Générale des Forêts, 1999, adjusted for inflation).

Overall, the direct use values of the different categories of forest outputs add up to an annual sum of US$100.8 million. Grazing is the major economic activity in most forest areas (Ben Mansoura *et al.*, 2001) and accounts for 71% of the total. Cork constitutes the second major output with 9%, although cork oak trees occupy only 5.5% of the total forest area (Direction Générale des Forêts, 1995a). In contrast, wood generates a low contribution with only 4%, due to its low quality for industrial use.

Indirect use values

Notwithstanding the limited extent of Tunisian forests and their degradation, they provide substantial services in terms of soil erosion prevention, soil fixation and stability. In Tunisia, soil erosion is a major ecological threat affecting 56% of all the country's area, the highest proportion in North Africa (Marcoux, 1996). Concern for the spread of soil erosion processes is particularly high in the northern part of the country, where nearly 1.5 Mha show more or less advanced signs of soil degradation (Direction de la Conservation des Eaux et des Sols, 1996) and where the most important water storage reservoirs are present.[6] Forests in northern Tunisia are therefore expected to provide a significant contribution to the protection of the region's dams and of the surrounding agricultural lands, which constitute the main source of food for the country's population. However, the existing literature offers no estimates of this contribution in monetary value terms (Direction Générale des Forêts, 1994).

The estimate for the economic value of watershed protection is based on the avoided losses to water storage capacity and agricultural yield in the areas protected by the presence of forest cover.

Reduced sedimentation

Forest cover contributes to reducing erosion in watersheds, thereby decreasing sedimentation in reservoirs that are mainly used for irrigation. This benefit was valued using the cost avoided technique, assuming the cost of replacing water storage capacity, which would be lost in the absence of forest cover. In Tunisia, there are 22 dam reservoirs located in watersheds with erosion rates varying between 1.9 m^3/ha/year (e.g. in Bezirk and Cap-Bon) and 16.2 m^3/ha/year (e.g. in Joumine, North) (Direction Générale des Etudes des Travaux Hydrauliques, 1998). The lowest observed erosion rate of 1.9 m^3/ha occurs in a watershed with high levels of forest cover. Conversely, the highest observed erosion rate of 16.2 m^3/ha is found in a watershed with poor vegetation cover, where land uses such as agriculture and pasture prevail. Of course, erosion rates are affected by various other factors in addition to forest cover, including rainfall patterns, soil pedology and slopes. However, in the absence of better information, it was generally assumed that forests reduce erosion, and consequently dam sedimentation, from the higher to the lower bound, i.e. by 14.3 m^3/ha/year. The average cost of constructing water storage facilities between 1990 and 2001 was US$1.2/$m^3$ (World Bank, 1994). Applying this value to the avoided erosion in the total forest area of 942,800 ha results in an estimated benefit of US$16.2 million.

Agricultural soil conservation

In addition to purifying water, forests also prevent erosion damage in nearby cultivated lands of around 1 Mha. The value of this damage is estimated at between US$4.7 and US$14.2/ha of cultivated land[7] (World Bank, 2003). Applying a mean value (US$9.4/ha), the avoided economic loss amounts to US$9.4 million.

Based on these estimates, the economic value of watershed protection amounts to about US$25.6 million. However, it should be remembered that there is considerable variability of the protective value, depending on the type of forests, levels of degradation and other factors.

Valuation of carbon sequestration is based on the methodology proposed by IPCC (Intergovernmental Panel on Climate Change) and OECD (Organization for Economic Cooperation and Development) for valuing the emissions and absorptions of greenhouse gases. Accordingly, in 1994, forests sequestered 811,600 tCO_2 (Ministère de l'Environnement et de l'Aménagement du Territoire, 2001). Update and conversion[8] into carbon quantity yields an absorption of 220,890 tC in 1998. Valued at a price of

US$20/tC (Fankhauser, 1995), the economic value of carbon would be US$4.4 million. The economic value of carbon sequestration is rather low compared with that of water and soil conservation due to low levels of forest productivity. Other woody vegetation ecosystems, such as the country's orchards (2 Mha), may lead to a carbon fixation quantity six times greater than that of forests.

Option, bequest and existence values

There is a total lack of statistics and only a small amount of literature dealing with option, bequest and existence values of Tunisian forests. Nevertheless, some option values may very well be linked to recreation in parks and forests, as well as the extraction of medicinal and aromatic plants as personal future benefits. It is likely that greater economic development and improved life conditions, over time, will increase the social demand for environmental conservation and outdoor leisure activities.

In addition, several plant species are gaining increasing interest for their essential oils and secondary substances that are useful for processing perfumes and medicinal drugs with high potential for export. This is counterbalanced by an increasing disaffection with traditional medicine due to the dissemination of modern health centres among both urban and rural inhabitants, including those living in remote areas. Trained doctors and nurses supplying basic health care products and family planning means, at affordable cost or free of charge for the poor, helped to improve the living conditions for all Tunisians and contributed to reduced reliance on traditional herbal medicine.

Regarding the existence and bequest values, no attempts have been so far made for their appraisal using the contingent valuation method (CVM), although the national parks offer pertinent examples for such analyses. For instance, investigations of these values in parks sheltering endangered fauna, such as buffaloes in Ichkeul park and deer in El-Feija park, may be of interest to both present and future generations. Presently, the costs of conservation and management may represent a good indicator – certainly a minimum estimate – of existence and bequest values. For all Tunisian parks and reserves, these costs amounted to an annual average of US$2.768 million during 1997–2001 (Direction Générale des Forêts, 2001b), reflecting the importance of forest contribution to the preservation of biodiversity in the country. However, species richness, for both fauna and flora, cannot be fully considered without including the costs incurred for ensuring the proper surveillance and enforcement means, which in 1998 amounted to US$3.662 million. Theoretically, the value of all expected benefits from forest resources is supposed to be far greater than the sum of all expenses in forest conservation.

Negative externalities linked to forests

Illegal acts and forest fires are the major causes of negative externalities affecting Tunisian forests. For example, deforestation causes a yearly loss of about 300 ha of forest and woody vegetation stands (Direction Générale des Forêts, 2001b). The economic value of the losses due to illegal activities of stocking, logging, clearing and poaching is based on the monetary value of the fines paid for illegal acts committed against forests.[9] Obviously, any damage appraisal in this manner bears an inherent underestimation because: (i) many other indirect benefits are lost and overlooked in every transgression case; (ii) the fine itself is a nominal amount which might only partially reflect the full value of forest loss; and (iii) it is very likely that many illegal forest users are not caught and, therefore, do not pay the fines. In 1998, there were around 3900 transgressions of forest law for which the total value of fines paid amounted to US$161,200. In addition, there were 250 offences transmitted to courts for judgement, with damage estimated at US$82,000. It is assumed that the damage caused is more important than fines levied, therefore its unit value was assumed as twice that of the mean of fines paid for all forest offences.

Damage due to forest fires is valued on the basis of losses of products (wood, grazing, cork and other NWFPs) and services (watershed management and carbon sequestration) with the addition of forest administration expenses for fire prevention and control. The cost of reforestation of burnt areas, which is usually included in similar analyses, was overlooked because incinerated areas of conifers are often left to natural regeneration. Advocates of natural ecosystem conservation prefer to rely on natural recovery rather than turning to reforestation that requires the clearing of native vegetation for successful establishment of newly planted seedlings. The value of burnt wood and other NWFP losses is calculated in relation to the annual burnt area (2215 ha) (Ministère de l'Agriculture, de l'Environnement et des Ressources Hydrauliques–Observatoire du Sahel et du Sahara, 2002), mean growing stock of wood (35 m^3/ha), a price of US\$8.8/m^3 and a mean value for other products and services (US\$131.7/ha), leading to US\$0.97 million. The State expenses for fire prevention and firefighting amount to US\$2.4 million (Direction Générale des Forêts, 1998b).

The value of damage due to overgrazing was not estimated due to complete lack of quantitative data dealing with its specific impacts on Tunisian forests. Overgrazing, as a major cause of soil erosion in Tunisia (Marcoux, 1996), usually results in reduced forage production levels and induces undesirable shifts in the botanical composition of the under-storey vegetation. It is well known that overstocking causes the invasion of unpalatable woody plant species and reduces the presence of herbaceous ones, which are more valuable in terms of forage quality and animal productivity. Overgrazing can lead to a loss of biodiversity as illustrated by the total extinction of the annual *Atriplex* species due to their higher palatability compared with perennial and woody species of the same genus. Overgrazing and other human activities cause problems of regeneration in the ageing cork oak forests, the area of which has diminished by about 1000 ha/year during recent decades (Souayah *et al.*, 2001). Loss of biodiversity in both the flora and fauna, and reduced species richness are still difficult to quantify in terms of their monetary value.

Towards the Total Economic Value of Tunisian Forests

The attempt to produce a comprehensive assessment of the most important economic

values shows that forest production in Tunisia generated a total monetary value of US\$133.7 million in 1998. Figure 7.2 presents the estimated TEV, of which direct use values form the biggest contributor. Comparison among the different TEV components is, however, difficult, as the magnitude of the values of the other categories is underestimated, due to the scarce information available for their valuation. For example, the estimates of option, bequest and existence values tend to underestimate the true value of this category due to the lack of relevant studies. Also valuation of negative externalities does not refer to all aspects linked to forest depletion in Tunisia, such as damage due to overgrazing.

Bearing in mind these limitations, the research sheds light on the most significant forest benefits in Tunisia. Livestock grazing represents more than 54% of the TEV (Fig. 7.3), the benefits of which were collected by local forest inhabitants in accordance with their usage rights. The second major forest output was soil and water conservation (19.4%), followed by cork (7%) and other NWFPs (10%), biodiversity conservation function (4.9%), carbon sequestration (3.3%), wood (3.2%) and others.

These results reflect that Tunisian forests are essentially devoted to provide shelter and to meet the basic needs of Tunisian local communities, in addition to their environmental protection role. This is illustrated by the fact that nearly 65% of the estimated total economic benefits of forests were channelled into local populations in various forms of income and sources for livelihoods (Table 7.5). Overall, each household benefited by an average sum of US\$483 in 1998 (see Box 7.2 for results of a case study investigation). Grazing and other NWFPs

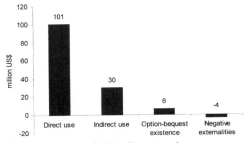

Fig. 7.2. Estimated TEV of Tunisian forests.

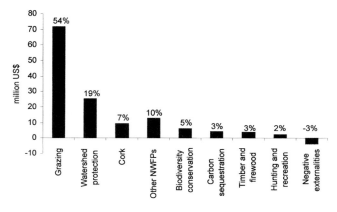

Fig. 7.3. Main components of the TEV of Tunisian forests.

Table 7.5. Components of the estimated TEV of forest and recipients (1998) (US$000).

| Recipients | Direct use values | | | Indirect use values | Non-use (conservation) values | Total | % |
	Grazing, cork and other NWFPs	Wood	Hunting and recreation				
State[a]	9,109	769	935			10,813	8
Local communities[b]	84,983	3,449	1,067			89,500	65
General public[c]			532	30,041	6,430	37,003	27
Total	94,093	4,218	2,534	30,041	6,430	137,316	100
%	69	3	2	21	5	100	
Social costs[d]					3,636	3,636	
Total economic value						133,680	

[a]The value of forest products sold by the State, plus the value of licences, taxes and harvest fees, minus the cost of harvest.
[b]The value of forest products sold or consumed by local users, plus the cost of harvest, plus the premium for herding out wild boars to hunters and the game value.
[c]The values of recreation, watershed protection, carbon storage and biodiversity conservation.
[d]The values of losses due to forest fires and illegal acts.

constitute the two main components with US$387 and US$67, respectively. Nevertheless, household income remains quite variable, as does the share of forest resources in the annual revenue.

Table 7.6 shows the importance of forage, watershed protection, cork and other NWFPs in the estimated production value of 1 ha of forest. Hence, cost–benefit analysis (CBA) should not be limited solely to the value of timber production, as has been dealt with so far. The alternative of extended economic analysis, that takes into account all other benefits that can be estimated, offers a better economic evaluation of forest investments. Table 7.6 suggests that plantations with the aim of soil conservation as well as grazing areas would be highly

valuable and might justify investments in these areas.

Perspectives and Crucial Issues

Forest investments constitute 0.9% of all capital assets channelled into the country's efforts for development in the various economic and social sectors. The slight difference between the forest sector contribution to the GDP (0.1%) and to the total investments may be explained by the fact that Tunisian forests are orientated mainly towards environmental conservation rather than production. Forests and degraded woodlands also play a prominent

Box 7.2. Forest contribution to seasonal employment and income of local inhabitants.

In remote and less favoured rural areas, forests bring a significant contribution to the seasonal employment of local inhabitants and ensure a substantial part of their revenues through numerous work opportunities in tree nurseries, plantations, logging and cork harvesting. Field investigations show that most forest workers receive low wages, have little education and are mostly elderly people seeking part-time employment in forestry as a secondary activity for additional income.

In the area of Skhira, Sejnane, the contribution of the employment in the forestry sector amounted to 27% of all household earnings, attaining US$1400/household or US$261/*capita* in 1999. Raising domestic livestock created 52% of the families' annual revenue, while honey bees, tobacco cropping, charcoal making and handcrafting accounted for 7, 6 and 8% of the income, respectively.

Studies also indicate diffuse hardship and poverty in remote forest areas. For instance, there are 41,000 inhabitants living in scattered and isolated groups of households under predominantly rural settings covering 89% of the area of Sejnane county where forest covers 53% of the total area. In other areas – Zilia and Jbel Essmâa – about 58% of local households benefited from forest employment. Households were categorized further into non-farm holders (28%), small-size farm holders (31%) and medium-size farm owners (41%). The first two categories were more affected by economic hardship as they relied primarily upon forest products for subsistence, whether these products were obtained legally or illegally. They were also affected by rural outmigration for better living conditions. In contrast, farmers of the third category showed a greater diversity in their sources of income and utilized forest resources more correcty. In addition, they demonstrated greater capability and willingness to change, as they could be more easily integrated into increasingly modern lifestyles.

In conclusion, the improvement of living conditions in remote forest areas remains largely linked to greater agro-silvo-pastoral development and urbanization efforts in order to diversify people's economic activities. The participation of local inhabitants, which is limited currently to seasonal employment and livestock grazing, must be widened to include other forest activities. Common Interest Forest Associations may support the forest administration in logging operations and trade of forest products, without jeopardizing their sustainability. Planned grazing schemes, promotion of handcrafting and greater forest product transformation could enhance the revenue of local inhabitants further and improve their living conditions.

Table 7.6. Summary of the estimated production values/ha of forest (1998).

	Grazing	Watershed protection	Cork, other NWFPs	Biodiversity conservation	Carbon sequestration	Wood	Hunting, recreation	Negative externalities	TEV
Value (US$/ha)	76.3	27.2	23.5	6.8	4.7	4.5	2.7	−3.9	141.8

socio-economic role in offering shelter for local inhabitants and in providing them with grazing and seasonal working opportunities. Despite their limited area, forests continue to ensure the employment of 2.5% of the country's total active population in a wide array of seasonal jobs. They also offer grazing and fodder opportunities for nearly half of Tunisia's domestic livestock resources.

Indirectly, forests contribute to the productivity of the surrounding agricultural lands by filtering runoff waters and preventing soil erosion, salt deposits and sedimentation. In various regions of the country, forests promote a wide array of industrial activities, such as

cork processing, wood transformation, extraction of essential oils from aromatic plants, and handcrafting activities resulting from the production and processing of wool, leather and mohair. Multiple use of forest resources may have greater perspectives for expansion in the future, particularly after the country's adherence to free exchange and open market policies. Increased output of manufactured goods using NWFPs such as perfumes, fragrances, drugs and essential oils might be expected and may result in enhancing the value of the aromatic and medicinal plants. However, fierce international competition and reduced size of local enterprises can lead to a further underutilization of Tunisian timber and other raw materials.

Continued urbanization accompanied by greater job opportunities in a diversified economy are likely to reduce the anthropogenic pressure to which Tunisian forests have been exposed. Widespread overgrazing and local users' free access to forest resources constitute a serious threat to the sustainability of the Tunisian forest cover. However, with improved living conditions and enhanced income, local populations are expected to move to higher levels of hierarchical needs. It is only then that awareness about forest conservation, biodiversity and recreation would overtake users' short-term self-interest.

Concerns about the necessity of forest conservation and the aim of achieving greater forest coverage in both rural and urban areas are illustrated by the importance of the investments in the forest sector. The areas of fenced parks and reserves where endangered species find a natural refuge in unique and diverse habitats are increasing in different regions of the country. Financial incentives, such as 50% subsidy of plantation costs, are offered to private owners to encourage plantations, especially of species with potential for fodder production, in order to reduce the chronic summer forage deficits. Nevertheless, changes in the forest legislation are required in order to improve the flexibility in the management of forests and afforested areas, and the trade and handling of forest products. These changes must include the local inhabitants whose active participation is essential for improving the sustainable development in forest areas.

Notes

[1]   Between 1992 and 1996, a nationwide annual average of 5900 illegal actions against forests was reported. Only 18% of all forest offences were transmitted to local courts for penal judgement. The remaining cases were settled out of court after the payment of moderate fines amounting to an average fee of US$26/illegal action (Direction Générale des Forêts, 2001b).

[2]   The greatest extent of soil erosion in the country occurred in the districts with the highest ratios of standard livestock unit (SLU) per capita: Siliana (64% of total area) with 0.7 livestock/capita and le Kef

(59%) with 0.6 SLU/capita (Direction de la Conservation des Eaux et des Sols, 1996; Ben Mansoura and Garchi, 2001).

[3]   Not including the Djebil desert reserve, situated in the southern district of Kebili and covering 150,000 ha.

[4]   In fact, forage production is variable due to erratic and unpredictable rainfall patterns and depends on the geographic location and its aridity. Most often, forage yields vary between 600 FU/ha in the north to 50 FU/ha in the south (Ben Mansoura et al., 2001). Because most of the Tunisian forest is located in the north (Fig. 7.1), this geographic region constitutes the largest contributor to forest grazing and fodder production in woodlands.

[5]   The harvest fees are paid to the administration for collecting forest products.

[6]   Silting control operations for 13 dams carried out in 1975 and 1991 showed an average loss rate of 1.1% from their total initial holding capacity. Control operations were undertaken under the responsibility of the Direction Générale des Etudes des Travaux Hydrauliques, Ministry of Agriculture. Soil loss varied between 2.4 and 14.2 m^3/ha/year (Direction Générale des Etudes des Travaux Hydrauliques, 1998).

[7]   In Tunisia, soil erosion and desertification cause the loss of agricultural land of an area estimated between 10,000 and 30,000 ha/year. The damage cost is valued between US$26.6 million and US$80 million in 1999. This amount is higher than the annual replacement cost for erosion prevention of US$20.1 million (World Bank, 2003).

[8]   The update considers an annual increment of 0.2% in the woody biomass, hence in carbon fixation. In addition, the carbon atomic mass corresponds to 0.27/molecular mass of CO_2.

[9]   It should be recognized that illegal actions provide both private benefits to local users and social costs of environmental degradation. Private benefits have been only partially valued, for example related to overgrazing and included within the 'grazing' valuation in the direct use values. For valuing the social costs, only the amount of fines paid to the State by the illegal users is available and considered a rough proxy for valuation.

References

Aouididi, A. (1996) Evaluation Économique du Parc National de l'Ichkeul. Ministère de l'Environnement et de l'Aménagement du Territoire, Tunis.

Ben Mansoura, A. and Garchi, S. (2001) Intégration de l'Élevage en Fonction du Type de Couvert ou d'Utilisation des Terres et de l'Aridité en Tunisie.

Projet   PNM/SERST96/BIRD07,   INRGREF, Tunis.

Ben Mansoura, A., Garchi, S. and Daly, H. (2001) Analyzing forest users' destructive behavior in Northern Tunisia. *Land Use Policy* 18, 153–163.

Boudy, P. (1948) *Economie Forestière Nord-Africaine. Tome Premier: Milieu Physique et Milieu Humain.* Editions Larose, Paris.

Daly-Hassen, H. (1998) Les perspectives de l'offre et de la demande de bois rond industriel en Tunisie à l'horizon 2015. *Annales de l'INRAT, Tunis* 71, 275–296.

Daly-Hassen, H. and Gader, G. (2001) *Rapport Annuel de Prospective du Secteur Forestier en Tunisie.* INRGREF/DGF, Ministère de l'Agriculture, Tunis.

Daly-Hassen, H., Ben Mansoura, A. and Mhadhbi, F. (2003) *Analyse de la Filière des Plantes Aromatiques et Médicinales en Tunisie.* Mémoire présenté au XIIème Congrès Forestier Mondial, Québec, Canada, 21–28 Septembre 2003.

Direction de la Conservation des Eaux et des Sols (1996) *National Strategy Plan for Soil and Water Conservation.* Ministry of Agriculture, Tunis [in Arabic].

Direction Générale des Etudes des Travaux Hydrauliques (1998) *Envasement des Barrages en Tunisie.* Ministère de l'Agriculture, Tunis.

Direction Générale de la Planification, du Développement et des Investissements Agricoles (1997) *Etude sur la Stratégie des Ressources Naturelles.* SCET/BDPA-SCETAGRI, Ministère de l'Agriculture, Tunis, Partie 2, Vol. 1, pp. 91–116.

Direction Générale des Forêts (1992) *Plan National de Protection des Forêts Contre les Incendies.* Ministère de l'Agriculture, Tunis.

Direction Générale des Forêts (1994) *Second Forestry Development Project, Annex 9: Financial and Economic Analysis,* Ministry of Agriculture, Tunis.

Direction Générale des Forêts (1995a) *Résultats du Premier Inventaire Forestier National en Tunisie.* Ministère de l'Agriculture, Tunis.

Direction Générale des Forêts (1995b) *Résultats du Premier Inventaire Pastoral National en Tunisie.* Ministère de l'Agriculture, Tunis.

Direction Générale des Forêts (1998a) *Analyse du Bilan du Bois d'Énergie et Identification d'un Plan d'Action.* SCET-Tunisie/Scandia Consult Natura AB, Tunis.

Direction Générale des Forêts (1998b) *Rapport sur la Protection des Forêts Contre les Incendies.* Ministère de l'Agriculture, Tunis (in Arabic).

Direction Générale des Forêts (1999) *Recensement et Valorisation des Produits Forestiers Non Ligneux.* Jaakopoyry and Extra Consult/ Ministère de l'Agriculture, Tunis.

Direction Générale des Forêts (2001a) *A Report on the Planning of the Development of the Forestry Sector for 2002–2011.* Ministry of Agriculture, Tunis (in Arabic).

Direction Générale des Forêts (2001b) *Stratégie Nationale de Développement Forestier et Pastoral.* Ministère de l'Agriculture, Tunis.

Direction Générale des Forêts (2002) *Etude sur les Indicateurs de Gestion Durable des Forêts en Tunisie.* Ministère de l'Agriculture, Tunis.

Fankhauser, S. (1995) *Valuing Climate Change. The Economics of the Greenhouse.* Earthscan, London.

Food and Agriculture Organisation (2002) *Yearbook of Forest Products.* FAO, Rome.

Hamzaoui, S. (1993) Sociologie des délits en milieu forestier. In: *Problèmes Socio-économiques de la Forêt du Nord-ouest de la Tunisie: la Kroumirie. Cahiers CERES, Série Géographique* 8, 73–95.

Hardin, G. (1968) The tragedy of the commons. *Science* 162, 1243–1248.

Institut National de la Statistique (1994) *Premiers Résultats du Recensement Général de la Population et de l'Habitat.* INS, Ministère du Plan et du Développement Régional, Tunis.

Institut National de la Statistique (2002) *Annuaire du Commerce Extérieur de la Tunisie.* INS, Ministère du Plan et du Développement Régional, Tunis.

Kayouli, C. and Buldgen, A. (2001) *Elevage Durable dans les Petites Exploitations du Nord-ouest de la Tunisie.* Faculté Universitaire des Sciences Agronomiques de Gembloux (FUSAGx), Belgium.

Marcoux, A. (1996) *Population Change – Natural Resources – Environmental Linkages in the Arab States Region.* POPIN-DESIPA Report, Population Program Service, FAO, Rome.

Ministère de l'Agriculture, de l'Environnement et des Ressources Hydrauliques–Observatoire du Sahel et du Sahara (2002) *Etude sur les Indicateurs de Gestion Durable des Formations Forestières et Steppiques.* Projet de suivi-évaluation du PAN-LCD, Tunis.

Ministère de l'Environnement et de l'Aménagement du Territoire (1997) *Etude Nationale de la Diversité Biologique de la Tunisie,* Tome 4. MEAT, Tunis.

Ministère de l'Environnement et de l'Aménagement du Territoire (2001) *Communication Initiale de la Tunisie à la Convention Cadre des Nations Unies sur les Changements Climatiques.* MEAT, Tunis.

Régie d'Exploitation Forestière (2002) *Rapport Annuel d'Activités – Gestion 1998.* Ministère de l'Agriculture, Tunis.

République Tunisenne (1993) *Code Forestier.* Imprimerie officielle, Tunis.

Sghaier, T., Khaldi, A., Khouja, M.L. and Nsibi, R. (1997) Estimation du rendement en cônes et en graines du pin d'alep de la forêt de Ouergha (Sakiet Sidi Youssef-Tunisie). *Annales de Recherches Forestières du Maroc* 30, 84–89.

Snane, M.H. (1993) La dégradation de la forêt de la Kroumirie: causes et effets. In: *Problèmes Socio-économiques de la Forêt du Nord-ouest de la Tunisie: la Kroumirie. Cahiers CERES, Série Géographique* 8, 47–72.

Souayah, N., Khouja, M.L. and Nsibi, R. (2001) *Etude de la Variabilité Morphométrique Chez le Chêne Liège au Stade Juvénile.* Les Journées Scientifiques de l'INRGREF, 19–20 Décembre 2001, Nabeul, Tunisia.

World Bank (1994) *Analyse Économique des Projets Concernant les Ressources Naturelles en Tunisie.* ITALECO, Rome.

World Bank (2003) *Evaluation du Coût de la Dégradation de l'Environnement en Tunisie.* METAP/ANPE, Tunis.

8 Egypt

Lelia Croitoru

University of Padova, Centre for Environmental Accounting and Management in Agriculture and Forestry (CONTAGRAF), Via Roma 34, Corte Benedettina, 35020 Legnaro (PD), Italy

Introduction

Egypt is an arid country, covering the north-eastern corner of Africa and the Sinai Peninsula in Asia. It covers almost 1 Mkm^2, of which around 96% is desert. The other 4% is represented by the Nile valleys, the most inhabited and cultivated areas in the country. The main country regions are (United Nations Environment Programme/Mediterranean Action Plan/Plan Bleu, 1996):

- The Nile basin – including the valley in the south (Upper Egypt) and the Delta in the north (Lower Egypt), where farmlands are particularly concentrated. Agriculture depends largely on irrigation, for which the Nile meets around 84% of the sector needs.
- The Western Desert – essentially a flat plateau with various depressions, except for the mountain of Gebel Uweinat (1907 m) on the extreme west of the the border with Sudan. The area is characterized by ranges of parallel belts of sand dunes extending from north to south.
- The Eastern Desert – comprising a chain of rugged mountains running parallel to the Red Sea coast and dividing the coastal plain to the east and the main stretch of desert plateau to the west.
- The Sinai Peninsula – of which Mediterranean coastal belt includes sand areas, Lake Bardawil and salt marshes.

In contrast to the arid zones characterizing most of the country, the coastal land is considered Mediterranean, especially due to the influence of the sea. This explains the concentration of Mediterranean vegetation (taxa) along the Nile valleys and Delta. Actual woodlands practically do not exist, except as tree savannah in the steppe habitats and relic woodlands on mountain peaks (United Nations Environment Programme/Mediterranean Action Plan/Plan Bleu, 1996). Since most of the tree plantations are in the form of windbreaks and linear plantations, Egypt is one of the very few countries stating its forest resources as a number of individual trees: 60.5 million trees in 1992/1993, of which about 40.6 million trees are *Casuarina* sp. (Ministry of Agriculture and Land Reclamation, 2000).

In the Mediterranean part of Egypt, the climate is characterized by a hot dry summer and a mild winter, whereas in the south, the climate is more continental, with large variations of temperature between day and night. Rainfall is very scarce and occurs mainly in the winter, with an annual average of 10 mm. It is concentrated in the north (150–200 mm) and decreases gradually towards the south (24 mm in Cairo and 1.5 mm on Aswan) (Riad, 2000).

With a population of 67.8 million (United Nations, 2001), Egypt is the second most populated country in the Mediterranean region after Turkey. The vast majority (99%) of the population live in the area of the Nile valley and Delta.

The remainder are scattered among the oases of the Western Desert, the Mediterranean coastal regions (where rainfall allows for rainfed agriculture), the Sinai peninsula and the Red Sea coast (Food and Agriculture Organization, 2001).

Forest Resources

Area

Historical records certify that forests in Egypt have always been scarce and the wood quality of many native tree species left much to be desired (Thirgood, 1981). It is, however, argued that ancient times witnessed the presence of a more extended forest cover and a different landscape from the one of today, which has been diminished continuously by human impact.

Presently, Egyptian forest cover is formed mainly by plantations, covering only 71,600 ha, or less than 0.1% of the country's area (Food and Agriculture Organization, 2003a). They mainly occur in the form of windbreaks and linear plantations, along fields, roads and canals and concentrated around the Nile valley and Delta. During 1990–2000, forest area increased by an annual rate of 3.3%, due to the planting rate of about 2000 ha/year. The growing stock is estimated at about 108 m^3/ha, with a biomass of 106 t/ha (Undersecretariat for Afforestation, 1999).

Natural forests are almost non-existent and only a very few native species are still growing. In the desert areas, natural vegetation is widely dispersed, providing the local people (around 564,000 inhabitants) with important resources for living: fuelwood, charcoal and medicinal herbs as cash crops.

In the last decade, an extensive programme of afforestation and land reclamation has started. The Ministry of Agriculture created several tree nurseries. The trees produced are supposed to be used further as roadside and agricultural windbreaks, protecting crops in reclaimed lands and creating parks and green spaces. Often, they are used in establishing plantations irrigated with wastewater, or in the creation of man-made forests. Several of these man-made forest systems are operating on small-scale sites at Luxor, Sadat City and Ismailia.

Typologies

According to the Food and Agriculture Organization (2003a), forest plantations started in the 19th century, when *Casuarina* and *Eucalyptus* seeds were introduced. At present, *Casuarina* spp. (*Casuarina glauca*, *C. cunninghamian* and *C. equisetifolia*) are the most common trees in the country (57% of forest area), followed by *Eucalyptus camaldulensis* (10%) (Table 8.1). Other commonly grown trees include *Dalbergia sissoo*, *Acacia saligna*, *Albizia lebbeck*, *Tipuana speciosa*, *Tamarix aphylla* and *Populus nigra*.

It is argued that in Upper Egypt the species that grow most satisfactorily include *Tamarindus indica*, *Terminalia arjuna*, *Khaya senegalensis* and *Swietenia mahogoni*. From the indigenous species, *Ficus sycamores* and *Morus* spp. are planted along the canals for shade and for their edible fruits. Other native species are *Tamarix aphylla*, *Acacia arabica*, *Parkinsonia aculeata* and *Palatines aegyptiaca*.

It has been estimated that the Mediterranean coast holds around 1095 vegetation species, which make up around 53% of the total number of native species in Egypt (Boulos, 1975, cited by Food and Agriculture Organization, 2003a). The coastal area has been divided into three sectors (Zahran *et al.*, 1985, cited by Food and Agriculture Organization, 2003a): western (the Mareotis, extending for 550 km between Sallum and Alexandria); middle (Deltaic, extending for 180 km between Alexandria and Port Said); and eastern (Sinaitic, extending for 220 km between Port Said and Rafah). It was estimated that Mediterranean species account for around 50% of the total number of species in the Deltaic sector (Marshaly, 1987, cited by Food and Agriculture Organization, 2003a) and for 45% in the Sinaitic sector (Gibaly, 1988, cited by Food and Agriculture Organization, 2003a).

Functions

The majority of Egyptian forest plantations perform protective functions, such as protecting soils, watercourses and farms from winds and storms. They also act as windbreaks in the newly reclaimed areas, as shelterbelts, and as

Table 8.1. Plantation areas by ownership type.

Species	Area					
	Total		Public		Private	
	ha	%	ha	%	ha	%
Casuarina spp.	45,600	63.7	21,979	48	23,621	52
Eucalyptus	8,300	11.6	4,233	51	4,067	49
Acacia spp.	2,000	2.8	1,280	64	720	34
Dalbergia	4,600	6.4	3,680	80	920	20
Mahoganies	1,800	2.5	1,800	100	0	0
Terminalia	50	0.1	NA	NA	NA	NA
Other broadleaved	8,700	12.2	4,914	56	3,786	44
Pinus spp.	50	0.1	NA	NA	NA	NA
Other coniferous	500	0.7	NA	NA	NA	NA
Total	81,600	100.0	37,886	53	33,114	47

The separation between private/public ownership has been made with reference to a total of 70,000 ha of forests, i.e. the total forest area except for the one planted with *Terminalia*, *Pinus* spp. and other coniferous, for which data are not available.
Source: adapted from Food and Agriculture Organization (2003a); Undersecretariat for Afforestation (1999).

purifiers of polluted and contaminated sewage waters. In addition, a small part of the forest area has been planted with fast-growing species for productive reasons, in order to bridge the gap between the domestic demand and supply of timber.

Nature protection[1] in Egypt dates back to the early times of Greek and Roman civilizations, which followed basic management principles. Traditional forms of nature protection can still be found in the 'sacred areas' such as Mount Sinai, where hunting was permanently banned by local Bedouin tribes (Hobbs, 1989; Morrow, 1990). Additionally, in the Gebel Elba Mountains, restrictions on the exploitation of natural resources have been imposed by nomadic communities through tribal and religious dictates (Goodman, 1985). Community protection also exists, such as the 'lineage preserves' (*Sayaal*) where patrilineal descendants have the responsibility to protect resources in certain *wadis*: for example, the preserve set up in 1999 to protect *Acacia* in 1900 near Jbel Gataar in the Western Desert (Hobbs, 1989).

Under Ottoman law, the environmental legislation was covered under the civil code of which Art. 1243 states that land and associated trees growing wild in the mountains could not be possessed and should remain ownerless. The first conservation legislation during the 20th century came into force with the creation of a Royal hunting reserve at Wadi Rishrash in 1900, while the first protected site was established at El Omayed. However, the legal framework for protected areas was created by Law No. 102/83, referring to the definition and delineation of the 'natural protectorate'. The main body responsible for the enforcement of environmental protection and conservation is the Egyptian Environmental Affairs Agency (EEAA).

The Egyptian Wildlife Service (EWS), acting under the responsibility of the Ministry of Agriculture, is in charge of the management of natural protectorates and wildlife research. However, Law 102/83 has been regarded as being ineffective, due to the lack of legislative provisions for management plans, operational budgets and demarcation of protected area boundaries. Since 1991, the reorganization of the EEAA has been undertaken in line with a new attitude towards the protected areas that no longer have been considered as isolated units, but rather within a multi-disciplinary approach of resource management.

Institutional Aspects

Ownership and size of properties

From the total forest area, about 53% is publicly owned and generally serves as windbreaks, shelterbelts or strip plantings. Private forests account for the remaining 47% and are mainly in the form of field windbreaks (Table 8.1). Only a few woodlots have been established by some companies with private financial aid and are mostly oriented towards wood industries (Riad, 2000).

Concerning the ownership and tenure of the forest plantations, the land's owner is the trees' owner. Consequently, the trees planted on private land are owned by the private sector, while the trees planted along the roads, irrigation and drainage canals are owned by the government (Omran, 2000).

Administration and policies

The main public institution responsible for the afforestation activities is the Undersecretariat of Afforestation, under the Ministry of Agriculture and Land Reclamation (MALR). It has various departments acting in all governorates. It is responsible for carrying out various works of afforestation, supervising private nurseries, supplying tree seeds and seedlings, introducing new tree varieties and tackling the reclaimed desert areas. It also offers training services and creates job opportunities in the field of afforestation. Other institutions contributing to forestry activities include the Ministry of Irrigation, the General Authority for Roads and Bridges and the Agency for Environmental Affairs (Ministry of Agriculture and Land Reclamation, 2000).

Forest policy is part of the agricultural policy and is enforced without an official law. The strategies of the formulated forest policy are: rational use and preservation of the existing trees; increasing planted areas through afforestation programmes based on rotation and sustainable wood supply; encouraging the private sector to share in the afforestation activities; and raising public awareness of the socio-economic benefits of trees and commitment to regional and global processes. National forestry laws and regulations are prepared and implemented by MALR. However, so far, existing laws govern the rights for tree cutting and replacement: the government gives the owner the right to cut an adult tree on condition that it is replaced (Omran, 2000).

The national forestry action plan is carried out by MALR and the above-mentioned institutions. The general programme includes (Ministry of Agriculture and Land Reclamation, 2000): planting trees along roads and canals, stabilizing sand dunes and combating desertification through afforestation programmes; expanding man-made forests in the newly reclaimed areas; and establishing green belts around large cities for the protection of these areas.

Contribution of the Forest to the National Economy

Gross domestic product and employment

In contrast to the strong weight of agriculture in the overall economy (17%), the contribution of forestry is practically nil. This is linked to the scarcity of forest and the strong orientation towards other economic sectors. Employment in the forestry sector is estimated at around 171,000 people, representing just 1.2% of the total labour force (Riad, 2000). Of these, around 14,000 are personnel engaged in forestry works by the MALR and the Directorate of Agriculture at district levels.

Timber-based industries are mostly represented by wood construction and furniture and carpentry industries, while there are no sawmilling enterprises (Riad, 2000). The furniture industry is labour intensive, based on local labour, which in some areas of the country boasts a long tradition in woodworking and furniture manufacturing (Centro Studi Industria Leggera, 1997). Most of the timber-based enterprises are very small scaled, employing around 9.6% of the total small-scale industries workforce in Egypt (Fisseha, 1987), which represent around 6240 personnel.[2] However, in terms of employment, non-wood forest-based activities are much more diffuse in rural areas, compared with wood-based ones. All (wood and non-wood) forest-based industries contribute 18%

to the value added and 19% to the value of the production of all small-scale industries in the country.

International trade

The scarcity of forest in Egypt makes it a net importer of forest products as a main input for the furniture and wood construction industries. According to the Food and Agriculture Organization on-line database, timber production is almost entirely offset by imports that constitute 6.7 Mm^3, whereas production of firewood (16,200 m^3) covers all the country's needs. Due to the lack of a sawmilling industry, most of the softwood needs are met by imported timber, 70% of which is consumed by the construction industry and the rest by the furniture/carpentry sector. There is an increasing trend towards using more softwood in the furniture industry due to its relatively low price compared with hardwood.

From the total softwood imports, the share of the private sector is around 75% and the only public sector enterprise is Societé Commerciale Des Bois (FABAS) that imports the balance. Prior to 1992, about 60% of the FABAS imports were through private agreements with the former USSR, in exchange for agricultural and other petroleum products. After 1993, with the economic structural reform, FABAS has diminished its position significantly. Therefore, given the State's low priority for wood imports, the private sector role in the international timber trade is expected to increase (Riad, 2000).

International trade in furniture (or other forest value-added products) is hardly relevant, both because local consumers tend to purchase less expensive local products and because Egyptian companies are not very competitive on foreign markets: only 2% of Egyptian furniture consumers purchase imported products. The national production of home furniture is industrial, of medium and top craftsmanship, which satisfies a total of 40% of the population; the remaining demand is satisfied by low-level craftsmanship, second-hand goods and other inferior substitutes (Centro Studi Industria Leggera, 1997). Overall, imports of forest products account for about €700 million, while exports stand at just about €8 million (Food and Agriculture Organization, 2000).

Forest-based industries

Forest-based industries are mostly dominated by small-scale enterprises located in rural areas and specializing in carpentry/furniture (23%) and basket/mat/hat weaving (70.4%) (Fisseha, 1987). Even though the country is almost devoid of wood resources of its own, the manufacture of wooden furniture is the third largest industry outside large towns and cities (Mead, 1982). The furniture industry is generally labour intensive, while at the technological level, the sector is rather backward. Local supply is oriented towards satisfying demand from within the country rather than finding outlets on foreign markets: in fact, only 3% of national production is exported. Locally produced furniture is still strongly influenced by traditional styles, and modern designs are totally absent (Centro Studi Industria Leggera, 1997).

The small forest-based processing enterprises are generally very small, with 1.9 workers on average. They rely heavily on the entrepreneurs' family for labour, which accounts for around 76% of production. Technologically, the operations are mostly household based and of a high level of simplicity. Eighty per cent of the forest-based enterprises are located in the rural environment and they employ around 65% of the rural labour force. Women's share in total ownership accounts for around 65% mostly due to the prevalence of weaving activities (Arnold et al., 1987).

The non-wood product-based activities constitute an increasing share in the total small-scale industries (14.8%), much higher than that of the timber-based industry.

The Values of Egyptian Forests

Forest scarcity in Egypt and the lack of comprehensive statistics linked to forest values are the basic reasons for deriving only an incomplete picture of the values of the forest products in this country. Forest plantations contribute only modest quantities of non-wood forest products

(NWFPs), of which edible fruits, medical and aromatic oils and fodder are the most important (Food and Agriculture Organization, 2003a). Only a brief account of forest values based on international-level publications is given; some of the values are just rough estimates, due to a lack of more accurate data. Whenever relevant, the picture is completed with results of a local level survey undertaken in Wadi Allaqi Biosphere Reserve located in the Nubian desert of Egypt (Belal et al., 1999).

Direct use values

The role of national wood forest products (WFPs) is negligible, especially if compared with other forest goods and services such as grazing, honey and medicinal plants. Actually, the United States Department of Agriculture (2002) reports a nil national production of WFPs for Egypt, while according to the Food and Agriculture Organization on-line database, timber and firewood production rises to an irrelevant 268 m³ and 16,182 m³ respectively. Valued at roadside prices of €45/m³ and €27/m³, respectively (T. Omran, University of Alexandria, 2003, personal communication), it gives a monetary value of €12,000 for timber and €436,900 for firewood. Firewood and charcoal production are significant sources of livelihood, especially for the close-to-desert and oasis inhabitants (Box 8.1).

Grazing is a traditional activity in Egypt and animals have generally both economic (subsistence and commercial) and social value. Sheep, goats and camels are among the most common grazing animals, at least in the arid and hyper-arid zones. The social and symbolic value of animals is reflected on different occasions, for instance by the rituals of slaughtering sheep and goats to mark births, weddings or funerals, or other Islamic feasts. Camels are rarely slaughtered for food; they are mostly desired as a means of transport. However, no data regarding the number of animals grazing in the forests were found, therefore, no monetary valuation of this activity could be undertaken.

Egypt is the primary honey producer in Africa. According to the latest statistics of the Arab Organization for Agricultural Development (AOAD), the annual honey production in Egypt is about 16,000 t, which is more than half the total amount of honey produced in all the Arab countries (El Shehawi, 1998). The valuation assumes very roughly that in 2001, the quantity of honey was at least the same as a decade ago. At a market price of around €4/kg (T. Omran, University of Alexandria, 2003, personal communication), the total monetary value reaches €64,000. It is difficult to estimate the proportion of this value attributed to forestland and other natural vegetation.

Medicinal and aromatic plants, such as rosemary, wild thyme, sweet basil and henna, grow on around 17,000 ha and represent the primary export product. Some of these have a strong potential for pharmaceutical use, such as Cymbopogon proximus (Box 8.1).

Other direct uses of forests are related to silkworm breeding and glue production. Some of the trees and shrubs are used for sand dune fixation as well as windbreaks. A number of Acacia species are used as forage and pasture

---

Box 8.1.   The uses and economic value of forest products in Wadi Allaqi (case study).[a]

Belal et al. (1999) report a local level survey conducted with the aim of valuing in economic terms the actual and potential uses of desert resources in Egypt, with the scope of suggesting instruments for environmentally sound management and policies in this area. It was carried out in Wadi Allaqi Biosphere Reserve, in the Nubian desert of Egypt, which is characterized by a 'hyper-arid environment'.

After the Lake Nasser flooding in 1967, new vegetation types emerged to which local people adapted. The area's residents – numbering around 218 in 1986 – are the Bedouins, who lead a semi-sedentary life, mostly based on natural resource exploitation: charcoal production, pastoralism and medicinal plants. Valuation of all these products has been undertaken on the basis of their market price and, when not available, the surrogate market price. All the values are updated to 2001 euro prices.

Charcoal production. Charcoal is generally produced from Acacia trees and their dead branches and is marketed in Aswan at a market price ranging between €0.4 and €0.6/kg, depending on the product quality. Annual charcoal production in 1996/1997 was 18 t and the average market price was €0.6/kg. Therefore, the total charcoal value equalled €10,200, i.e. a very significant income source.

Charcoal production is based on the controlled burning of wood in the sand and it takes several weeks, or months if carried out in more than one location. It is a winter activity, generally undertaken in conjunction with small animal grazing in distant hills and wadis. Production is generally carried out by the wealthier families, as they are in a position to provide food and camel transportation for the labourers. The price of charcoal depends on its quality. Lumps which are well burnt, black in colour and large sized are favoured.

Fodder. Fodder is an important forest product, as grazing of camels, goats and sheep is the dominant economic activity in the area. Animals can graze more than half of the existing species in the area. The desert pasture plants have been divided into three groups, according to their availability: (i) perennial plants (trees and shrubs) – form the permanent, but limited source of fodder for livestock, i.e. *Acacia* spp., *Balanites aegyptiaca* and *Tamarix nilotica*; (ii) perennials whose life-span depends on the water availability stored in the wadi-filled deposits after rainfall, i.e. *Sena alexandrina*, *Aerva javanica* and *Citrullus colocintus*; and (iii) annual and ephemeral plants that provide temporary (a few months) but abundant (high biomass and high nutrition) pasture for livestock, i.e. *Eragrostis aegyptiaca* and *Fimbrystillis bis-umbrellata*.

The fodder shadow price was estimated in two ways, both giving rise to similar results. First, the surrogate market approach was applied, using the prices per kilogram of energy in dry matter of crude protein in sorghum (~€215/t) and 'Teben' (~€145/t). Therefore, it can be considered that the market price of fodder varies between €145 and €215/t. Alternatively, the productivity change method was used, considering the conversion of fodder into meat and milk, which have observable market prices. The implicit fodder price obtained was about €147/t. This was, however, considered an underestimate by not including the value of other livestock outputs, such as wool and hide.

Around 100,000 camels annually graze in the Allaqi region, having a daily requirement of about 6.3 kg of fodder/camel, resulting in a total of 630 Mt of fodder consumed annually. On this basis, the annual value of camel grazing is €92,600.

Medicinal plants and pharmaceuticals. Nearly half of the recorded species (56) in the region are of medicinal value or are under investigation for their potential use in medicinal compounds. From these, the value of ten indigenous species has been estimated, considering those collected and used by the Bedouin and those sold in shops. The price of these plants ranges according to species, but all have a market value and their sale represents an importance source of income for Bedouin: *Balanites* (€2.5/kg), *Citrullus* (€0.2/fruit), *Cassia senna* (€1.2/kg), *Cleome* (€3.8/kg), *Cymbopogon* (€1.2/kg), *Haplophyllum* (€3.4/kg), *Hyoscyamus* (€1.3/kg), *Pulicaria* (€4.5/kg), *Salvadora* (€0.1/piece) and *Solenostemma* (€2/kg).

Another important medicinal plant is *Cymbopogon proximus*, the chief constituent of the pharmaceutical proximol (Halpha-bar). In 1997, 20 t of this plant were traded through Allaqi. Depending on the level of processing, its price ranges from €950–1300/t to about €1600–1900/t. In pharmacological applications, proximol is produced from *C. proximus* at a rate of 1.5%. The final product has a selling price of around €1 per 60 mg of proximol (which contains 18.5 mg of extract). Table 8.2 shows that the value of *C. proximus* as a dry plant exceeds the market value of the processed extract; this is due to the influence of the State-controlled pharmaceutical industries in the country.

Table 8.2. Market value of *Cymbopogon proximus*.

	Unit	Market price (€ 2001)	Market value of 20 t (€)
Plant (dry)	t	1,100	22,000
Plant (crushed)	t	1,730	34,700
Extract (proximol)	60 mg	1	17,400

[a]This is the author's own summary of the case study taken from Belal *et al.* (1999). The author takes full responsibility for all possible errors and shortcomings in this summary.

crops, especially on the north-western coast and Sinai peninsula (Riad, 2000).

Indirect use values

Indirect use values are even more difficult to estimate at a national level. However, the positive role played by forests in a country with such harsh climatic conditions is very important.

The man-made forests established in Egypt are of paramount importance in improving the climatic conditions in the afforested desert areas. They diminish the intensity of the desertification process and, as such, protect agricultural fields. They also create an environmental balance by offering the opportunity for birds and animals to live and reproduce. Forests help to decrease pollution by excessive quantities of sewage water. They also provide work opportunities for the citizens living in the nearby areas (Riad, 2000). One important species for its multi-purpose use is acacia (*Acacia saligna*), serving mainly for sand dune fixation, soil erosion and as a fuelwood source. More than 1 million acacia seedlings were transplanted along the Mediterranean coasts of Egypt for native range rehabilitation (El Shaer, 2000).

Only the estimate of carbon sequestration in (non-forest) trees is undertaken in monetary terms. Based on the methodology developed by the Intergovernmental Panel on Climate Change (IPCC) in 1995, it was calculated that in Egypt the trees store in their annual growth increment 2.7 MtC (United Nations Framework Convention on Climate Change, 1999). Valued at a shadow price of €20/tC (Fankhauser, 1995), the total value of carbon sequestration is €54 million. This estimate does not take into account the losses due to removals, as this amount is minimal because nearly all timber is imported. If the quantity of actual harvested WFPs (even if minor) is taken into account, the total value of carbon sequestration would be slightly less.

Conclusions

Despite the scarcity of forests and of other vegetation in the desert areas of Egypt, their

role is of considerable economic importance for society as a whole and in particular for rural and indigenous people who utilize this resource.

From the direct use values, WFPs are generally represented by fuelwood and serve subsistence purposes of cooking and heating, but are also used as an input for producing charcoal. Of outstanding value are the NWFPs, such as medicinal plants – with high potential for pharmaceutical production – and forage for livestock, given the importance of grazing throughout the country. Egypt is the primary honey producer in Africa; however, existing valuation does not allow the identification of the actual contribution of forests to this value.

Indirect use values are represented by the protective functions of forests, such as windbreaks, shelterbelts, soil conservation, climate balance, water purificators and reducing desertification. Despite the scarce estimates, these functions could be considered as the most important, given the adverse climate conditions and the dominance of the desert in this country.

Acknowledgements

The author would like to give special thanks to Professor Talat Abdel-Hamid Omran, Alexandria University, Egypt, for the careful revision, suggestions and improvements of the chapter.

Notes

[1]   The following paragraphs of this section are based on Food and Agriculture Organization (2003b).
[2]   The data refer only to the small-scaled timber-based industries.

References

Arnold, J.E.M., Chipeta, M.E. and Fisseha, Y. (1987) The importance of small forest-based processing enterprises in developing countries. *Unasylva* 39, 3–4.

Belal, A.E., Leith, B., Solway, J. and Springuel I. (eds) (1999) *Environmental Valuation and Management of Plants in Wadi Allaqi, Egypt, Final Report*. International Development Research Centre (IDRC).

Centro Studi Industria Leggera (1997) The furniture sector in Egypt. In: *World Furniture Industry: the Middle East and Africa.* http://web.tin.it/csil-furnitureWorldFurniture/apr98/art2n2.html (accessed September 2002).

El Shaer, H.M. (2000) *Utilization of* Acacia saligna *as Livestock Fodder in Arid and Semi-arid Areas in Egypt.* CIHEAM, Options Méditerranéennes.

El Shehawi, M. (1998) *The Future of Bees and Honey Production in Arab Countries.* www.beekeeping.com/articles/us/arab_countries.htm (accessed September 2002).

Fankhauser, S. (1995) *Valuing Climate Change. The Economics of the Greenhouse.* Earthscan, London.

Fisseha, Y. (1987) *Basic Features of Rural Small-scale Forest-based Processing Enterprises in Developing Countries.* FAO, Rome.

Food and Agriculture Organization (2001) *State of the World's Forests.* FAO, Rome.

Food and Agriculture Organization (2003a) Country profiles: Egypt, www.fao.org/forestry/fo/country Accessed February 2003.

Food and Agriculture Organization (2003b) *Protected areas – Policy and Legislation.* FAO, Rome.

Goodma, S.M. (1985) *Natural Resources and Management Consideration.* Gebel Elba Conservation Area. WWF/IUCN Project No. 3612.

Hobbs, J.J. (1989) *Bedouin Life in the Egyptian Wilderness.* University of Texas Press, Austin, Texas.

Mead, D.C. (1982) Small industries in Egypt: an exploration of the economics of small-furniture producers. *International Journal of Middle East Studies* 14, 159–171.

Ministry of Agriculture and Land Reclamation (2000) *The National Afforestation and Forestry Action Plan.* Country report. MALR, Undersecretary for Afforestation, 3–6 June 2000.

Morrow, L. (1990) Trashing Mount Sinai. *Time* 129, p. 26.

Omran, T. (2000) *Country Forestry Brief for Egypt.* FAO, Rome.

Riad, M. (2000) *Egypt,* Forestry Outlook Studies in Africa (FOSA). FAO, Rome.

Thirgood, J.V. (1981) *Man and the Mediterranean Forest – a History of Resource Depletion.* Academic Press, New York.

Undersecretariat for Afforestation (1999) *FRA 2000 for Egypt.* The Ministry of Agriculture and Land Reclamation, Undersecretariat for Afforestation, Cairo.

United Nations (2001) *World Population Prospects: the 2000 Revision.* Population Division and Social Affairs, United Nations Secretariat.

United Nations Environment Programme/Mediterranean Action Plan/Plan Bleu (1996) *Egypt, Series of Mediterranean Country Profiles: Institutions – Environment – Development.* Blue Plan, Regional Activity Centre, Sophia Antipolis, France.

United Nations Framework Convention on Climate Change (1999) *The Arab Republic of Egypt: Initial National Communication on Climate Change.* UNFCC.

United States Department of Agriculture (2002) *Egypt – Solid Wood Products 2002.* Prepared by Abdi, A. and Ibrahim, S., Foreign Agricultural Service, GAIN Report.

9   Palestine

Roubina Ghattas, Nader Hrimat and Jad Isaac

Applied Research Institute Jerusalem (ARIJ), PO Box 860, Caritas St, Bethlehem, West Bank

Introduction

Palestine, as part of the eastern Mediterranean region, is the meeting ground for plant species originating from far-flung world regions, as far apart as Western Europe, Central Asia and Eastern Africa. The area's proud history as the cradle of civilization and a focal point of the world's three monotheistic religions has long given it a global influence despite its small size. The extent to which historical Palestine has attracted the combined attention and often fervent intentions of world civilization has been a source even for its ecological richness. It represents a rich base for flora and fauna, where the natural biota is composed of an estimated 2483 plant species, 470 avian species, 95 mammalian species, seven amphibian species and 93 reptilian species (Shmida, 1995).

The country extends over two geographical areas: the West Bank, covering 5820 km², which includes East Jerusalem, and the Gaza Strip, extending over 365 km² (Applied Research Institute Jerusalem, 2000). The population is currently at 3.2 million inhabitants, representing a high growth rate of about 4% (Palestinian Central Bureau for Statistics, 2000). The high demographic and urbanization rates have a large impact on natural resources and their possibility of meeting market demand and satisfying increasing human needs. In addition, Palestinians face serious problems in generating sufficient cash income to meet the most basic necessities, especially in the context of the decreased area due to Israeli constraints, confiscations and continuous land degradation. At best, the overall results are static crop yields and widespread poverty, especially during 2001–2002 and during the current *Al Aqsa Intifada* when 64.9% of households fell bellow the poverty line (Palestinian Central Bureau for Statistics, 2000, 2001).

The climate of Palestine as a whole, and the West Bank in particular, is of a Mediterranean type – although with semi-desertic influences in the south – marked by a mild, rainy winter and a prolonged dry and hot summer. From north to south, the annual amount of rainfall decreases, while the temperatures increase. From west to east, annual rainfall and mean temperatures undergo similar, but less regular changes (Isaac *et al.*, 1997). The ecosystem in the West Bank, scarcely changed over time, is divided into four longitudinal belts: the Semi Coast, Central Highlands, Eastern Slopes and the Jordan Valley (Fig. 9.1). They are well marked by differences in geomorphological features, climatic and soil conditions as well as plant life. The Semi Coast and the Central Highlands constitute most of the West Bank land and lie completely under the semi-humid Mediterranean climate. They receive adequate rainfall and constitute a favourable environment for plant growth. The main ecosystem of the Gaza Strip is the Coastal Plain.

The country's location at the crossroads between Eurasia and Africa also nurtures its biological diversity through the abruptness with which climatic zones, desert, steppe,

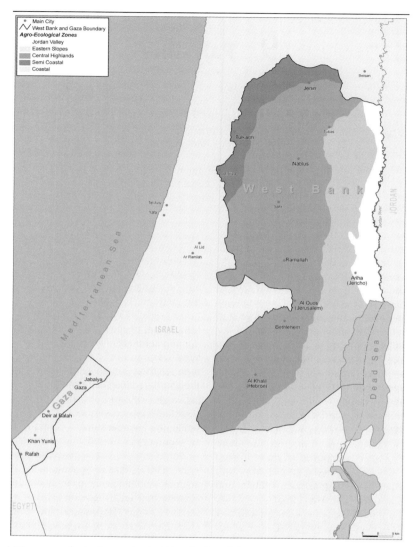

Fig. 9.1. Main agroecological zones in the West Bank and Gaza Strip.

Mediterranean woodland and even oasis join one another. The vegetation includes a variety of plant formations ranging from dense forests to thin patches of desert herbs, passing through different forms of woodland such as *maquis*, *garrigue* and *batha*, with diverse genetic backgrounds. There are 60 species of natural trees and 90 of bush plants distributed throughout the country (Breghiet, 1995). If, historically, this land was famous for its rich vegetal cover, the landscape, ecosystems and vegetation have been subjected for thousands of years to change on a large scale. The rate of destruction of nature is even much higher today, especially as a result of the uprooting of trees and forest overgrazing; yet, most of the Palestinian plant species have managed to survive, albeit some of them in small populations.

Forest Resources

Area and people

The forest area covers 23,159 ha, or 3.7% of the country's surface area. It is formed by 93 major forest sites extending on 22,959 ha of the West Bank and 13 forest sites covering 200 ha of the Gaza Strip.

Afforestation programmes in the West Bank were first implemented during the British Mandate (1918–1948). They concerned the planting of around 230 ha of mountainous and sloping land with *Cupressus* and *Pinus* spp. and the establishment of nurseries in order to produce seedlings for local governments and people as part of a Grand National Afforestation scheme. During the Jordanian administration (1948–1967), plantations were made on an area of 3535 ha, with the main species being *Pinus* spp. (*P. pinea*, *P. halepensis*, *P. brutia* and *P. canariensis*), *Eucalyptus camaldulensis*, *Cupressus* spp. and *Acacia* spp. (Ministry of Agriculture, 1999).

In the Gaza Strip, before 1948, the areas with a naturalistic value comprised mainly shifting sand dunes along the coast. During the Egyptian Administration (1948–1967), plantations were undertaken on 4200 ha, with the purpose of stabilizing the dunes, therefore protecting the hinterland. The main species planted were *Eucalyptus* spp., *Acacia* spp., *Tamarix* spp. and *Atriplex* spp. Other abundant natural perennial plants included *Retama raetam* and *Artemisia monosperma* (Euroconsult/Iwaco, 1994).

Between 1971 and 1999, the forest area decreased by 23%, mostly in the plantations of the Gaza Strip and, to a lesser extent, in the natural forest area (Fig. 9.2).

Around 80% of the lost forest area can be attributed to the construction of Israeli settlements (77%), military camps (2%) and by-pass roads (1%), where, for example, 670,000 fruit and forestry trees were uprooted during 2001 alone by the Israelis.[1] In fact, all forestry activities were prohibited and forest nurseries were closed in most districts of the West Bank except for Wadi Al-Quof Nursery in the Hebron district (Box 9.1). The remaining 20% decrease is due to woodcutting by local people, overgrazing, conversion of private forests to agriculture or urban uses and, to a small extent, natural factors.

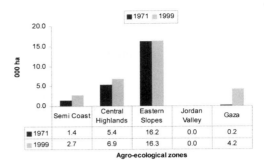

Fig. 9.2. Officially designated forest area in 1971 and 1999.

	Semi Coast	Central Highlands	Eastern Slopes	Jordan Valley	Gaza
■ 1971	1.4	5.4	16.2	0.0	0.2
▨ 1999	2.7	6.9	16.3	0.0	4.2

Agro-ecological zones

Typologies

Of the total forest area, around 79% comprises natural forest, 12% plantations and the remaining 9% is bare land with sparse vegetation (Fig. 9.3).

Most of the natural forest area is concentrated on the Eastern Slopes. It consists of a very open pseudo-savanna type with sparse large trees of *Ceratonia siliqua* and small shrubs such as *Pistacia lentiscus* and *Rhamnus palaestinus*. The dry areas of the eastern slopes contain species such as *Ziziphus lotus* and *Retama raetam*, while the dwarf shrubs *Sarcopoterium spinosum* are located between the central highlands area and grasses.

In the Central Highlands, natural forests[2] are represented by Aleppo pine and evergreen oak *maquis*. The principle tree and shrub species include *Quercus caliprinos*, *C. siliqua*, *Pistacia palaestina* and *P. lentiscus*. The open *garrigue* and *batha* are mostly represented by *S. spinosum*, *Cistus villosus*, *Phlomis viscosa* and *Thymus capitatus*. These species also grow on the Semi Coast, where, additionally, species such as *Euphorbia perelis*, *Senecio vernalis*, *Thymelaea hirsutum* and *Lupinus palaestinus* can be found.

The Jordan Valley does not contain any officially designated forests. However, there is a large area of natural forests, partly protected as Israel declared them nature reserves. Along the River Jordan and the Dead Sea, there is a large area of riparian forest and wetland – considered military land since 1970 – with closed reed trees, such as *Tamarix jordanica*, and shrubs, such as *Atriplex* spp., *Lycium* spp. and *Nitraria retusa*.

Box 9.1.   Tree nurseries in Palestine.

The first nurseries in Palestine were established as early as 1927. Until 1970, a significant reforestation programme was implemented in the West Bank. Forest seedlings were produced in five nurseries, of which four were located in agricultural stations. Total production of seedlings amounted to 2.1 million in 1971. From 1970 until 1995, all nurseries in the West Bank, except for Wadi Al-Quoff and Gaza, ceased operating, and the reforestation programme was reduced to the production of only 120,000 seedlings. All the nurseries were handed over from Israel to the Palestinian Authority in December 1995. They have been developed and rehabilitated since the creation of the Palestinian Authority, and in 1999 they produced 850,000 seedlings (Table 9.1). Species produced in the governmental nurseries consist of: conifers, such as *Pinus* spp. and *Cupressus* spp. (40% of the total amount), indigenous broadleaved species, such as *Pistacia palaestina*, *Quercus calliprinos* and *Ceratonia siliqua* (30%), exotic and ornamental trees, such as *Bauhinia variegata*, *Dodonea viscosa* and *Eucalyptus* spp. (20%), and dry area species for rangeland rehabilitation, such as *Atriplex* spp. and *Acacia* spp. (10%).

Table 9.1.   Nurseries between 1971 and 1999.

	Annual production of seedlings		
Governorate/nursery	1971	1971–1994	1999
Jenin	500,000	0	100,000
Tulkarem	250,000	0	0
Nablus	0	0	0
Ramallah	250,000	0	0
Jerusalem	0	0	0
Hebron-Wadi Al Quof	200,000	60,000	250,000
Hebron-Arroub	500,000	0	100,000
Gaza-Green Dam	0	0	250,000
Gaza-El Shati	400,000	60,000	150,000
Total	2,100,000	120,000	850,000

Source: Ministry of Agriculture (1999).

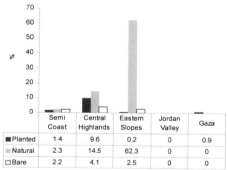

	Semi Coast	Central Highlands	Eastern Slopes	Jordan Valley	Gaza
■ Planted	1.4	9.6	0.2	0	0.9
▨ Natural	2.3	14.5	62.3	0	0
□ Bare	2.2	4.1	2.5	0	0

Fig. 9.3.   Forest types according to agroecological zones.

The planted forests are mainly located in the Central Highlands (Fig. 9.5). To a small extent, they can also be found in the Coastal Plain of Gaza where plantations were undertaken at a very low density with species including

Acacia spp., *E. camaldulensis* and *Tamarix* spp. The main sand dune fixation species in Gaza are *Suaeda splendens*, *Salsola soda*, *Aster tripolium*, *Atriplex hasitatata*, *Ipomaea stolonifera*, *Salsola kali*, *Euphorbia peplis*, *Tamarix nilotica*, *Artemisia monosperma* and *Ammopila arenaria*. Most of these forests, though 'naturalized', are still classified as planted forests (Euroconsult/Iwaco, 1994).

Officially designated bare land with sparse vegetation is concentrated in the Central Highlands and Semi Coast. It should be stressed that actually most of the natural forest area in the Eastern Slopes is currently bare and consists of sparse vegetation.

According to available knowledge at Palestinian level, a map was produced in order to distinguish between natural and planted forests, excluding nature reserves. A link can be established between the Quézel classification and this map, bearing in mind that natural forests in Palestine are represented mainly by the carob–lentisk *maquis*, the deciduous oak forest and the evergreen oak *maquis* and forest, whereas the pine forests are represented by plantations (Box 9.2).

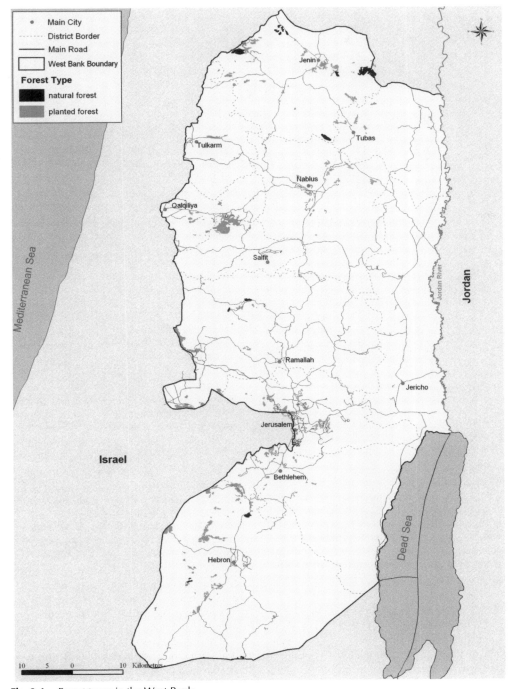

Fig. 9.4. Forest types in the West Bank.

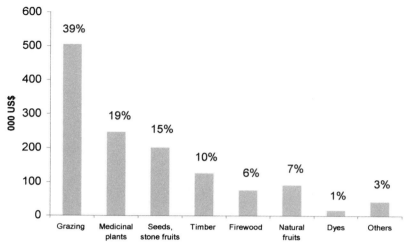

Fig. 9.5.   Main components of the TEV of Palestinian forests.

Box 9.2.   Palestinian forest types according to the Quézel classification.

According to the Quézel classification, the Palestinian forests can be divided into the following:

(i) Carob–lentisk *maquis* corresponding to thermophilic wild olive and pistachio scrubs. This is a rather dense carpet of low shrubs, consisting of *Pistacia lentiscus* and other associated species. It occurs scattered together with carob trees, which often attain a height of 4 m or more. Both of the leading plants are evergreen. The carob–lentisk *maquis* occupies large stretches in Palestine. It is widespread on the western foothills of the mountain belt, on the slopes of Galilee and Nablus, and on dunes and kurkar hills on the Coastal Plain. The soil varies from *terra rossa* to rendzina and kurkar sandstone. In this association, numerous Mediterranean chamaephytes, such as *Cistus villosus*, *C. salvifolius*, *Calycotome villosa* and *Phlomis viscose* are found. In the north, the association comprises Mediterranean shrubs such as *Olea europaea* and *Amygdalus communis*. On sand dunes, the association includes two leading species, *Ceratonia* and *Pistacia*, together with a series of shrubs, such as *Retama raetam*, *Artemisia monosperma* and *Lycium europeaum* (Zohary, 1962).

(ii) Pine forest corresponding to the Mediterranen conifer forests of Aleppo pine, brutia pine, stone pine and Phoenician juniper. This type of forest is dominated by Aleppo pine (*Pinus halepensis*) and is often accompanied by shrubs and trees of *maquis* and *garrigue*, such as *Quercus calliprinos*, *Pistacia lentiscus*, *P. palaestina*, *Arbutus andrachne*, *Juniperus oxycedrus*, *Cistus salvifolius*, *Salvia fruticosa*, *Calycotome villosa* and many other perennial and annual species. The Aleppo pine forest extends from the sea level in Lebanon to an altitude of more than 800 m. It is confined to and scattered on rendzina soils all over the Mediterranean mountain range, clearly indicating its former sphere of distribution, especially considering that rendzina soil moisture content is sufficient to support pine seedlings during the summer. Larger stands have been preserved on Mount Carmel, in the mountains of southern Nablus. Pine forests are very susceptible to fire; unlike other forest and *maquis* trees, the Aleppo pine is not able to renew growth from its stump and propagates from seed only. In addition, due to its high quality timber, it has been clearcut over large areas (Zohary, 1962).

(iii) Evergreen oak *maquis* and forest corresponding to sclerophyllous evergreen oak forest of holm oak, cork oak and kermes oak. This is the most typical and common forest and *maquis* formation of the Mediterranean part of Palestine. The dominant type of association is the *Quercus calliprinos-Pistacia palaestina* association. This occurs generally in the form of *maquis* and comprises, apart from the dominating *Quercus* and *Pistacia*, a series of other Mediterranean evergreen trees and shrubs such as *Laurus nobilis*, *Arbutus andrachne* and *Phillyrea media*; and, in addition, *Styrax officinalis*, *Rhamnus palaestina* and *Crataegus azarolus*. The most typical climbers of the *maquis* are *Clematis cirrhosa*,

Tamus communis and *Lonicera etruca*. The *maquis* gives shelter to a large number of beautiful bulb and tuber plants such as the species of *Tulipa* (Liliaceae), *Allium* (Liliaceae), *Colchicum* (Liliaceae), *Crocus* (Iridaceae), *Orchis* (Orchidaceae), *Ophrys* (Orchidaceae) and shade-demanding ferns. Where the *maquis* is less dense, it offers optimal growth conditions for a wealth of annual and perennial herbs. This type of *maquis* is common throughout the western mountain belt, from the foot of the Lebanese hills in the north, up to the Judean Mountains (Jerusalem and Hebron) in the south. It is most characteristic of the Mediterranean *terra rossa*, but it also occurs on certain variants of rendzina (Zohary, 1962).

(iv) Deciduous oak forest corresponding to deciduous forests of zeen oak, afares oak, Lebanese oak, tauzin oak, hornbeam, ash and occasionally beech. This type belongs to a large group of broadleaved, deciduous forests. It reaches its southern limit of distribution and has different forms of association. It can be found together with a grass community dominated by *Desmostachya bipinnata*. However, this association has been almost totally destroyed by man and citriculture. On the other hand, a typical oak forest is that which is associated with and accompanied by *Styrax officinalis* and, under favourable ecological conditions, also by *P. palaestina*, *Crataegus azarolus*, *Phillyrea media*, *C. cirrhosa*, *Anemone coronaria*, *Cyclamen persicum* and *Arum palaestinum* (Zohary, 1966).

(v) Savannah forests (not included in Quézel classification). This type largely consists of thorny acacia species (such as *A. raddiana*), *Ziziphus spina-christi*, *Salvadora persica* and other tropical trees and shrubs which are distributed throughout the Jordan Valley, Dead Sea shore and in the southern Coastal Plain. *Z. spina-christi* is widely spread in the Gaza Strip and other places characterized by high temperatures. It is considered an important series of plant communities for the environmental balance in the valleys, coast and Gaza Strip.

(vi) Riparian forests (not included in Quézel classification). Consisting mainly of various species of *Salix* spp. (such as *S. acmophylla*), *Tamarix* spp. (such as *T. jordanus*) and *Populus* spp. (such as *P. euphratica*), these predominate near rivers in warm areas. At the same time, forests of *Platanus* (such as *P. orientalis*), *Fraxinus* (such as *F. syriaca*) and *Ulmus* (such as *U. canescens*) occupy cold areas near water sources.

Functions

Among the principle forest functions, production of wood plays a minor role. Wood is used mainly as fuelwood and as an input for tourist and decorative crafts. *Cupressus* spp., *Quercus* spp., *Acacia* spp., *Pinus* spp., *Eucalyptus* spp. and *Tamarix* spp. are those species which have a major economic value in Palestine. There is a wide range of non-wood forest products (NWFPs) such a fruits (food and/or feed), resins and fodder that contribute to a diversified economy. In addition, small natural elements, such as pools and streams, patches and lines of ruderal herbs, trees and shrubs, line plantations along roads and planted fences, serve multiple purposes by integrating biodiversity values with production functions.

A very important use of the forest is grazing. It has a positive effect by increasing the open area in the forests, resulting in a higher level of biodiversity. However, due to forest overgrazing, negative effects have appeared: the whole cover of woody plants is being removed, resulting in another ecosystem of dwarf shrubs and grasses (*batha*), which is currently too common (Kaplan and Gutman, 1996).

Perhaps most important are the ecological functions of trees and forests, including microclimate regulation, filtration of air pollutants generated from urban areas, fixation of mobile sand dunes and soils, protection of crops against wind and regulation of surface water flow through forests and woodlands. Afforestation is one of the principal means of controlling soil erosion. In the hills, forests prevent the topsoil from being washed away by the winter rains. Shelterbelts of trees moderate the onslaught of winter storms on adjoining fields and control wind erosion. Recreation-tourism has a high potential, but is currently underused as most of the nature reserves are in Israeli hands and facilities for tourists are barely available in other forests and nature areas.

Institutional Aspects

Ownership and size of properties

In Palestine, the systems of forest ownership and management are peculiar. The reserve system is applicable where the State is the major forest owner, whereas private ownership is limited. This regime was created and implemented mainly during the colonial period of the British Mandate. Ninety-one per cent of the forest area is State owned and divided into natural forests (73%), bare lands (8%) and planted forests (10%). Private forests occupy 8% of the forest area and are planted forests. It should be noted that in addition to the officially designated forest area, a further 1000 ha of unregistered private forests and line plantations are estimated to exist.

Administration and policies

The history of the legal system in Palestine is unique. For centuries, different regimes have been enforced by various powers: the Ottoman Empire, the British Mandate, the Jordanian and Egyptian Administrations and the Israeli military occupation. Each administration adopted extant laws that were applied and later modified. As such, the Palestinian National Authority (PNA) inherited a jungle of laws, with various applications according to the area. However, the PNA has formulated strategies concerning all issues, including forests, that would finally adjust the previous legislation.

The Forest Law issued in March 1926 and its Amendments No. 8/1928, No. 30/1937 and No. 7/1942 are still in force in the Gaza Strip. In the West Bank, however, this was replaced with Law No. 81/1951. In addition, another new law was declared, entitled 'The United Law of Forests', and was implemented throughout the land under Jordanian control on both banks of the River Jordan. The two laws functional in the West Bank included general principles related to the protection of existing forests, the establishment of new protected forests and their management. They also include the rules for rangers and the appropriate regulations and methods for obtaining a licence to benefit from the forest products or grazing resources.

During the Israeli occupation, several laws were issued for the protection of natural resources. However, they gave Israel full control, mainly over land for security reasons (especially for building colonies). The outcome of these laws can be assessed by the loss of 65 and 50% of the total areas of the West Bank and Gaza, respectively, due to confiscations up to 1991. The applied policies and military regulations led to an increased rate of destruction and reduced biodiversity throughout the country. In addition to the closure of grazing areas, several military bases have been established, causing changes in the topography and natural stream flow routing, and increased soil erosion.

Several agreements were held between Israel and Palestine to aid cooperation in implementing principles and standards. They should conform with internationally accepted standards concerning the protection of endangered species and of wild fauna and flora, including the restriction of trade, the conservation of wildlife and migratory species, and the preservation of existing forests and nature reserves. According to the Oslo II Interim Agreement, 'powers and responsibilities in the sphere of Forests in the West Bank and Gaza Strip shall be transferred from the military government and its Civil Administration to the Palestinian side'. This sphere includes, *inter alia*, the establishment, administration, supervision, protection and preservation of all forests (planted and unplanted).

Currently, there are major weaknesses in the policies, methods and mechanisms of enforcement in place to support and develop the multi-functional role of forests. However, a Palestinian Forest Policy, the Strategic Options and Scenarios, was formulated based on the analysis of the situation and constraints with respect to nature and forests in Palestine and in compliance with the existing agricultural and other relevant policies, mainly the National Biodiversity Strategy and Action Plan, and the National Policies for Physical Development.

The administration of forests is under the supervision of the Forestry Department of the Ministry of Agriculture, in both the West Bank and Gaza Strip. The long-term policy objectives are:

● Preserving natural and forestlands, increasing their areas, conserving wild plants and animals and organizing their exploitation.

- Promoting and strengthening the institutional structure and human capacity for forest and nature.
- Developing an adequate legislative framework for nature conservation, sustainable forest management and protection of plant and animal species, consistent with regional and international legislations, agreements and acts and integrated in different laws.
- Undertaking of an inventory, monitoring and research of nature and forest areas in order to obtain the basis for their planning and management.
- Increasing public awareness and enhancing local knowledge and skills in the field of conservation and sustainable uses of nature and forests.
- Developing and increasing the regional and international cooperation in forest and nature conservation.

Contribution of the Forest to the National Economy

Forests contributed 0.03% to the gross domestic product (GDP) in 2000 (Palestinian Central Bureau for Statistics, 2000, 2003). Employment in agriculture and forestry is about 13.4% of the total labour force. Over the last century, employment in forestry has decreased, in terms of both the number of jobs and the number of working hours. This was particularly true during recent decades, when most of the tree nurseries were closed. Nowadays, forest operations are performed under the responsibility of the Ministry of Agriculture (Forestry Department) and more forest rangers have been reinstated. It should be stressed that not only are research and studies dedicated to forestry issues limited, but also those analysing the agricultural sector have failed to address this nationally vital issue realistically.

The Values of Palestinian Forests

Forests in the West Bank and Gaza are limited in area and are therefore not expected to be of significant commercial use in the short or medium term. Annual wood production of natural forest ranges from 1.0 to 3.3 m^3/ha for oak stands and around 4.3 m^3/ha for moderately dense pine forests. The annual growth rate of natural forest amounts to 0.2 m^3/ha, which is much less than the threshold for commercial forests (1 m^3/ha). Artificially planted pine forests grow on average 3 m^3/ha annually.

A major reason behind the reluctance and slow rate of forestation is the relatively high investment required (at least US$2500–3000/ha) and risks in terms of uprooting or overgrazing, which add new costs to forest maintenance. In addition, forests cannot withstand competition with other agricultural crops, even with those of a low economic return in the short term.

In open areas, forest expansion is significant and sustainable only when implemented in marginal zones, which need reclamation. This requires relatively high infrastructure rehabilitation, upgrading operations and labour force training. It is essential that central government, non-governmental organizations (NGOs), other international private organizations and local communities step in to carry out such development projects.

Direct and indirect use values

Estimates of direct use values are based on data collected from traders and people in the market (Breghiet and Qanam, 1998). Timber and firewood valuation is based on the estimated amount of annual removals and their roadside prices. The value of seeds, stone fruits, medicinal plants and other natural fruits collected from the forests is estimated by using the commercialized quantities and their average market price. Dyes and other colouring items refer to the products made from natural plants growing in the forests and, therefore, have a value of transformation (Table 9.2).

Grazing occurs in agricultural lands, pastures and forests as a free activity, unconstrained by tax or other regulations as it mainly takes place on unoccupied land. During normal seasons, with average annual rainfalls varying between 100 and 250 mm, the production

Table 9.2.   Values of Palestinian forests.

Valuation method/output	Quantity	Value (US$ 1998)	Value (€ 2001)
Market price valuation			
Timber (m³)	1,500	125,000	121,250
Firewood (m³)	1,500	75,000	72,750
Seeds, stone fruits (t)	500	200,000	194,000
Medicinal plants (t)	700	245,000	237,650
Natural fruits (t)	300	90,000	87,300
Dyes and other colouring items (t)	50	15,000	14,550
Others	—	40,000	38,800
Substitute goods pricing			
Grazing (t of fodder)	5,040	504,000	488,880
Total direct use values		1,294,000	1,255,180

Source: Breghiet and Qanam (1998).

capability of pastures does not exceed 120 kg of dry matter/ha, taking into account that the moisture content of naturally dried forages is 12%. Therefore, it can be deduced that the West Bank's natural pastures – including those restricted to Palestinians – have the capacity to support 27,000–30,000 livestock annually. At present, only 10% of this area is accessible to grazing.

The value of grazing is estimated according to the amount of fodder grazed yearly in the forests and its market price. In Palestine, around 200,000 head of livestock graze in forests and pastures for a period of 180 days/year. Based on the daily consumption equivalent to 1.4 kg of fodder/day, it results in a total consumption of 50,400 t of fodder. At a price of US$100/t of fodder, the value of grazing in forests and pastures equals US$5.04 million. Assuming that only 10% of total fodder consumption comes from woodlands, the value of forest grazing equals US$504,000. It is worth mentioning that livestock is increasingly dependent on concentrates (0.7 kg/day) as a supplement to grazing due to the thinning vegetation cover in pasture areas and the increasing scarcity of many plants.

In total, the value of forest products reaches US$1.3 million. This low estimate can be explained by the lack or even absence of information regarding many forest products, either those which are commercialized or, in particular, those which are self-consumed. Accurate estimates are often difficult to obtain, also due to the impossibility in distinguishing between those products collected either in the forest,

agricultural land or other ecosystems. However, some authors argue that forests are expected to have a higher annual production and economic value, close to US$9 million (Breghiet and Qanam, 1998).

Of the indirect use values, only valuation of carbon sequestration is attempted. It can be approximately assumed that on the 23,159 ha of forests there are around 7,719,600 trees, and each tree annually sequesters on average some 6 kg of CO_2. Applying a carbon transformation coefficient of 0.27, this gives a quantity of 4406 tC. Valued at US$20/tC, the monetary value of carbon sequestration is US$88,120. However, this figure does not account for the carbon released through deforestation and other losses, thus the valuation is very partial.

Negative externalities linked to forests

Landscape, vegetation and flora of Palestine have been subjected for thousands of years to large-scale change. Deforestation, transhumance, grazing, agriculture, fire, plantation forestry, introduction of exotic species, urban and industrial development, tourism, population growth and movement, and land confiscation have dramatically altered the face of Palestine. The biomass is now comprised of remnants of natural and semi-natural vegetation in a mosaic of agricultural land, planted forest areas, wasteland, roads, industrial landscapes and urban settlements. Today many wild plant species occur in small, fluctuating and poorly dispersed populations as a result of habitat fragmentation, and face extinction or severe genetic loss.

The people who arrived in the ancient geographical area of Palestine were hunter–gatherers. Their habits and numbers were such that they interfered little with their host ecosystem. However, once they began settling and farming the fertile valleys and hillsides, they encroached quite radically upon the ecological *status quo*. Cultivated crops replaced native vegetation, as vines and orchards substituted native forest and shrub land, maintained with terracing, man-made irrigation channels and drainage ditches. A balance was maintained providing that this network of human alteration of the native vegetation composition and landscape received the attentive human maintenance it required.

Unfortunately, a continuity of maintenance was prevented by the succession of wars brought to the area by subsequent waves of conquerors who killed or drove out the farmers. The so-called human improvements and cultivated vegetation were left abandoned and fell into ruin, leaving the soil, now deprived of even its native cover, subject to extensive erosion. This was especially true at the higher elevation where the scarcity of soil, combined with the erratic rainfall of the region, rendered it difficult for the original vegetation (mainly forests) to reassert itself. The vegetation was depleted further by the cutting of trees for several purposes, mainly for the construction of colonies, which has been undertaken intensively especially during the last 50 years.

Israeli military activities are probably the most environmentally damaging. A total of 2975 ha of land was destroyed in the Gaza Strip between September 2000 and March 2001. In addition, clearcutting of around 112,900 trees was reported during the years 2000–2003 (Ministry of Agriculture, 2003). Overgrazing is another factor damaging the forests and causing negative externalities, in terms of soil erosion, desertification and species loss. Estimating these negative externalities in monetary terms is not possible due to insufficient data related to the many variables and, more particularly, the complete lack of quantitative information.

Towards the Total Economic Value of Palestinian forests

The contribution of forests to the total economic value (TEV) is measured not only by the tangible products they provide, but also by the services they offer, which in this case is difficult to quantify. A significant proportion of forest products is consumed by those who collect them, in amounts varying according to seasonality, access and options. Most of the available information related to the contribution of forests to the national economy is descriptive and often site specific. No studies quantifying the proportion of household inputs, labour allocation, income and costs attributable to forest products have been conducted.

However, despite the shortage of information related to the other TEV components, the available estimates of direct uses of forests in Palestine show that grazing is the most significant forest use (39%), followed by collection of medicinal plants (19%) and other NWFPs (15%). Timber and firewood are less important with just 10 and 6%, respectively.

Perspective and Crucial Issues

Change in the current situation can be brought about in the long term by means of education that needs to engender a significant change in attitudes. However, the complexity of pressures threatening the Palestinian forest ecosystems also demands an immediate response. The most rapid and significant change can be achieved by means of a clear, comprehensive and effective legislation, which will be readily enforced to carry out the forest policy recommendations prepared by the Ministry of Agriculture.

Palestine has few institutions whose aim is to record and study its biodiversity. In most countries, national museums of natural history and herbaria assume this role. Very few data have been collected regarding forests and their value. There is no formal biological survey and the studies undertaken by Israeli scientists many years ago are no longer accurate. The Ministry of Agriculture's Forestry Department and a few Palestinian NGOs have been working on preliminary studies concerning plants in general and woody plants in particular. However, no comprehensive or systematic research of forests and forest species has been conducted (Orni, 1978).

Although, generally speaking, there is a growing awareness in Palestine of the need to conserve natural resources, there is a great lack of formal institutions for studying, recording and monitoring forests. It is imperative that a biological survey is undertaken in order to monitor changes in forest ecosystems. However, national and international funding for these activities is crucial and national guidance is required in order to ensure the continued monitoring of the country's natural resources.

Efficient planning for new forests and sound management of existing ones both require detailed surveying and mapping of relevant areas. Appropriate development organizations should encourage research in ethno-biology to identify plant and animal species used by local people living in the surrounding forests. The specialized knowledge accumulated at a local level regarding the economically useful plant and animal species as well as the manner in which the ecosystem functions (including the likely effect of certain human disturbances) can be of use to modern society. Such research would prevent the forests from being irretrievably lost (Ledec and Goodlands, 1988).

Notes

[1] Overall, more than 0.37 Mha were confiscated, including almost 93% of the total forest area and rangelands – most of which were declared as Israeli closed military areas and bases.

[2] Most of the natural forest in the Central Highlands and the Semi Coast is concentrated in Jenin district, with well-developed dense evergreen oak stands as in Um Rehan and evergreen oak open woodland as at Um Al-Tut.

References

Applied Research Institute Jerusalem (2000) *Geographic Information System and Remote Sensing Unit*. GIS database, ARIJ, Bethlehem, West Bank.

Breghiet, A. (1995) *Forest and Woodland in Palestine from 1950 to 1995*. PNA, Ministry of Agriculture, Ramallah Governorate, West Bank.

Breghiet, A. and Qanam, K. (1998) *Implementation of Modern Technology in the Development of Forest Resources*. Ministry of Agriculture, Department of Forestry, Ramallah Governorate, West Bank.

Brown, S. (2001) Measuring carbon in forests and future challenges. *Environmental Pollution* 116, 363–372.

Euroconsult/Iwaco (1994) *Gaza Environmental Profile*, Parts 1, 2 and 3. Mopic, EPD, Gaza Palestinian Territories.

Isaac, J., Qumsieh, V., Owewi, M., Hrimat, N., Sabbah, W., Sha'lan, B., Hosh, L., Bassous, R., Al Hodali, D., Al-Dajani, N., Abu Amrieh, M., Al-Junaidi, F., Neiroukh, F., Sleibi, O., Al-Halaykah, A., Quttosh, N., Al-A'raj, I. and Zboun, I. (1997) *The Status of the Environment in the West Bank*. ARIJ, Bethlehem, West Bank.

Kaplan, D. and Gutman, M. (1996) Effect of thinning and grazing on tree development and the visual aspect of an oak forest on the Golan heights. *Israel Journal of Plant Sciences* 44, 381–386.

Ledec, G. and Goodland, R. (1988) *Wildlands, Their Protection and Management in Economic Development*. The World Bank, Washington, DC.

Ministry of Agriculture (1999) *Forest and Nature Strategy*. Fourth Draft. Palestinian National Authority, Ministry of Agriculture, Ramallah, West Bank.

Ministry of Agriculture (2003) *The Palestinian Agricultural Losses Due to Israeli Aggressions (September 29, 2000–January 31, 2003)*. Palestinian National Authority, Ministry of Agriculture, Ramallah, West Bank.

Orni, E. (1978) *Afforestation in Israel*. Ahva Press, Jewish National Fund, Jerusalem.

Palestinian Central Bureau for Statistics (2000) *Population Housing Establishment Census*. PCBS, Palestine National Authority, January 2000, Ramallah, West Bank.

Palestinian Central Bureau for Statistics (2001) *Impact of the Israeli Measures on the Economic Conditions of Palestine Households*. PCBS, Palestine National Authority, Ramallah, West Bank.

Palestinian Central Bureau for Statistics (2003) *National Accounts at Current Constant Prices (1994–2000)*. PCBS, Ramallah Governorate, West Bank.

Palestinian Environmental Authority (1999) *National Biodiversity Strategy and Action Plan for Palestine*. Palestinian Environmental Authority, Ramallah Governorate, West Bank.

Shmida, A. (1995) *General References on Biodiversity and Theory of Ecological Richness, Especially Concerned with Arid and Mediterranean Ecosystems*. The Hebrew University, Jerusalem.

Zohary, M. (1962) *Plant Life in Palestine*. Ronald Press, New York.

Zohary, M. (1966) *Flora Palaestina*, Vols 1 and 2. Israel Academy of Science and Humanities, Jerusalem.

10 Israel

Avi Gafni

Forest Division, Land Development Authority, Jewish National Fund (JNF),
Mobile Post Shimshon, KKL Eshtaol, Israel 99775

Introduction

Israel is located on the eastern shore of the Mediterranean Sea, with a total area of 21,920 km². Its southernmost tip extends to the Gulf of Aqaba, an arm of the Red Sea. The country can be divided into five physiographic regions: the highlands and valleys of Galilee; the Judean and Samaria hills; the Mediterranean coastal plains; the Negev; and the Great Rift Valley. The Golan Heights, east of the upper Jordan River, is a natural region under the control of Israel.

The mountains and valleys of Galilee dominate the northern section of Israel, extending about 40 km from the narrow coastal plain across to Lake Kinneret (also known as Lake Tiberias or the Sea of Galilee). Excluding Mount Hermon at the northern tip of the Golan Heights, the recognized highest point in the country is Mount Meiron (1208 m) at the very centre of Upper Galilee. Galilee is the most naturally forested region in the country, enjoying an average annual precipitation of around 900–1000 mm. The mountains of Galilee are separated from the Judean and Samarian hills to the south by the plain of the Yizre'el Valley – the most fertile and agriculturally productive area. The Judean and Samarian hills run north to south, forming the mountainous backbone of central Israel west of the Jordan River, with an annual precipitation of around 600 mm (Encyclopedia Britannica, 1980).

The coastal plains consist of the Plain of Zevulun, the Plain of Sharon and the Plain of Judea. They contain most of Israel's large cities, industry and commerce. The central and southern sections of the Great Rift Valley and the Negev are the most arid parts of the country, forming some 70% of its total area. While most of Israel falls under the definition of a semi-arid country with long, dry summers and short wet winters, the Negev and the Great Rift Valley are particularly low-rainfall regions, with less than 200 mm/year. They constitute the greatest challenge for greening the country efforts, either by farming or by afforestation. Part of the solution for both these nationally important activities (agriculture and forestry) lies in large-scale water-conveying systems from the north to the south and exploiting marginal water resources in the Negev itself.

From the very inception of modern Israel, afforestation of the hillsides, or restoring the old-time vegetative appearance of the country, as vividly described in the Bible, was an integral element of developing and settling the country throughout its length and breadth. The plantation efforts, initiated at the beginning of the 20th century, resulted in tens of thousands of hectares planted in central and northern Israel. Some of the afforestation endeavours in Israel are unique in the sense of a 'lack' of environmental logic in planting on semi-arid lands, such as the Yatir forest in the northern Negev at the fringe of the desert. Yatir forest constitutes an interesting case study in understanding the capacity of semi-deserts to support pine forests, along with the impact of those forests on the environmental conditions: climate, hydrology and the atmospheric carbon sequestration status.

Forest Resources

Area

The total forest area in Israel is 193,100 ha, divided into natural forests (96,000 ha) and planted forests (97,100 ha) (Fig. 10.1). The total forestland area under the jurisdiction of the Jewish National Fund (JNF), as of February 2004 (Tauber, 2004), encompasses 98,348 ha, of which 52,222 ha (53.1%) are conifers, 26,198 ha (26.6%) are broadleaved forests, 5385 ha (5.5%) are other species including sparsely planted trees and experimental plots and 14,543 ha (14.8%) comprise forest roads, forest firebreaks, areas designated for future afforestation and other non-used lands.

Of the broadleaved forests, 8207 ha (8.3% of JNF forests) are planted with various eucalyptus species, 11,219 ha (11.4%) are made up of other local and introduced broadleaves, 2761 ha (2.8%) are covered by fruit trees, 3963 ha (4%) are natural woodlands and the remaining 47 ha (0.0%) are covered by exotic ornamental bushes. Of the total conifer forests, 41,552 ha (42.3% of JNF forests) are pines and mixed conifers, 3041 ha (3.1%) are cypress, 7435 ha (7.6%) are mixed conifers and broadleaved, and 194 ha (0.2%) are other conifers.

Thousands of hectares of forests annually are destroyed by fires, which average 1000/year. Among the various reasons for the frequent forest fires – to the degree that the actual causes are traceable – arson is estimated to account for 60% of fires occurring on JNF-managed forests (603 in 2003, affecting 2077 ha). JNF-planted forests bear the greatest brunt of the forest fire catastrophes; although it varies considerably from year to year, 3400 ha of damage per year is a typical figure.

Typologies

The forests of Israel are distributed among the different areas of the country as follows.

Mountainous region

Most of the woods in the mountainous region, which can reach the considerable height of 1000 m, consist of the common oak group (*Quercus calliprinos*) and the Palestine terebinth (*Pistacia palaestina*). They are usually shrubby as a result of having been cut or browsed by sheep and particularly by goats (Feliks, 1973). This bush grows extensively on mountains of altitudes between 300 and 1000 m above sea level. There is also the gall oak (*Quercus infectoria*), a deciduous tree with a tall trunk, alongside which grows the hawthorn (*Crataegus azarolus*). Due to generations of intensive agriculture on the mountain slopes, natural forests of Jerusalem pine – the Aleppo pine (*Pinus halepensis*), the tallest forest tree in Israel – have been severely curtailed and have survived in only a few places. Nowadays, Jerusalem pine, well suited to the local natural conditions, as well as brutia pine (*P. brutia*), are the main planted trees in the mountainous region.

Associated with the pine are the common oak (*Q. calliprinos* Webb), the Palestine terebinth (*P. palaestina* Boiss.), the mastic terebinth (*P. lentiscus*), the carob (*Ceratonia siliqua*), the arbutus (*Arbutus andrachne* L.) and the rhamnus (*Rhamnus alaternus* L.), as well as many shrubs and wild grasses. Under favourable humid conditions, the sweet bay (*Laurus nobilis*) and the Judas tree (*Cercis siliquastrum*) also grow in this region. On the western ridges of Carmel and western Galilee and on the western slopes of the Judean Mountains, there is *maquis*, where groups of carob (*C. siliqua*) and the mastic terebinth (*P. lentiscus*) grow, along with many species of shrubs, climbers, annuals and perennials. A third species of oak – the Tabor oak (*Quercus ithaburensis*) – predominates on the western ranges of the Lower Galilean mountains, accompanied by the storax tree (*Styrax officialis*). In the northern Huleh Valley, it grows alongside the Atlantic terebinth (*Pistacia atlantica*). These two species of trees are the largest in Israel, some in the vicinity of Dan (Upper Hula Valley) having trunks 6 m in circumference and a height of 20 m.

Another genus of Mediterranean plants comprises flora groups called *garrigue*, which in Israel consist predominantly of shrubs and dwarf

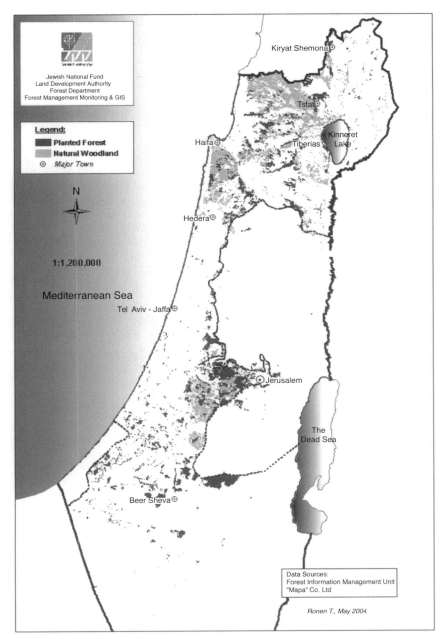

Fig. 10.1. Map of planted forests and natural woodlands in Israel. Source: Ronen (2004).

shrubs no taller than a man. The main plants of the *garrigue* are the calycotome thorn bush (*Calycotome villosa*) and sage (*Salvia tribola*). Characteristic of the unforested Mediterranean landscape are dwarf shrubs, of which the most widespread is the rium thorny burnet (*Sarcopoterium spinosum*). Reaching a height of less than 0.5 m, it grows densely and is one of the principal factors in preventing the erosion of mountain soil.

The coastal region

The soil in this region is sandy or a mixture of sandy chalk and sandy clay. It is poor in organic matter and has a low water-holding capacity. Plants typically develop deep roots and are adapted to occasional water shortage. There are also annual plants that sprout and ripen during the rainy winter months. Here flora of Israel's three phyto-geographic areas can be found, as well as that of the Sudano-Deccanic, such as the sycamore fig (*Ficus sycomorus*) and the wild jujube (*Zizyphus spina-christi*).

Near the sea, where the winds carry sea spray on to the flora, plants resistant to salty sea water grow, for example the Russian thistle (*Salsola kali*) and species of marigold fig (*Mesembryanthemum* sp.). In the recent past, most of the plains along the coast were planted with citrus groves. These are gradually disappearing. In places characterized by hard sandy–chalk soil, groups of carob (*C. siliqua* L.) and mastic (*P. lentiscus* L.) can also be found.

The semi-arid regions

These regions extend on the loess of the northern Negev, the Judean and Binyamin Desert (north, southern and eastern Israel). Here the climate is dry, with rainfall varying from 200 to 300 mm. In this area, there are practically no natural forests but only sparse trees, such as the plant association of the Atlantic terebinth (*P. atlantica*) and the lotus jujube (*Zizyphus lotus*). Plants of the *Retama duriaei* association characterize the slopes bordering the Judean and Beit-Shean Valleys, whereas the most important plant association is of a species of wormwood (*Artemisia herba-alba*) that grows extensively in the Negev and the Judean Desert.

The arid region

This area, which extends over most of Israel (central and southern Negev and the Dead Sea region), has the poorest flora. Its annual rainfall amounts to less than 200 mm (usually much less) and varies considerably from year to year. The soil is infertile and includes Hamada, desert, gravel and rocks with some saline tracts of land bare of vegetation of any kind. Trees grow only in the dry stream beds. The most

typical plant in the Hamada is the small shrub *Zygophyllum dumosum*, which is capable of surviving in areas with a rainfall of less than 50 mm. Most species here spring up and flower quickly after a shower of rain; some scatter their seeds only a few weeks after germinating and are capable of preserving their germination potential for many years. Many species of desert flora have a great ability to absorb groundwater; one species, the *Reaumaria palaestina*, develops an osmotic capacity of more than 200 atmospheres. In the sandy desert regions, the flora is usually more abundant, the predominant species here being the haloxylon (*Haloxylon*) and the broom (*Retama roetam*). In the Arava and in the lower Jordan Valley, where there is widespread salinity, saline flora, including species of atriplex (*Atriplex*) and salicornia (*Salicornia herbacea* L.), grow densely.

Swampy areas in the northern valleys and the northern coastal plains

Hydrophylic flora grows near expanses of water in all areas of Israel. Large numbers of poplar (*Populus euphratica*), as well as willow (*Salix* sp.) and tamarisk species (*Tamarix* sp.) grow on riverbanks, as does plane (*Platanus orientalis*) and Syrian ash (*Fraxinus syriaca*) in the north. Alongside these trees, oleander (*Nerium oleander*) usually grows, together with numerous species of annuals and perennials. The reed and the cattail are found near almost every expanse of water.

Afforestation

The present forest mantle of Israel is due primarily to over 120 years of tree planting efforts, since the early return of the Jewish settlers to Israel during the 1880s (Box 10.1). It is widely accepted by scholars (Zohary, 1959; von Forbrig and Grov-Umstadt, 1989; Bonneh, 2000; Liphschitz and Biger, 2000) that the natural woody vegetation of the country was exploited and practically eliminated during centuries of overgrazing, felling, land preparation for crop cultivation, development and use as an energy source. Early plantation during the late

Box 10.1. Afforestation in the Negev: combating desertification.

A great emphasis in the past three decades has been placed on afforestation in the more arid parts of the country – the Negev region, which constitutes about 60% of Israel. The total area planted in the Negev amounts to 25,067 ha (Tauber, 2004). The Negev comprises three out of the four dry-land types (hyper-arid, arid and semi-arid). The 'greening' of the Negev and combating desertification trends is of utmost national and environmental importance, for which significant resources are allocated annually by the JNF and the government of Israel through its various ministries.

Sustainable forest plantations[1] in the Negev are first and foremost an issue of efficient water harvest purported to ensure sufficient moisture for each planted tree (in excess of the direct natural rainfall inputs).To achieve this goal, various innovative techniques were adopted, some having roots in the Roman–Byzantine period (Moshe, 1989). In principle, micro-walled catchments (limans), elongated cross trenches or simple pits, dammed on the down-slope side (*molehills*), contour stripes are constructed, so as to collect the runoff developing on the larger, up-slope contributing area, during the few, short-lived rain storms (Kligler, 1989).

Various models of water harvest and runoff manipulations were experimented with over the years to utilize every drop of water that occurs in the region, down to the 90 mm rainfall isohyets (the average annual rainfall). In addition, research and development efforts, employing more intensive means, such as 'Brush Blankets', protecting sleeves (tree shelters), hydro-gels, mycorrhiza, etc., were also gradually introduced to increase the rate of sapling establishment and survival. Selection of the tree species for planting in the plains and the barren slopes of the Negev hilly territory was undertaken after long trials and when sound evidence of the species' ability to survive under the characteristic extreme Negev conditions had been obtained.

Despite the frequent droughts, great variance in the rainfall amounts from year to year, the deep groundwater resources, typical saline/sodified soils and other adverse conditions, planted forests in the Negev region are largely successful. The Yatir Forest (2800 ha, 35 years old) on the sloping hills of southern Judea (350 mm rainfall) is a striking example of the capacity of the Jerusalem pine (*Pinus halepensis*) to adapt to the local conditions by adjusting its physiological function, conserving soil water and by manipulating the carbon sequestration processes (Grunzweig *et al.*, 2003). Other planted, well-functioning forests in the Negev region (Lahav, Meitar, Lehavim, Syeret Shaked and in the periphery of Beer Sheva) as well as the eucalyptus groves in the western Negev indicate the feasibility of sustainable afforestation in the semi-arid terrains.

Plantations in the Negev serve also as a means to enhance biodiversity, provide shade for the nomads, avail parks and a green setting for recreational facilities, improve the microclimate, screen out dust storms and 'green' the landscape. The afforestation activities in the Negev are steadily supported by research based on field trials conducted by various academic research institutions, as well as by the JNF applied field research unit. Thus, more than 200 exotic species of eucalyptus were introduced and trialed under the real Negev conditions in the past 40 years.

The genetic mapping of elite, stress-tolerant trees was performed to select the most suitable strains for propagation; probabilistic hydrological studies were undertaken to determine the most prospective sites, timing of planting and the optimal irrigation techniques, to ensure the establishment of young saplings. The above is just a sample of the ongoing research that accompanies the ever-expanding afforestation of the Negev. Being a vast 'last frontier' of Israel, the northern Negev region constitutes the only territory reserved for future afforestation, whereas in central and northern Israel, the high demand for housing, construction and development has left virtually no areas for new forest plantations, except for deserted agricultural lands converted to farm forestry.

Ottoman regime was not for afforestation *per se* but for the purposes of drying out swamps and marshes, employing eucalyptus species imported from Australia. In 1901, the JNF, presently the Israeli forestry authority, was founded; its ultimate goals were to purchase land, plant trees and settle newcomers to the country. In 1909, the first olive plantation was planted by the JNF on 200 ha in central Israel (Hulda) and then on the barren slopes west of the Sea of Galilee. This pioneering effort to plant fruit trees failed due to low profitability and limited professional knowledge. The policy then changed and the new plantations employed non-fruit trees, such as pines and eucalyptus.

Up to the beginning of the British Mandate (1918–1948), only local, isolated plantations were initiated in the then few settled parts of the country. A countrywide survey conducted in 1920 (cited by Liphschitz and Biger, 2000) showed that there were some 60,000 ha of natural degraded forests and natural woodlands in the Galilee, Judea and Samaria (then Palestine). The existing forests were tended while the afforestation efforts commenced. In 1926, the British Government issued The Forestry Ordinance aimed at protecting the forests and defining the basic laws concerning their management. In this framework, 430 forest reserves were demarcated throughout Mandatory Palestine, totalling 83,000 ha. During the British Mandate, an organized and professional afforestation scheme was carried out by the combined efforts of the governmental Department of Agriculture and Forests and the JNF. Afforestation in various parts of the country was undertaken on land considered unsuitable for agriculture, being under the control (ownership) of the JNF. Nurseries were also created that ensured the supply of about half a million seedlings during the British Mandate. The major trees planted included pine (*Pinus halepensis*, *P. pinea*) and eucalyptus (*Eucalyptus camaldulensis*), although other prospective species were introduced and trialled.

Statistically, at the end of 1945, of the 70,000 ha of land then under JNF control, only 1750 ha were afforested: about 2.5%. One of the early planting objectives of the British Forestry Department was the stabilization of the sand dunes along the coastal plain. In terms of long-term planning, by 1945 only 12% of the designated area for afforestation was actually realized (Liphschitz and Biger, 2000). According to Kaplan (1999), by 1948, 8000 ha of lands in various parts in central and northern Israel were afforested, of which the British Government planted 4100 ha and the JNF planted 3900 ha.

Upon the establishment of the State of Israel in 1948, the rate of forest planting was significantly increased to 1000–2000 ha/year. Motivated by the ideals of greening the country, all land not suitable for agriculture was, in principle, subjected to afforestation. Protection against soil erosion and dust storms, providing rural employment, as well as climate moderation, were also important rationales. Since 1959, the JNF has been the sole agency in charge of planting and managing the forests and the natural woodlands in Israel. The largest planted forests in Israel were initiated in the 1950s and 1960s, when a total of 40,000 ha were planted. Since the 1970s, the rate of forest planting declined due to decreasing land availability and the need to allocate a large proportion of the financial resources to maintain the existing forests. In the past three decades, around 57,100 ha were added, making up the present total of afforested lands in Israel of 97,100 ha.

Despite more favourable environmental conditions in central and northern Israel, the afforested areas are roughly evenly distributed among the central, northern and southern major regions of Israel. However, forest establishment activities are by necessity more intense in the south, as is the investment. Survival rates in new plantations are constantly improving with adoption of new techniques (site preparation, propagation material, planting methods and others). In addition, innovative supportive irrigation regimes, conducted by the research and development teams, are steadily tested and applied, if proved effective. Monitoring survival rates of new plantations is performed annually as a matter of routine.

A tree survival survey was undertaken on a planted area of 789 ha in 2003. It showed that the overall (countrywide) survival rate was 76%. The highest success occurred in the northern region (83%), whereas the lowest rate was in the central region (52%).

Institutional Aspects

The Palestine Forest Ordinance of 1926 was the first legislation aimed at protecting the natural woodlands and set out the laws concerning new forest plantations. Pre-dating the founding of modern Israel by more than 20 years, it provided both framework and precedent for progressive legislation in the country, which took the form of new edicts to extend the amount of protection to the forest and its trees (Weitz, 1970). The Forest Edict of 1956 (promulgating a list of protected trees) contains three supplements detailing the various species[2] covered by the protection law.

To help enforce these few ordinances and edicts, inspectors and wardens were required. In 1960, new Forestry Regulations were introduced under the 1926 Forest Ordinance Art. 26, according to which 'forest officers' were to be appointed with the power to grant licences for tree cutting. Upon the establishment of the State of Israel (1948), forest laws came within the domain of the Ministry of Agriculture. Since 1959, all forestry operations, in both natural and planted woods, have been in the care of the JNF Forest Department, which has also assumed the responsibilities of wardenship (Weitz, 1970).

In 1961, a covenant was signed between the Government of Israel and the JNF, detailing the authority, functions and responsibilities delegated to the JNF with respect to Israel's land development and afforestation. Clause 10 of the Covenant reads: 'The reclamation and afforestation of Israel's lands shall be concentrated in the hands of JNF, which shall establish a *Land Development Administration* for that purpose'.

Clause 12 states that the Ministry of Agriculture (ministry in charge) 'shall henceforth engage in afforestation research only. However, the Minister of Agriculture shall continue to be charged with the implementation of the 1926 Forestry Ordinance, through JNF – Land Development Administration'. Administration of all State lands including JNF land (17% of Israel's land mass) was handed over to Israel Lands Administration (ILA). More specifically, ILA is charged with leasing out the State lands, including those owned by JNF – which make up 17% of the public (State) lands. The revenue obtained by ILA from leasing out the JNF lands constitutes a major source of income for the JNF (60% of the budget).

Recognizing the weakness of the Forestry Ordinance of 1926 in directing the evolving forestry needs and the ambiguity of the authorities, the Forestry Department of the JNF has initiated several amendments and, even better, proposed a new forestry law. Under this tentative law, the JNF would be empowered to be the sole forest agency for the management of forests and trees. However, these amendments and the long-due foresty law have not yet been implemented. Obscure areas of authoritative dichotomy still exist between the JNF and the Ministry of Agriculture.

In 1959, JNF was mandated by the Government of Israel to manage the State forests and woodlands. As such, the JNF is referred to as the authorized Forest Service of Israel, with all its derived implications: professional, statutory, planning, maintaining, utilizing in-forest resources, protecting, expanding and making available the goods and services of the managed forests to the public.

The operative goals of the JNF Forestry Department are established every year in the annual plans and are given budget allocations. These have changed over the years as each time period brings new priorities and preferences. For instance, the growing importance of recreation and outdoor leisure time by the 1970s became a central issue in the strategic as well as in the operational plans of JNF forests. In recent years, this function within the JNF forest plans has been significantly expanded, extended and accordingly heavily funded (Avni, 2004).

Another initiative, known as National Master Plan 22 (NMP22) (Kaplan, 1999), originating from the late 1980s, aims at assisting the Israeli government with its future country-wide master development plans. The rationales of NMP22 are twofold: (i) to conserve existing planted forests and natural woodlands; and (ii) to ensure sufficient forestlands and green open spaces, depending on the population distribution and the particular needs of the region. NMP22 was submitted and approved by the Israeli government in November 1995.

Besides providing some measure of protection to the existing and future forests (although, as yet, not an absolute protection), NMP22 has helped re-map, designate and logically organize all forest areas (the green natural resources) in the country. NMP22 clearly defines the 'green' needs of the public and turns them into an official, policy-directing, forestry document that each land designer as well as any State developing planners should consider. For the JNF, NMP22 is the most important statutory document developed and approved in the past 30 years (Box 10.2). NMP22 has extended the forests' recognition and jurisdiction, as well as presenting them with new challenges and opportunities. Meeting these challenges will undoubtedly benefit both residents and visitors.

Box 10.2.   The National Master Plan for Afforestation (NMP22).

NMP22 embraces a total area of 162,000 ha, mostly north of Beersheba – some 18% of the State territory in the northern region. The plan classifies and specifies eight types of forests and afforestations
(existing planted, suggested planted, existing forest park, suggested forest park, natural woodlands for
management, forest for conservation, coastal forest parks and riverbank plantings) and stipulates guidelines for their conservation, management and incorporation in planning according to the needs of the
State and society.

    NMP22 lends statutory protection to Israel's forests, whether natural or planted, existing or
planned, in order to safeguard open spaces from the accelerated pressures of development and construction. The country's forests are multi-faceted, serving important ecological and social functions. Forests
not only allay the greenhouse effect by absorbing carbon dioxide and supplying oxygen but, in Israel,
they also act as buffers against congested built-up areas, filter out pollutants, dust and noise, and
conserve soil against erosion and depletion. NMP22 was devised with the following objectives in mind:

- To conserve forests and natural woodlands as ecologically integral to Israel's landscapes.
- To allocate forestland according to population distribution, providing the public with nearby,
  green, open spaces suitably adapted to outdoor leisure and recreation.

Thus, NMP22 views the country's varied forests and woodlands as a haven and natural habitat for
numerous forms of plant and animal life. It strives to preserve Israel's unique biodiversity and cultivates
forests and woodlands typical of each and every region.

    In recent years, forests have also been developed for recreation and tourism purposes, in addition
to their role in soil and environmental conservation. Israel's outdoor leisure facilities rely heavily on
the infrastructure of planted forests, parks and recreation areas established by the JNF. NMP22 seeks
to earmark green space for rest and relaxation in both forest blocks and green belts planted around
communities.

    Finally, NMP22 relates to the economic value of forests: creating jobs in recreation, tourism and
forestry; appreciating land for residential areas and tourism sites; increasing timber supply for local
industry, and preventing erosion and soil loss.

Contribution of the Forest to the National Economy

Employment

Forestry has played a central role during
various stages of Israel's development as a
source of immediately available work that
requires relatively minimal investments in
tools and supply (Weitz, 1970; Kaplan, 1999;
Ginsberg, 2000). During the 1950s and 1960s,
when waves of newcomers arrived in the country, afforestation was a significant employer
(6000 workers) resulting in a tenfold increase
in the forest cover (from 3300 ha to 34,600 ha).
In the early 1990s, when another 1 million
Russian immigrants arrived in Israel, many
were temporarily employed in various
afforestation and land development projects.
At present, the JNF alone employs 1400
workers, but its myriad of countrywide land
management activities support many other
entrepreneurs.

Wood products from the JNF forests

Timber utilization from the JNF forests was
never a primary goal in managing the national
forests, as other more environmentally oriented
priorities were at stake. Wood products rather
were by-products of the silvicultural activities. If
the traditional uses of natural forests since biblical times, such as building materials, charcoal,
fuel and agricultural tools, are not taken into
account, the 1950s might be considered as the
start of timber utilization from the JNF forests
– primarily for the fibreboard industry. In the
1950s, agriculture was in its prime and the
demand for poles, stakes, hand tools, and fruit
and vegetable packing boxes was constantly on
the rise, using coniferous (pines and cypress)
timber (Spetter, 1989). However, all these uses
were low-value products, resulting from thinning the existing forests, or felling mature or
burnt woodlots. While in the 1950s the annual
wood yield from conifers (soft wood) amounted
to some 2200 m^3, it increased to 37,300 m^3 in

the late 1980s. Utilization of broadleaved wood (mostly eucalyptus) began in the 1960s, when mature eucalyptus forests were clearcut at the end of the first coppice rotation. Eucalyptus forests of the north show very good growth owing to the deep soils in the plains, high groundwater and reasonable amounts of precipitation. Oak woodlands in the northern mountain ranges were used principally for charcoal production on a scale of 1000 m³/year. All sources of timber taken together added up in the 1990s to an annual wood yield of 100,000–120,000 m³ from the JNF forests.

Medium-return, commercial forestry, either by companies or by individual farmers, was not initiated in Israel until the mid-1990s, when the medium density fibreboard (MDF) processing line was built in the Lower Galilee. The annual timber consumption for MDF (with the intention of using local eucalyptus and pines) was on the scale of 100,000 m³. The MDF plant functioned for a while, both as a market and as a catalyst for the emerging small-scale forest tracts for production. None the less, the 10-year supply contract signed between the plant and the JNF, instead of promoting the embryonic wood industry in Israel, turned out to be an excuse for a failure, the reasons being neither economic nor reflecting the potential of a solid commercial forestry in Israel. Initiation of the plant and the concurrent decline of the cultivation of traditional field crops prompted several farmers to consider planting eucalyptus for profit, despite the non-attractive returns expected from the MDF plant (€20/m³ at the factory gate). The MDF plant was sold and disassembled in 2002.

The demise of the MDF plant did not discourage other entrepreneurs – a new commercial forestry-related venture appeared called EUCALYPTOP. This new firm attempts to establish commercial afforestation along with the wood production industry in Israel, based on genetically improved, economically promising eucalyptus ecotypes, developed in Israel or as imported seeds from Australia. Because the JNF refrains from any commercial engagement, EUCALYPTOP resorted to the Ministry of Agriculture to seek recognition and professional assistance with a view to establish the entire forestry timber production line: from a modern tree nursery to a range of sawn timber end-products, such as home, kitchen and garden furniture.

Some 500 ha of eucalyptus plantation plots have been developed so far in various parts of the country, mostly north of the 500 mm isohyets. The new venture is based on a long rotation of up to 16 years, while shorter interim rotations will allow the farmer-forester to obtain some cash flow along the way. This new initiative leans on the results from 10 years of intensive research in the northern Yizre'el Valley, which tested the viability of growing selected eucalyptus trees, for attaining both environmental and commercial goals (Gafni and Zohar, 2004). This approach, known as biological drainage–commercial forestry, could prove viable if fundamental conditions of deep soils, high groundwater, sufficient rainfall and long hours of radiation are met. In Israel, in the plains north of the 400 mm isohyets, many hectares of abandoned agricultural lands are available that satisfy these requisites. Biomass yield of selected ecotypes of *E. camaldulensis* during the 10-year study in the Yizre'el Valley was around 30 t/ha/year. This yield is believed to be economic, but with seed-source improvement, along with stricter, more intensive crop management, the yield is expected to rise appreciably.

Wood industry and trade

The wood industry in Israel (wood production and wood products) includes 500 enterprises, excluding furniture manufacturers. This comprises sawmills, the plywood industry, wood products for house construction and others. Total sales in 2002 (for the local market and for export) amounted to €32 million. The plywood products serve the furniture industry primarily for kitchens and doors, but also as construction casts, fences and railings. Since most sales of wood products are to the housing industry (98% for the local market), which has been in deep crisis over the past 4 years, the whole wood industry is depressed. The number of manufacturers dropped from 1000 in 1996 to 500 in 2002, while the workforce decreased from 7000 in 1995 to 4000 in 2002 (Dan and Bredstreet, 2004). The plywood industry underwent deregulation twice in the 1990s (1992 and 1998). All raw material import limits were lifted and absolute free wood trade was established. This resulted in the rapid replacement of the locally produced plywood by cheaper imported raw material, causing a total collapse of the plywood industry and mass layoffs.

In 2002, imported wood and plywood (mostly MDF) totalled €230 million, a reduction of 16% compared with the year 2000. The export of that year amounted to only €6 million. As a share in the total value of wood imports, Israel's imports come from the following countries: Italy (12%), the USA (6.6%), Germany (6.1%), Spain (5.7%), Portugal (4.5%), Sweden (2.9%), France (2.6%) and Belgium (1.5%), i.e. 35.3% of the wood imports are from European Union countries.

Until recently, only a small number of companies controlled the importers of wood to Israel. Presently, new wholesalers, such as ACE and Home Center, are directly importing wood, thus increasing the competition and causing price reduction. Due to the current depression and privatization of the wood trade, many companies are at high risk – 45% of the 500 listed are at very high risk.

The Values of Israeli Forests

In many aspects, analysis of the values of the Israeli forests should not be separated from a wider evaluation of the so-called 'open spaces' – the non-built-upon lands in the country. The high level of importance of 'open spaces' is reflected by the approved and ongoing NMP[3] aimed at their protection; the constant increase in the number of governmental and non-governmental organizations (NGOs) dedicated to environmental protection; and the research and academic programmes currently dedicated to the subject. Yet, efforts to translate the values of the 'open spaces' into concrete monetary terms are embryonic, controversial and believed to be underestimated, since many goods and services are either ignored or unmeasurable (Eliyahu and Becker, 2004).

Direct use values

Timber production

Wood production for profit is not a goal in managing the JNF forests. The marginal income from selling felled or thinned wood logs is neither sought after nor managed for, rather the income is traded off and retuned to the contractor to carry out needed forest activities (thinning, and forest fire prevention at difficult-access high-terrain locales). The new initiative by the private sector aiming at high-yield private forestry might bring about a dramatic change in this challenging area. In 2003, timber production from JNF forests accounted for 30,000 t, with a market value of €420,000.

Honey

There is a natural lack of pasture in Israel due to water shortage, limiting nectar-rich crops. Uprooting of many citrus groves and roadside eucalyptus trees further aggravated the situation. As one alternative, beekeepers place their beehives in eucalyptus, carob and almond forests/groves, or resort to planting their own farm forestry for this particular end. Honey production ranges between 20 and 60 kg/beehive, depending on the scale of operation. Annual production is around 3500 t of honey, 55% of which comes from citrus blossoms and the remaining 45% is produced from wild flowers, herbs, eucalyptus trees, orchards and legumes (Israel Export and International Cooperation Institute, 2003). The total value of honey and pollen of the JNF planted forests amounts to €2.7 million.

Mushroom picking

There is no commercial mushroom production in the Israeli forests. None the less, as a winter hobby, it is becoming increasingly popular to pick mushrooms in the mature pine forests right after the occurrence of rain. This new pastime gained momentum after the wave of Russian emigration to Israel in the early 1990s.

Forest grazing

The forests and other woodlands in Israel provide grazing grounds for livestock, notably during the winter–early springtime. There are about 70,000 head of livestock, of which 60,000 are sheep and goats and 10,000 are cows. Grazing occurs on an area of 25,000 ha of forestlands in the Negev region alone. There is practically no charge for these herds entering the forests. Actually, the positive impact of grazing on forests and other woodlands is highly appreciated by the foresters, by reducing the high risk of forest fires and by opening up the landscape for the public and enhancing biodiversity (Henkin *et al.*, 2002).

Recreation

Of all benefits that the forest resources provide for the public, recreation should be listed as number one. The pressurized day-to-day life in Israel, combined with the still-afforded leisure times for most citizens, causes virtually a mass exodus to the forests and open spaces during weekends and holidays, somewhat less on weekdays. A countrywide survey (Fleischer, 1993) estimates roughly 10 million visitors a year in the JNF forests, with large variations among the 20 studied forest sites. Another travel cost survey estimates a total travel expenditure to JNF forests of about €16.5 million, based on the average fuel expenditure per car of about €5.5/visit and a total number of 3 million visits per year (Eliyahu and Becker, 2004).

Indirect use values

Watershed protection

In the planting operations conducted by the JNF, soil conservation, erosion control and watershed protection are important objectives. In several monitoring studies designed to evaluate the impact of the measures applied on soil erosion prevention, the results were satisfactory – gullies development was brought under control and soil transport from up-slopes significantly reduced. In a recent study designed to evaluate the hydrological regime under the growing Yatir Forest in the northern Negev (Yakir *et al.*, 2002), it was found that practically no runoff was generated during moderate rain storms – the forest contained all rain inputs.

Carbon sequestration

In detailed research conducted on the Yatir Forest (2400 ha) in recent years, the daily, weekly and annual CO_2 sequestration was measured (Yakir *et al.*, 2002). The study showed that the annual absorption of carbon by the forest amounts to around 2.7 tC/ha, similar to the world average. Despite the aridity of the region, the forest functioned exceptionally well, by adapting its internal mechanism to the prevailing conditions.

Aesthetics and cultural values

The forests, woodlands and other open spaces encompassing huge tracts of land rich in culture, archeological remains, heritage sites, biblical playgrounds, battlefields as well as serene and secluded forest spots provide endless inspiration, recollections and spiritual stimulation to the hikers/visitors. Among other activities, forests and woodlands serve for educational and field study day programmes catering for all grade levels. Do-it-yourself plant-a-tree programmes, notably during the Jewish holiday Tu B'Shvat, are representative of this culture–landscape–nature appreciation connection. Memory too is preserved in the forest through a unique Israeli tradition of dedicating groves to beloved or honoured individuals by way of a financial donation to the JNF. The forest becomes for some a sanctuary of remembrance (Ginsberg, 2000).

Conclusions and Perspectives

Four nationally important factors will surely determine the future of Israel's forests and woodlands. These are:

- An ever-increasing demand for land to be made available for construction, commerce and housing, in the more populated, central regions of the country.
- The decline of agriculture, particularly field crops, leading to the abandonment of previously extensive cultivated land.
- The increasing political strength of environmental groups, who are criticizing the JNF-planted forests of containing allegedly exotic (and controversial) species.
- The present severe financial crisis of the JNF, which will greatly impact its ability to manage the forests under its responsibility adequately.

The latter is of critical importance as many JNF forest stands are nearing their maturity and soon need to be replanted. In contrast, the role of forests and their function as 'green lungs', in both urban and peripheral areas, is now well appreciated and recognized by the public (Ginsberg, 2000; Bronshtein, 2002). The number of visitors to JNF forests and woodlands, as well as to nature reserves and national parks, is steadily increasing and small enterprises are emerging that assist in the alleviation of the poor economic situation in remote, peripheral settlements. However, the only significant areas of land reserved for future plantations are located in the semi-arid southern zones of Israel. Unfortunately, afforestation in the south is a very costly undertaking, and requires high-power research and development to support the risky endeavours. Excluding the south, it appears that the major future silvicultural efforts, in central and northern Israel, will focus on maintaining the existing forests/woodlands, by thinning, pruning and renewing the ageing stands.

The drought spells of the late 1990s and the early 2000s, particularly in the Negev, urged researchers to identify and propagate the best suited genetic sources for future plantations in the south. The unequivocal demand by environmentalists for more authentic, mixed 'Israeli' forests is met with sincere and bold efforts by JNF foresters to comply with such understandable wishes. However, this is often difficult to realize. For instance, the most desired species, the oak, is very hard to establish, let alone for its growth to be accelerated. None the less, efforts to plant mixed forests, with a stronger emphasis on broadleaved species, are of high priority in the JNF strategic planning and operations. The old 'loyal' Jerusalem pine (*Pinus halepensis*) is undoubtedly, as agreed among most researchers, the species most suitable to address the prevailing natural conditions of low rainfall, shallow, calcareous soils and lack of groundwater resources. However, due to its sensitivity to fatal diseases, it is gradually being replaced with *Pinus brutia*.

To solidify the status of the JNF as the sole authority to manage the forests of Israel independently and protect them against 'clear-cutting for construction' options – quite likely under the dynamic life in the country – the JNF is contemplating the proposal of a new law,

to be termed 'The Forest Law'. Being a territorial agency, belonging to the entire Jewish nation worldwide, the JNF is struggling to maintain its independence from possible arbitrary decisions by the Israeli government. The worst threat is the nationalization of the JNF. This would end its authority as the national forest service, along with its financial instruments.

As for the forests, the future trend is obvious – the public will become more involved in the planning and decision-making processes. For its part, the JNF is becoming ever increasingly more publicly oriented. This finds expression in many different ways: allocation of funds for parks and recreational facilities; the establishment of new planning and designing units within the JNF; the establishment of special teams to involve, interact with and guide the public; the allocation of research money to survey and identify the public needs; the public's participation in various statutory and planning committees; and other community engagement efforts.

A major challenge for the JNF is to convince the government of Israel to share the financial cost of creating new forests/woodlands/parks and to maintain them professionally. In the current 'privatization' atmosphere prevailing in the country, this might not be a simple task. It is difficult at this stage to evaluate the prospects for commercial forestry or, on a smaller scale, tree farms. More than 100 years of growing eucalyptus in the country have proved its suitability to the local conditions, in particular, in the northern plains that are characterized by deep soils, shallow groundwater and minimal rainfall. The fact that the largest sawmill in Israel (Kelet Afikim) was privatized and re-operated recently, this time by a purely profit-oriented firm, presents a practically unlimited market for locally supplied logs – it can potentially save Israel hard currency for otherwise imported wood. Nevertheless, neither the related government sectors nor the JNF are willing to commit themselves, as yet, in any serious way. Despite the hesitation of the official establishment, some farmers and municipalities have ventured into this prospective economic enterprise, which may signify the beginning of private forestry in Israel, replacing non-profit traditional farm crops. According to initial business plans, the returns expected vary from €35/m³ for 3-year-old eucalyptus poles, to over €85/m³ for mature eucalyptus (12- to 16-year-old) at the end of the first rotation. When this chapter was written (May 2004), there were some signs of official recognition of tree farms by the Ministry of Agriculture as a legitimate agricultural branch.

Notes

[1]   The trees and shrubs planted for afforestation in the Negev are: *Acacia albida, A. cyanophylla (A. saligna), A. cyclopes, A. etbaica, A. gerrardii* var. *negevensis, A. horrida, A. pendula, A. raddiana, A. salicina, A. sclerosperma, A. tortilis, A. victoriae, Callitris quadrivalvis (Tetraclinis articulata), C. verucosa, Cassia artemisioides, C. eremophila, Eucalyptus astringens, E. brockway, E. camaldulensis, E. cornuta, E. dundasii, E. ebbanoensis, E. gomphocephala, E. lesoueffi, E. leucoxylon, E. occidentalis, E. populnea, E. salubris, E. sargenti, E. spathulata, E. torquata, E. woodwardi, Ficus carica, F. sycomorus, Olea europaea* var. *communis, Parkinsonia aculeata, Phoenix dactylifera, Pinus brutia, P. canariensis, P. halepensis, P. pinea, Pistacia atlantica, P. palaestina, Prosopis alba, P. juliflora, P. nigra, Retama raetam, Schinus terebinthifolius, Tamarix articulata, T. articulata* var. *erecta, Washingtonia filifera* and *Zizyphus spina-christi* (Kligler, 1989).

[2]   These are: oak (*Quercus* sp.), terebinth (*Pistacia terebinthus*), pine (*Pinus* sp.), tamarisk (*Tamarix* sp.), arbutus (*Arbutus unedo*), sage (*Salvia*), cistus (*Cistus* sp.), Judas tree (*Cercis siliquastrum*), Storax (*Styrax officinalis*), phillyrea (*Phillyrea media*), poplar (*Populus* sp.), willow (*Salix* sp.), acacia (*Acacia* sp.), eucalyptus (*Eucalyptus* sp.), casuarina (*Casuarina*), grevillea (*Grevillea robusta*), cypress (*Cupressus* sp.), laurel (*Laurus nobilis*), thorny burnet (*Sarcopoterium spinosum*), Christ's-thorn (*Zizyphus spina christi*), lotus (*Zizyphus lotus*), retama (*Retama* sp.) and myrtle (*Myrtus communis*).

[3]   NMP8 for National Parks and Nature Reserves, NMP13 for Coasts, NMP22 for Forests and Afforestation, NMP12 for Tourism and Recreation and NMP2020 for Israel (Bronshtein, 2002).

References

Avni, Z. (2004) The Effect of Institutions on Open Space Landscapes and Preservation Patterns in Israel. PhD Thesis, The Hebrew University, Jerusalem, Israel [in Hebrew].

Bonneh, O. (2000) Management of planted pine forests in Israel: past, present and future. In: Ne'eman, G. and Trabaud, L. (eds) *Ecology, Biogeography and Management of* Pinus halepensis *and* P. brutia *Forest Ecosystems in the Mediterranean Basin.* Backhuys Publishers, Leiden, The Netherlands, pp. 377–390.

Bronshtein, E. (2002) The Attitude of Carmel Coast Residents to Open Spaces – Nature Reserves, National Parks and Planted Forests and Their Implications for Regional Sustainable Development. Research Thesis, Technion-Israeli Institute of Technology, Haifa, Israel [in Hebrew, abstract in English].

Dan and Bredstreet (2004) *Wood Industry and Wood Products in Israel: an Economic Analysis.* Economics Department, Bank Leumi of Israel [in Hebrew].

Eliyahu, O. and Becker, N. (2004) The economics of the Agmon. In: Gofen, M. (ed.) *The Agmon: History, Policy, Implementation and Economics.* Kluwer Publishing, Dordrecht, The Netherlands.

Encyclopedia Britannica (1980) *Israel.* Vol. 9, pp. 1059–1067.

Feliks, J. (1973) *Geography: Flora and Fauna.* Keter Publishing House, Jerusalem, Israel, pp. 137–150.

Fleischer, A. (1993) *Forests and Parks – Demand Survey.* JNF publication.

Gafni, A. and Zohar, Y. (2004) Biological drainage – commercial forestry, a green alternative to utilize marginal lands. *Water and Irrigation* 449, 24–28.

Ginsberg, P. (2000) Afforestation in Israel: a source of social goods and services. *Journal of Forestry* 98, 32–36.

Grunzweig, J., Lin, T., Rotenberg, E., Schwartz, A. and Yakir, D. (2003) Carbon sequestration in arid-land forest. *Global Change Biology* 9, 791–799.

Henkin, Z., Atsmon, N., Gutman, M., De-Kvinik, H. and Karni, Y. (2002) *Forestation in Open Herbaceous Natural Pasture Lands.* Final report submitted to the JNF.

Israel Export and International Cooperation Institute (2003) *Israel's Agriculture.* The Israel Export and International Cooperation Institute.

Jewish National Fund (2004) *Forestry Inventory.* Statistical information of JNF forests [in Hebrew].

Kaplan, M (ed.) (1999) *National Master Plan 22: Forests and Forestry.* JNF, Ministry of Interior, Israel Lands Administration [in Hebrew].

Kligler, E. (1989) Planting techniques in the semi-arid Negev regions. *Allgemeine Forest Zeitschrift, Munchen-Afforestation in Israel* 24–26, 636–637.

Liphschitz, N. and Biger, G. (2000) *Green Dress for a Country, Afforestation in Eretz-Israel, the First Hundred Years 1850–1950.* Ariel, Jerusalem [in Hebrew].

Moshe, Y. (1989) Plantings by runoff harvesting in the Negev. *Allgemeine Forest Zeitschrift, Munchen-Afforstation in Israel*. 24–26, 640–641.

Ronen, T. (2004) *Map of Planted Forests and Woodlands in Israel*. JNF, Forestry Inventory Annual Report, 2004.

Spetter, E. (1989) *Timber utilization in Israel. Allgemeine Forest Zeitschrift, Munchen,- Afforstation in Israel*. 24–26, 652–653.

Tauber, I. (2004) *Forestry Inventory Annual Report, 2004*. Statistical information on JNF forests, JNF [in Hebrew].

von Forbrig, A. and Grov-Umstadt. (1989) Past afforestation and present natural forests in Israel. *Allgemeine Forest Zeitschritf, Munchen,- Afforstation in Israel*. Vols 24–26, pp. 614–617.

Weitz, J. (1970) *Forests and Afforestation in Israel*. Massada Press, Jerusalem.

Yakir, D., Berliner, P. and Schartz, A. (2002) *Evaluation of Water, Energy and CO_2 Budgets in the Yatir Forest*. Final research report submitted to JNF [in Hebrew].

Zohary, M. (1959) *Geobotany*. Hapoalim Library Publishing [in Hebrew].

11 Lebanon

Elsa Sattout[1], Salma Talhouk[1] and Nader Kabbani[2]

[1]*Department of Plant Sciences, Faculty of Food and Agricultural Sciences;*
[2]*Department of Economics, American University of Beirut,*
PO Box 11-0236-Beirut, Lebanon

Introduction

Historical records, dating back to Phoenician times, bear witness to the important economic role that Lebanese forests have played in supplying the Mediterranean people with high quality timber (Mikesell, 1969; Meiggs, 1982; Alptekin *et al.*, 1997). However, systematic woodcutting over several millennia, followed by the expansion of agropastoral activities and subsequent urbanization, has left the mountains of Lebanon with relic forest patches and scrub vegetation.

Lebanon extends over 10,450 km^2 on the eastern shore of the Mediterranean Sea. The country is mostly mountainous, with 73% of the total area consisting of two mountain ranges. The Mount Lebanon range is adjacent to the sea and is separated from the inner Anti-Lebanon range by an elevated fertile plain. Both ranges run parallel to the Mediterranean seashore, causing marked climatic variability over short distances. This diverse topography gives rise to the existence of many microclimates. A transect of 50 km includes a narrow coastline with a subtropical climate, middle elevation slopes that are typically Mediterranean, high mountain peaks covered with snow for most of the year and subdesertic plains.

Lebanese forests harbour high species richness, with an estimated total of 2600 plant species for the country as a whole (Zohary,

1973). Moreover, the share of endemic plant species of the total (12%) is quite high when compared with other Mediterranean countries (Davis *et al.*, 1994; Quézel and Medail, 1995; Hamadeh *et al.*, 1996). This noted diversity is the result of the country's physiography and its location at a crossroads between continents.

Forest Resources

Area and people

The current forest cover in Lebanon is estimated at 134,300 ha, i.e. 12.8% of the country's area (Fig. 11.1). It is formed by dense forests (4.7% of the country's area) and open forestlands (8.1%) (Lichaa El-Khoury and Bakhos, 2003). The total population, of about 3.8 million, is unevenly distributed throughout the country, with about 60% clustered in the narrow coastal zone. This spatial distribution has major implications for the environment, in terms of increased demand for both land resources and environmental services.

The multi-purpose nature of Lebanese forests makes them an important source of income for local communities. Juniper and oak forests were, and are still in part, exploited intensively for the production of resin, charcoal

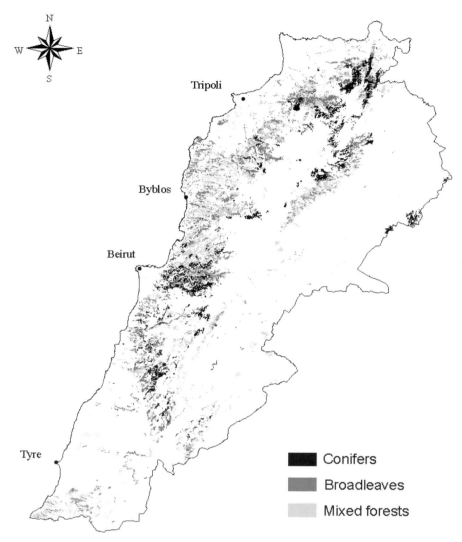

Fig. 11.1. Forest types in Lebanon according to the land cover map. Source: Lichaa El-Khoury and Bakhos (2003).

and firewood. Stone pine forests are harvested for pine kernels by local villagers, and southern coastal stands are important for molasses production. Forests are also used for the collection of edible, aromatic and medicinal plants, upon which many villagers rely for their livelihood during the spring. To some extent, in the mountainous regions and remote areas, local communities have been using the natural resources encountered in the forests in a sustainable way.

Typologies

Ten vegetation zones and 22 vegetation associations falling under the Mediterranean and pre-steppe areas have been identified in Lebanon. The ten bioclimatic ranges are described briefly below (Chouchani, 1972; Zohary, 1973; Abi-Saleh *et al.*, 1996) and shown in Fig. 11.2.

The thermo-Mediterranean zone extends from the coast to an altitude of 500 m and is occupied predominantly by carob (*Ceratonia siliqua*),

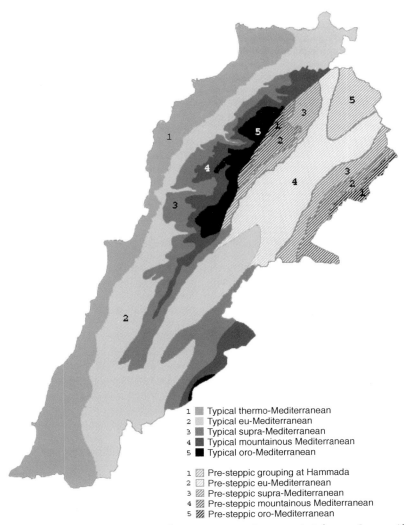

1 ▨ Typical thermo-Mediterranean
2 ▨ Typical eu-Mediterranean
3 ▨ Typical supra-Mediterranean
4 ▨ Typical mountainous Mediterranean
5 ▨ Typical oro-Mediterranean

1 ▨ Pre-steppic grouping at Hammada
2 ▨ Pre-steppic eu-Mediterranean
3 ▨ Pre-steppic supra-Mediterranean
4 ▨ Pre-steppic mountainous Mediterranean
5 ▨ Pre-steppic oro-Mediterranean

Fig. 11.2. Map of typical and pre-steppe Mediterranean vegetation zones in Lebanon. Source: Abi-Saleh and Safi (1998).

pistachio (*Pistacia palaestina* Boiss and *P. lentiscus* L.), myrtle (*Myrtus communis* L.) and oak (*Quercus calliprinos* Webb.). The eu-Mediterranean zone, consisting of the upper western slopes of the Mount Lebanon range (500–1000 m), comprises open woodlands dominated by oak, Calabrian pine (*Pinus brutia* Ten.), Aleppo pine (*Pinus halepensis*), pistachio (*P. palaestina*), carob (*C. siliqua* Miller), laurel (*Laurus nobilis* L.) and Judas tree (*Cercis siliquastrum* L.).

The oro-Mediterranean zone is found at higher altitudes and consists of a lower supra-

Mediterranean zone extending from 1000 to 1500 m, and a mountainous zone at an altitude of 1500–2000 m. Pine and oak forests characterize the former zone, while the latter includes relic patches of cedar (*Cedrus libani* A. Rich), fir (*Abies cilicica* Ant. & Kotschy) and juniper (*Juniper excelsa* M. Bieb and *J. oxycedrus* L.) forests.

The high summits (above 2000 m) are limited to the northern end of the Mount Lebanon range and consist of alpine zone vegetation typically inhabited by low-growing perennials such as milk-vetch (*Astragalus* spp.), prickly thrift

(*Acantholimon libanoticum* Boiss.) and stumpy spurge (*Euphorbia caudiculosa* Boiss.). This zone extends to similar altitudes on the eastern slopes of the Mount Lebanon range and is then gradually replaced by the mountainous oro-Mediterranean pre-steppic zone, which extends down to 1800 m. In this latter zone, degraded juniper forests dominate the landscape.

The supra-Mediterranean pre-steppic vegetation occurs at altitudes ranging from 1400 to 1800 m and consists mostly of mixed plant communities dominated by kermes oak (*Quercus calliprinos*) and cypress oak (*Q. infectoria* Oliv.), with sporadic populations of wild almond, wild pear and maple.

The Mediterranean pre-steppic zone, consisting mostly of degraded oak stands, ranges from 1000 to 1400 m and constitutes the foothills of the eastern slopes of the Mount Lebanon range leading to the agricultural Bekaa plain (900 m). On the other side of the plain and parallel to the Mount Lebanon range lie the western slopes of the Anti-Lebanon mountain chain which display similar vegetation zones to those described for the eastern slopes of Mount Lebanon, except that the land is more degraded and barren.

Functions

Oak woodlands constitute the major part of the Lebanese forests with 53.3% (Table 11.1). They consist of oak species growing at different altitudes and climatic zones. Most of them are, however, neglected, overgrazed and used for firewood and charcoal production. Pine forests occupy 15% of the total forest cover. Half of these forests are plantations consisting of *Pinus pinea*, a naturalized tree with economic potential due to the high market value of its pine nuts. The remaining pine forest areas consist of two species, *Pinus halepensis* and *Pinus brutia*, with no reported economic use in Lebanon. Pine forests are poorly managed and have therefore become highly susceptible to fire events.[1] The pine forests at higher altitudes have been targeted by developers for the construction of summer houses and resorts. Urban expansion within these forests, initially intended to take advantage of this natural habitat, unfortunately is leading to its destruction.

In contrast to oak and pine forests, cedar forests, which constitute less than 1% of the total forest cover, have received national, regional and international attention. The value of the cedar tree in Lebanon is connected to its historical and symbolic value and the rich flora and fauna associated with the remnant stands. The recreational income-generating activities related to these forests have been limited mostly to visits to the Bsharre stand and the associated trade of small item cedar woodcrafts. More recently, Ehden and Al-shouf cedar forests, declared as nature reserves, are being managed where nature-based tourism activities are planned.

As for the remaining conifer forests, namely the fir, cypress and juniper forests, they grow as

Table 11.1.   Dense and open forest areas and their functions.

Forest type	Dense forests (ha)	Open forests (ha)	Total ha	Total %	Forest functions
Cedar	626	510	1,136	0.85	Protection
Fir	717	886	1,603	1.2	Protection
Cypress	7	197	204	0.15	Protection
Juniper	0	11,917	11,917	8.87	Protection and production (firewood, charcoal, resin)
Pine	9,512	10,832	20,344	15.15	Protection and production of firewood and pine kernels
Oak	26,588	44,943	71,531	53.27	Production (firewood and charcoal)
Mixed forests	11,374	13,145	24,519	18.26	Production (firewood and charcoal)
Other broadleaved	729	2,293	3,022	2.25	Molasses and fuelwood
Total forest cover	49,553	84,723	134,276	100	

Source: El-Khoury and Bakhos (2003).

sparse and degraded populations. These have been neglected consistently as they spread over remote areas of Lebanon where agropastoral practices still prevail.

Institutional Aspects

Ownership and size of properties

About 77.3% of the Lebanese forests are public property belonging either to the State (51.5%) or to municipalities (25.8%), while the remaining 22.7% are privately owned lands (M. Bassim, Ministry of Agriculture, 2001, personal communication). Regardless of ownership, management and utilization of any resource within Lebanese forests require prior authorization from the Ministry of Agriculture.

Administration and policies

Following independence (1943), Lebanon adopted forestry laws pertinent to the use of forestlands. The laws and decrees were based on Turkish forestry law and were later amended according to changes in forest conditions and other factors. The 1949 Forest Code Art. 93 recognizes the perennial nature of forests and guarantees their protection and conservation. Accordingly, all activities related to forest use, such as cutting of trees, selling wood, clearing land, grazing, extraction of stone, sand, earth, grass and plants, natural manure, seeds, fruits and gathering forest products, require authorization.

The Forest Code prohibits the cutting of conifer species in both public and private forests, while the pruning of conifer species requires permits (Art. 93). Similarly, it regulates grazing by defining the size of flocks that have access to the forests and the duration of the land lease. Grazing practices are forbidden for a period of 10 years in burned forests (Art. 109). While, during the civil war, the law was not fully enforced, nowadays, the procedures are taken seriously and monitored.

The key institutions responsible for forest management are the Ministries of Agriculture and Environment, the Caza and the Department of Antiquities (Darwish *et al.*, 1996). The Ministry of Agriculture is responsible for managing all forest areas, except for the nature reserves[2], which fall under the responsibility of the Ministry of Environment, Directorate of Protection and Urban Development. The Department of Antiquities is involved in situations where forests are considered as part of the cultural heritage. The approval of the Administrative Heads of the Caza (District) is needed to proceed with the implementation of management and conservation activities. However, protection and management of reserves have never been performed by ministry employees. In the initial stages following their establishment, the management was subcontracted to local non-governmental organizations (NGOs), to which the ministry allocated €173,700 in 2000 (ECODIT, 2001). More recently, however, the government has appointed a committee of representatives of municipalities, state agencies and scientific institutions to take on the responsibility of managing strategies for nature reserves. The government is currently investigating options for selecting an implementing body.

While the ministerial decisions for the establishment of these reserves was the founding block for conservation and management, the enforcement of these decisions has generated some conflicts between the State and forest communities. Even though the enforcement of the law has led to a notable increase in natural regeneration in the Al-shouf and Ehden reserves, it has also alienated local communities (ECODIT, 2001). After being the natural harvesters and acquiring local knowledge that helped guide them through conservation of these natural resources, many local communities feel that their traditional rights to use and develop local forests have been stripped away from them.

In the late 1960s, national reforestation plans and forest management strategies were developed with the financial and technical support of the Food and Agriculture Organization (FAO). These were halted with the onset of the civil war in the 1970s and have slowly resumed since the 1990s, with limited funding and local personnel. In recent years, the Ministry of Agriculture received 7–10% of the government budget (Combating Desertification Lebanon, 2002); however, forestry practices did not improve substantially.

Contribution of the Forest to the National Economy

Gross domestic product and employment

The contribution of the forest sector to Lebanon's gross domestic product (GDP) was 0.93% in 2001. This figure is not available from official sources. Instead, it is generated by estimating the value added from all forest-related activities, relying on official and 'grey' data given by State agencies and other organizations (Table 11.2). As such, the value added from forest-related activities estimated at market price value accounts for €155.6 million, of which 22.7% comes from wood-based industries and 77.3% accrues to other forest-based industries: pine

nut production, grazing, honey making, hunting and medicinal plant industries (Darwish *et al.*, 1996; Hamze *et al.*, 1996; Ministry of Finance, 2002). This finding represents the minimum estimate of the economic value for the items considered at market price values.

Forestry contributes 0.02% to the total labour force. This share is represented by some 370 employees at the Forestry Service under the Ministries of Agriculture and Environment (G. Akl, Ministry of Agriculture, 2001, personal communication) that includes 221 forest guards and members of two protected area management teams (ECODIT, 2001). Consequently, public forest-related activities have a negligible contribution to the national employment.

The wood products industry (except furniture and paper/pulp) employs around 7000

Table 11.2. Values of Lebanese forests (2001 prices).

Valuation method/output	Quantity	Value (000 €)
Direct use values		
Market price valuation		
Firewood (m³)	82,300	1,890
Charcoal (Mt)	11,400	1,890
Pine kernels (t)	2,592	52,488
Medicinal and aromatic plants (subsistence and commercial)	—	16,650
Honey and wax	1,028	12,150
Substitute goods pricing		
Grazing (million FU)	9.6	960
Cost of harvest and other raw materials		
Carob (t)	2,000	587
Seeds	1,200	480
Molasses	800	107
Permit price		
Hunting (no. of hunters)	600,000	12,000
Legal hunting	200,000	6,000
Illegal hunting (poaching)	400,000	6,000
Entrance fees		
Recreation in forest reserves (no. visits)	—	270
Total direct use values		98,885
Option, bequest and existence values		
Cost-based method: biodiversity conservation (ha)	—	864
Total option, bequest and existence values		864
Negative externalities		
Carbon emissions (tC)	72,138	1,443
Losses due to forest fires (ha)	2,215	4,770
Damage due to pest infestation	—	72
Total negative externalities		6,285
TEV		93,464

workers and the furniture industry employs some 11,000 workers (Ministry of Industry, 2000). Overall, this represents 1.2% of the Lebanese workforce, a lower figure compared with the 1.7% as reported for 1995 (Mediterranean Environmental Technical Assistance Program, 1995). However, since most timber is imported, this employment cannot be attributed to the existence of Lebanese forests. In contrast, most firewood is obtained locally, generating employment in woodcutting and selling activities. However, this is not considered in official figures partly due to the unlawful nature of the activity: woodcutting is illegal in Lebanon without prior consent from the Ministry of Agriculture.

International trade

The small relic forests and the degraded state of the Lebanese woodland do not allow their exploitation for the timber industry (Baltaxe, 1965). Accordingly, most timber[3] is imported (96,600 m^3) and its value is estimated at €1.9 million (Food and Agriculture Organization Statistics, 2000). In contrast, the national needs for firewood and charcoal are covered by local production, amounting to 82,300 m^3 and 11,400 Mt, respectively, with an estimated value of €1.9 million each.

Other forest-based industries: tourism and green belts

The high elevation zone (800–1800 m) of the Mount Lebanon range, which runs parallel with the Mediterranean Sea, was sought historically as a refuge by persecuted minorities in the region that today constitute a sizeable portion of the Lebanese population. This zone has also acted traditionally as a refuge from the hot and humid summers of the coastal cities.

The forests throughout this zone, consisting mostly of cedar trees (*C. libani*) and pine trees (*P. pinea* and *P. brutia*), have acquired a special value among the Lebanese. The cedar tree symbolizes the glorious past and the current national identity and is reproduced on the Lebanese flag, while the pine forests are closely

associated with the modern history of Lebanon and its traditional village communities.

Lebanese forests have always been managed and protected to some extent by local communities, relying on indigenous knowledge (local sciences) and cultural values. Over the decades, harvesting of non-wood forest products (NWFPs) and production of artisan products were handled mostly by women. Thus, gender involvement has played a key role in the development of local microindustries based on NWFPs, such as production of jams, syrups, tea mixtures and dried fruit. However, in recent years, increased urbanization trends led to the breakdown of traditional socio-economic activities, which has altered the communities' relationship with the forests.

Awareness of the need for forest preservation led governmental organizations and NGOs to take action in promoting ecotourism and developing markets for traditional NWFPs. The income generated by these organizations and local communities, either through local or international funding or through other fund-raising activities, should be recognized officially especially in cases where positive externalities are generated: preserving biological diversity, promoting the landscape assets and developing gender involvement within the forest users groups. An example of issues that need to be addressed is the ability of local communities or NGOs to collect revenue from the use of communal forests for recreational or other purposes. At present, the collection of such fees is illegal and awaits a ministerial decree or decision (Boxes 11.1 and 11.2).

The Values of Lebanese Forests

Despite widespread interest from many national groups and institutions, economic valuation of Lebanese forests is practically non-existent. Indeed, official data often fall short of providing the required indicators for generating adequate estimates. The estimates subsequently reported rely partly on official data and partly on information derived from 'grey' literature collected from public and private Lebanese sources. Case studies are also used to illustrate the economic potential of these forests.

Box 11.1. Case study of Al-Shouf Cedar reserve.

The Al-Shouf Cedar forest covers 16,000 ha in Mount Lebanon and was designated by Forest Law No. 532/1996 as a nature reserve. It is managed by a local NGO that is subcontracted by the Ministry of Environment within the framework of a Global Environmental Facility/United Nations Development Programme (GEF/UNDP)-funded project designed to assess the effectiveness and feasibility of protecting natural areas (United Nations Development Programme/Ministry of Environment, 1996). The project aimed to promote the social and environmental sustainability of the reserve.

During the 5-year project, the reserve has attracted €500,000 through various fund-raising activities as a major source of income. The project contributed to the creation of local jobs, including: 60 women involved in the production of artisan and local products by using cultural practices and indigenous knowledge, 13 permanent employees and five seasonal employees involved in the management body of the reserve. Job creation, especially through gender involvement, has been a success; it has widened the scope for new economic opportunities and generated additional income for 53 households.

In 2001, the revenues generated by the project amounted to €180,800, originating from:

Sales of NWFPs, artisan handicrafts and promotional material: €16,500
Revenues from recreational activities (hiking[a], etc.): €16,800
Contribution fees: €42,500
Other fund-raising activities[b]: €105,000.

The reasonable success of the Al-Shouf protected area is basically related to the integration of local communities and gender involvement in the forest management and planning strategies. Future plans for income generation by the NGOs include: establishment of a cedar nursery and sale of seedlings, opening of campsites and biking trails, and launching an educational centre for increasing the awareness of the need for forest and environmental protection.

The total income pertaining to the services and goods provided by the Al-Shouf reserve is increasing every year, as reported by the official data. Therefore, it is believed that establishment of new nature reserves might help to preserve forestlands and increase local communities' welfare. The generation of new job opportunities in the forest villages will contribute to some extent to a decrease in the rural migration of youth and will initiate gender involvement in the traditional microindustries where income generated contributes to an increase in their household well-being.

[a]The clubs for hiking and nature-related activities drew an average of 1500 visitors each year during the past 2 years, which generated an average of €10,000/year.
[b]Includes the Biodiversity Conservation grants, which total €105,000 over the 5-year project.

Box 11.2. Case study of Bsharre forest.

The cedar grove in Bsharre consists of only 11 ha and a total of 376 trees, and yet it is the most famous in the country because of the age and beauty of some of the tree specimens. Revenues from entrance fees to the grove increased by 580% during 1993–1994 and by 307% during 1994–1995 (Darwish *et al.*, 1996). The entrance fees accounted for €31,533 in 2001 (Kozhaya Hanna, Municipality of Bsharre, 2002, personal communication). In compliance with national regulations, collection of entrance fees was stopped in the late 1990s and, instead, donations and in-kind contributions were solicited in order to continue the activities for the preservation of the reserve. A survey on the willingness to pay to conserve and protect the Bsharre cedar grove indicated an average value of €7.5/visit (Darwish *et al.*, 1996). It also revealed that the Lebanese are willing to pay similar rates to visit other forest reserves providing recreational services in Lebanon.

Direct use values

Lebanese forests have been used traditionally in small-scale industry and subsistence, and very rarely for major commercial purposes. However, estimates of subsistence uses are not included in the country's GDP. The forests support the national demand for firewood and charcoal. Valuation is based on their market price, resulting in an estimate of €1.9 million for firewood and €1.9 million for charcoal pro-duction (Food and Agriculture Organization Statistics, 2000).

The amount of NWFPs is much higher than that of firewood and charcoal. Pine kernels are the most important NWFP. Quantitative valuation takes into account the production area (5400 ha) and the corresponding yield (480 kg/ha). The local market price averaging €20.3/kg is higher than that of imported kernels (Darwish *et al.*, 1996) and leads to a total mone-tary value of €52.5 million. It is believed that revenues could be increased further with better stand management that could increase produc-tivity to 1200 kg/ha. In cases where stands are publicly owned, the land is leased to contractors for harvesting and selling pine kernels as well as managing the pine trees (concession).

Medicinal, aromatic, and other edible plants, which are an integral part of Lebanese cuisine, are used extensively for both subsistence and income generation. The monetary value of medicinal and aromatic plants is estimated at €16.7 million, based on the export and the local consumption values (Darwish *et al.*, 1996). Examples of plants harvested from the wild include commonly used species such as camomile (*Matricaria* sp.), oregano (*Origanum* sp.), sage (*Salvia* sp.), fennel (*Foeniculum* sp.) and Syrian eryngo (*Eryngium* sp.). Other spe-cies, such as the A'kkoub (*Gundelia* sp.) are less well known. No official data are available, as all plants harvested from the wild are either traded in informal markets or are considered under a pooled category in agricultural statistics referred to as medicinal plants. It is estimated that around 6000 t of *Salvia* and 300 kg of *Origanum syriacum* L. are exported every year (G. Akl, Ministry of Agriculture, 2001, personal commu-nication). These plants are usually harvested by women who typically receive a *per diem* rate of approximately €4–12 per 50–60 kg of sage (Rask, 2002).

Valuation of honey is based on the reported revenues from honey-making activities, giving a total of €12.1 million (Food and Agriculture Organization/Ministry of Agriculture, 2000). The income generated from the beekeeping by-product royal jelly is estimated at €86,400. The common practice in honeybee management in Lebanon is transporting the beehives in the winter to lower elevation locations, where they

are usually placed in fruit orchards and other cultivated lands (H. Hammoud, Ministry of Agri-culture, 2001, personal communication). There-fore, assessing the direct values of forest areas with respect to honey and other by-products requires the separation of production in summer and winter periods, since in the former the bees feed on wild flowers in woodlands while in the latter they most probably feed on cultivated trees.

Grazing is one of the most controversial forest uses in Lebanon. Continuous grazing has prevented the regeneration of forests, especially the slow-growing conifers, and compounded the effects of deforestation. Grazing by goats has been blamed as the main reason for much habi-tat destruction despite the presence of laws that organize grazing practices by defining the size of flocks entering the forests and limiting the annual lease period to 2 months. Compliance rates are estimated at 60% and involve paying for leases that are renewed on an annual basis (S. Hammadeh, Department of Animal Sciences, American University of Beirut, 2001, personal communication).

Valuation of grazing is based on substitute goods pricing, assuming that the nutritional value of one forage unit (FU) is equivalent to that of 1 kg of barley. There are about 400,000 head grazing in leased woodland and rangelands over a period of 2 months of the year. Assuming that 40% of their diet derives from grazing and the average daily consumption is 2 FU, the total annual consumption is 19.2 million FU. Valued at a cost of €0.1/FU, a total value of €1.92 mil-lion is obtained. It is assumed that only half of this value can be attributed to grazing in forests, while the other half occurs in open rangelands. Therefore, the monetary value of grazing in forests is probably closer to €0.96 million. The area of forests grazed is difficult to estimate, as it varies from one region to another accord-ing to the management and local/municipal arrangements.

The value of the carob is estimated based on the net revenue from molasses production and the revenue generated from the export of carob seeds to European countries. The quantity of harvested carob is 2000 t, of which 800 t are mollasses and 1200 t are seeds. The net revenue generated from molasses production (€0.5 mil-lion) is equal to the revenue (€1.2 million) minus labour costs (800,000 kg × €0.7/kg) and costs of raw materials (800,000 kg × €0.2/kg). Additionally, 1 t of carob fruits gives approximately 80 kg of seeds. At a price of €1.12/kg, the total value of seeds is €0.1 million. Adding up the two estimates results in a total value of €0.6 million (G. Zein, Greenland SARL, 2002, personal communication).

Hunting is valued based on the price of permits issued by the National Council for Wild Hunting. These permits are seasonal; they do not limit the number of game animals hunted, but they forbid their sale. Permits to own hunting guns are issued once and do not need renewal. The hunting law has been updated recently to decrease the annual hunting period to protect migratory and native bird species. In addition, the prices of both hunting and gun-owning permits have been increased. The total number of forest hunters is estimated at around 600,000, of which two-thirds have no hunting licences. However, both legal and illegal hunting bring private benefits, and are therefore accounted as positive values. It is recognized that illegal hunting is often responsible for negative externalities in terms of threatening many bird species and other wildlife (ECODIT, 2001). The value of legal hunting is based on the number of hunters having licences (200,000) and the permit fee (€30), giving a figure of €6 million. The value of illegal hunting is based on the number of hunters without permits (400,000) and half the permit fee (€15), resulting in €6 million. Overall, the value of hunting is €12 million.

Only a few surveys have been undertaken to obtain the consumers' willingness to pay for recreational services in Lebanon and they refer mainly to forest reserves (Box 11.2). Therefore, the estimate of forest recreation services is based on the net revenue generated by environmental clubs that organize recreational activities in Lebanese forests, including the donations collected at the entrance of the three forest reserves of Al-shouf, Bsharre and Ehden. The income generated from nature-based activities within the forests by three NGOs and six private societies[4] is estimated at €0.3 million. This amount includes the donations obtained from the three forest reserves (€0.08 million). This value corresponds to only a small part of recreational services in the Lebanese forests.

Indirect use values

Precipitation in Lebanon results in an average flow of 8600 Mm^3 yearly. Many activities affect the water cycle, one of which is deforestation. It alters the conditions of water replenishment and induces soil erosion, which in turn diminishes the groundwater recharge. Thus, forests can play a major role in watershed management in relation to the topographical nature of the Lebanese territories where valleys confine 17 perennial rivers with an average annual flow ranging from 11 to 793 Mm^3. The budget allocated for wastewater and water management over 10 years is €850 million, where 5.1% is allocated for the assessment of river basins and their protection from pollution and flooding (ECODIT, 2001). However, it is not known what share of this value would truly correspond to the watershed protection function that is actually performed by forests.

Option, bequest and existence values

The high biodiversity richness resulting from the country's location, physiography and climate makes Lebanon a reservoir for plant genetic resources. Gene diversity is a hidden resource, which, if protected and conserved, could generate income for generations to come. Between 1996 and 2005, international funding agencies allocated €4.3 million for protecting Lebanese forests and an additional amount of €4.4 million for biodiversity conservation, where forest conservation and sustainable uses were included directly and indirectly within the project proposals (ECODIT, 2001). For 2001, international funding for forest conservation amounted to €0.86 million and this reflects at least partially the international community interest in preserving biodiversity in Lebanon. This can be considered just a partial estimate for the option, bequest and existence values of Lebanese forests.

Negative externalities linked to forests

Among the negative externalities, only carbon losses due to wood harvest and fires, other damage caused by forest fires and the decline in forest health due to pest infestations are estimated in monetary terms.

Valuation of carbon losses caused by wood harvest and fires is based on the methodology used within the Lebanese National Communication on Climate Change (Ministry of Environment, 1999) applied to the entire country's forest area (134,400 ha). The net annual quantity of biomass (tonnes of dry matter, tdm) is equal to the difference between the biomass stored in forest increment and that lost by wood harvest and forest fires:

- Forest increment biomass = (2.5 tdm/ha × 106,735 ha) + (1.5 tdm/ha × 27,541 ha) = 308,149 tdm.
- Biomass loss due to wood harvest (firewood, industrial wood, worked wood) = 417,298 tdm.
- Biomass loss due to forest fires = 35,127 tdm.

Accordingly, it results in an annual loss of 144,276 tdm, i.e. 72,138 tC, after applying a conversion factor of 0.5. Valued at €20/tC (Fankhauser, 1995), the monetary value of carbon loss is €1.4 million.[5]

The damage caused by forest fires is estimated by using the replacement costs approach. The Forestry Service indicates that 1200 ha of natural forests are burnt every year, inducing further degradation of the vegetation cover, causing landslides and soil erosion (Lebanese Environment and Development Observatory (LEDO), 2000). The annual cost of combating fires and restoring the burnt sites is €4.8 million (M. Bou Ghanem, Association for Forest Development and Conservation, 2001, personal communication).

The decline of Mediterranean conifers may be attributed to the complex and cumulative effects of urban expansion and associated stresses such as air pollution (Bussotti and Ferretti, 1998). More specifically, the decline in the Lebanese pine forests is associated with a noted increase in insect attacks, causing a widespread dieback syndrome in the managed and unmanaged forests and allergy in local communities. Also, the cedar forests have been attacked by a complex insect (*Hymenoptera*). Valuation of the damage caused by pest infestation is estimated with reference to Tannourine's cedar site, which currently is attacked by *Cephalsia tannourenensis* Chevin. The annual cost for controlling this pest is €18,000 and includes expenses related to use of a plane, a pilot, a mechanic, experts as well as insecticides and related products necessary for spraying the pests (N. Nemer, American University of Beirut, 2000, personal communication). In addition, the processionary moth (*Thaumetopea wilkinsoni* Tams.) currently is attacking pine forests, and the annual costs for its control (insecticides) are estimated at €54,000 (G. Akl, Ministry of Agriculture, 2000, personal communication). Overall, the actual cost of treating Lebanese forests for pest infestation is €72,000.

The implications resulting in the loss of Lebanese forests due to anthropogenic activities often go unnoticed. Box 11.3 presents an economic cost analysis of a landslide and forest fire in Shouf pine forest.

Towards the Total Economic Value of Lebanese Forests

The calculation of the Lebanese forest values revealed that stone pine forests occupy the first rank in the contribution to the TEV (Fig. 11.3). Therefore, plantation forestry with these naturalized tree species is beneficial wherever the soil and altitude parameters are adequate.

On the other hand, grazing practices, which have been a major contributor to the degradation of forest coverage in Lebanon, have registered one of the lowest direct values. Therefore, this practice should be either strictly controlled or forbidden in certain forest areas. The law totally bans grazing activities in coniferous forests. In other forest types and rangelands, the Ministry of Agriculture set strict terms of condition for grazing permits, but still the law is not fully enforced.

NWFPs, such as medicinal, aromatic plants and honey have also registered relatively high values. However, the country has not yet developed laws and regulations for the sustainable utilization of its genetic resources, nor has it developed strategies to identify uses and develop industries and markets for them. Presently, production and marketing of NWFPs in Lebanon are just a minor industry, despite the great potential in cultivating marginal lands. It is believed that if properly managed and utilized, NWFPs

---

Box 11.3. Economic cost analysis of a landslide and forest fire in Shouf pine forest.

A study was conducted in the locality of Shouf, Deir Baba (Lebanese Environment and Development Observatory (LEDO), 2000) where a 20 ha pine forest (*P. pinea*) was severely damaged due to the opening of a new road and a forest fire. A landslide followed, with major impacts on the main road and on agricultural fields downhill. The damage caused by the landslide and forest fire was evaluated using a variety of approaches.

The direct impacts of the landslide were assessed by using the restoration costs (€15,000 to clear the soil from the road and €41,400 to rebuild the terraces destroyed by the landslide) and additional preventive expenditures (€5000 to build a retaining wall) for a total of €61,400.

The indirect impacts of the landslide were valued in a variety of ways. The loss in property value was estimated by using the property value approach, based on its tourist and residential potential rather than its agricultural potential. This approach is typically used to place a value on the deterioration in the environmental quality of land. It indirectly uses the market price and is based on the assumption that the real estate market is competitive. The impact of the landslide was high since the monetary value of land in the affected area dropped from €8–10/m^2 to around €4–5/m^2. For a total affected area of 500,000 m^2, the loss was estimated at €2–2.5 million.

The increase in health care expenses was based on a doubling of expenditure on medicines (normally 2.85% of family expenditure) and a tripling of expenditure on cleaning materials (normally 2.57% of family expenditure) observed at the pharmacy nearest to Deir Baba. This increased demand lasted for 6 months, until the threat of more mass movements had passed. With a monthly average family income of around €200–300, and given that the hazard effects lasted for 6 months, the total loss in income was estimated at around €48,300. The landslide also led to at least 20 days of work lost by workers and employees who were delayed and even prevented from going to work by the landslide, but it was not possible to determine the loss of earnings.

The direct costs of the fire could not be calculated as too much time had passed. However, it is known that nearly all families of Deir Baba derive part of their income from gathering pine kernels. The revenue losses caused by the fire were calculated based on the number of trees lost and the revenue from each tree, resulting in €20,000/year.

The uphill land (planted with 4 ha of olives and 4 ha of plum) affected by the slide lost 80% of its productivity, leading to a loss of €38,400 in 1993. Downhill land (85 ha, 17.6% planted with olive trees and the rest mainly with plums) lost around 20–40% of its productivity according to local farmers' estimates. This loss is estimated at around €180,000.

The Deir Baba watershed feeds the Damour River, a major source of irrigation for the southern coastal region. The sediment discharged into the river from Deir Baba and from other similar watersheds will probably seriously affect water quality.

---

could raise the economic opportunities and improve the local communities' welfare.

One initiative towards achieving these goals was undertaken recently by the Ministry of Agriculture by issuing a ministerial decision to regulate the wild harvesting of *Origanum* sp. and *Salvia* sp. This is a direct response to the increased exploitation of these species for which export markets have been developed. The decision allows collection only with permits issued (for free) by the Ministry of Agriculture. It also forbids export of unprocessed *O. syriacum* in order to promote local processing activities. Export is allowed only after it is processed and mixed with other ingredients to form a herbal mixture commonly consumed in the region.

The value of recreational services is lower than that of NWFPs, mainly due to insufficient valuation studies. However, it can be argued that the potential of the Lebanese forests for nature-based tourism is promising. Strategic planning for such activities and more extensive valuation studies are necessary.

Reforestation programmes have been one of the major activities in the Forestry Department at the Ministry of Agriculture for several decades. However, these programmes were not always successful due to the selection of species unsuitable for the corresponding vegetation zones and the inadequate provenance of seeds, showing low adaptation rates. This resulted in a halt to reforestation programmes since their early phases of implementation.

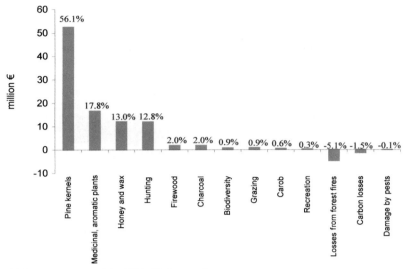

Fig. 11.3. Main components of the TEV of Lebanese forests.

Perspectives and Crucial Issues

Considering the size of Lebanon, the current status of its woodlands and the increasing rate of urbanization, a comprehensive assessment of the TEV of forest resources is certainly needed to ensure a sustainable conservation and a development action plan for the country. The economic valuation of national forests can play a major role in leading state agencies to decide on the best uses of the country's forests and natural resources. Additionally, an assessment of the TEV is a reliable tool for influencing decision making regarding the conservation of forests and even in developing policies for their management as well as elaboration of strategies for restoration programmes. Apart from the direct use values, the economic value of national forests also includes indirect services such as watershed protection, biodiversity and landscape quality, all of them contributing to the well-being of Lebanon, the Mediterranean and the arid regions surrounding the country.

It is worth mentioning that throughout the study, commercial use values (i.e. the direct use values) are estimated based on market prices and represent the minimum estimate of the real TEV provided by the forests. In the absence of sufficient data for other benefits linked to indirect use and non-use values, qualitative descriptions

are provided. However, attempts are made through case studies to shed some light on the magnitude of various forest benefits and their significance for the rural communities.

Acknowledgements

We wish to thank G. Akl, M. Bassil, R. Darwish, S. Hamadeh, H. Hammoud, D. Lichaa El-khoury, M. Moufarej, N. Nemer, R. Zurayk and all others who contributed to the completion of this study.

Notes

[1] In 1992, 60 fire events were reported in Lebanon and these were mostly in the pine forest zone (Lebanese Environment and Development Observatory (LEDO), 2000).

[2] There are seven nature reserves established in 1996 and extending over 14.8% of forest area. They include four cedar sites: Al-shouf, Ehden, Bsharre and Tannourine. In addition, five new sites are under investigation for their designation as forest reserves (L. Samaha, Ministry of Environment, personal communication, 2001).

[3] Except for a small national production of plywood and paper for which data are not available.

[4] Many of the 122 environmentally focused NGOs are directly involved in forest management,

ecotourism activities and the marketing of NWFPs (ECODIT, 2001). The establishment of nature reserves has gained the general approval of local communities, being a source of improving sustainable development, by providing new job opportunities, developing microindustries and preserving genetic resources for present and future generations.

[5]   It should be stressed, however, that the quantity of carbon actually released into the atmosphere as greenhouse gases as a result of firewood burning and charcoal production ($201,535$ tdm $\times 0.5 = 100,768$ tC) and forest fires ($35,127$ tdm $\times 0.5 = 17,564$ tC) is less than that stored in the annual forest increment ($308,149$ tdm $\times 0.5 = 154,075$ tC). If other wood harvest, such as industrial wood, is not taken into account, one can argue that the Lebanese forests act as a carbon sink.

References

Abi-Saleh, B. and Safi, S. (1998) Carte de la végétation du Liban (1/2,000,000). *Ecologia Mediterranea* 14, 123–141.

Abi-Saleh, B., Nasser, N., Hanna, R., Safi, N., Safi, S. and Tohme, H. (1996) *Lebanon Country Study on Biological Diversity. Terrestrial Flora.* Republic of Lebanon, Ministry of Agriculture and UNDP, Vol. 3.

Alptekin, C.U., Bariteau, M. and Fabre, J.P. (1997) Le cèdre de Turquie: aire naturelle, insectes ravageurs, perspectives d'utilisation pour les reboisements en France. *Biologie et Forêt* 49, 19–31.

Baltaxe, R. (1965) *Projet de Bonification Integrale de la Montagne Libanaise.* Report on Mapping the Forests of Lebanon. FAO and Green Plan.

Bussotti, F. and Ferretti, M. (1998) Air pollution, forest condition and forest decline in Southern Europe: an overview. *Environmental Pollution* 101, 49–65.

Chouchani, B. (1972) Le Liban: Contribution à son Étude Climatique et Phytogeographique. Mémoire du Doctorat de 3ème cycle, Université de Toulouse, Toulouse. France.

Combatting Desertification Lebanon (2002) *Lebanese National Action Plan.* United Nations Convention for Combating Desertification, Lebanon.

Davis, S.D., Heywood, V.H. and Hamilton, A.C. (1994) *Centre of Plant Diversity: a Guide and Strategy for their Conservation*, Vol. 1. The World Wide Fund for Nature (WWF) and The World Conservation Union (IUCN), Switzerland.

Darwish, R., Zakhia A., Hamadeh, S., Hayek, A. and Serhal, J. (1996) *Lebanon Country Study on Biological Diversity.* National Current Capacity and Economic Valuation, Republic of Lebanon, Ministry of Agriculture and UNDP, Vol. 8.

ECODIT (2001) State of the Environment Report (SOER). Lebanese Environment and Development Observatory, Ministry of Environment, Lebanon. Website: http://www.moe.gov.lb/ledo/soer2001.html Access date: May, 2003.

Food and Agriculture Organization/Ministry of Agriculture (2000) *Agriculture Statistics Project in Lebanon.* FAO.

Food and Agriculture Organization Statistics (2000) http://www.fao.org/forestry/fo/country   Access date: January, 2001.

Fankhauser, S. (1995) *Valuing Climate Change. The Economics of the Greenhouse.* Earthscan, London.

Hamadeh, S., Khouzami, M. and Tohmé, G. (1996) *Lebanon Country Study on Biological Diversity.* Country Study Report, Ministry of Agriculture and UNDP, Lebanon, Vol. 9.

Hamze, M., Abi Antoun, M., El-Hajj, S., Hamadeh, S. and Tohme, H. (1996) *Biological Diversity of Lebanon.* Agricultural and Livestock Habitats and Nature Reserve, Ministry of Agriculture and UNDP, Lebanon, Vol. 7.

Lebanese Environment and Development Observatory (LEDO) (2000) *Biodiversity Indicators*, EC LIFE program and Ministry of Environment, Lebanon.

Lichaa El-Khoury, D. and Bakhos, W. (2003) *Land Cover Land Use Map of Lebanon (1/20000) – Technical Report.* LEDO, Ministry of Environment/UNDP, Ministry of Agriculture/FAO and National Centre for Remote Sensing, Lebanon.

Mediterranean Environmental Technical Assistance Program (1995) *Lebanon: Assessment of the State of the Environment.* World Bank, Ministry of Environment, Lebanon.

Meiggs, R. (1982) *Trees and Timber in the Ancient Mediterranean World.* Oxford University Press, Oxford, UK.

Mikesell, M.W. (1969) The deforestation of Mount Lebanon. *Geographical Review* 19, 1–28.

Ministry of Environment (1999) Lebanon First National Communication under the United Nations Framework Convention on Climate Change (UNFCC), Republic of Lebanon, Ministry of Environment, Website: www.moe.gov.lb/climate/new_index.htm, Access Date: April 2003.

Ministry of Finance (2002) *The Lebanese Republic Country Profile.* Ministry of Finance, Republic of Lebanon.

Ministry of Industry (2000) *A Report on Industry in Lebanon 1998–1999: Statistics and Findings.* Sponsored by GTZ.

Quézel, P. and Medail, P. (1995) La region Circum-Meditérranéene, Centre Mondial Majeur de Biodiversité Végétale. *Actes des 6 èmes Rencontres de l'Agence Regionale pour l'Environnement Provence-Alpes-Cote d'Azur.* Colloque Scientifique Internationale Bio'Mes, Gap, pp. 152–160.

Rask, N. (2002) *Marketing of Native Plants in Lebanon: a Preliminary Assessment.* Report prepared under the project on Bioprospection: an alternative for sustainable agricultural development in Lebanon, American University of Beirut, Beirut, Lebanon.

United Nations Development Program/Ministry of Environment (1996) *Strengthening of National Capacity and Grassroots* – In Situ *Conservation for Sustainable Biodiversity Protection.* Lebanon. UNDP/Ministry of Environment.

Zohary, M. (1973) *Geobotanical Foundation of the Middle East*, Vols 1–2. Gustav Fisher Verlag, Stuttgart, Germany.

12 Syria

Ibrahim Nahal[1] and Salim Zahoueh[2]

[1]Aleppo University, Department of Forestry and Ecology, PO Box 5008,
Aleppo, Syria; [2]FAO, Syrian Arab Republic, Damascus, Syria

Introduction

Syria is a Middle East country that covers an area of 185,180 km². It overlooks the Mediterranean in the west, the Taurus Mountains in the north, and embraces part of the Arabian steppe (desert) in the south-east. From west to east, four geographical regions can be identified:

- The coastal region, which is sandy and narrow and runs more or less alongside a twin chain of mountains separated by a rift valley.
- The mountainous region, which runs from the north down to the south of the country, and includes mountains and hills that are parallel to the Mediterranean Sea.
- The interior (plains) region, which comprises the areas situated to the east of the mountainous region towards the Iraqi and Jordanian borders.
- The steppe region, with elevations between 400 and 600 m, which accounts for about 45% of the total land area and includes much of the central and south-eastern sections of the country.

Elevations range from 0 to 200 m along the Mediterranean shore, part of the Ghab Valley, along the Euphrates River and the foot of the Golan Heights, accounting for some 5–6% of the country area, to more than 1000 m in the mountains, covering about the same proportion of the total land area. Separate mountains and high plateaux, including those in the south-west,

cover about one-third of the country, with elevations ranging from 600 to more than 1500 m. Land with elevations between 400 and 600 m accounts for about 60% of the total land area and includes much of the central and south-eastern sections of the country.

The estimated population reached 18.6 million in 2003 (Central Bureau of Statistics, 2003), with 45% under the age of 15 years. The urban population is 50.1% compared with 49.9% rural. The average growth rate is actually estimated at 2.4% (United Nations Population Fund, 2003). The urban population is registering continuous growth, as demonstrated by the fact that 40% of the population is concentrated in only three cities: Damascus, Aleppo and Homs.

According to land use and soil type, the country's area is classified into cultivable land (32.5%), uncultivable land (20%), pastureland and steppe (45%) and forests (2.5%) (Central Bureau of Statistics, 2003). The contribution of forestry production, including industrial wood, firewood and charcoal, has a minor significance in the gross domestic product (GDP) of the country (Central Bureau of Statistics, 2002). However, the importance of forest should not be underestimated in a country with such a high degree of aridity as Syria. Forests provide a constant supply of fodder for a large number of grazing animals. They play a significant role in water catchments, in desertification control, in prevention of soil erosion and flooding, and in conservation of biodiversity. Moreover, Syrian forests constitute a substantial source of food security for poor rural households.

Forest Resources

Area

According to the Ministry of Agriculture and Agrarian Reform (2003b) and the Food and Agriculture Organization (2001), forestlands cover approximately 461,000 ha, of which 232,000 ha are natural and 229,000 ha are man-made plantations. The forested areas have

gradually degenerated mainly because of over-grazing, fire, overcutting for fuel and charcoal production, clearing for cultivation and general lack of management. A map of forests and other land uses of the country is shown in Fig. 12.1.

The estimate of forest area is based on the definition of the forest cover in botanical terms, adopted by the Law of 1953: it includes trees, shrubs, bushes and plants, as well as their roots and fruits. Their names are expressly mentioned in the Law, for example: *Quercus calliprinos, Pinus brutia, Cedrus libani, Laurus nobilis, Ceratonia siliqua, Pistacia atlantica, Myrtus communis* and *Phylleria media*. This definition creates serious ambiguities. On the one hand,

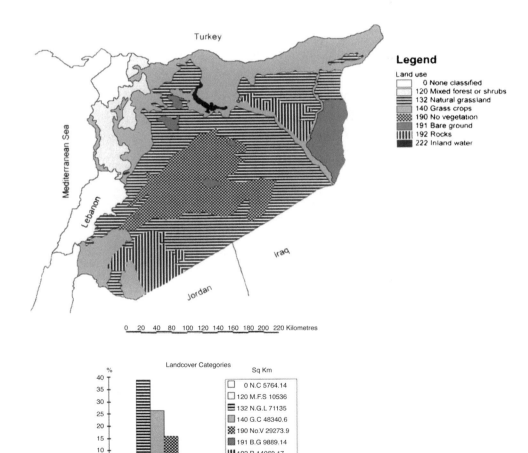

Fig. 12.1. Map of land cover of Syria. Source: adapted from General Organization for Remote Sensing (2001).

the list may prove non-exhaustive with respect to species that are omitted, either because they were unknown when the Forestry Law was drafted, or because they were introduced into the country at a later date. On the other hand, this list may, conversely, prove too broad with regard to species that are now very rare or may become extinct in the future.

Forestlands are broadly classified into three categories of land: (i) land covered with forest vegetation; (ii) bare, vacant land, the afforestation of which is believed to be beneficial for protection purposes; and (iii) agricultural land, in which forest trees occupy more than 10% of the total area (Forestry Law enacted by Legislative Decree No. 66/1953). While the first and second categories of land are quite commonly deemed to be forestlands in comparative forestry law, the third one normally is not. Usually, where forest trees are not predominant in cultivated land, they may be afforded some kind of protection, especially if they are endangered. Nevertheless, they are not placed under the statutory forest regime as part of forestlands, nor is their use necessarily subject to the specific rules governing forest exploitation.

Moreover, the data regarding forestland should be verified for a better reconciliation between the natural and artificial areas. In fact, the official sources indicated that in 1970, the area of natural forest covered a surface of 420,000 ha (Central Bureau of Statistics, 1975). The High Commission for Afforestation (HCA) was established in 1977, with an annual work plan of 20,000 ha for afforestation development. If this is the case, a significant area of natural forest disappeared during the last three decades, which is approximately equivalent in size to the afforested areas. A detailed analysis (Zahoueh, 1994a,b) shows that the expansion of cultivable areas in the coastal mountainous regions was undertaken to the detriment of the forest cover. Additionally, urban expansion is also contributing to the shrinking of forest areas.

Most of the plantations realized after 1977 cover roadsides and bare hills in areas of low rainfall and are therefore of ecological value and aesthetic significance. The Greenbelt Project, with a length of 1100 km and a width of 10–20 km, has greatly contributed to the nationwide forests, fodder and fruit trees. Moreover,

forest road construction has also increased noticeably from about 200 km/year in the early 1980s to an annual average of some 700 km in the last decade.

Typologies

Syrian forests can be classified according to five altitudinal vegetation zones. These are as follows.

The coastal mountains (Alaouites Mountains)

These are divided into the following (Nahal, 1961b):

THE WESTERN EXPOSURE

- 0–200 m: *C. siliqua* and *Pistacia lentiscus* zone, belonging to the subhumid bioclimate.
- 200–750 m: *Q. calliprinos* and *Pistacia palaestina* zone. In this zone, *P. brutia* is localized on marl and marly limestone, belonging to the upper temperate subhumid bioclimate.
- 750–850 m: *Quercus infectoria* zone, belonging to the middle temperate humid bioclimate.
- 850–1200 m: *Quercus cerris* subsp. *pseudocerris* zone, belonging to the upper cold humid bioclimate.
- 1200–1570 m: *Abies cilicica* zone, belonging to the lower cold per-humid bioclimate.

THE EASTERN EXPOSURE

- 300–900 m: *Q. calliprinos* and *P. palaestina* zone. In this zone, *P. brutia* and *Cupressus sempervirens* are localized on marl.
- 900–1000 m: *Q. infectoria* zone.
- 1000–1300 m: *Q. cerris* subsp. *pseudocerris*.
- 1300–1570 m: *C. libani* zone.

It should be stressed that the *Q. infectoria* zone in the Western and Eastern exposure is considerably degraded and this species is replaced by many heliophilous woody species. At the same time, *C. libani* is found as scattered trees at lower altitude than in its potential zone.

The Baer-Bassit Mountains
(Nahal, 1961b, 1974)

- 0–100 m: *C. siliqua* and *P. lentiscus* zone.
- 100–400 m: *P. brutia* zone, with *Q. calliprinos* in competition, according to the nature of the soil.
- 400–900 m: *Q. cerris* subsp. *pseudocerris* zone, on soil derived from *gabbros*. *P. brutia* is in competition and is localized on serpentine and pyroxenic peridodite.

Aleppo Mountain (Kurds Mountain)
(Nahal, 1975)

- 400–900 m: *Q. calliprinos* zone. In this zone, *P. brutia* is found on marl and marly limestone. *Quercus aegilops* is present with *Q. calliprinos* on *terra rossa* and basalt. It belongs to the upper subhumid and cold lower humid bioclimates.
- 900–1050 m: *Q. infectoria* zone, belonging to the cold lower humid bioclimate.
- 1050–1200 m: *Q. cerris* subsp. *pseudocerris* zone, belonging to cold middle humid bioclimate.

Hermon Mountain (Pabot, 1957; Abou Zakhm, 1997b; Abido and Kurbaisa, 2003)

The vegetation in this mountain is mostly degraded with various phases of degradation.

- Less than 800 m: this zone has lost its natural plant cover and is desertified or planted with fruit trees.
- 800–1200 m: *Crataegus azorolus* and *Poterium spinosum* are the dominant species.
- 1200–1600 m: broadleaved forests zone, with *Q. calliprinos*, *Q. infectoria*, *Pyrus syriaca* and *Amygdalus orientalis* as dominant species.
- 1600–2000 m: *Juniperus excelsa* zone.
- 2000–2500 m: almost degraded zone, with *Ferula hermonis* and *Astragalus hermoneum*.

Al Arab Mountain (Druzes Mountain)
(Mouterde, 1953)

- 850–1050 m: *P. atlantica* zone.
- 1050–1400 m: *Q. calliprinos* zone.

- 1400–1700 m: caducous-leaved oaks zone: *Q. infectoria*, *Q. look* and *Q. cerris* subsp. *pseudocerris*. Soils on this mountain are derived from basalts.

Syria represents a remarkable phytogeographic entity in relation to its forest vegetation due to the great variety of geographic, geological and climatic conditions. In fact, important mountain chains in the coastal zone and in the inlands, a variety of parent materials (hard limestone, dolomitic limestone, marls, basalts, ultra basic green rocks such as pyroxenic perododites, gabbros, diorites and amphibolites) and various bioclimatic zones, from the per-humid up to the extremely arid (rainfall varying between 1900 mm in the coastal mountains and < 100 mm in the steppe area), create conditions for a rich diversity of the forest vegetation. The series of forest vegetation in Syria are presented in Box 12.1.

Functions

Almost 60% of the forested areas in Syria are severely affected by fires, overgrazing, overcutting and encroachment from agriculture. As a consequence, large tracts of natural forests have degenerated into secondary plant communities of low economic and environmental value (United Nations Evironment Program, 2000). Nevertheless, Syrian forests represent an important source of livelihood. They provide a wide range of non-wood forest products (NWFPs), such as food, medicinal and aromatic plants, and fodder materials, which constitute a substantial source of food security for poor rural households.

According to the main functions, natural forests (*sensu lato*) can be classified into: protection forests (85% of forest area), multi-purpose forests (10%) and protected areas for biodiversity conservation (5%). Of the total area covered by plantations, 23% is managed for wood production[1] and 77% for protection against soil and wind erosion.

Institutional Aspects

As a matter of principle, forestlands 'falling under the control of the State' or 'pertaining to

Box 12.1. Series of forest vegetation in Syria.

According to Nahal (1961b, 1962, 2003), Abi-Saleh *et al.* (1976), Barbero *et al.* (1976), Chalabi (1980) and Martini (1989, 1999), the series of forest vegetation in Syria, in relation to the bioclimatic zones, is as follows.

1. Thermo-Mediterranean zone

1.1. *Thermophilous series of* Ceratonia siliqua *and* Pistacia lentiscus
This series represents the climatic vegetation (*Ceratonieto-Lentiscetum*) and occupies the coastal range: 0–300 m above sea level. Deforestation began in the second millennium BC by the Cananean–Phoenicians. Only scattered small groves and trees of *C. siliqua* still persist. This series of vegetation is represented by three subseries:

The normal subseries: a shrubby formation composed mainly of *C. siliqua*, *P. lentiscus*, *Myrtus communis*, *Oryzopsis coerulensis* and *Olea europaea*.

The subseries of *Pinus brutia*: in this subseries, *P. brutia* is an invader and constitutes an important element in addition to the above-cited species, which can be found on the Bassit coast. This series is found on various substrata: green rocks, hard limestone, *terra rossa* and red clay.

The subseries of *Pinus halepensis* such as in Cirestane at 150–250 m on calcareous substratum with silex kidneys.

1.2. *Thermophilous series of* Quercus calliprinos
This series covers a large zone in the coastal mountains and Baer-Bassit Mountains and is based on hard limestone, *terra rossa*, green rocks, red clay and other substrata. This series is degraded and consists of shrubby *garrigue* where *Q. calliprinos* and *Pistacia palaestina* are dominant. It represents the thermophilous variant of the association *Pistacieto-Quercetum calliprini alaouitaea* (Nahal, 1962). Two facies could be distinguished:

The normal facies: a *maquis* where *Q. calliprinos* is dominant, but *P. lentiscus*, *M. communis* and *C. siliqua* are still present. Silvatic species of *Quercion calliprini* are also well represented.

The facies with *P. brutia*: *P. brutia* is associated with *Q. calliprinos*, *P. lentiscus*, *M. communis* and occasionally *C. siliqua*.

1.3. *Thermophilous series of* Pinus brutia
On marl and marly limestone: *P. brutia* represents the dominant species. *Q. calliprinos* is still present, but its role is reduced. Differential species are found in this series such as: *Gonocytisus pterocladus*, *Cytisopsis dorycnifolium*, *Satureia thymbra* and *Putoria calabrica* that are linked to this substratum. However, the indicator species of the thermo-Mediterranean zone are present: *P. lentiscus* and *C. siliqua*.

On serpentine and pyroxenic peridodite: this series is found in Baer and Bassit on soils very rich in magnesium derived from serpentine and pyroxenic periododite. The following species are indicators of this substratum: *Ptosimopapus bracteatus*, *Quercus microphylla*, *Salvia aramiensis*, *Genista cassia*, *Scorzonera kotschyi* and *Centaurea cataonica*.

2. Eu-Mediterranean zone

2.1. *Eu-Mediterranean series of* Quercus calliprinos
This series occupies a large area of hard limestone, *terra rossa* and dolomitic limestone, in the coastal mountains, at altitudes from 300 up to 700–800 m above sea level and also in the Al Qousair mountain. Phyto-sociologically, it represents the *Pistacieto-Quercetum calliprini alaouitaea*, a garrigue very rich in plant species, such as: *Pi. palaestina*, *Phillyrea media*, *Rhamnus palaestina*, *Styrax officinalis*, *Arbutus andrachne*, *Laurus nobilis*, *Cercis siliquastrum*, *Olea europaea*, *Spartium junceum* and *Quercus infectoria* (in the upper part). This series is represented by three facies:

The normal facies where *Q. calliprinos* is dominant. It is the most widespread facies.
The facies with *Pi. brutia*. *P. brutia* is an invader species in this facies.
The facies with *Quercus aegilops*, in the Al Qousair Mountain and Aleppo Mountain.

2.2. *Eu-Mediterranean series of* Quercus infectoria

continued

Box 12.1. *Continued.*

2.3. *Eu-Mediterranean series of* Pinus brutia
This series is found on marl and marly limestone in the coastal mountains where it occupies a large area. The main indicator species are: *S. officinalis, Juniperus oxycedrus, Rhus cotinus, Dorycnium hirsutum, Cercis siliquastrum, Ruscus aculeatus, Fontanesia phillyreoides* and *Aristolochia altissima*, in addition to the species linked to the marls and marly limestones.

2.4 *Eu-Mediterranean series of* Pinus halepensis *and* Cupressus sempervirens
This series is found on marls and marly limestones in Qadmous. The indicator species are: *G. pterocladus, C. dorycniifolia, Genista acanthoclada, Linum aroanium, Syphonostegia syriaca* and *Lygia aucheri*.

2.5 *Eu-Mediterranean series of* Pinus brutia *on serpentine and pyroxenic peridotite*
This series occupies a large area in the Baer-Bassit Mountains in the coastal zone. The same indicator species of the magnesium-rich soils cited in the thermo-Mediterranean zone are found here: *P. bracteatus, Q. microphylla, S. aramiensis, S. kotschyi* and *C. cataonica*. Excellent stands of *P. brutia* exist in this zone.

2.6 *Eu-Mediterranean series of* Pinus brutia *and* Quercus cerris *subsp.* pseudocerris
This series is found at 400–500 m, on western and northern slopes in the Baer-Bassit, on soils rich in magnesium derived from serpentine and pyroxenic peridodites. It is characterized by the following species: *Tamus communis, Cephalantra ensifolia, Sorbus torminalis, Rubia aucheri, Asperula libanotica, Brachypodium sylvaticum* and *Melica uniflora*, in addition to special species of the region such as: *Glycirrhiza flavescens, Cytisus cassius, Genista cassia, Euphorbia biglandulosa, Onosma cassia* and *Trifolium cassium*. The indicator species of the magnesium-rich soils, as previously cited, is present (*P. bracteatus*).

3. Supra-Mediterranean zone
3.1. *Series of* Quercus cerris *subsp.* pseudocerris *on gabbros, diorites and amphibolites*
This series is found at an altitude of 500–700 m above sea level in the Baer-Bassit on deep soils derived from gabbros, diorites and amphibolites, and is characterized by a high water reserve which can reach 200–300 mm/ha, compared with 30–50 mm/ha in the soils derived from serpentine and pyroxenic peridotites where *P. brutia* is dominant. The following species are present: *Chrysanthemum cilicicum, Doronicum caucasicum, Cyclamen coum, Lecokia cretica, Luzula forsteri, Potentilla micrantha, Geranium asphodeloides, Circea lutetiana, Orobus hirsutus, Ostrya carpinifolia* and *Carpinus orientalis* – in the degraded phases.

3.2. *Series of* Quercus infectoria
This series is now rare due to deforestation for the establishment of fruit tree plantations. The original forest has disappeared. A few groves of shrubby formation exist at 800–1000 m in the coastal mountains, Aleppo and Al Qousair Mountains, on hard limestones and dolomitic limestones.

3.3. *Series of* Quercus cerris *subsp.* pseudocerris *on hard limestones and dolomitic limestones*
This series is well represented at 1000–1200 m on the western and eastern slopes of the coastal mountains (particularly in Slenfeh), in the Aleppo and Al Qousair mountains with the following main species: *C. azarolus, Daphne oleoides, Juniperus drupacea, J. oxycedrus, Origanum capitatum, Phillyrea media, Q. infectoria, R. aculeatus* and *S. officinalis*.

4. Mountain Mediterranean zone

4.1. *Series of* Abies cilicica
This series occupies the western exposure of the coastal mountains, between 1200 and 1500 m. *Abies cilicica* is accompanied by many woody species such as: *Quercus brantii, Q. cerris* subsp. *pseudocerris, Q. libani, Q. infectoria, Acer monspessulanum, A. hyrcanum, Sorbus torminalis, S. umbellata* var. *flabellifolia, Fraxinus ornus* and *J. drupacea*. This series has two subseries corresponding to two plant associations:

Subseries of *A. cilicica* and *Anthriscus lamprocarpa* with the following plant characteristics: *Orchis laxiflora, Scleranthus verticillatus, Sedum stedelli, Scila sibirensis, Scrophularia nusairensis, Astragalus eriophylloides, Trifolium dichroanthoides, Euphorbia herniariipholia, Saxifraga scotophila, Arum gratum* and *Saponaria bargyliana.*

Subseries of *Abies cilicica* and *Rhamnus cathartica* with the following plant characteristics: *Sorbus aria, Sorbus torminalis, Mercurialis ovata, A. monspessulanum, Cotoneaster nummularia, R. cathartica, Rubia aucheri, Acer hyrcanum* and *Orchis compariana.*

4.2. *Series of* Cedrus libani
This series is found on the eastern exposure of the upper zone of the coastal mountains. *Cedrus libani* is accompanied by many woody species of the *Abies cilicica* series. It is represented by two subseries:

Subseries of *C. libani* and *Cytisus depranolobus* with the following characteristics: *Genista pestalozza, Ferulago cassia, Centaurea cassia* and *Daphne libanotica.*

Subseries of *C. libani* and *Q. infectoria* with the following plant characteristics: *Prunus ursina, Pyretrum cilicicum, Quercus cedrorum, Crocus kotshyanus, Geranium libani, Fraxinus ornus, Rosa glutinosa, Ferulago cassia* and *Ulmus montana.* This subseries represents degraded stands.

4.3. *Mountain* garrigue *series*
This series is found in the coastal mountains between 1200 and 1500 m, spreading on the various summits, where the vegetation is considerably degraded and the soil highly eroded, leaving the hard limestone at the surface, particularly in Nabi Younis and Nabi Matta. The main species composing the *garrigue* are: *D. oleoides, Hedera helix, J. oxycedrus, J. drupacea, Q. cerris* subsp. *pseudocerris, Q. calliprinos* and *S. officinalis.*

5. Oro-Mediterranean zone
The forest vegetation is represented by the series of *J. excelsa* in Assal Al-Ward ranges of the Antilebanon Mountains between 1880 and 2200 m. It is degraded vegetation where the soil is highly eroded, leaving the hard limestone on the surface. Well-preserved stands exist in Assal Al-Ward with the following main species: *Amygdalus orientalis, C. azarolus, Pyrus syriaca, Cotoneaster nummularia, Berberis cretica, R. palaestina, Astragalus hermoneus, A. angustifolius, A. exiguus, Asphodeline taurica, Prumus ursina* and *Prunus microcarpa.* This series is represented by two subseries:

A normal series of *Juniperus excelsa.*

A series of xerophytic spiny species in the form of small cushions – in particular *Astragalus* sp. This subseries could be classified within the pre-steppic series.

6. Xerophytic forest formations
These are formations with silvatic aspects in arid and semi-arid bioclimatic zones. They are very rare in Syria and represent the remnant of the original forest vegetation before its past destruction, such as the remnant forests of *Pistacia altantica* in Abou-Rigmen in the Palmyrean chain, on limestone and in Qanawat in Djebel Al Arab, on basalt. In Qanawat, *P. atlantica* is accompanied by: *Q. calliprinos, C. azarolus, Crataegus sinaica, Pyrus syriaca* and *R. palaestina.*

7. Pre-steppic forest series
The term pre-steppic forest series is taken to mean a certain number of arboreous landscapes made up of scattered trees or shrubs, with very degraded soils, due to overexploitation and erosion. The possibility of their evolution to real silvatic landscapes could not be considered. They are found mainly in semi-arid bioclimatic zones, but also exist in the lower subhumid zone.

7.1. *Pre-steppic series of* Quercus calliprinos
This series is found in the Antilebanon chain on limestone, with the following species: *Amygdalus orientalis, Crataegus azarolus, Q. infectoria, P. microcarpa, P. atlantica, P. syriaca, Poterium spinosum, Arum dioscoridis, Asphodelus microcarpus, Astragalus spinosum, Phlomis viscosa, Euphorbia macroclada, Ononis spinosa* and *Rosa dumetorum.*

7.2. *Pre-steppic series of* Pistacia atlantica
This series is found in Djebel Al Arab, Djebel Abdul Aziz, Djebel Bilaas and Bichri on limestone, with the following main species (in Bilaas): *A. orientalis, R. palaestina, P. microcarpa, Capparis spinosa, Poa bulbosa, Poa sinaica, Hordeum bulbosum, Artimisia herba alba, Anabasis Hausknechtii* and *Peganum harmala.* The floristic composition of this series reflects the severe degradation of the vegetation in semi-arid and arid zones.

continued

Box 12.1. *Continued.*

8. Riparian and hydrophilous forest vegetation

8.1. *Riparian forest vegetation of* Populus euphratica

This is found along the Euphrates and Tigris Rivers and on the islands. *Populus euphratica* is accompanied by *Salix persica* and *Tamarix tigrensis*. This vegetation is becoming rare due to overexploitation.

8.2. *Riparian forest vegetation of* Platanus orientalis *and* Alnus orientalis

This is found along the water courses in the coastal zone where *Platanus orientalis* and *Alnus orientalis* are accompanied by *Tamarix pentendra* and *Salix alba*.

8.3. *Hydrophilous forest formation of* Fraxinus syriaca

This formation of *Fraxinus syriaca* is found in Al Ghab valley, as a remnant of the original forest that once covered this valley before it was drained. *F. syriaca* is accompanied by *Iris pseudacorus*.

9. Productive plantations

9.1. *Artificial plantations of* Populus sp.

These plantations are concentrated in the Euphrates valley and in Ghouta, near Damascus. They are irrigated and composed of local clones of *Populus alba* var. *roumi* and *Populus nigra* var. *hamoui*.

9.2. *Artificial plantations of* Castanea sativa

These are found in the Al Qousair Mountain on basalt and in the humid bioclimatic zone over 900 m.

9.3. *Artificial plantations of* Juglans regia

These are found along the water courses and irrigated canals in many regions.

10. Protective plantations

The main species used for protective plantations are as follows.

10.1. *Protection against soil erosion*

Acacia cyanophylla, Cupressus sempervirens, P. brutia, Pinus halepensis, Eucalyptus camaldulensis and Eucalyptus gomphocephala.

10.2. *Protection against wind*

Casuarina cunninghamiana, Cupressus arizonica and Cupressus sempervirens.

---

the State', as better termed in the draft forestry act, are State forests. Therefore, with the exception of forests that occur on privately owned land, the legal presumption is that all forests belong to the State. Provision is made, however, for the possibility of establishing private and communal forests.

Ownership and size of properties

In actual fact, nearly 99% of existing forests are owned by the State, with most (~90%) having been classified as forests of the State's private domain.[2] As such, they have been recorded in the Land Cadastre in the name of the Forestry Department. They are therefore granted a high degree of legal protection, as lands of the private domain may not be forfeited, nor may

any title thereto be acquired by prescription. The Forestry Law of 1953 does not provide, however, for the procedures to be followed in establishing State forests. In particular, it makes no mention of prior public enquiries allowing for existing rights of ownership or use to be preserved. In fact, the said Law lacks clear procedural and organizational arrangements for clarifying measures to be taken to consolidate ownership and user rights. Similarly, it is mute with respect to changing procedures of recorded forests to such an extent that the transfer of property rights' ownership is almost non-existent.

Recently, as a matter of both policy and law, the role of private forestry has been officially acknowledged in Syria. Afforestation and reforestation are among the top priorities of the forestry sector, and forest plants and seedlings are widely distributed to interested people, free of charge, to encourage tree planting. Privately owned forests are, in principle, to be managed and preserved by their individual or corporate body holders/owners, under the technical supervision of the Forestry Department. However, there are a number of legal constraints that presently obstruct the development of private forestry.

Administration

The major central government institution responsible for forest management, protection and exploitation is the Forestry Department under the Ministry of Agriculture and Agrarian Reform. As stated in Forest Law No. 7/1994, the Forestry Department seeks to 'conserve and manage forest resources and their protection, conserve biological diversity and establish environmental protected areas'. Headed by a Director who reports to the Minister, the Department consists of four divisions, namely: Production (seedlings and afforestation); Silviculture (management); Protection (fire control and forest police/guards); and Exploitation (public and private forests). In the various Provinces, as part of the Provincial Directorates of Agriculture, there are the Forestry Services, the organizational structure of which is generally modelled after that of the Department. Service Chiefs report to the Forestry Headquarters through the Provincial Directors of Agriculture.

Impact of past policies on forestry

No policy has been, so far, officially adopted by the Government to prescribe long-term objectives in the forestry sector. Hence, there is no structured, detailed national strategy formally laid down for the development and conservation of the country's forest. The forestry programme consists of no more than targets to be achieved and budgets to undertake them.

The first Forestry Law enacted by Legislative Decree No. 66/1953 was the basic enactment governing forestry in Syria. A second law, much more limited in scope, was the Forest Police Law[3] enacted by Legislative Decree No. 86/1953 (amended in 1962, 1969 and 1970). As a strictly regulatory law focused mainly on

prohibitions, limitations and sanctions, it was both restrictive and repressive. It neglected the 'social dimension' of forests, for example by prohibiting the traditional customary rights of forest users. It did not contain any specific provisions regarding forest policy, administration, inventory, management plans, private forestry, research, training and extension, social (community) forestry and public participation, and environmental impact assessments. Because of the repressive character of this law, the forest areas in Syria have seriously declined during the past four decades and many of them have lost their socio-economic and environmental values.

This law was replaced by a new Forest Law of 1994. It brings certain improvements to the old one, such as guaranteeing free user rights to people living in forest villages. These rights include: the use of dead wood found on the ground, wood for repair of dwellings and the making of agricultural tools, fuelwood, and grazing activities, except for goats and sheep. The procedures to be pursued for the effective exercise of these rights are, however, highly complex and constraining. This law still conserves the main weakness of the previous one concerning policy, social forestry, public participation and involvement, planning, private forestry, research and extension.

Another Law on Environmental Protection was adopted in 1994. It provides, in general terms, the protection of flora, fauna, soil and natural resources. It empowers competent authorities to issue 'standards, specification and regulations for the protection of flora and fauna and for the sites of protected areas in order to ensure environmental balance and the conservation of living organisms'. It is believed that this law will affect forestry in Syria in a positive way.

Presently, there is a general policy direction for preserving existing natural forests and creating protective forest plantations on around 15% of the country's area. These activities initially were entrusted to HCA, a forestry body established by Presidential Decree No. 108/1977. HCA was abolished in 2001 and all responsibilities under its mandate were transferred to the Ministry of Agriculture.

Additionally, there are two bodies, established in 2001, dealing with the environment, that are also concerned with forestry. One is the General Commission for Environmental Affairs,

specially charged with preparation of environmental plans and law, the assessment of environmental problems, the prevention and control
of ecologically harmful activities, and the promotion of environmental public awareness. The
second is the Supreme Council for Environmental Safety, a prominent decision-making
organ, that has the power to adopt environmental policies, regulations and standards,
as well as to prohibit any environmentally
damaging activities.

Contribution of the Forest to the National Economy

Gross domestic product and employment

In the absence of a real forest inventory, and
therefore of reliable data regarding forest
resources, it is difficult to assess the role of
forestry thoroughly – in terms of market and
non-market goods and services – in the Syrian
economy. However, recent rough estimates
suggest that the contribution of forest production (industrial wood, firewood and charcoal)
to the GDP is of minor significance, of around
0.01% (Central Bureau of Statistics, 2002).
Nevertheless, the environmental role of forests
is much more important than its productive
function.

Statistics regarding forestry employment
show 12,000 workers, corresponding to 0.25%
of the total workforce in 2002 (Ministry of Agriculture and Agrarian Reform, 2003a). This figure
mainly accounts for the workforce employed
in the 'Forest Silviculture and Development
Project', established in 1987 for conducting
environmentally oriented cuts, managing
nurseries, providing fire guards and undertaking
afforestation activities. Some of this employment, which is classified within public sector
services, is on a casual labour basis. Unfortunately, reliable statistics are not available on
employment in timber-based industries.

International trade

Syria is a net importer of wood and wood products. Indeed, the country only produces 5–6%

of its wood consumption, with the larger part of
domestic demand for wood and paper products
being met by imports. Syria produces very
modest volumes of sawn timber, veneer, plywood and particleboard, which mostly originate
from environmentally oriented fellings, applied
to a very small extent in certain regions by
the Forestry Department of the Ministry of
Agriculture and Agrarian Reform.

Other forest-based industries: tourism and green belts

Over recent decades, forests have been assuming an increasing role in promoting tourist and
recreational activities, and their contribution to
the promotion of better quality of life is acknowledged at the level of both society in general and
decision makers. These forests take advantage
of their proximity to the Mediterranean Sea
and their location in mountainous areas.
Special recreation areas in zones reserved for
recreational activities have been envisaged.

The Values of Syrian Forests

As already mentioned, the forestry sector
contributes very little to the national economy,
compared with other sectors. However, the
importance of the forest services, though more
difficult to quantify, should not be underestimated. It should be stressed that in a country as
arid as Syria, forests provide a constant supply
of fodder for a large number of grazing animals
and play a significant role in water catchment
protection, desertification control and prevention of soil erosion and flooding. They represent
a rich source of medicinal and edible herbs
and perform a crucial function in preserving
biological diversity. A preliminary estimation of
Syrian forest values is shown in Table 12.1,
based on 2003 quantities and prices.

Direct use values

Syrian forests produce very modest volumes of
wood, which are mainly used to satisfy small
industrial and domestic needs, firewood and

Table 12.1. Values of Syrian forests.

Valuation method/output	Quantity	Value (€)
Direct use values		
Roadside price valuation[a]		
Timber for industrial use[b] (t)	5,096	267,960
Firewood (t)	2,489	102,700
Charcoal (t)	115	2,470
Medicinal and aromatic plants (t)	3,167	1,453,800
Chestnuts, small fruits (t)	6	14,850
Market price valuation		
Net growth of standing timber stock (m³)[b]	91,000	3,417,900
Firewood, illegal cuts (t)	750	38,250
Charcoal, illegal cuts (t)	40	850
Other NWFPs[c] (t)	4,300	1,973,000
Honey (kg)	30,000	177,000
Surrogate market pricing		
Grazing (000 FU)	1,500	173,300
Total direct use values		7,622,080
Indirect use values		
Cost-avoided method		
Watershed protection (ha)	200,000	42,500,000
Shadow pricing		
Carbon sequestration (tC)	130,500	2,610,000
Total indirect use values		45,110,000
Negative externalities		
Cost-based method		
Soil erosion due to lack of management (m³/ha)	200,000	−7,140,000
Losses produced by forest fire (ha)	1,500	−2,550,000
Total negative externalities		−9,690,000
TEV		43,042,080

[a]Quantity sold by licences according to the statistics of the Forestry Department (2003).
[b]The conversion rate applied by the Forestry Department is 1t = 0.7 m³.
[c]Quantities collected for self-consumption (free rights of local people) and sold in the local markets (main species *Laurus nobilis*, *Myrus communis*, *Thymus syriaca*, *Capparis spinisa* and *Rhus coriaria*).

charcoal, and a wide range of NWFPs. The majority of these products – timber, firewood, charcoal, medicinal and aromatic plants, chestnuts and small fruits – are estimated based on the quantities sold at public auction, through permits or licences issued by the Forestry Department. These products are valued according to their roadside prices (Table 12.1).

However, the valuation of medicinal and aromatic plants takes into account both the quantities collected by private vendors (professionals), through official permits or licences issued by Forestry Department, and those collected for free by forest villagers. For the quantities commercialized by the Forestry Department,

the average roadside prices are used, thus giving the income accruing to the treasury. The forest outputs' returns appropriated by individuals living in forestry villages are estimated by using their average market prices. Overall, the monetary value of medicinal and aromatic plants is €1.4 million.

Net yearly growth of standing timber stock is on average 1 m³/ha of wood in mature accessible stands. This corresponds to a total of 130,000 t (or ~91,000 m³), if the major natural forests – chiefly situated in foothills and on high ground in the western part of the country – are excluded, due to the fact that sylvicultural activities largely depend on a site's accessibility. Based on half of the stumpage prices, the total value of the net growth of standing timber is estimated at €3.4 million.

Illegal firewood collection and charcoal production are common practices in Syria, mainly due to the limited numbers of forest guards and lack of available means to carry out proper surveillance. Physical valuation is the difference between the quantity sold in local markets and that collected through official permits issued by the Forestry Department. Monetary valuation is based on the roadside prices and gives a value of €38,250 for firewood and €850 for charcoal.

The valuation of honey is based on the estimated quantity produced by beehives in and around forests. The forests' degradation and limited accessibility are responsible for the low honey productivity, of about 0.5 kg/ha, over a surface of 60,000 ha. At an average market price of €5.9/kg (Abed, 1998), the total value of honey is €177,000. It should be noted that beekeeping activities depend, among other factors, on the density of forest, canopy cover, dominant species, climate and forest accessibility.

The value of grazing takes into account the number of forage units (FU) consumed and the value of 1 FU. According to the Ministry of Agriculture (Zahoueh, 1994c), there are 50,000 ha of grazed forest, with an average forage consumption of approximately 30 FU/ha (Abou Zakhm, 1997a). The market price of 1 FU is derived from the market price of barley (€0.12/kg), based on the surrogate market technique. Accordingly, total grazing value is around €173,300.

There are no available estimates relating to hunting and recreation in Syrian forests. However, it should be mentioned that all hunting activities in forest areas are illegally practised by professionals, despite the fact that national environmental regulations impede hunting on all territory. The State, as the sole forest owner, has not yet defined hunting rights due to the threat encountered by endangered species. Some recreational areas have been newly established for developing tourist activities. However, at this stage, it is difficult to estimate the number of day-visits/year and consumer surplus/visit in the forest.

Indirect use values

Despite the minor importance of wood production, Syrian forests play a significant role in the protection of water catchments, desertification control, prevention of soil erosion and flooding, and biodiversity conservation. Moreover, they constitute a substantial source of food security for poor rural households. Other indirect use values provided by forests include landscape quality and carbon sequestration. Monetary valuation of these functions is not an easy task and involves many large assumptions.

Water-related services are valued on the basis of the costs avoided method. Accordingly, it is assumed that the magnitude of the water protection function is proportional to the costs of floods, erosion and other impacts avoided by the existence of forest cover (*within* and *without* forest situation). The western and north-western parts of Syria (the coastal region, Idleb, Al-Ghab and parts of Hama and Aleppo), where the major natural forests are chiefly situated in foothills and on high ground with a prevailing Mediterranean climate, are subject to high risk of water-related hazards (floods and erosion). These high-risk areas cover approximately 200,000 ha. In addition, other areas of lowland (with the same magnitude of surface, 200,000 ha) with intensive land uses in agriculture, principally located in coastal and interior plains, are well protected by the well-managed forest stands. It is estimated that the average national public expenses for soil conservation and hydraulic works maintenance (flows) amount to €210/ha. Consequently, it can be roughly assumed that the well-managed forests perform a watershed protection function worth €42.5 million. This value allows calculation of a benefit of about €92/ha for forests in Syria.

Information required to support estimates of forest volume and wood biomass, which are important indicators of the forests' potential to sequester carbon, is not satisfactory in Syria. This means that assumptions and extrapolations have to be used and the results are therefore dubious. Based on estimates provided by the Food and Agriculture Organization (2000) in similar conditions, the net quantity of carbon fixed yearly in the Syrian forest biomass (28 m^3/ha), estimated according to forest nature and vitality, was around 130,500 tC for 2002.

The application of a shadow price of €20/tC results in an annual value of €2.6 million for carbon sequestration.

Option, bequest and existence values

In this respect, no data can be reported. Studies and research should be conducted to explore the impact of these indicators on the TEV of the Syrian forests. The results obtained in other countries cannot be simply extrapolated and used in the Syrian forests due to their specific context.

Negative externalities linked to forests

The total forested area in Syria is affected by forest fire, overgrazing, overcutting, encroachment from agriculture, negligence and lack of management. As a consequence, large tracts of original forests have degenerated into secondary plant communities of low economic and environmental value (United Nations Evironment Program, 2000). These externalities are likely to reduce the TEV of the forests substantially.

Erosion in hilly and mountainous areas represents the main consequence of the degradation of vegetative cover, and its partial or complete destruction. The soil erosion rate varies in relation to several factors, of which the nature of soil, vegetation, slope and precipitations are the most important. This rate is estimated to be up to 20 m³/ha/year, equivalent to 2 mm/year, after a destructive fire in coastal slopes. On average, soil losses in degraded forest area of 200,000 ha could be 3.5 m³/ha/year with a minimum cost of €10.2/m³. This gives a total value of €7.1 million.

The losses caused by forest fires are considerable (for more details refer to Zahoueh, 1994a,b; Samman, 2000). For various reasons, forest fires in Syria are primarily a 'social problem', and the rural people living in and around forests are the main agent for causing fires. In fact, more than 95% of the annual fires, ravaging more than 1500 ha/year (average over the period 1995–2000) are started directly by people and/or indirectly by their related activities:

grazing within forest stands, conversion of forestlands to agriculture, house construction and negligence. The value of damage caused by forest fire based on the restoration costs is €2.6 million.

In conclusion, the estimated components of the TEV give an overview of Syrian forests as a whole. The dominant magnitude of the water issues (watershed protection) is attributed mainly to the well-managed forests, most of which are located in the foothills and on high land. They protect around 200,000 ha of lowland, with intensive uses (agriculture) located on the coastal and inner plains. This clearly indicates that the sustainable management plans should capture the water-related issues at watershed level and/or at geographically similar regions.

Towards the Total Economic Value of Syrian Forests

Although the contribution of forestry production to the national economy in Syria is very little compared with other sectors, calculation of the TEV shows that their environmental role is paramount for developing sustainable plans and strategies at country level. Assigning monetary values to these functions is difficult, and the results are sometimes dubious. However, they are meant to encourage the consideration of the entire range of forest functions within sustainable management plans and new forestry policies.

As an outcome of this preliminary valuation, Fig. 12.2 highlights the estimated values of the TEV categories: direct and indirect use value and negative externalities. Despite the scarce

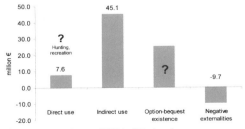

Fig. 12.2. Estimated TEV of Syrian forests.

information on several components, it appears that the indirect use values represented by the environmental functions have the major importance, being €45.1 million. Direct use values, of only €7.6 million, are underestimated, by not including values such as hunting and recreation. Negative externalities account for €9.7 million and are represented mainly by erosion and losses due to forest fires.

Figure 12.3 shows that Syrian forests play a significant role in water catchments and in prevention of soil erosion and flooding. While the results should be strengthened through more research in fields related to water issues, they confirm the importance of forests in conserving soils in mountainous areas and protecting agriculture production in lowland regions. The estimated value of forests in this respect is around €42.5 million. Conservation and sustainable development of these forests should be a priority in order to reconcile the role of forests in meeting both the socio-economic and environmental objectives. Therefore, ecological considerations should be viewed not as subordinate, but as an integral part of economic policy and planning.

In addition, the estimated production values were calculated on a per hectare basis as shown in Fig. 12.2. The results show the high importance of watershed protection, with €92/ha of forests. Therefore, in many of the steeply sloped mountainous areas, the need for sustaining forest stands as well as agricultural and rural development should be seriously considered.

Perspectives and Crucial Issues: Strategy and Policy for Sustainable Forest Development

1.   Forests have always played an important role in Syria. Presently, the country's forests are generally degraded, due to overexploitation, overgrazing, deforestation and fires. They should be rehabilitated and managed on a sustainable basis.

The demand for their numerous functions and outputs is increasing with the expanding population. For this reason, conservation and sustainable development of these forests have now emerged as priority items in the context of conservation and sustainable development of renewable natural resources. The challenge ahead is to reconcile the role of forests in meeting both the socio-economic and environmental objectives at national level. It should be stressed that ecological considerations should be viewed not as subordinate, but as an integral part of economic policy and planning.

2.   The formulation of strategy approaches to sustainable forest development requires the harmonization of human activities with the biological and physical aspects of forest ecosystems. Human activities, forest ecosystems and the inter-relationships between the two are dynamic and change over time and space. Consequently, monitoring the two systems and their interaction is crucial for improving sustainable forest development and involves a number of ecological, socio-economic, technological and political considerations.

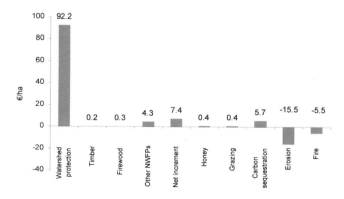

Fig. 12.3.   Average estimates of Syrian forest values/ha.

The challenge of reaching sustainable forest development may be pursued through a number of specific actions, including research and extension, education, legislation, forest and environmental policy, forestry practices and management.

A basic principle of the strategy for sustainable forestry development could be the incorporation of the social dimension into the forestry sector through the use of social analysis, which describes and analyses the potential effects of planned interventions upon people living in forest areas. The traditional rights of the local people, mainly referring to the collection of firewood and other NWFPs, are historic and very important for the preservation of forests.

The Forestry Department should encourage the active participation of the local people in the social analysis of projects in the forestry sector: forest management, protected areas, plantations, combating forest fire and forest degradation.

3. Although present policy reflected in the forest legislation provides the basis for the protection of forestlands, there is no well-defined and officially announced/adopted National Forest Policy in Syria. A policy oriented towards a sustainable development of forestry is greatly and urgently needed. The following objectives could be taken into consideration in its development.

- To dedicate the present forested areas as a Permanent Forest Estate, strategically located in the different ecological zones of the country, in accordance with the concept of rational land use. This should ensure: (i) the sound climatic and physical conditions of the country, i.e. the safeguarding of water supplies, soil fertility and environmental quality and the minimization of damage by floods and erosion to rivers and agricultural land (*Protective Forests*); (ii) the continuous supply at reasonable rates of all forms of forest produce which can be produced economically within the country and are required for agricultural, domestic and industrial purposes, and for export (*Productive Forests*); and (iii) the conservation of adequate forest areas for recreation, education, research and protection of the country's

unique flora and fauna (*Amenity Forests and Protected Forests*).

- To manage the Permanent Forest Estate with the objective of maximizing social, economic and environmental benefits for the nation and its people in accordance with the principles of sound forest management.

- To pursue a sound programme of forest development through regeneration and rehabilitation operations in accordance with approved silvicultural practices in order to achieve maximum productivity from the Permanent Forest Estate.

- To promote efficient harvesting and utilization of all forms of forest produce and to stimulate the development of appropriate wood-based industries with determined capacities commensurate with the resource flow in order to achieve maximum resource utilization and to create employment opportunities.

- To undertake and support an intensive research programme in forest development aimed at improving yield from the Productive Forests and enhance the direct and indirect benefits linked to harvesting, protection and conservation of the resource base.

- To undertake and support a comprehensive programme of forestry training at all levels in the public sector in order to ensure an adequate supply of trained labour to meet the requirements of forestry and wood-based industries.

The National Forestry Policy should be supplemented by specific guidelines to ensure sustainable forest management and development.

Notes

[1] Especially irrigated poplars in the Euphrates valley.
[2] Under Syrian land legislation (1926 Law, as subsequently amended), State holdings are part of either the public domain of the State (building, infrastructures, roads and other stands meant for public use), or the private domain of the State (public holdings privately managed by Government institutions).
[3] The Forest Law was complemented further by three other statutes, namely: (i) Legislative Decree No.

128/1958 on the protection of trees and plantations and damage caused by goats; (ii) Legislative Decree No. 65/1966 on the prohibition of grazing of certain animals in certain lands; and (iii) Legislative Decree No. 169/1959 on the distribution of arid lands in the forested regions of Syria.

References

Abed, T. (1998) *Etude de Potentialité Apicole en Vue de Développement de l'Apiculture en Zone Forestière Littorale.* Rapport de consultation dans le cadre du projet GCP/INT/539/ITA, FAO.

Abido, M.S. and Kurbaisa, M.S. (2003) *The Present Status of Syrian Juniper Forests on the East Lebanon Mountain Chain.* Al Khalij Review, Arabian Gulf University, Bahrain, vol. 1, no. 21, pp. 64–70 [in Arabic with summary in English].

Abi-Saleh, B., Barbero, M., Nahal, I. and Quézel, P. (1976) Les séries forestières de végétation au Liban: essai d'interprétation schématique. *Bulletin de la Société Botanique de France* 123, 541–560.

Abou Zakhm, A. (1997a) Investigation sur l'Application du Pâturage Contrôlé en Plantation Artificielle du Pinus brutia. Rapport de consultation dans le cadre du projet GCP/INT/539/ITA, FAO.

Abou Zakhm, A. (1997b) The renewable natural resources in the south of Syria. *37th Science Week*, Damascus, Syria, pp. 51–77 [in Arabic].

Barbero, M., Chalabi, N., Nahal, I. and Quézel, P. (1976) Les formations à conifères méditerranéens en Syrie littorale. *Ecologia Mediterranea* 2, 87–99.

Central Bureau of Statistics (1975) *Statistical Abstract.* CBS, Office of the Prime Minister, Syrian Arab Republic.

Central Bureau of Statistics (2002) *Statistical Abstract.* CBS, Office of the Prime Minister, Syrian Arab Republic.

Central Bureau of Statistics (2003) *Statistical Abstract.* CBS, Office of the Prime Minister, Syrian Arab Republic.

Chalabi, M.N. (1980) Analyse Phytosociologique, Phytoécologique, Dendrométrique des Forêts de *Quercus cerris* ssp. *pseudocerris*, et Contribution à l'Étude Taxonomique du Genre *Quercus* L. en Syrie. Thèse de Docteur dès science. Faculté des Science et Techniques, St Jérôme, Canada.

Food and Agriculture Organization (2000) *Global Forest Resources Assessment 2000.* FAO Forestry Paper No. 140.

Food and Agriculture Organization (2001) Forestry country profiles navigation page: www.fao.org/ forestry/fo/country/nav_world.jsp. Accessed May 2003.

Forestry Department (2003) *Annual Report of Forestry Department Activities.* Ministry of Agriculture and Agrarian Reform, Damascus.

General Organization for Remote Sensing (2001) *Mosaique Map of Syria.* Damascus, Office of the Prime Minister, Syrian Arab Republic.

Martini, G. (1989) Etude Écologique de Réserve Forestière de Jabal Matta (Chaîne Montagneuse Syrienne). Thèse de Master, Faculté d'Agronomie, Université d'Alep, Syria.

Martini, G. (1999) Analyse Écologique et Phytosociologique su Versant Est de la Chaîne Montagneuse Syrienne. Thèse de Doctorat en Science Forestière et Ecologique, Faculté d'Agronomie, Université d'Alep, Syria.

Ministry of Agriculture and Agrarian Reform (2003a) *Annual Agricultural Statistical Abstract.* Ministry of Agriculture and Agrarian Reform Publications, Damascus.

Ministry of Agriculture and Agrarian Reform (2003b) *National Forestry Programme.* Ministry of Agriculture and Agrarian Reform, Forestry Department, Damascus.

Mouterde, P. (1953) *La Flore du Djebel Druze.* Paul Lechevalier, Paris.

Nahal, I. (1961a) La garrigue à *Quercus calliprinos* et *Pistacia palaestina* du Djebel Alaouite de Syrie. *Annales d'ENEFT* 18, 409–420.

Nahal, I. (1961b) La végétation forestière naturelle dans le Nord-ouest de la Syrie. *Revue Forestière Française* 3, 90–101.

Nahal, I. (1962) Contribution à l'étude de la végétation dans le Baer-Bassit et le Djebel Alaouite de Syrie. *Webbia* 16, 477–641.

Nahal, I. (1974) Réflexions et recherches sur la notion de climax de la végétation sous le climat méditerranéen oriental. *Revue de Biologie et d'Écologie Méditerranéenne* 1, 1–10.

Nahal, I. (1975) *Etude de Végétation dans le Kurdagh (Jabal de Kurd).* Aleppo University, Faculty of Agriculture, Aleppo, Syria.

Nahal, I. (2003) *Science de l'Ecologie Forestière.* Publication de l'Université d'Alep, Syria.

Pabot, H. (1957) *Rapport au Gouvernement de Syrie sur l'Écologie Végétal et ses Applications.* FAO, Rome, Rapport No. 663.

Samman, G. (2000) Effect of Fire on Soil Chemical and Physical Properties and Vegetation Characteristics in the Mountains Forests of the Syrian Coast. MSc Thesis in Forestry, Faculty of Agriculture, University of Aleppo, Syria.

United Nations Environment Program (2000) *Global Environment Outlook (GEO).* UNEP, Geneva.

United Nations Population Fund (2003) *Population and Demograhic Aspects of the Syrian Arab Republic*. UNFPA, Damascus.

Zahoueh, S. (1994a) Integrated forest management plan for fire prevention in Syria (pilot area of Om-Altoyour, Lattakia). In: Specialized international training course on '*Soil and Water Conservation in Arid Environment*', 29 May–12 June, jointly organized by IAM Chania, IAM Bari, CIHEAM, and FAO, Damascus, Syria.

Zahoueh, S. (1994b) *Plan d'Aménagement Intégré Anti-incendie pour Mieux Préserver la Forêt en Syrie (Zone Pilote Om-Altoyour, Lattakia)*. Rapport de consultation dans le cadre du projet GCP/INT/539/ITA, FAO.

Zahoueh, S. (1994c) *Primary Study on Controlled Grazing in Natural Forests and Pine Plantations in Kessaibya (Aleppo)*. Consultancy report within FAO project GCP/INT/539/ITA, FAO.

13 Turkey

Mustafa Fehmi Türker[1], Mehmet Pak[2] and Atakan Öztürk[3]

[1]Karadeniz Technical University, Faculty of Forestry, Division of Forest Economics in the Department of Forest Engineering, Trabzon, Turkey; [2]Kahramanmaraş Sütçüimam University, Faculty of Forestry, Division of Forest Economics in the Department of Forest Engineering, Kahramanmaraş, Turkey; [3]Kafkas University, Faculty of Forestry, Division of Forest Economics in the Department of Forest Engineering, Artvin, Turkey

Introduction

Turkey occupies a unique geographical and cultural position at the crossroads between Europe and Asia. It covers 80 million ha, one-quarter of which is devoted to – or at least designated as – forest. The country's topography is extremely varied, with an unusual diversity of agroecological conditions. Mountain ranges generally run parallel to the northern and southern coasts, surrounding the central Anatolian Plain, which rises from 500 m above sea level in the west to over 2000 m in the east. Soils are also variable: on gentle slopes, they tend to be deep, moderately fertile and slightly alkaline, while on steeper slopes, they are usually shallow, rocky and infertile. About 80% of the country's soils suffer from moderate to severe sheet and gully erosion, and most rivers carry heavy loads of sediment.

Turkey is divided into seven agroecological zones, defined largely by major crops and climatic conditions. In the western and southern coastal zones, a subtropical Mediterranean climate predominates, with short, mild, wet winters and long, hot, dry summers. Arid and semi-arid continental climates prevail in central regions, where winter is often extremely harsh, with frequent and heavy snowfall in the higher parts of the Anatolian Plain. There is a remarkably even distribution of productive forests among these zones. However, the greatest concentration is found in the regions with higher rainfall, such as the Black Sea coast, where the average annual rainfall exceeds 1000 mm, and around the Marmara Sea. Almost all the country's major broadleaved forests are located in these regions (Muthoo, 2001).

The country has a sensitive ecosystem due to the irregular rainfall regime, steep slopes and a geographic location situated between the arid areas in the south and the humid belt in the north (Ministry of Forestry, 1997). There is a rich structure and variety of tree species, among which the main conifers are *Pinus* sp., *Abies* sp., *Cedrus* sp., *Picea orientalis* and *Juniperus* sp., and the non-conifers are *Fagus* sp., *Quercus* sp., *Castanea* sp., *Alnus* sp., *Carpinus* sp., *Fraxinus* sp. and *Eucalyptus* sp. (Muthoo, 2001).

Turkish forest management is mostly aimed at wood forest products (WFPs) and only partly at non-wood forest products (NWFPs). Other non-market forest values linked to indirect uses, option and non-uses have been ignored to a certain extent, or at least not yet recognized explicitly by forest management. New approaches for a sustainable forest management are needed in order to capture the multiple products and functions of forests and to reduce management costs, while giving a clear emphasis on social objectives.

Forest Resources

Area and people

In Turkey, the designated forest area[1] covers 20.7 Mha, or 26.6% of the country's surface (United Nations Economic Commission for Europe-Food and Agriculture Organization, 2000). Around 91.6% is natural forest and the remaining 8.4% consists of plantations.

The country's total population at the 1991 census was 56.5 million. Just under 60% lived in towns, a percentage that rose significantly during the following 10 years. Approximately 40% of the working population was engaged in agriculture, a very high proportion by European standards, but the figure fell by 10% in the following decade, and continues to do so. In 2000, the total population in Turkey was 67.8 million (Muthoo, 2001; State Institute of Statistics, 2002). An estimated 7.6 million people live in or around the forests, distributed among about 20,080 villages. Many of these villages are situated in remote areas, far from markets and hindered by generally outdated means of transport and inadequate communications, health care and education services. At around US$200/*capita*, the yearly income of these villagers is well below the national average.

Forest villages have received the lowest proportion of the government investment budget for infrastructure in all the subsequent national 5-year development plans. At the same time, forest villagers are largely dependent on natural resources, generally combining forestry activities with small-scale subsistence farming (Muthoo, 2001).

Typologies

Turkish forests contain a diversity of ecosystems ranging from well-stocked forests to degraded, if not depleted, areas subject to severe natural limitations (Fig. 13.1). Only 51% of the forest area, or 10.5 Mha, is represented by productive forests. Of this, around 8 Mha are classified as productive high forest and about 2.5 Mha as productive coppice forest (World Bank, 2001).

Conifers make up more than two-thirds of the high forest, while broadleaved species represent about 20% and are often also predominant in the mixed high forest (Table 13.1). The coppice woodlands, a third of the forest area, are dominated by a large variety of broadleaved species, of which oaks are prominent. Broadleaves are not of good industrial quality, except for limited areas of beech and oak along the

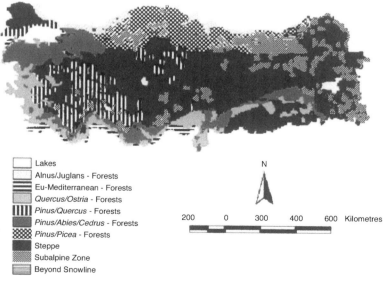

Lakes
Alnus/Juglans - Forests
Eu-Mediterranean - Forests
Quercus/Ostria - Forests
Pinus/Quercus - Forests
Pinus/Abies/Cedrus - Forests
Pinus/Picea - Forests
Steppe
Subalpine Zone
Beyond Snowline

N

200   0   300   400   600   Kilometres

Fig. 13.1. Map of Turkish forests. Source: Ernst (1997) (redrawn).

Table 13.1. Area, growing stock and increment of Turkish forests by major forest types

Forest type	Tree species	Forest condition	Area		Growing stock		Current annual increment	
			Mha	%	Mm³	%	Mm³	%
High forest	Coniferous forest	Normal	5.9	28.8	720.9	61.8	18.9	58.6
		Degraded	3.9	19.0	45.1	3.9	0.9	2.9
		Total	9.9	47.8	766.1	65.7	19.9	61.5
	Broadleaved forest	Normal	1,4	6.8	272.6	23.4	6.5	20.1
		Degraded	1.2	5.7	16.4	1.4	0.3	1.2
		Total	2.6	12.5	289.1	24.8	6.9	21.3
	Mixed forest	Normal	0.6	3.1	—	—	—	—
		Degraded	0.7	3.5	—	—	—	—
		Total	1.3	6.6	—	—	—	—
	Total	Normal	8.0	38.7	993.6	85.2	25.5	78.7
		Degraded	5.8	28.2	61.6	5.3	1.3	4.1
		Total	13.8	66.9	1,055.2	90.5	26.8	82.8
Coppice forest		Normal	2.5	12.3	83.9	7.2	4.9	15.0
		Degraded	4.3	20.9	27.2	2.3	0.7	2.2
		Total	6.8	33.2	111.2	9.5	5.6	17.2
Grand total		Normal	10.5	51.0	1,077.6	92.4	30.4	93.7
		Degraded	10.1	49.0	88.9	7.6	2.0	6.3
		Total	20.7	100	1,166.5	100	32.4	100

Black Sea coast. The regenerative capacity of major species, especially *Pinus brutia*, is very good, so that natural high forest can easily be maintained through natural regeneration, provided appropriate silvicultural practices are used. Dense broadleaved forests comprise only 7% of the forest area and 14% of the high forest. *Fagus orientalis* (42%) and various oak species (25%) are dominant. Some 5% of the forest area consists of mixed high forest where broadleaves often play an important role. The coppice woodlands are dominated by oaks (46%), but contain a large variety of other broadleaved species (Muthoo, 2001).

Figure 13.2 illustrates some broad trends of Turkish forest area over the last three decades. Part of the change is due to revised definitions of forest type; nevertheless, several important observations can be made (World Bank, 2001):

- The total designated forest area increased by 5%, or 0.5 Mha, mainly due to the planting of forest clearings and natural regeneration of abandoned land as a result of outmigration.
- The area of productive and degraded high forest increased by 27%, while the area of

coppices has decreased correspondingly. This is explained by the enrichment planting of coppice areas, conversion of beech coppice to high forest by improvement cuttings[2] and afforestation of forest clearings.

- The degraded forest area fell by 10%, as a cumulative result of a decrease in the area of degraded coppice and an increase in the degraded high forest areas.

Nearly half of the total forest area is degraded, mainly due to forest fires, overgrazing, land clearance and illegal settlement (Ministry of Forestry, 2000). Between 1937 and 1999, forest fires affected about 1.5 Mha (State Planning Organization, 2001). Grazing is a major threat to forests, causing serious degradation, particularly to forest regeneration, and increases soil erosion especially on steep slopes. This is due to the fact that the total pasture area of 1.5 Mha is far from adequate to meet grazing requirements. Controlled grazing, range improvement, fodder production and stall-feeding efforts and practices are lacking. Illicit woodcutting and encroachment for farming are other important causes of degradation and productivity decrease.

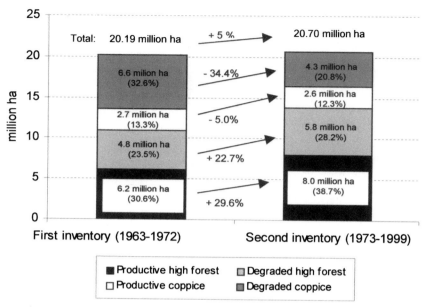

Fig. 13.2. Breakdown of total forest area by forest type at two inventories.

Silvicultural aspects

By the end of 1999, the total growing stock was estimated at 1.17 billion m^3 (Table 13.1). The average growing stock was 56 m^3/ha, which appears low because of the large area of degraded, understocked forest including scrub *maquis* vegetation of typical Mediterranean forest formations. In the productive high forest, growing stock was 125 m^3/ha, just below the European average, while stocking in productive coppice averaged 33 m^3/ha. The estimated current annual growth is 1.5 m^3/ha. Along the coastal areas, especially of the Black Sea region, characterized by favourable timber-growing conditions, fast-growing plantations can grow up to 20 m^3/ha/year. Meanwhile, in the central zones, large areas of oak register an increment of 2 m^3/ha/year. A comparison between the first forest inventory (1963–1972) and the second (1973–1999) shows an increase of about 25% in total growing stock and of around 15% in total current growth.

Despite this trend, the Ministry of Forestry reduced the Allowable Annual Cut (AAC) by 25% over the same period – from 22.8 to 17.2 Mm^3 – to allow the growing stock to increase in understocked stands and to improve

the conservation of protected forest areas. Production from State-owned forests has stabilized recently at about 7.6 Mm^3 of industrial roundwood and 8 Mm^3 of fuelwood. If a minimum 5–6 Mm^3 of illegal or unrecorded removal is estimated,[3] total annual production reaches 21 Mm^3. This represents about 11 Mm^3 below the annual increment, but some 4 Mm^3 above the AAC (World Bank, 2001).

In addition to State forest production, some 3.5–4 Mm^3/year comes from 200,000 ha of fast-growing, short-rotation private plantations. Agroforestry sites are mostly planted with poplars and, to a lesser extent, fruit trees, and play an important role in meeting domestic wood demand and reducing pressures on natural forests (Doğru, 2001).

Functions

Forests are renewable and complex ecosystems capable of providing a wide range of environmental, economic, social and cultural benefits. They supply various products and services that contribute directly to the social well-being and are vital to the economic and environmental conditions of the country. While the essential roles of forests are increasingly recognized by Turkish society as a whole, their benefits and functions are valued differently amongst different people and segments of society. Moreover, such valuation continues to change over time, due to changing needs and expectations of society (Doğru, 2001).

Turkish forests are divided into three categories by Forest Law 6831/1956, Art. 4, according to their functions: protection forests; national parks; and production forests. According to the same Law (Art. 23), protection forests cover around 355,545 ha, being dispersed over 52 different sites. They perform various functions, such as hydrological, erosion prevention, climatic, society health and national defence (Savaşan, 1999).

Konukçu (2001) states that 80.2% of the total forest area is managed for productive purposes; about 15.8% is allocated as protected area, including recreation and protected sites.[4] The remaining 4.0% is allocated to biodiversity conservation comprising national parks,[5] nature parks, nature conservation areas, natural monuments, seed stands, gene conservation forests, clonal seed orchards and specially protected areas.

Afforestation operations have been undertaken on degraded forestland and unplanted areas under the provisions of the Turkish forest regime. About 48% of the afforested areas are orientated principally towards wood production, while the remainder has purely protection and environmental roles. Productive high forests cover about 39.7% of the total forest area (Konukçu, 1998; Ministry of Forestry, 2000).

Institutional Aspects

Forest policy in Turkey began with the political reforms made under the Ottoman State in 1839. The state's financial crisis prompted the establishment of the General Directorate of Forests (GDF) in order to collect taxes from fuelwood, timber extraction and charcoal production. Regional Forest Directorates were also established (Bingöl, 1990). The 1870 Forest Regulation (*Orman Nizamnamesi*) was the first legal regulation concerning forest ownership and management (Özdönmez *et al.*, 1996).

The first Forest Law No. 3116/1937 classified forestland into state, public, foundation and private forests. It considered forests as an important source of national income. State forests became subject to protection, expansion, planning and management by the state for the common good. In 1945, in an attempt to protect and utilize forestland more effectively, almost all forests were nationalized under Law No. 4785/1945. However, because of negative public reactions, Law No. 5658/1950 approved a partial restitution to their previous owners, provided that specific conditions were verified. Finally, Forest Law No. 6831/1956, which is currently in force, states the various social and cultural forest benefits, suggesting the need for forest protection, and introduced the concept of the national park (Muthoo, 2001).[6]

The main forestry policy concepts that have been commonly accepted in different documents and implemented during the last half century can be summarized as follows (Doğru, 2001):

- Protecting forests against damage caused by anthropogenic and natural factors.
- Improving forest conditions and productivity so as to provide a sufficient supply of WFPs to meet domestic demands from forest industries, local communities, urban wood consumers and others on a sustainable basis (sustained yield).
- Increasing provision of NWFPs, and specifically of protective and environmental services.
- Reforesting and rehabilitating degraded forestland to enhance their potential for producing WFPs and NWFPs, controlling erosion, and providing other functions and services.
- Expanding and improving management of parks, nature and wildlife reserves and forest recreation areas.
- Supporting forest-dependent local populations, by meeting their needs for WFPs and NWFPs, providing employment in forest operations and credit assistance for small-scale income-generating activities, to improve local living standards and to reduce pressures on forestlands and other natural resources.
- Accepting that most forests are public goods, retaining them under State ownership, with

planning and management undertaken exclusively by State forestry organizations.

- Using forest revenues, such as those obtained from timber sales, to finance forestry programmes and expenditures related to forest protection, planning, silviculture, reforestation of degraded lands, and support of small-scale income-generating activities in forest villages.

These traditional forest policies are still in force, but their relative importance has evolved over time. For example, the increasing awareness of environmental and social forest values and the impacts of global and regional initiatives, related to the environment and forests, contribute to the recent changes in the national forest policies and strategies.

Ownership and scale of management units

Currently, 99.9% of the forest area is owned by the State, with the remainder belonging to 277 private owners and 51 public organizations (State Planning Organization, 2001). The management units of Turkish forests are the State Forests Enterprises (SFEs). Their size can be considered as the average size of a forest property (Türker *et al.*, 2001a). With an average size of 83,000 ha and some 190 workers, SFEs are large enterprises according to the Dietrich classification of forest enterprises in German conditions (Fırat, 1971). Their large size is explained mostly by their ownership and management by the State (Türker, 2000). Each SFE is divided into Forest Administration Chief Offices (FACOs), whose average size varies between 10,000 and 100,000 ha. Over time, the number and size of SFEs have changed[7] due to political decisions and changing physical–environmental, social and economic objectives. According to Yazıcı (1991), one of the main reasons for their frequent closing and opening is that economic criteria have not always been adequately taken into account when establishing SFEs.

Administration and policies

The Ministry of Forestry is responsible for administering and managing almost all forest areas. Most of their staff are employed in four General Directorates: Forests; National Parks and Hunting–Wildlife; Afforestation and Erosion Control; and Forest and Rural Relations. The GDF is responsible for the overall economic aspects of state forest management and has an autonomous revolving fund with a supplementary budget. Its duties and responsibilities were set out by Law No. 3234/1985 (Muthoo, 2001). The other three General Directorates are the main service units, and receive direct budget funding. The organization of the Ministry of Forestry comprises nine Regional Directorates who oversee operations in their geographical areas, including nurseries and afforestation activities (Muthoo, 2001). State forests are divided into 27 Regional Forest Directorates, 241 SFEs and 1328 FACOs.

The Environment and Culture ministries are also given responsibilities that may touch on forest management by applicable legislation. These overlapping responsibilities can result in institutional conflicts, as the legislation does not provide for their coordination.

Management and participation

Since the early 1960s, State forests have been managed according to traditional forest management plans, prepared and revised on a 10-year cycle. Each plan typically covers one or two watershed areas. They mainly concentrate on silviculture and wood use, and aim mainly at obtaining maximum wood production. The GDF is developing a new approach to management plans that incorporates planning for silvicultural and forest protection and is intended to be both more intensive and environmentally friendly. Though a major step towards sustainable forest management in Turkey, it does not yet address the issue of livestock and forest grazing.

Management plans have paid relatively little attention to the social conditions of villagers as actual and potential co-managers of forests and rangelands (Muthoo, 2001) – what is now called 'joint management' is still outside the management principles of Turkish forests. Management and planning of forests and protected areas is regarded in the relevant legislation as a technocratic discipline, being reserved exclusively for the responsible authorities. There is limited consultation with other institutions, affected local residents or other stakeholders. The people concerned are not consulted prior to the declaration of protected areas or to the adoption of forest management plans.

The lack of a legal framework for integrated, participatory agro-silvo-pastoral management is partly the result of constitutional provisions that give the State the exclusive right to manage and exploit natural resources (Art. 169). The Constitution does not prohibit the delegation of these functions; this may be specified by an act of Parliament and, in fact, forest villagers have already been involved in forest harvesting and have had other limited use rights. However, many interpret the Constitution's provision as excluding involvement of parties other than the State in forest management. This interpretation hampers the involvement of forest communities and other stakeholders in forest management, meanwhile preventing privatization of the forestry sector (World Bank, 2001).

Several non-governmental organizations (NGOs) are actively participating in forest management. For example, an NGO of the Foundation for Protection and Promotion of the Environmental and Cultural Heritage (ÇEKUL) focuses on local concerns and brings together youth for reforestation and other related activities. The Turkish Development Foundation (TKV) is concerned with overall policy and is very active in land use planning, sustainable agricultural and rural development and agroforestry at the national level. Also, the Turkish Foundation for Reforestation, Protection of Natural Habitats and Combating Soil Erosion (TEMA) promotes public awareness regarding the seriousness of the environmental problems in the country, especially soil erosion. Current projects include preparing a detailed 'erosion map', establishing a documentation centre and rehabilitating rangelands in several areas. The TEMA has also been actively supporting projects for biodiversity, forest and soil conservation, tree planting, building self-reliance among villages, and the employment of young people and women (Muthoo, 2001). All these NGOs are located in İstanbul.

Contribution of the Forest to the National Economy

Gross domestic product and employment

The contribution of forestry to the gross domestic product (GDP) is 0.5% on average (Türker, 1999). This figure includes mainly the value of raw materials sold by forest enterprises and of small processing operations. If processed WFPs and NWFPs such as sawnwood, chips and particles, mine poles, wire poles, pulp and paper, and styrax (liquidambar oil) were included, the contribution of forestry to the national economy would rise to 1.76% (Çakır, 1984). Moreover, forestry provides services such as erosion control, protection of water quality and flows, and recreation opportunities, which are not always valued in monetary terms. If these goods and services were included in the calculation, the share of forestry in the national economy would be much higher.

Employment in forestry contributes 0.2% to the total national employment (State Institute of Statistics, 2002). Under the Ministry of Forestry, there are 3813 forest engineers, 8196 forest guards, 11,513 professional and administrative staff, and a further 3097 permanent and 16,900 temporary staff (World Bank, 2001). In addition, a considerable number of villagers are hired for forest-related works: in 1999, 270,000 personnel were employed for harvesting work and 6000 workers were hired for afforestation. Between 1975 and 1999, about 13,000 people were employed to implement 466 projects supported by the General Directorate of Forest and Rural Relations (State Planning Organization, 2001).

International trade

Wood exports and imports account for a negligible share of the Turkish foreign trade, with just 0.05 and 0.06%, respectively (Türker, 1999). National wood production is insufficient to meet local demand that is partially satisfied through imports. Also, the export opportunities are very limited and this tendency might be changed only in the long term (Geray, 1986). A more detailed overview of the situation can be found in Box 13.1.

The Values of Turkish Forests

Forest outputs, including the public goods and externalities, were identified according to the total economic value (TEV) framework and estimated on the basis of a wide range of valuation methods. The results, reported in Table 13.2, are national level estimates of annual flows, except for recreation and angling which are local level results. The table shows the results of the monetary valuations in US$ for 1998 as originally estimated (third column) and also updated to 2001 euro prices (fourth column).

Direct use values

Direct use values are represented mainly by WFPs and a wide variety of NWFPs, such as

resin, styrax, mushrooms, pine nuts, bay leaves and many types of medicinal and aromatic herbs. Other important NWFPs include animal fodder, in the form of leaves, grass and other vegetation, particularly from those areas designated as rangelands. Water from forested catchment areas, the benefits of urban green-belts and recreational amenities should also be included under this heading (Muthoo, 2001).

Table 13.2 shows the value of timber and firewood for sale, estimated according to the roadside prices, as reported by the official statistics of the Ministry of Forestry (General Directorate of Forests, 1999; State Planning Organization, 2001). In addition, illegally harvested firewood represents roughly half of Turkey's annual firewood production and passes through no market whatsoever. Even though it is a result of an illicit activity, it represents a use value by meeting the fuel needs of many low-income forest villagers depending on forests. According to the Ministry of Forestry, average auction sale prices for fuelwood in 1997 were US$20.1/m^3. The fuelwood prices

---

Box 13.1.   Contribution of forestry to the national economy according to input–output tables.

Input–output models analytically show the economic structure of a country. Relationships between the various sectors and branches of the economy can be described, analysed and simulated, showing effects on production, income and employment.

Türker (1999) presents a model of the Turkish economy in which forestry is one of 64 branches. Overall, forestry contributes only 0.5% of the total GDP and a negligible amount of internationally traded products. However, the share of intermediate demand in total supply of forest products is 82%, much higher than the average for all sectors, which is about 39%, thus demonstrating that forestry is a strategic sector that supplies a considerable amount of raw/processed materials to other sectors. Conversely, the share of other industries' inputs to the forestry sector is 14%, compared with 38% of the average national economy, showing that forestry does not depend greatly on inputs provided by other sectors.

The share of gross added value (i.e. output less external expenses) in total output is around 76% for forestry, compared with around 57% in all branches. The difference between these figures is due to the high involvement of internal factors: capital, organization and, in particular, intensive labour.

From the input–output analysis, it can be seen that:

- the contribution of forestry to the national economy is small – although if free public services were included, it would have a greater weight;
- the share in national exports is low as the sector is mainly devoted to other internal intermediate industries and final consumption;
- forestry might be used as an important tool for decreasing the income differences and increasing employment, especially in underdeveloped rural areas.

Within the 64 sectors of the national economy, forestry ranks 17th due to the labour-intensive technology. At the same time, timber-based industries occupy 34th place (Türker, 1994).

Table 13.2. Values of Turkish forests.

Valuation method/output	Quantity	Value (000 US$)	Value (000 €) 2001 prices[a]
Direct use values			
Market price valuation			
Timber for sale[b] (m³)	10,316,000	435,030	421,979
Firewood for sale[b] (m³)	13,620,000	14,785	14,341
Firewood illegally harvested [c] (m³)	10,000,000	40,000	38,800
Resin[b] (t)	391	1,898	1,841
Mushrooms[b] (t)	11.4	11,482	11,138
Medicinal and aromatic plants[b] (t)	—	8,642	8,383
Truffles[b] (t)	395	0.5	0.5
Styrax (liquidambar oil)[b] (t)	5.9	56	54
Sticks and twigs[b] (t)	3,711	22	21
Bay leaves[b] (t)	4,221	9,253	8,975
Carob (fruit)[b] (t)	12	6	6
Chestnuts[b] (t)	262	262	254
Pine kernels[b] (t)	541	7,172	6,957
Snowdrop, cyclamen and other bulbous plants[b] (t)	180	1,087	1,054
Thyme, oregano[b] (*Thymus*, *Oreganium*)	6,038	13,237	12,840
Others[b] (t)	—	32,927	31,939
Substitute goods			
Grazing[c] (t of fodder)	2,300,000	225,000	218,250
Permit price, licences, fees			
Hunting[c] (no. hunters)	350,000	15,800	15,326
Angling[c]	—	20,148	19,544
Recreation[c] (day-visits)	93,400	2,000	1,940
Total direct use values		838,807	813,643
Indirect use values			
Shadow pricing			
Carbon sequestration[d] (tC)	7,920,000	158,400	153,648
Total indirect use values		158,400	153,648
Option, bequest and existence values			
Production function approach			
Pharmaceuticals[c] (no. of plant species)	9,000	112,500	109,125
Cost-based method			
Biodiversity conservation	—	1,380	1,339
Total option, bequest and existence values		113,880	110,464
Negative externalities			
Cost-based method			
Erosion, floods and landslides[c] (t of nutrients)	110,000	−125,000	−121,250
Losses due to forest fires[e] (ha)	5,800	−8,607	−8,349
Total negative externalities		−133,607	−129,599
TEV		977,480	948,156

[a]Adjustment to 2001 prices based on US$ inflation and exchange rates (IMF, 2003).
[b]General Directorate of Forestry (1999).
[c]Bann and Clemens (2001).
[d]United Nations Economic Commission for Europe/Food and Agriculture Organization (2000) quantitative evaluation, and Fankhauser (1995) monetary valuation.
[e]Ministry of Forestry (1999).

subsidized by the GDF to forest villagers and cooperatives averaged US$7.7/m³, suggesting a possible revealed willingness to pay. It would not be reasonable, however, to value the quantity of illicit fuelwood consumption at even the subsidized price, since firewood demand would most probably decrease if users who currently obtain their firewood for free were to actually pay. Conservatively assuming that the true current willingness to pay for illicit firewood is just half of the subsidized price (US$4/m³), then the annual value of illicit fuelwood production is US$40 million.

It should be stressed that forest planning decisions that incorporate only the interests of those utilizing formal firewood distribution channels run the risk of imposing a negative externality on millions of firewood users. In other words, the damage of the unrecorded exploitation of the forest resource produces considerable negative externalities in Turkey.

Several NWFPs, such as resin, mushrooms, medicinal and aromatic plants, and truffles, are valued according to the quantities traded on the market and their average price (Table 13.2).

Grazing is estimated on the basis of substitute goods (hay) and their average price on the market. Forestland fodder represents a large and valuable free input to the economic life and welfare of forest communities. According to the State Planning Organization (2001), there are 5.6 million cattle, 10.7 million sheep, 11.8 million goats and 1.6 million horses grazing in forest areas. While grazing patterns vary significantly between regions, animals commonly graze freely in the forest for 8 months of the year, i.e. roughly from April to November. Quantitative valuation is based on the grazed forest area of 5.8 Mha, from which 2.3 Mt of fodder are consumed annually. According to the Ministry of Forestry, the market price of hay cut from meadows is around US$0.23/kg. Conservatively assuming that the value of forest pasture is less than half of this figure, i.e. US$0.098/kg, forest fodder can be valued at around US$225 million annually.

Hunting is estimated by summing the revenue obtained from the issue of hunting permits and licences (US$15.8 million). Hunting permits are issued by the Ministry of Forestry for bear, ibex, wild goat and wild boar, and provided an income of around US$0.2 million in 1997.

Revenue obtained from hunting licences was around US$15.6 million.

The value of angling is assessed by summing the market value of captured fish species and other fees and it refers to freshwater fishing, trout breeding and sport fishing. Freshwater fishing is undertaken in 69 lakes and ponds, some of which pertain to forestland. Freshwater fish production totals 37,500 t and is valued at around US$79 million, on the basis of the market prices of captured species (State Planning Organization, 1996). It is assumed that only one-quarter of this revenue (US$20 million) is obtained from the forest areas. In addition, there are government-managed trout breeding stations, with a capacity ranging from 50,000 to 2 million fish per station, located in forest areas. According to the General Directorate of National Parks, Game and Wildlife (1997), gross income from fish sales at just six of these stations totals over US$147,000. Moreover, 700 anglers visited Yedigoller National Park, generating US$1500 in revenue. The aggregate value obtained from these estimates, however, only partially reflects the true value of this activity, due to unavailability of data at a national level.

Recreation is estimated by means of the revenue derived from the entrance fees paid by visitors for forest recreational sites and facilities. There are 428 forest recreation sites covering 15,950 ha, which meet an important part of the demand for recreation in Turkey. According to the General Directorate of National Parks, Game and Wildlife (1997), revenue obtained from the forest recreational areas calculated on the basis of entrance fees is US$2 million.

Indirect use values

No serious attempt has been undertaken to estimate the indirect use values of forests in Turkey until 1998, when the Ministry of Forestry launched a Forestry Sector Review Project assisted technically and financially by The World Bank. The results of this study related to valuation of soil protection, avalanche prevention, water quality and purification are provided at local level and reported in Box 13.2.

Carbon sequestration is the only indirect use value that could be estimated at a national level. Quantitative valuation made reference to the annual net forest increment as proposed by the United Nations Economic Commission for Europe-Food and Agriculture Organization (2000). Accordingly, the amount of carbon annually sequestered by forests is estimated at 7.92 MtC. The economic value is derived by applying a shadow price of US$20/tC (Fankhauser, 1995).

Option, bequest and existence values

The option value is estimated only for pharmaceuticals derived from genetic pools linked to forests. Valuation is based on the following sample model (Bann and Clemens, 2001):

$$V_p = (N \times p \times r \times a \times V/n)/H$$

where: V_p = the pharmaceutical value of 1 ha of forest (US$/ha/year), N = the number of plant species in the forest, p = the probability of a hit, r = the royalty rate, a = the appropriation rate, or rent capture, V/n = the average value of drugs developed (US$/year) and H = the area of forest (ha).

Based on best guess factors for Turkey, estimates of the potential value of pharmaceuticals based on three scenarios are given in Table 13.3. The present valuation adopts the medium scenario reporting an option value of US$6.3/ha; based on this, the total value is US$112.5 million.

The bequest–existence value of biodiversity is valued on the basis of the willingness to pay by international organizations for biodiversity

---

Box 13.2. The values of soil protection, avalanche prevention, water quality and purification in Altındere Valley National Park (1988).

The values of soil protection, avalanche prevention, water quality and purification and therapy functions of forests were calculated at a local level. These values may give indicators that could be applied in order to estimate the value of these functions and services throughout the country. Calculations were made, generally using the cost avoided method, for Altındere Valley National Park located around Trabzon city, which was established in 1987 covering 4800 ha of forest area.

The value of soil protection against erosion is US$53,800, derived from summing up the costs of dam, bench, sleeper and dry walls that would need to be constructed if forests did not exist. The value of avalanche prevention is US$13,500, referring to the cost of technical precautions that would need to be taken to protect the roads and villages if forests were not in place. Water quality and purification services were valued at US$606,500 with reference to the costs of afforestation and additional means (fence, terrace, afforestation and barbed wire fences) that would need to be implemented in the catchment area if forests did not exist. The value of landscape and nature protection services provided by forests was estimated at US$44,900, with reference to costs for establishing a similar national park if forest cover did not actually exist in Altındere Valley. Therapy value attains US$181,600, calculated on the basis of the income generated by national, and especially foreign, visitors to 'Meryemana Monastery' located in Altındere Valley National Park.

Source: Yavuz *et al.* (1988).

Table 13.3. The option value of pharmaceuticals derived from Turkey's forest resources.

Scenario	Values (US$)		Assumptions
	Unit value (US$/ha/year)	Total value (million US$)	
Low	0.05	0.9	Appropriation rate: 0.1; drug value = US$0.39 million
Medium	6.30	112.5	Appropriation rate: 0.5; drug value = US$1 million
High	87	1575.0	Appropriation rate: 1.0; drug value = US$7 million

General assumptions: n = 9000; probability of a hit = 0.0005[a]; royalty rate = 0.05; H = 17.8 Mha.

[a]A hit rate of 0.0005 has been estimated for tropical forest ecosystems and is applied here on the assumption that Turkey's biodiversity is comparable in richness.
Source: Bann and Clemens (2001).

conservation. The total is a one-off payment of almost US$14 million, or US$0.70/ha of forest. Annualized at a discount rate of 10%, this is financially equivalent to a perpetual recurring payment of US$1.38 million/year (Bann and Clemens, 2001). This should be regarded as a minimum estimate of biodiversity conservation value as: (i) the overall sum (US$14 million) refers only to the protected forest area, which is much less than the total forest; and (ii) the willingness to pay itself by international organizations represents a conservative measure of estimating biodiversity conservation.

Negative externalities linked to forests

From this category, the erosion due to poor forest management and the damage caused by forest fires are estimated at national level by means of the replacement cost method. Box 13.3 presents a local level valuation of the damage caused by deforestation in 16 catchment areas of Turkey.

Estimates of soil loss vary from 500 million to 1 billion t for the whole country. Based on estimates at the lower end of this scale, this is equivalent to the loss of 2.2 Mt of plant nutrients, assuming that the soil contains on average 0.1% N, 0.15% P_2O_5 and 0.154% K_2O. The price of fertilizers ranges from US$0.45 to US$2/kg. Using an average cost figure of around US$1.25/kg, the annual cost of replacing lost nutrients due to all erosion can be estimated at US$2.75 billion. Of course, available data do not indicate how much of the nationwide total

nutrient loss is provoked by degraded/depleted forests and, therefore, imputable to poor forest management or even neglect. However, if, for example, just 5% of total erosion were induced by anthropogenic poor forest management, this would still represent over US$125 million in replacement costs per year. The replacement cost at current prices is not, however, equivalent to damage values, as the manner in which prices would change if the 'replacement' of nutrients had literally to take place is unknown. However, this figure can be taken as a very conservative estimate or minimum value. This is because: (i) it is a very conservative assumption to accept that about 5% of all erosion is due to forest-related causes, given that roughly half of the country's forests have been depleted and destroyed; and (ii) this valuation takes into account only nutrient losses, without any effort to estimate the soil material which is lost and practically impossible to regain (Bann and Clemens, 2001).

Forest fires annually affect an area of 5804 ha. The damage provoked by forest fires is valued by summing the value of burnt wood (US$2.2 million), the additional reforestation costs exceeding the ordinary costs (US$4.5 million) and other extinguishing costs (US$1.8 million). The damage of NWFPs and especially endemic species destroyed completely by forest fires, and the opportunity cost of the labour force used to extinguish the forest fires have not been accounted for in the calculation. Neither the revenue, which could be obtained from the land during the years subsequent to fires (from other land uses), nor the general administration costs for this area are taken into account (Türker, 2000; Türker et al., 2001b).

Box 13.3.   The value of watershed damage caused by deforestation in 16 catchment areas (1997).

Watershed management functions performed by forests increase the economic life of reservoirs due to reduced sedimentation. In Turkey, there are 175 reservoirs in operation, and 99 under construction. Measurements of the sediment accumulated within a reservoir per year are available for 16 critical catchment areas.

Assuming that the original economic life of the dam was 100 years (based on typical engineering standards), the value of forgone production for these 16 dams is estimated at US$64 million (present value, r = 10%). This is financially equivalent to a perpetual annual damage of US$5.8 million. Additional assumptions of this analysis include the supposition that all sediment accumulation in these reservoirs, above and beyond that accounted for in the original lifetime estimate of 100 years, is attributable to deforestation.

Source: Bann and Clemens (2001).

Towards the Total Economic Value of Turkish Forests

Figure 13.3 shows the estimated values of the TEV components: direct and indirect use values, option/bequest/existence values and the negative externalities linked to the Turkish forests. The estimated TEV is about US$977.4 million, of which direct use values are distinguished as the main component with US$838.8 million, mainly due to the weight of timber and grazing. Substantially lower values accrue to indirect uses (US$158.4 million) and option/bequest/existence values (US$113.8 million) for which minimum estimates were provided, due to data scarcity at the national level. For example, only local level estimates were calculated concerning soil conservation, avalanche prevention, landscape and therapy. Negative externalities, totalling US$133.6 million, are represented mainly by the impacts of erosion on the agricultural production in the proximity of

forests. It should be remembered that valuation of negative effects, such as those provoked by illegal cutting or overgrazing, is lacking due to scarcity of data.

The contribution of the main forest outputs, including public goods and externalities, to the TEV is shown in Fig. 13.4. The highest share is given by timber (45%), followed by grazing (23%), carbon sequestration (16%), pharmaceuticals (12%) and firewood (6%). Erosion due to poor forest management reduces the positive TEV by 13%, whereas forest fires decrease it by 1%.

Perspectives and Crucial Issues

The research leads to a TEV estimate of US$977.4 million, which can be considered to a large extent a minimum estimate, particularly due to the scarce information on indirect use values, option, bequest and existence values. It appears that WFPs comprise a considerable share (51%) of the TEV, which confirms the wood-based approach of Turkish forest management. Other tangible NWFPs such as mushrooms, medicinal plants, styrax and oregano contribute only 8% to the TEV. This falls in line with the primary objective of Turkish forest policy, 'to provide WFPs and NWFPs since the beginning' (Özdönmez *et al.*, 1996). The values of recreation, hunting and angling total 4% of the TEV, much less than the previous values. This should be regarded in the context of

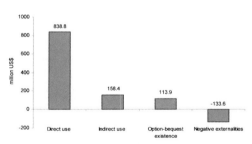

Fig. 13.3. Estimated TEV of Turkish forests.

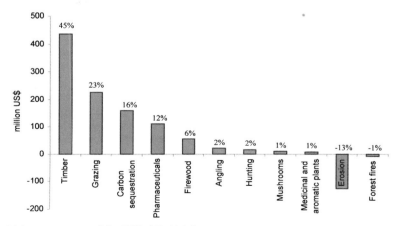

Fig. 13.4. Main components of the TEV of Turkish forests.

Turkish forest management, for which recreational purposes are less significant objectives than WFPs and NWFPs. Other forest values such as grazing (23%), carbon storage (16%) and pharmaceuticals (12%), though significant, have not yet been adequately considered in forest management and policy.

Notwithstanding the data scarcity and valuation limitations, the research shows that the 0.5% forestry contribution to the GDP is highly underestimated. The national balance sheets account only for the values of WFPs, and partly for the NWFPs, i.e. hunting and recreation. These values comprise 62% of the TEV, while the remainder, provided mainly by grazing, carbon sequestration and pharmaceuticals, are not accounted for in balance sheets.

Moving towards sustainable forest management requires the consideration of the TEV categories and particularly the non-market public goods and externalities. This will lead to more comprehensive approaches that incorporate the forest's multiple uses and internalize its externalities (Türker et al., 2001c,d) and render the objectives of forest management more consistent with those of the national economy.

The issue of sustainable development should be seen in the context of high population density in the mountain and forest villages, where income per capita and education levels are quite low (Türker and Ayaz, 1997). The local communities' dependence on forests, often driven by poverty, can give rise to negative externalities, such as erosion due to illegal cutting, which threaten forest sustainability (Türker et al., 2001c). These negative effects, together with damage provoked by forest fires and erosion, need to be addressed in a consistent forest policy.

The role of the TEV estimates is also significant as it makes the incorporation of true values of non-market forest products and services into a realistic cost–benefit analysis (CBA) of forestry projects possible. Furthermore, research of this nature will contribute to the development of a new perspective based on functional planning by replacing the current management system based on wood production with a new system, which considers the forests together with other natural resource ecosystems both on the surface and underground. To achieve this, the conversion of the current extensive forestry applications to intensive forestry practices is necessary.

Overall, it can be seen that the total value of NWFPs and other positive goods and externalities is greater than US$620 million/year, while wood sale revenue from State forests totalled less than US$450 million. This demonstrates that the current contribution of Turkish forests to the national economy lies not only in wood products, but also elsewhere. These values remain external to forestry planning that focuses intensively upon timber, leaving open the possibility of large negative externalities when non-wood resources are significantly affected by forestry planning decisions. This can cast doubt on the degree to which those decisions provide benefit to Turkish society. It is therefore desirable that forestry planners take certain actions. These can be broadly classified according to three themes: (i) multi-functional management of State forests; (ii) joint management of State forests; and (iii) improved management of protected areas (Bann and Clemens, 2001).

Multi-purpose forest management is the most important step for Turkish forestry, allowing an increase in the TEV, by raising the positive externalities and reducing the negative ones. With this aim, the following actions can be proposed:

1. Attention should be paid to participatory management and production of NWFPs, including grazing. To achieve this, principles, targets and strategies related to sustainable forest management and NWFP use should be set in the regional forestry plans. Inventory, research, control and evaluation programmes and institutional capacity related to production, economy, marketing and social issues linked to NWFPs should be improved, as well as ownership and use rights.

2. The energy system in forest villages needs to be developed and fuelwood price subsidies should be eliminated, to enable the reduction of illicit fuelwood consumption, which is responsible for a large degree of forest degradation. To achieve these aims, several initiatives can be taken. First, forest management plans must be adjusted by also taking into account the needs of local people. Secondly, programmes and precautions that will provide fuelwood saving should be supported. Thirdly, inventory and research on the present and future fuelwood production and consumption (legal and illegal) should be developed. Finally, the utilization of coppice forests by local villagers should be advised and fast-growing species stands should be established as a main fuelwood source.

3. The contribution of local people and the private sector to hunting and fishing management should be supported. Wildlife should be taken into account in the forest management plans, and the revenues obtained by the State should be spent to improve this sector.

4. The revenue obtained from the recreational areas should be used for their management. Presently, about 0.045% of total forest area in Turkey is used for recreational purposes, which is a very low share compared with the average European share (2%). Ecotourism plans considering the carrying capacity of forest areas should be prepared by involving the participation of local people.

The specific actions required to move towards the elimination of the negative externalities associated with Turkish forestry decisions are numerous and diversified. However, the adoption of a very few basic principles by the forestry planning process – multi-functional, joint and ecologically sound management of protected areas – can help to reduce the potential for a wide variety of major negative externalities. The reward for these efforts is no less than the knowledge that Turkey's forest sector is advancing the interests of the largest possible social segment, rather than those of certain groups to the detriment of everyone else (Bann and Clemens, 2001).

In conclusion, the following issues are outstanding in relation to the values of goods, services and externalities of Turkish forestry:

- The increasing importance of forests is revealed by their real share in the national economy.
- More monetary support for investments should be allocated to the forestry sector.
- A good strategy and policy for improving the effectiveness of forest resources should be determined based on sustainable development principles.

Notes

[1] In addition to the designated forests, which are almost entirely under the control of the Ministry of Forestry, there are also 25 Mha of rangeland and shrub land, mostly controlled by the Ministry of Agriculture.

[2] Some of this apparent change is due, however, to changes in forest area definitions.

[3] Illegal wood cutting is common throughout Turkey and is dominant in the eastern Black Sea region. Since 1985, there has been no serious survey on illegal wood removal. Several estimates of the annual unrecorded cut show some 5–6 Mm^3 (General Directorate of Forests, 1999) or 10 Mm^3 (World Bank, 2001).

[4] Within the protected forest area, an estimated 856,653 ha are given to nature conservation, aesthetic values and recreation, including hunting reserves (Muthoo, 2001). An additional 92,188 ha of forests are allocated to scientific and educational roles and are given to forest faculties, forestry research institutions and the like (Savaşan, 1999).

[5] There are now 33 national parks under the National Park category in Turkey.

[6] The concept of forest sustainability and the interaction between forests and the public are also addressed in the Constitution, in addition to other Turkish laws, which states that 'the protection of forests and the development of forest villages is a responsibility of the Turkish Republic'.

[7] As a matter of fact, before 1980, 292 SFEs were established, of which several were affected by closure and re-opening, so that by 1992 there were 207 SFEs (Geray, 1992).

References

Bann, C. and Clemens, M. (2001) *Turkey Forest Sector Review-Global Environmental Overlays Program Final Report*. April 1999, İksir Tanıtım Ltd.Sti, Ankara [in Turkish].

Bingöl, I.H. (1990) Our forests and forestry from the past to today. *The Foundation of Forestry Culture and Education Publication 3*, Vol. 1, İstanbul [in Turkish].

Çakır, M. (1984) The importance of the forestry sector for the national income. *Forest Engineers Association Publications* No. 9, Ankara [in Turkish].

Doğru, M. (2001) *Planning and Management of Forest Resources In Turkey (Draft)*. Assistance for the Preparation of a National Programme for Turkey, Ankara.

Ernst, F. (1997) *Reconstruction of Vegetation Development in Turkey Using Geographic Information Systems*. Turkey 4th Arc-Info and ERDAS Users Meeting Papers, 10–2 June 1997, Ankara. Website: http://www.islem.com.tr/BILDIRILER/FRED_ERNST/quarter4.htm Access date: March 1998.

Fankhauser, S. (1995) *Valuing Climate Change. The Economics of the Greenhouse*. Earthscan, London.

Firat, F. (1971) *The Economics of Forest Management*. İstanbul University, Faculty of Forestry Publications, No. 155, Istanbul [in Turkish].

General Directorate of Forests (1999) *General Directorate of Forests' Records*. GDF, Ankara. Website: http://www.ogm.gov.tr Access date: April 2001 [in Turkish].

General Directorate of National Parks, Game and Wildlife (1997) *The General Directorate of National Parks and Game-Wildlife's Records*. GDNPGW, Ankara [in Turkish].

Geray, U. (1986) *Planning Lecturer's Notes*. İstanbul University, Forestry Faculty, İstanbul [in Turkish].

Geray, U. (1992) *Mediterranean Region Forest Enterprises and Organization Problems*. İstanbul University, Faculty of Forestry, Forestry Research and Application Centre Publication No. 1, İstanbul [in Turkish].

International Monetary Fund (2003) *International Financial Statistics*. January 2004, IMF, Washington, DC.

Konukçu, M. (1998) *Statistical Profile of Turkish Forestry*. State Planning Organization, Ankara.

Konukçu M. (2001) *Forests and Turkish Forestry*. State Planning Organization, Ankara.

Ministry of Forestry (1997) *Forestry in Turkey*. Leaflet, XI.World Forestry Congress, Ministry of Forestry, Antalya, Turkey [in Turkish].

Ministry of Forestry (1999) *The Evaluation Report of Fighting Activities with Forest Fires*. Ministry of Forestry [in Turkish].

Ministry of Forestry (2000) *Facts and Figures Forestry in Turkey*. Ministry of Forestry, Ankara.

Muthoo, M.K. (2001) *Forests and Forestry in Turkey*, 2nd edn. Ministry of Forestry, Ankara.

Özdönmez M., İstanbullu T., Akesen A. and Ekizoğlu, A. (1996) *Forestry Policy*. İstanbul University, Faculty of Forestry Publication No. 3968, İstanbul [in Turkish].

Savaşan, T. (1999) *Functional Planning, Forest Management Meeting in Fethiye*. Forest Administration and Management, Department of General Directorate of Forests, Fethiye, Turkey [in Turkish].

State Institute of Statistics (2002) Website: www.die.gov.tr Access date: May 2003.

State Planning Organization (1996) *The Production Price of Water Products*. State Planning Organization, Ankara.

State Planning Organization (2001) *VIII. Five Year Development Plan (2001–2005) – Forestry Special Impression* Commission Report, State Planning Organization Publication No. 2531, Ankara [in Turkish].

Türker, M.F. (1994) The economic analysis of forestry sectors using input–output model (country and eastern Black Sea region samples). *Regional Development Symposium*, 13–15 October 1994, Trabzon, Turkey [in Turkish].

Türker, M.F. (1999) The determination of the importance of the forestry sector in national economy using input–output analysis, *Turkish Journal of Agriculture and Forestry* 23, Supplement 1, 229–237 [in Turkish].

Türker, M.F. (2000) *Forest Management Lecturer Notes*. Karadeniz Technical University, Faculty of Forestry, Lecturer's Notes No. 59, Trabzon, Turkey [in Turkish].

Türker, M.F. and Ayaz, H. (1997) The socio-economic profile of forest villages and destruction of forest resources. Paper presented at the *Third International Ecology Summer School*, 21–27 July 1997, Trabzon, Turkey [in Turkish].

Türker, M.F., Öztürk, A. and Pak, M. (2001a) Ownership scale size and economic analysis of the state forest enterprises constitute the frame of forestry sector. *The 31st Southern Forest Economics Workers (SOFEW) Annual Meeting 'Forest Law and Economics'*, 27–28 March 2001, Sheraton Buckhead Hotel, Atlanta, Georgia.

Türker, M.F., Öztürk, A. and Pak, M. (2001b) Evaluation of the method used in the determination of economic loss arisen from forest fires: forest economics perspective. *The Economics of Natural Hazards in Forestry*, 7–10 June 2001, Padua University Press, Solsona, Catalonia.

Türker, M.F., Öztürk, A. and Pak, M. (2001c) Evaluation of externality concept with regard to the Turkish forest resources and forest management. *First National Forestry Congress*, 19–20 March 2001, Turkey Foresters Association, Publication Series No. 1, Ankara, pp. 154–170 [in Turkish].

Türker, M.F., Pak, M. and Öztürk, A. (2001d) Total economic value and evaluation of the outputs supplied by the forest resources in this scope. *First National Forestry Congress*, 19–20 March 2001, Turkey Foresters Association, Publication Series No. 1, Ankara, pp. 171–182 [in Turkish].

United Nations Economic Commission for Europe/Food and Agriculture Organization (2000) *Global Forest Resources Assessment 2000*. Main Report, Geneva Timber and Forest Study Papers No. 17, United Nations, Geneva.

World Bank (2001) *Turkey Forestry Sector Review*. Report No. 22458-TU, World Bank, Washington, DC.

Yavuz, H., Türker, M.F. and Gül, A.U. (1988) The Calculation of Functional Values of Altındere Valley National Park. Postgraduate seminar study for 'The determination of management objectives in forestry course', unpublished, Trabzon, Turkey [in Turkish].

Yazıcı, K. (1991) The determination of minimum economical size for a state enterprise in the eastern Black Sea region, Istanbul University. *Journal of Forest Faculty, A Series* 41, 2 [in Turkish].

14 Cyprus

Department of Forests

*Ministry of Agriculture, Natural Resources and Environment, Louki Akrita 26,
PC 1414 Nicosia, Cyprus*

Introduction

Cyprus is the third largest Mediterranean island after Sicily and Sardinia, with a population of 793,100.[1] It occupies 925,100 ha, of which 41.7% are forests and the remainder natural vegetation. Four geographical regions can be identified, with different geological formations and other interesting features:

1. *The Troodos range* is a dome-shaped highland of mainly infertile igneous rocks situated in the central-western part of the island, rising to 1951 m at Mount Olympus. The forests that cover the Troodos massif, combined with its steep slopes, precipices, narrow valleys and crevices, create a magnificent landscape. A lower belt of dome-shaped pillow lava, a most infertile area, which levels off gradually towards the coast, surrounds the hard igneous rocks.

2. *The Pendadaktylos range*, situated in the north, is mainly composed of limestone and rises to 1024 m. Part of this range consists of a finger-shaped mountain known as *Pendadaktylos*. It is a beautiful mountain composed of a succession of mostly allochthonous sedimentary formations ranging from the Permian to the Middle Miocene in age.

3. *The 'Messaoria'* or *central plain* is situated between the Troodos and Pendadaktylos Mountain ranges and has a low relief, not exceeding 180 m, near Nicosia, the capital of Cyprus. This plain is composed of flysch-type rocks carried by rivers from the Troodos and Pendadaktylos ranges and was formed during the Holocene period.

4. *The coastlands* form valleys that almost entirely surround the country: Kyrenia valley with its narrow coasts in the north; Larnaka and Limassol valleys in the south; Pafos and Chrysochou valleys in the west; and Famagusta valley in the east. The soils are alluvial and fertile and are suitable for agriculture.

Cyprus has an intense Mediterranean climate characterized by hot, dry summers and rainy, rather changeable winters, separated by short autumn and spring seasons. The central Troodos massif, and, to a lesser extent, the long narrow Pendadaktylos mountain range, play an important part in the country's meteorology. The average annual rainfall increases ascending the south-western windward slopes from 450 mm to nearly 1100 mm at the top of the central Troodos massif. On the leeward slopes, amounts decrease steadily northwards and eastwards to between 300 and 350 mm in the central plain and the flat south-eastern parts of the island. The narrow ridge of the Keryneia (Kyrenia) range, stretching 160 km from east to west along the extreme north of the island, produces a relatively small increase of rainfall to nearly 550 mm along its ridge at about 1000 m.

Forest Resources

Area and people

Forests and other wooded land cover 385,600 ha, or 41.7% of the country's area, according to the 1999 vegetation map, based on the Food and Agriculture Organization definition of forest area (Pantelas *et al.*, 1999). However, the United Nations Economic Commission for Europe/Food and Agriculture Organization (2000) provides a less accurate estimate of forest area and other wooded land, of about 280,000 ha. The difference between these two figures can be explained by the fact that reliable estimates regarding private forests and other wooded land became available only after the 1999 vegetation map was produced. Before that, the reported estimates varied significantly among sources. At the same time, it should be stressed that data regarding State forests have always been accurate.

In Cyprus, very few people depend exclusively on forestry. The high urbanization rate (70%), the insignificant contribution of forestry and forest-related activities to the national economy (0.03%) and the overwhelming deficit in wood trade balance – 97% imports, equal to 1% of the gross domestic product (GDP) – are indisputable indicators reflecting the actual picture of the forests in Cyprus. Additionally, the urbanization trend persists and the rural population has been leaving the forest and migrating to large villages on the plain and coastal areas, where the service sector (such as tourism and public services) is growing.

Urbanization induces increasing demand for recreational services and option values rather than for timber products (Alexandrou, 1999). The results of an economic valuation study show that the overall social and environmental value of forests greatly exceeds the commercial use value of trees as wood material (Costantinides, 1999). Tourism–recreation possesses a dominant share in the valuation, corresponding to around 25% of the total economy of the island. It is therefore expected that the pressure of recreational and tourism services and infrastructure (such as housing, hotels and roads) will be increasingly directed on to forestland, with positive and negative effects on the forests.

Typologies

Almost all forests (99%) are either natural or semi-natural. Of the total forest area, natural vegetation includes high forests (44.6%), *maquis* (32.6%) and lower vegetation types such as scrub and *phrygana* (22.8%). There are no productive plantations and although in the past many areas have been planted for fuelwood production, today most plantations are managed for amenity and other non-marketable benefits.

Forest degradation is mainly the result of overgrazing, fires in forest and rural land, quarrying and mining as well as land clearance for agricultural purposes (Hadjikyriakos and Pantelas, 1998). Furthermore, development for tourism frequently leads to degradation of natural vegetation.

The first reference regarding the natural vegetation of Cyprus, in a phyto-sociological perspective, was made by Zohary (1973). He gave an account of the country's vegetation and described some associations, especially for the central plain of Mesaoria, which has been overlooked by subsequent researchers. Later on, the literature was enriched with comprehensive descriptions of characteristic climax or potential vegetation (Barberó and Quézel, 1979; Quézel and Barberó, 1985) and of coastal communities (Arnold *et al.*, 1984; Géhu *et al.*, 1984, 1990). These studies greatly facilitated the description and mapping of the country's vegetation according to the classification proposed by Quézel (1976, cited by M'Hirit, 1999). Some deviations from the original classification were necessary in order to accommodate the peculiarities of certain vegetation types and achieve a better representation (Fig. 14.1, Box 14.1).

Functions

Forestry in Cyprus has for many years been based on multiple uses. In ancient times, the forest area was more extensive and the population smaller; timber was felled for building houses and ships, and forests were a source of food, fodder and other basic needs. Their capacity to meet peoples' requirements was not questioned and sustainability was not an issue.

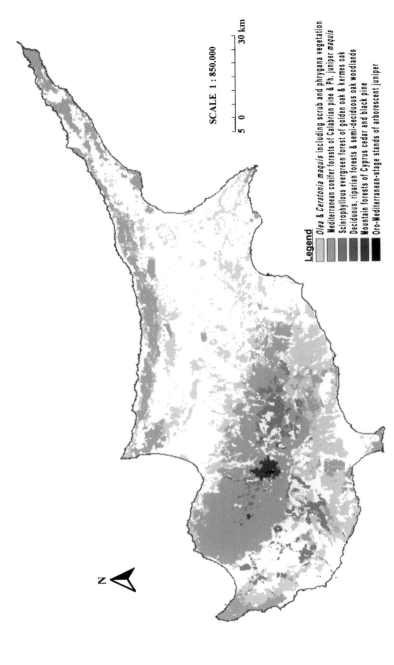

Legend

Olea & Ceratonia maquis including scrub and phrygana vegetation
Mediterranean conifer forests of Calabrian pine & Ph. juniper *maquis*
Sclerophyllous evergreen forest of golden oak & kermes oak
Deciduous, riparian forests & semi-deciduous oak woodlands
Mountain forests of Cyprus cedar and black pine
Oro-Mediterranean-stage stands of arborescent juniper

SCALE 1 : 850.000

5   0   30 km

N

Fig. 14.1. Map of Cyprus forest vegetation according to adapted Quézel classification.

Box 14.1.   Cyprus forest types according to the Quézel classification.

(i) *Olea* and *Ceratonia maquis* including scrub and phrygana vegetation – corresponding to thermophilic wild olive and pistachio scrubs (total estimated area 226,671 ha)
This category occurs in the thermo-Mediterranean zone from sea level up to about 600 m, but it may also be found up to 800 m depending on the slope aspect and the local climatic conditions. It is widely distributed from the Akamas peninsula, in the west, to the foothills of the Troodos (southern) range where it is particularly common on the southern flanks, and all along the Pentadaktylos (northern) range up to Cape Apostolos Andreas to the east.

   This vegetation type is mainly represented by *Olea* and *Ceratonia* forests, corresponding to *Oleo-Ceratonion* (= *Ceratonio-Rhamnion*) (European Commission, 2003); however, only a small proportion is sufficiently tall to be classified as forest, whereas most formations are *maquis* attaining a height of about 4 m. Generally they are composed of evergreen, sclerophyllous woody species, such as *Olea europaea*, *Ceratonia siliqua*, *Pistacia lentiscus*, *Rhamnus oleoides* subsp. *graecus* and *Myrtus communis*, and they are dominated either by *O. europaea* or by *C. siliqua*; whereas along the coastline *P. lentiscus* frequently dominates locally. Other woody species occurring in these communities include *Pistacia terebinthus*, *Quercus coccifera* subsp. *calliprinos* and *Crataegus azarolus*, which may have a considerable participation especially at the higher altitudes or moister situations such as streambeds.

   The vegetation of this category belongs to the class *Quercetea ilicis*, order *Pistacio-Rhamnietalia*, alliance *Ceratonio-Rhamnion*, and two associations have been described: namely *Ceratonio siliquae–Genistetum sphacellatae* and *C. siliquae–Ballotetum integrifoliae* (Barberó and Quézel, 1979).

   In areas where the *Oleo-Ceratonion* (= *Ceratonio-Rhamnion*) *maquis* has been degraded by repeated fires and/or overgrazing, there is a lower, often dense scrub vegetation (up to 2 m) dominated by *Genista fasselata* subsp. *fasselata*, including also many species of the *Cisto-Micromerietea* and often of the *Pistacio-Rhamnetalia*. These communities are classified under the *Cisto-Micromerietea*: *Sarcopoterietalia spinosi* and two alliances have been proposed for Cyprus: *Sarcopoterio-Genistion sphacelatae* and *Genistion fassellatae* with the association *Genistetum fasselatae cyprium* that includes all the formations characterized by *G. fasselata*. These formations are considered as pre-forest communities (Costa *et al.*, 1984; Brullo *et al.*, 1997).

   Phrygana vegetation, corresponding to *Sarcopoterium spinosum* (European Commission, 2003), occurs at the forest degradation stage, most often in large openings of the *maquis*, or as climax stage in sites with adverse conditions (aridity, poor soils, exposure and grazing). These communities belong to the *Cisto-Micromerietea* and they are normally dominated by *Sarcopoterium spinosum* and *Thymus capitatus*, but other shrubs such as *Lithodora hispidula* subsp. *versicolor*, *Cistus* spp., *Onosma fruticosa*, *Fumana* spp. and *Phagnalon rupestre* are frequently common; also, *G. fasselata* subsp. *fasselata* and *Calycotome villosa* are sometimes present sporadically.

   Along rivers and temporary streams of this zone develops hydrophilic vegetation, composed of *Tamarix* spp. (mainly *Tamarix smyrnensis*, *T. tetragyna* and *T. tetrandra*), *Nerium oleander* and *Vitex agnus-castus*, corresponding to thermo-Mediterranean riparian galleries: *Nerio-Tamaricetea* (European Commission, 2003).

   Coastal halophytic wetlands are confined around the salt lakes of Akrotiri (Lemesos) and Larnaka and the marshy areas of Salamina (Ammochostos) and Morfou. The woody vegetation of these areas comprises halophilous scrubs, corresponding to Mediterranean and thermo-Atlantic halophilous scrubs (*Sarcocornetea fruticosi*) (European Commission, 2003), which are dominated by *Arthrocnemum macrostacyum*, *Halocnemum strobilaceum*, *Salicornia fruticosa* and *Inula crithmoides*. This vegetation type belongs to *Salicornietea fruticosae*: *Salicornietalia fruticosae*: *Arthrocnemion glauci* and has been named *Arthrocnemetum glauci* (Géhu *et al.*, 1990) or *Puccinelio convoluteae–Arthrocnemetum glauci*: *Inuletosum crithmoidis* (Arnold *et al.*, 1984).

   The central plain of Mesaoria, which represents about one-third of the island, is characterized as a semi-arid zone (Barber i Valles, 1995; Pantelas, 1996). This treeless plain consists generally of cereal fields (Meikle, 1977). The woody vegetation is mainly composed of *Ziziphus lotus* and *Crataegus azarolus*, which form the high shrub communities, and phryganic vegetation dominated by low shrubs, mainly *S. spinosum* and *T. capitatus*. These formations correspond to arborescent matorral with *Ziziphus* and to thermo-Mediterranean and pre-desert scrub, subtype xerophilous *Crataegus azarolus* var. *aronia* scrub (European Commission, 2003). However, *Z. lotus* communities do not clearly fit in the thermo-Mediterranean zone, as they are in part distinctively thermo-Mediterranean and in part intermediate between Mediterranean formations and open dry tropical woodland. The association *Ziziphetum loti cyprium* (*Quercetea ilicis*: *Hiparrhenietalia hirtae*: *Ziziphion loti*) describes communities of *Z. lotus* associated with segetal and phryganic vegetation (Zohary, 1973). The xerophilous *Crataegus azarolus* scrub also occurs within the same zone and is associated with herbaceous as well as phryganic vegetation.

(ii) Mediterranean conifer forests of Calabrian pine and Phoenician juniper *maquis* – corresponding to Mediterranean conifer forests of Aleppo pine, Brutia pine, stone pine, Barbary Thuya and Phoenician juniper (total estimated area 230,010 ha)

In the costal zone, up to about 450 m, Phoenician juniper (*Juniperus phoenicea*) *maquis* coexists with *Oleo-Ceratonion maquis*. This vegetation type corresponds to arborescent matorral with *Juniperus* spp., subtype: *J. phoenicea* arborescent matorral (European Commission, 2003). The best and most extended *maquis* are found in the Akamas peninsula (west), Episkopi and Akrotiri peninsula (south), Cape Kormakiti (north-west), Cape Gkreko (south-east), and the Karpasia promontory where they form extensive thickets around Apostolos Andreas (Meikle, 1977). These *maquis* are dominated by *J. phoenicea*, while *Pistacia lentiscus*, *Prasium majus* and *Rhamnus oleoides* subsp. *graecus* are common, and *O. europaea*, *C. siliqua* and sometimes *Pinus brutia* scarcely occur. In areas with large clearings, *G. fasselata* subsp. *fasselata*, *Calycotome villosa*, *Lithodora hispidula* subsp. *versicolor*, *S. spinosum*, *T. capitatus* and other small shrubs and subshrubs may also be present along with a wide range of herbaceous plants.

The association *Ephedro campylopodae-Juniperetum lyciae* (*Quercetea ilicis*: *Pistacio-Rhamnietalia*: *Ceratonio-Rhamnion*) is typical of this zone; among the characteristic species are *J. phoenicea*, *P. brutia*, *P. lentiscus* and *Ephedra fragilis* subsp. *campylopoda* (Barberó and Quézel, 1979).

Calabrian pine (*P. brutia*) is the dominant forest species and occurs almost all over the island, except for the Mesaoria plain, from sea level up to 1400 m (from thermo-Mediterranean to meso-Mediterranean). This forest type corresponds to Mediterranean pine forests with endemic Mesogean pines (European Commission, 2003). In the coastal zone, small *P. brutia* forests are found in the Akamas peninsula, Episkopi (Lemesos), Akrotiri peninsula, Cape Kormakiti and Karpasia peninsula. Their composition varies from dense to sparse *P. brutia*, with usually a dense under-storey of *J. phoenicea*, *P. lentiscus*, *C. siliqua*, *O. europaea*, *P. terebinthus*, *R. oleoides* subsp. *graecus*, *G. fasselata* subsp. *fasselata*, *C. villosa*, *Cistus* spp. and many other small shrubs. On higher altitudes, along the Troodos (southern) and Pentadaktylos (northern) ranges, there are extensive *P. brutia* forests.

The Troodos range is well covered with thick pine forests, which attain their best development in Pafos forest (central and western parts of the Troodos range). In the lower pine zone, up to about 700–800 m, the forests have a thermo-Mediterranean sclerophyllous under-storey mainly composed of *O. europaea*, *P. lentiscus*, *P. terebinthus*, *Quercus coccifera* subsp. *calliprinos*, *Arbutus andrachne*, *R. oleoides* ssp. *graecus*, *Rhamnus alaternus* and *G. fasselata* subsp. *fasselata*. Furthermore, *Styrax officinalis*, *Acer obtusifolium* and, at disturbed sites, *Rhus coriaria*, participate significantly in the under-storey. Around 400 m, *Quercus alnifolia* appears, at first scattered but gradually becoming more abundant, and at around 700 m it often dominates the under-storey. In addition, the forest may have a phryganic under-storey chiefly dominated by *Cistus* spp. (mainly *C. creticus* and *C. salviifolius*) but also including a wide range of small shrubs. At the central part of the range (Tripylos), around the cedar valley, *P. brutia* forms mixed forests with the endemic tree *Cedrus brevifolia*.

The northern Pentadaktylos range is also well covered with *P. brutia* forests, which have a rich under-storey composed of high and low shrubs. Moreover, they form mixed stands with *Cupressus sempervirens*. Along the whole stretch of the Pentadaktylos range, Cypress forests (*C. sempervirens* var. *sempervirens*) colonize the limestone cliffs and slopes. This forest type corresponds to Cypress forest (*Acero-Cupression*) (European Commission, 2003). Apart from the northern range, smaller Cypress forests also occur in several places, from 100 to 1200 m, such as the Akamas peninsula, Vouni Panagias, Episkopi (Kourion), Kyparissia (Lemesos forest), Lagoudera and Saranti villages.

For the *P. brutia* forests, two associations have been described: (i) *Querco calliprini-Pinetum brutiae* (*Quercetea ilicis*: *Pistacio-Rhamnietalia*: *Quercion calliprini*), which is characterized by the dominance of *Q. coccifera* subsp. *calliprinos* and *P. brutia*; and (ii) *Querco alnifoliae-Pinetum brutiae* (*Quercetea ilicis*: *Quercetalia ilicis*: *Quercion alnifoliae*), which is restricted on the igneous formations

continued

Box 14.1. *Continued.*

of the Troodos range, in the meso-Mediterranean zone, and is dominated by *Q. alnifolia* and *P. brutia* (Barberó and Quézel, 1979).

(iii) Sclerophyllous evergreen forest of golden oak and Kermes oak – corresponding to sclerophyllous evergreen oak forest of holm oak, cork oak and Kermes oak (total estimated area 30,620 ha)
This category includes the evergreen, endemic forests of golden oak (*Q. alnifolia*), which occur exclusively on the ophiolite (igneous) rocks of the Troodos range in the meso-Mediterranean zone mainly within the subhumid ombrotype. This formation corresponds to the scrub and low forest vegetation with *Q. alnifolia* (European Commission, 2003). This endemic large shrub or small tree occurs from 400 to 1700 m, on steep slopes, typically on loose screes, offering soil stabilization against erosion.

The forests of *Q. alnifolia* are distinguished by two associations. (i) *Pino (brutiae)–Quercetum alnifloliae* (*Quercetea ilicis*: *Quercetalia ilicis*: *Quercion alnifoliae*: endemic alliance) (Barberó and Quézel, 1979); this association is clearly xerophilous and comprises mixed forests of *P. brutia* with *Q. alnifolia*, in the meso-Mediterranean zone, with the latter dominating the under-storey together with *Arbutus andrachne* and other sclerophyllous shrubs. Pure *Q. alnifolia* stands are rare in this association. (ii) *Crepido fraasii-Quercuetum anifoliae* (*Quercetea ilicis*: *Quercetalia ilicis*: *Quercion alnifoliae*), which is more mesophilous and is characterized by the dominance of *Q. alnifolia* that forms dense stands with only some sciophilous species in the under-storey such as *Crepis fraasii*, *Cyclamen cyprium* and *Lecokia cretica* (Barberó and Quézel, 1979). Furthermore, two subassociations have been described: the first represents pure stands of *Q. alnifolia* with frequent occurrence of *Acer obtusifolium*, and the second mixed communities of *Cedrus brevifolia* and *Q. alnifolia*, which dominate the under-storey.

This category also includes communities dominated by *Q. coccifera* subsp. *calliprinos*, whereas *Arbutus andrachne*, *P. terebinthus*, *M. communis* and *S. officinalis* participate considerably. They develop on both calcareous and igneous substrates from about 600 to 1200 m; however, very often they occur as low as 300 m. These formations have been classified under the alliance *Andrachno-Quercion cocciferae* (*Quercetea ilicis*: *Pistacio-Rhamnetalia*) (Barberó and Quézel, 1979; Quézel and Barberó, 1985; Barbéro *et al.*, 1991); however, no associations have been defined. In these formations, *Q. coccifera* subsp. *calliprinos*, *A. andrachne*, *P. terebinthus*, *M. communis* and *S. officinalis* are among the characteristic species.

On a small area of the Pentadaktylos range (~18 ha), meso-Mediterranean *Quercetalia ilicis*: *Adrachno-Quercion cocciferae* formations occur, dominated by *Laurus nobilis*, corresponding to matorral with *L. nobilis* (European Commission, 2003). Other species participating in these communities are *A. andrachne*, *C. siliqua*, *O. europaea*, *Phillyrea latifolia*, *Rubia tenuifolia* and *Smilax aspera*. In addition, small patches of *L. nobilis* occur on moist mountainsides of the Pafos forest.

(iv) Deciduous, riparian forests and semi-deciduous oak woodlands – corresponding to deciduous forests of Zeen oak, Afares oak, Lebanese oak, Tauzin oak, hornbeam, ash and occasionally beech (total estimated area 4235 ha)
In Cyprus, there are no real deciduous woodlands. Exceptions are the riparian forests of oriental plane (*Platanus orientalis*), oriental alder (*Alnus orientalis*) and white willow (*Salix alba*) galleries; however, they belong to the riparian azonal vegetation and are distributed from sea level up to 1400 m. These vegetation types correspond to *P. orientalis* woods (*Plantanion orientalis*), residual alluvial forests (*Alno-Pandion, Alnion incanae, Salicion albae*) and *S. alba* galleries, respectively (European Commission, 2003). The most common formations are those dominated by *P. orientalis* and usually have significant participation of *A. orientalis* along with *Nerium oleander*, *L. nobilis* and *S. alba*; they colonize streams, rivers and spring basins and are particularly common along the rivers traversing the Troodos range. Residual alluvial forests with *A. orientalis* are infrequent and colonize riverbeds with heavy soils; also, *S. alba* galleries are not so frequent and extended.

Woodlands with *Quercus infectoria* subsp. *veneris*,[2] corresponding to woodlands with *Q. infectoria* (*Anagyro foetidae-Quercetum infectoriae*) (European Commission, 2003) are distributed mainly from the higher elevations of the thermo-Mediterranean up to the higher altitudes of the supra-Mediterranean vegetation zone (~600–1100 m); they develop on deep soils of both calcareous and igneous formations, and are generally confined to valleys. Throughout the distribution range of *Q. infectoria* subsp. *veneris*, there are only few representative stands such as those around Panagia and Polemi villages, in the Pafos district, and near Agros on the Troodos range. The most common formations are represented by small stands or patches of trees and more often by scattered trees among *maquis* and isolated individuals along field margins or within cultivated land, which sometimes reach majestic proportions (Thirgood, 1987). These degraded formations are relics of larger and more widely distributed woodlands that have been degraded by human activities through deforestation to expand agricultural land, grazing and fires.

The association *Anagyro foetidiae–Quercetum* infectoriae describes *Q. infectoria* subsp. *veneris* woodlands with a characteristic under-storey composed of *Q. coccifera* subsp. *calliprinos*, *P. terebinthus*, *S. officinalis*, *Anagyris foedita* and *Crataegus* spp. (Barberó and Quézel, 1979).

(v) Mountain forests of Cyprus cedar and black pine – corresponding to mountain and high-altitude forests of cedar, black pine and firs (total estimated area 5101 ha)

This category comprises the endemic forest of Cyprus cedar (*Cedrus brevifolia*) and the black pine forest (*Pinus nigra* subsp. *pallasiana*). The Cyprus cedar forest is confined to the Tripylos area (Pafos forest), known as the cedar valley, of the Troodos range; however, Zohary (1973) suggests that *C. brevifolia* had a wider distribution in the past. This forest type corresponds to *C. brevifolia* forests (*Cedrosetum brevifoliae*) (European Commission, 2003). It is distributed from the upper limits of the meso-Mediterranean to the mid supra-Mediterranean zone, at altitudes between 900 and 1400 m, on igneous formations. *C. brevifolia* forms pure as well as mixed forests with *P. brutia* and *Q. alnifolia*, especially at the margins of Tripylos peak. Furthermore, scattered small stands of *C. brevifolia* occur at the surrounding peaks. It has also been planted extensively along the Troodos range and in many mountainous villages. The area of its natural distribution has been designated as a Nature Reserve (Tsintides *et al.*, 2002). According to Barberó and Quézel (1979), the Cyprus cedar forests have been attributed to the subassociation *Cedrosetum brevifoliae* (*Quercetea ilicis*: *Quercetalia ilicis*: *Quercion alnifoliae*: *Crepido fraasii–Quercetum alnifoliae*). Zohary (1973), on the other hand, classifies them as *Cedretum brevifoliae* under the east Mediterranean *Querco-Cedretea libani*: *Quedretea libani orientalia*.

The black pine forest (*Pinus nigra* subsp. *pallasiana*), corresponding to Pallas's pine forests (European, Commission, 2003), colonizes the highest slopes of Chionistra in the lower and mid-supra-Mediterranean zone. At about 1200–1400 m, depending on the slope aspect, *P. nigra* subsp. *pallasiana* forms mixed forests with *P. brutia*; whereas pure *P. nigra* subsp. *pallasiana* forests occur at around 1400 m (depending on the aspect: higher at south-facing slopes and lower at north-facing slopes) and reach almost the top of Chionistra (1952 m). The under-storey consists of *Q. alnifolia*, *Juniperus oxycedrus*, *J. foetidissima*, *Sorbus aria* subsp. *cretica*, *Berberis cretica*, *Arbutus andrachne*, *Rosa chionistrae* and *R. canina* as well as a wide range of smaller shrubs and herbs including many endemic and rare serpentinophilous species.

The association *Euphorbio rigoi–Pinetum pallasianae* (*Quercetea pubescentis*: *Querco-Cedretalia libani*: *Cephalorrhyncho-Pinion*) describes *P. nigra* subsp. *pallasiana* forests associated with the endemics *Euphorbia cassia* subsp. *rigoi* and *Salvia willeana* in the under-storey (Barberó and Quézel, 1979).

(vi) Oro-Mediterranean stage stands of arborescent junipers – corresponding to oro-Mediterranean stage stands of arborescent juniper and thorny xerophytes (total estimated area 177 ha)

This category occupies the highest peaks of the central Troodos range, from the higher supra-Mediterranean to oro-Mediterranean zones; generally, it comprises stinking juniper woods (*J. foetidissima*) on the top of Chionistra (1700–1952 m) and Grecian juniper (*Juniperetum excelsa*) arborescent matorral on the highest peaks of Madari and Papoutsa (1400–1650 m), whilst *J. oxycedrus* forms small patches in both areas. These formations correspond to endemic forests with *Juniperus* spp., subtype: stinking juniper woods; *J. foetidissima* and *J. excelsa* arborescent matorral; and: *J. oxycedrus* arborescent matorral (European Commission, 2003).

J. foetidissima appears at 1500 m, either as scattered individuals or as part of the *P. nigra* subsp. *pallasiana* under-storey; at about 1700 m, it forms small or large pure stands or mixed stands with *P. nigra* subsp. *pallasiana*, which reach up to the summit of Chionistra at 1952 m. At these elevations, where the bioclimatic zone is humid cold (Barber i Valles, 1995; Pantelas, 1996), the forest stands are usually open with an under-storey composed of *Sorbus aria* subsp. *cretica*, *Berberis cretica*, *Cotoneaster nummularia* and many dwarf shrubs and subshrubs such as *Genista fasselata* subsp. *crudelis*, *Alyssum troodi*, *Teucrium cyprium* subsp. *cyprium*, *Astragalus echinus* subsp. *chionistrae* and *S. willeana*, as well as many other herbaceous plants, which are predominantly local endemic species. The association *Sorbo oblongifoliae–Junipereteum foetidissimae* (*Quercetea pubescentis*: *Querco-Cedretalia libani*:

continued

Timber production based on the principle of 'sustained yield' was introduced by European foresters in the 19th century and was practised in Cyprus throughout the 20th century.

In recent years, the importance of forests as a rich source of benefits, including biodiversity, climate amelioration, water supply and purification, amenity and scenery, has been recognized and the sustainability concept has been expanded to cover them. Forest management is now focused on the protection of forests, so that their capacity to provide goods and services of all kinds – wood and non-wood forest products (WFPs and NWFPs), landscape quality, protection of water supplies, of rural life and village communities – is not impaired. With effective conservation and sound management, the forest resource offers multi-dimensional opportunities for socio-economic development, especially in rural areas.

Institutional Aspects

Ownership and size of properties

State forests account for 40.6% of the forest area, while forests growing on private and common land make up 59.4%. State forests and some forested common land[3] are managed exclusively by the Department of Forests, while private forests have never been under management due to the lack of any specific policies towards their regulation. It should be mentioned that 'private forest holdings' have never been recorded as an income-generating unit in agricultural censuses, at least during the second half of the 20th century. This confirms the trend of private forest abandonment and neglect by

owners as a result of low profitability and high costs of exploitation.

Private forests consist of usually very small holdings, having been acquired by inheritance from parents and even grandparents.[4] The forest vegetation grows almost exclusively on former agricultural lands, meadows and pasture, mainly situated on hilly and mountainous areas at elevations between 700 and 1200 m. Their average size ranges between 2.1 ha in mountainous areas and 4.4 ha in the dry land and vineyard zones (Philippides and Papayiannis, 1983; Papayiannis and Markou, 1999). They are scattered in many parts of the island and most of them occur along the delimitation line of the State forests (Philippides and Papayiannis, 1983). The unclear land tenure and the holdings' fragmentation have raised inherent problems regarding the establishment of a functional management system of private forests.

Administration and policies

The responsibility of forestry belongs to the Department of Forests, under the Ministry of Agriculture, Natural Resources and Environment. The Department of Forests administers the State forests, implements the Government policy with regard to forests and is responsible for implementing plans for forest development. Furthermore, it provides technical assistance and carries out reforestation on private and public lands.

Up to the 1980s, most of the work of the Department of Forests was concerned with protection and management of the State forests, although its remit is more general, according to official forest policy. In recent years, there is a close cooperation with the Cyprus Fire Brigade regarding firefighting, based on a common Action Plan, covering rural areas of the whole island.

Forest policy is given by the National Forest Policy Statement, adopted in 2002 as a tool for implementing the National Forest Programme (NFP), under a new forest strategy. It replaces the previous official policy statement, which was issued in 1950 and reconfirmed after independence in 1960. The new strategy, known as the Rural Betterment Strategy, is based on the multiple use of forest resources. It aims at improving the condition of forests, conserving soils and watersheds, protecting biodiversity and heritage sites, promoting ecotourism and sustainable production of WFPs and NWFPs.

The Forest Law has been amended from time to time since its first enactment in 1879; its last amendment was in 2003. The current Forest Law, No. 14/1967 to 78A(I)/2003, refers mainly to the State forests, but it also relates, although to a far lesser degree, to private forestry, to the Cyprus Forest Industries Co. Ltd and to Private Forest Industries (sawmills), i.e. the whole forestry sector.

The NFP for development of the forest sector in Cyprus was prepared by the Department of Forests, in 1999, with the full support of the Ministry of Agriculture, Natural Resources and Environment. External assistance was provided by the Food and Agriculture Organization (FAO) of the United Nations (UN), under the auspices of the Mediterranean Forestry Action Plan. It specifies, in general terms, the action required for the implementation of this strategy during 2000–2009. It is a flexible programme, which can evolve in response to future events and allows for uncertainty of timing and the availability of additional resources. A close collaboration was established with other government services and non-governmental organizations (NGOs) for the implementation of the NFP.

Contribution of the Forest to the National Economy

Gross domestic product and employment

The annual contribution of forestry to the economy of Cyprus is negligible, making up only 0.026% of the gross domestic product (GDP) in 2001. The share of timber-based industries is more significant, accounting for 1.1% of the GDP in the same period: 0.6% for wood-working industries (sawmilling, wood-based panels and wooden pallets) and 0.5% for wooden furniture. The contribution of the furniture industry has been reasonably stable during the last 5 years, despite the growing imports of furniture (Statistical Service, 2001).

In 2001, 651 people were employed in forestry, representing 0.2% of the total national employment. All these people are employed by the government for the general management and protection of the forests, and include foresters, workers and fireguards. It also includes a limited number of people who work in small exploitation enterprises. The workforce in the timber-based industry accounted for 5941 jobs representing 1.9% of the total national employment (Statistical Service, 2001).

International trade

In 2000, total wood consumption in Cyprus was 640,439 m^3 and the consumption *per capita* was 0.95 m^3 (United Nations Economic Commission for Europe/Food and Agriculture Organization, 2000). Timber-based industries are mostly based on imported timber, which accounts for 97.8% of timber consumption. Imported firewood makes up 92.7% of the total firewood consumption. Actually, Cyprus is a net importer of all wood-based products, making up 97.1% of the overall wood market, while exports are almost nil. These data confirm the present tendency of the wood market in shifting to imported final products, while at the same time, timber-based industries are gradually shrinking (Statistical Service, 2001; FAOSTAT, 2004). This is, however, a characteristic of the whole national economy during the last two decades, reflected by the decline of the primary and secondary sectors and the shift towards the tertiary and especially to the provision of tourism services.

Other forest-based industries: tourism and green belts

The contribution of other non-wood forest-based industries or activities and the indirect forest benefits to the GDP are not thoroughly acknowledged and there are very few relevant data available (Costantinides, 1999; Giorgio and Gabriilides, 1999). The honey-making and hunting industries are fully acknowledged and their contribution to the GDP is estimated at around 0.036 and 0.079%, respectively.[5] The tourism-recreation, watershed protection and agricultural benefits provided by forests are acknowledged, but not estimated in monetary terms. Similarly, other indirect use values, such as protection of soil and water resources, conservation of biological diversity, support to agricultural productivity, carbon sequestration and mitigation of global warming, combating of desertification, and protection of coastal resources and fisheries are not captured by statistics, but have become increasingly important to society (Costantinides, 1999).

The Values of Cypriot Forests

Table 14.1 presents the estimated values of Cypriot forests according to the total economic value (TEV) framework and the valuation methods applied. Estimates of direct use values are available mainly for products traded on the market. Valuation of other direct (recreation) and indirect uses (carbon sequestration) is based on both official statistics and other research results supported by conservative assumptions. For the option, bequest and existence values, data are completely lacking. The

Table 14.1. Values of Cypriot forests.

Valuation method/output	Quantity	Value (000 €)
Direct use values		
Market price valuation		
Timber (m³)	11,757	290.0
Firewood (m³)	6,555	56.5
Net growth of timber (m³)	90,166	901.6
Honey (t)	948	3,373.7
Aromatic plants (t)	181	286.5
Carob (t)	7,300	1,666.2
Permit price		
Hunting (no. of hunters)	19,330	1,150.1
Benefit transfer (TCM, CVM)		
Recreation (no. of visits)	734,250	1,835.6
Enjoyed by nationals	253,000	632.5
Enjoyed by foreigners	481,250	1,203.1
Total direct use values		9,560.3
Indirect use values		
Shadow pricing		
Carbon sequestration in forest stands and soils (tC)	120,000	2,400.0
Of which: carbon sequestration in stemwood (tC)	27,000	540.0
Total indirect use values		2,400.0
Negative externalities		
Cost-based methods		
Losses due to forest fires (ha)	190	−285.0
Allergy caused by *Thaumetopoea willkinsoni* (ha)	5,000	−255.0
Agricultural crop losses due to forest wildlife (no. of farmers)	450	−56.0
Total negative externalities		−596.0
TEV		11,364.3

valuation methods and indicators used are discussed in the subsequent sections.

Direct use values

Cypriot forests provide WFPs, i.e. mainly timber and firewood, together with a wide range of NWFPs, such as mushrooms, medicinal and aromatic plants, cones, acorns, resin and other minor forest products. Some of these products are traded in the market and therefore their value can be easily estimated.

Valuation of timber and firewood takes into account the quantities commercialized and their roadside prices (Department of Forests, 2001). The timber increment not yet exploited is valued based on the quantity of wood increment net of removals and half of the average stumpage price. Economic valuation of all NWFPs is based on their average market prices. Some details are worth mentioning:

- The quantity of honey is estimated on the basis of the number of beehives (46,932). Assuming that all beehives are placed in or around forests and other wooded land, with an average productivity of 20 kg/beehive (Department of Agriculture, 2001), the total value of honey is calculated as €3.3 million.
- Valuation of aromatic plants including sage, thyme, oregano, caper and laurel is based on the known quantities traded on the market (Giorgio and Gabriilides, 1999). The total monetary value is estimated at €0.3 million. This is certainly an underestimate, as it is believed that a greater number of aromatic plants are traded on local, national and international markets, part of which escape statistics.
- The estimate of carob fruits (€1.6 million) is the highest among them all, as carob trees are commonly found in agricultural and on other wooded land and exploited extensively due to the marketable value of their fruits (Sabra and Walter, 2001).

The economic value of hunting is assessed on the basis of the number of hunters in the forests (19,330) and the permit price (€59.5/year) (Department of Forests, 2001), resulting

in €1.15 million. Game is traditionally State property in Cyprus, as are hunting rights. The State issues hunting permits and designates the hunting areas.

The recreation value of forests is estimated by considering the number of day-visits of Cypriots (253,000) and foreigners (481,250) (Department of Forests, 2001). The national official statistics for Cyprus report only indicators for travel expenditure,[6] which is insufficient for calculating the recreational value of forests. In the lack of cost contingent valuation method (CVM) or travel cost method (TCM) applications, a consumer surplus of €2.5/day-visit is taken as a benefit transfer from Italian and Spanish valuations. It results in a total recreation value of €1.8 million.

Indirect use values

Forests in Cyprus provide important indirect benefits such as protection of soil and water resources, conservation of biological diversity, support to agricultural productivity, carbon sequestration and mitigation of global warming, combating desertification. Estimating these functions requires a large amount of information, which is lacking, and therefore only carbon sequestration could be valued.

The total carbon sequestered in forest soils and stands (private and State) equals 6 MtC (Intergovernmental Panel on Climate Change, 1996; Von Mirbach, 2000). Using a 2% discount rate, it results in an annual flow of 120,000 tC sequestered in forest soils and stands. Estimated at a price of €20/tC, this results in a value of €2.4 million.

Part of this value includes the carbon sequestered in woody biomass. This amount can be estimated separately, based on the annual quantity of carbon stored in gross stemwood increment of State productive forests (30,000 tC) net of carbon loss by fellings (3000 tC), i.e. 27,000 tC (Intergovernmental Panel on Climate Change, 1996; Von Mirbach, 2000). It corresponds to 0.5% of the total 1977 emissions of carbon in Cyprus (Georgopoulou and Kountenaki, 2001). Taking into account the international prices/compensations paid by international financial organizations aiming at stimulating carbon sequestration of €20/tC (Fankhauser, 1995), the value of carbon sequestered in woody biomass is €540,000.

Negative externalities linked to forests

The picture of the TEV of forests is completed only when negative externalities linked to forests are considered, for example erosion and floods due to poor forest management, damage caused by forest fires, loss of landscape and natural quality due to human impact on ecosystems, and damage caused by forest insects and wildlife. In Cyprus, data are available only for some of these externalities.

Valuation of damage caused by forest fires refers to fires in State forests only. The State forest area annually affected by fires is 190 ha (average for the period 1993–2002), while the average number of fires per year is 23 (Department of Forests, 2001). Monetary valuation of the damage considers only the restoration cost, which amounts to approximately €1500/ha and includes the cost of ground preparation (removal of debris/weeds and cultivation/terracing of the ground, if necessary) and application of planting or sowing on the prepared area. Other costs and expenses such as the forgone values of standing trees or fire extinguishing costs should be also taken into account but are not presently available.

Valuation of health problems of allergy (skin itching) caused by the pine defoliator *Thaumetopoea willkinsoni* was undertaken based on the defensive expenditure approach. An area of 5000 ha of pine forests is treated annually by applying insecticides. The cost of insecticide purchase, application and control measures is €255,000 (Department of Forests, 2001). In addition to reducing the allergic symptoms, a positive effect on the growth and health of forests and aesthetic improvement of the landscape is also achieved.

The loss of agricultural crops due to forest wildlife refers to the crop damage caused only by the wild moufflon. The monetary valuation is based on the compensatory payments that are paid by the State to farmers. There are about 450 farmers receiving compensatory payments,

the annual average[7] payment being €56,000 (Game Fund, 2001).

Towards the Total Economic Value of Cypriot forests

The valuation of the different forest goods and services in Cyprus posed problems due to the scarcity or lack of data for several groups of values. Estimates of forest direct uses such as WFPs and certain NWFPs are easily drawn from official statistics, based on traded quantities and average market prices. However, valuation of other NWFPs, for example aromatic plants, is confined to only limited quantities traded on markets, while for grazing, an important forest benefit for rural economies, estimates are not available. Estimates of indirect use values tend to be partial, while option and non-use values, even though of high importance for present and future generations, are completely lacking.

Due to these limitations, it is more appropriate to use the term 'aggregate value' rather than TEV. Figure 14.2 presents this aggregate value and its individual components grouped into the main TEV categories. It can be seen that direct use values are the most relevant component of the aggregate value, with €9.5 million. However, detailed comparisons among all TEV categories could be misleading, due to partial valuations. What can be safely concluded is that the annual estimated aggregate value represents a very small part of the TEV, consisting mainly of the direct and indirect use values.

Comparisons can be made between individual components, as presented in Fig. 14.3. NWFPs are the most important component with some 47% of the aggregate value, followed by carbon sequestration (21%), recreation (16%), hunting (10%) and net growth of timber (8%). Timber and firewood are of least importance, with 3 and 1%, respectively. Of course, negative externalities caused by forest fires, pests and wildlife should not be forgotten; however, overall, they reduce the TEV by less than 6%. It should be stressed that valuation limitations mask the existence of other negative externalities, such as degradation due to forest grazing,

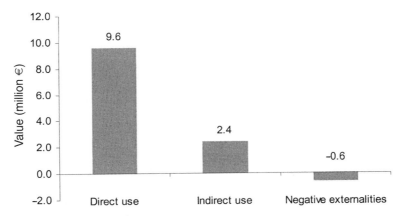

Fig. 14.2. Estimated TEV of Cypriot forests.

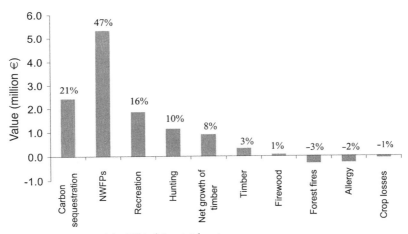

Fig. 14.3. Main components of the TEV of Cypriot forests.

which, if properly valued, might reduce the TEV of Cypriot forests even more.

Perspectives and Crucial Issues

This chapter attempts to draw a comprehensive picture of forest values, including public goods and externalities in Cyprus. Valuation efforts bear certain limitations due to the scarcity or lack of data regarding both quantitative and monetary aspects. As many environmental benefits are off-market and off-site, measurements are approximate and estimates are likely to be broad or partial, but this constraint does not limit the significance of valuation. Certainly,

further research is needed to improve and complete the whole picture of forest valuations.

The results show the high importance of NWFPs, carbon sequestration, recreation and hunting, relevant to increasing the welfare of local communities and society as a whole. At the same time, the role of WFPs is minor, if compared with all other goods and services provided by forests in Cyprus. This is actually in line with the multi-purpose forest policy, aimed at watershed protection, recreation and other services, rather than timber production.

It is not possible to capture all the categories of environmental values; there is widespread agreement on their significance despite the difficulty of expressing them in monetary terms. The greater the diversity of services attributed to the forest, the greater the scope of the Forestry Sector. Equally, the greater the emphasis on the multiple use of the forest, the greater the cluster of links with other sectors. This issue has far-reaching policy implications and justifies the need to apply a holistic approach to forestry planning and management.

Notes

[1]   Of which 80.7% are Greek Cypriots, 11.0% are Turkish Cypriots and the remaining 8.3% are foreign residents.

[2]   Although *Quercus infectoria* subsp. *veneris* is semi-deciduous, its woodlands are described in this category, which seems to be the nearest one.

[3]   In Cyprus, 'common land' referred to as 'hali land', consists of mostly unfertile land scattered throughout the island, not regularly exploited. It has never been claimed due to heavy property and farming taxes imposed by the Ottomans during the period from 1571 to 1878 BC (Thirgood, 1987). Later on, the British declared them as 'common lands' owned by the State who had availed them for the use of the people. This settlement still exists (Ioannou, 1991). Part of this land has gradually been naturally afforested.

[4]   The land tenure of a part of this land is unclear because the ownership has been split between many people, due to family inheritance, who are unknown to each other.

[5]   The estimates originate from Department of Agriculture, 2001 for honey; Statistical Service, 2000 for hunting.

[6]   Costantinides (1999) reports the expenditure for visiting forests of about €25.5/day-visit corresponding to 734,250 day-visits, while the expenditure for accommodation is €59.5/bed-night for 160,000 bed-nights. Overall, total expenditure for visiting forests reached €28.2 million in 1999.

[7]   Calculated for the period 1993–2002.

References

Alexandrou, C. (1999) A Plan to Improve the Valuation of Forest Goods and Services in Cyprus. A Project Submitted to the Cyprus International Institute of Management in Partial Fulfilment of the Requirements for the Degree of Master of Republic Sector Management, Nicosia.

Arnold, N., Biondi, E., Costa, M., Géhu, J.M. and Peris, J.B. (1984) Donees sur la vegetation maritime des cotes méridionales de l'île de Chypre (plages, dunes, lacs sales et falaises). *Documents Phytosociologiques* 8, 343–364.

Barber i Valles, A. (1995) Contribution to the Knowledge of the Bio-climate and Vegetation of the Island of Cyprus. Post Diploma Course in Forestry, Cyprus Forestry College, Nicosia.

Barbéro, M. and Quézel, P. (1979) Contribution à l'etude des groupments foréstiers de Chypre. *Documents Phytosociologiques* 4, 9–34.

Barbéro, M., Loisel, R. and Quézel, P. (1991) Sclerophyllous *Quercus* forests in the Eastern Mediterranean area: ethological significance. *Flora et Vegetation* 9, 189–198.

Brullo, S., Minissale, P. and Spampinato, G. (1997) La classe *Cisto-Micromerietea* nel Mediterraneo centrale e orientale. *Fitosociologia* 32, 29–60.

Costa, M., Géhu, J.M., Peris, J.B., Biondi, E. and Arnold, N. (1984) Sobre la vegetation thermomediterranea littoral le la Isla del Chipre. *Documents Phytosociologiques* 8, 365–376.

Costantinides, G. (1999) *Economic Report*. Prepared for the Department of Forests of the Ministry of Agriculture, Natural Resources and Environment in fulfillment of FAO Project (TCP/CYP 6712) on the Cyprus National Forestry Action Plan, Nicosia.

Department of Agriculture (2001) *Honey Census*. Internal Report, Department of Agriculture, Nicosia.

Department of Forests (2001) *Annual Report for 2001*. Department of Forests, Nicosia.

European Commission (2003) *Interpretation Manual of European Union Habitats – Version EUR 25*. European Commission DG Environment, Nature and Biodiversity.

Fankhauser, S. (1995) *Valuing Climate Change. The Economics of the Greenhouse*. Earthscan, London.

FAOSTAT (2004) Forestry Data. Agricultural Bulletin Board on Data Collection, Dissemination and Quality of Statistics. URL: http://faostat.fao.org/faostat/help-copyright/copyright-e.htm

Game Fund (2001) *Internal Report*. Game and Wildlife Service, Ministry of Interiors, Nicosia.

Géhu, J.M., Costa, M., Biondi, E., Peris, J.B. and Arnold, N. (1984) Donnes sur la vegetation maritime des cotes meridionales de l'ile de Chypre (Plages, dunes, lacs sales et falaises). *Documents Phytosociologiques* 8, 343–364.

Géhu, J.M., Costa, M. and Uslu, T. (1990) Analyse phytosociologioue de la vegetation littorale des cotes de la partie turque de l' ile de Chypre dans un souci conservatoire. *Documents Phytosociologiques* 12, 203–234.

Georgopoulou, E. and Kountenaki, D. (2001) *Inventory Results of the Greenhouse Gases Emissions in Cyprus*. National Observatory of Athens.

Giorgio, G. and Gabriilides, A. (1999) *Non Wood Forest Products in Cyprus*. Experts meeting on developing and coordinating the activities for non-wood forest products. FAO, Beirut.

Hadjikyriakos, X. and Pantelas, V. (1998) Cyprus and Desertification. *National Awareness Seminar on Desertification*, Nicosia, 21–22 September 1998. Department of Forests, Nicosia [in Greek].

Ioannou, C. (1991) *Land Survey Department. Functions Basic Legislation and Procedures*. I.J. Kasoulides, Nicosia.

Intergovernamental Panel on Climate Change (1996) *Guidelines for National Greenhouse Gas*. IPCC. Revised edition.

Meikle, R.D. (1977) *Flora of Cyprus*. Vol. 1. The Bentham–Moxon Trust, Royal Botanic Gardens, Kew, UK.

M'Hirit, O. (1999) Mediterranean forests: ecological space and economic and community wealth. *Unasylva* 197, 3–15.

Pantelas, V.S. (1996) Bioclimate and the climax plant associations in Cyprus. *Proceedings of the 6th Scientific Conference of Hellenic Botanical Society*, 6–11 April 1996, Paralimni, Cyprus, pp. 194–199 [in Greek].

Pantelas, V., Markides, L., Siamarias, N. and Aristokleous, G. (1999) Land use map of Cyprus. *Natural Vegetation and Agricultural Use*. Ministry of Agriculture, Natural Resources and Environment-Natural Resource Information and Remote Sensing Centre, Nicosia, Cyprus.

Papayiannis, C. and Markou, M. (1999) *A Structural and Economic Analysis of Farming in Cyprus*. Report 38, Agricultural Research Institute Ministry of Agriculture, Natural Resources and Environment, Agriculture Economics, Nicosia.

Philippides, P. and Papayiannis, C. (1983) *Agricultural Regions of Cyprus*. Comparative statistical and techno-economic analysis. Agricultural Economics Report 1. Agricultural Research Institute, Nicosia.

Quézel, P. (1976) Les forêts du pourtour méditerranéen. In: *Forêts et Maquis Méditerranéens: Écologie, Conservation et Aménagements*. Note technique MAB, 2, UNESCO, Paris, pp. 9–33.

Quézel, P. and Barberó, M. (1985) Carte de la végétation potentielle de la région méditerranéenne. Feuille No 1: *Mediterranee Orientale*. Éditions Du Centre National De La Recherche Scientifique, Paris.

Sabra, A. and Walter, S. (2001) *Non Wood Forest Products in the Near East: a Regional and National Overview*. Working Paper FOPW/01/2, FAO, Rome, pp. 23–27.

Statistical Service (2000) *Agriculture Statistics for 2000*. Ministry of Finance, Government Printing Office, Nicosia.

Statistical Service (2001) *Industrial Statistics for 2001*. Ministry of Finance, Government Printing Office, Nicosia.

Thirgood, J.V. (1987) *Cyprus, a Chronicle of its Forests, Land, and People*. University of British Columbia Press, Vancouver, Canada.

Tsintides, C.T., Hadjikyriakou, N.G. and Christodoulou, S.C. (2002) *Trees and Shrubs in Cyprus*. Foundation A.G. Leventis and Cyprus Forest Association, Nicosia.

United Nations Economic Commission for Europe/Food and Agriculture Organization (2000) *Global Forest Resources Assessment 2000*.

Von Mirbach, M. (2000) *Carbon Budget Accounting at the Forest Management Unit Level: an Overview of Issues and Methods*. Canadian Model Forest Network.

Zohary, M. (1973) *Geobotanical Foundations of the Middle East*. Vol. 2. Gustav Fischer Verlag, Stuttgart: Swets and Zeitlinger, Amsterdam.

15 Greece

Vassiliki Kazana[1] and Angelos Kazaklis[2]

[1]*Technological Education Institute of Kavala, Department of Forestry at Drama, 1st km Kalampaki-Drama, 66100 Drama, Greece;* [2]*Centre for Integrated Environmental Management (CIEM), 39 Androutsou Str., 55132 Kalamaria, Thessaloniki, Greece*

Introduction

Greece occupies the southern end of the Balkan Peninsula in the eastern Mediterranean with a population of 10.5 million people, 58% of whom live in urban areas, with the remaining 42% in semi-urban and rural areas (National Statistical Service of Greece, 1999). The country consists mainly of a hilly, mountainous and rocky mainland and a large number of islands scattered in the Ionian and Aegean Seas.

The climate is generally Mediterranean, with hot dry summers and rainy winters, varying greatly from region to region. It ranges from the continental type in the northern and north-western parts of the country with very cold winters and hot summers, to temperate in the southern and south-eastern parts of the country and the islands, with milder winters. Rainfall also varies greatly, from over 1200 mm along the western coast and in the Pindos Sierra, to about 600 mm along the Central Macedonia region and down to 300–400 mm along the eastern coast and the islands in the south-western Aegean sea.

As a result of the climatic variety, almost half of the country's land (49.4%) is covered by forests (Ministry of Agriculture, 1992; United Nations Economic Commission for Europe/Food and Agriculture Organization, 2000). However, only 25.5% of the country's forestland is classified as productive forestland, the remaining (23.9%) being other wooded land.

The majority of the forests extend over the northern and western part of the mainland.

Figure 15.1 shows six forest types of Greece, adjusted to the Quézel classification system (M'Hirit, 1999). In the coastal and plain zones, on altitudes ranging from 0 to 300 m above sea level, the typical Mediterranean low elevation natural vegetation (*Oleo ceratonion*) is predominant. Locust trees (*Ceratonia siliqua*), lentiscs (*Pistacia lentiscus*) and Olea oleasters (*Elaeagnus angustifolia*) are the most commonly diffused forest species of that zone, particularly in the middle and southern parts of the country. In the semi-mountainous zone on altitudes from 300 to 600 m above sea level, the *garrigue* and *maquis* types of vegetation (*Quercion ilicis*), resulting from thousands of years of grazing pressure, mainly from small animals (goats), are quite typical. Holm oaks (*Quercus ilex*) and holly oaks (*Q. coccifera*) are the most widely distributed evergreen oaks of this zone. According to the Quézel classification, the thermophilic wild olive and pistachio scrub class (i) and the sclerophyllous evergreen oak forest class (iii) fall into this zone.

At higher altitudes from 600 to 1200 m above sea level, in the mountainous zone (*Quercetalia pubescentis*), the growing spaces of Aleppo pines (*Pinus halepensis*), Calabrian pines (*P. brutia*) and all deciduous oak forests (*Quercus frainetto*, *Q. pubescens* and *Q. cerris*), are subordinated. The vegetation Quézel classes (ii) and (iv) fall into this zone.

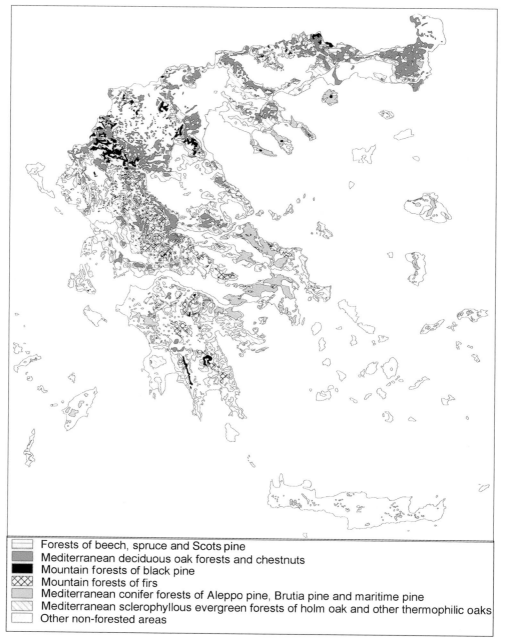

Fig. 15.1. Map of Greek forest types according to the adapted Quézel classification. Prepared by Angelos Kazaklis.

Boreal pines (*Pinus nigra, P. silvestris* and *P. heldreichii*), firs (*Abies borisii regis*), spruces (*Picea excelsa*) and beeches (*Fagus moesiaca* and *F. silvatica*) are typical species found at altitudes ranging from 1200 to 2000 m above sea level in the boreal conifer zone (*Fagetalia* and *Vaccinio picetalia*). The vegetation Quézel categories (v) and (vii) correspond to this zone.

No forests are found above 2000 m in the zone of Astragalo Acantholi monetalia, where vegetation is mainly subalpine.

Forest Resources

Area

About 6.5 Mha, or 49.4%, of Greece is classified as forest and other wooded land (Ministry of Agriculture, 1992; United Nations Economic Commission for Europe/Food and Agriculture Organization, 2000). The largest portion of this land (98%) consists of natural forests, while the remainder has plantations mainly of poplar species.

In the First National Forest Inventory, forestland included land parcels of more than 0.5 ha (5000 m^2) with over 10% tree crown cover (or equivalent stocking level) and shelterbelts of trees over 30 m wide. Also, stands with crown densities less than 10% and unstocked areas due to either human intervention or natural causes, yet expected to revert to forest, were considered as forestland. In the National Forest Inventory, final forestland included all the reforested areas and areas under the Mediterranean type of vegetation, such as evergreen broadleaved trees, shrubs, bushes, *maquis* and *garrigue*.

No discrepancy in the basic forest resource statistics of Greece was detected between the First National Forest Inventory report and other more recent sources, such as the Global Forest Resource Assessment (United Nations Economic Commission for Europe/Food and Agriculture Organization, 2000). Most probably, the latter was based to a great extent on the First National Forest Inventory results, as many parts of this report regarding Greek forests note 1992 as the reference year for the information cited. Indeed, the most reliable information on the forest resources of Greece to date can be derived from the results of the First National Forest Inventory, except for plantation forestry, which was not recorded specifically in the inventory. Most of the information related to plantation forestry in the context of this chapter has been based solely on the United Nations Economic

Commission for Europe/Food and Agriculture Organization report.

About 56% of the country's communities are located in mountainous or semi-mountainous areas, according to the National Population Census (National Statistical Service of Greece, 1995). Almost 80% of these communities have forests in their territory and 32% of the country's population lives in these areas. Therefore, forest resources play an important role in the economy of the mountainous and semi-mountainous areas of the country.

Typologies

The majority of the Greek forests, i.e. 48%, are *maquis* and *garrigue* types of forests, 34% are coppices and only 16% of the forests have been classified as high forests. The average total growing stock of the industrial forestland amounts to only around 44.2 m^3/ha, a rather low value when compared with that of other European countries. However, this value is not quite representative of the true production capability of many forests. There are a substantial number of forests with much higher yields, but also a large number of forests with a much lower yield. The total and per hectare growing stock values in the First National Inventory were calculated with reference to forest species rather than forest management types. Therefore, the growing stock of the Aleppo and Calabrian pine forests is about 26 m^3/ha, the Austrian pine forests 54 m^3/ha, the Scots pine forests 122.83 m^3/ha and all the other conifer forests about 267 m^3/ha. The spruce forests in the north of the country have an even higher growing stock of 342 m^3/ha.

The First National Forest Inventory of 1992 categorizes around 910,000 ha, or 14%, of the total forest area as damaged and/or degraded forestland. The main causes of damage/degradation according to this report are attributed to insects and diseases (27%), illegal cutting (12%) and various reasons (29%). Damage due to grazing accounts for only 1.7% and forest fire 1.2%. However, these grazing and forest fire figures and therefore the total degraded area seem rather unrealistic. The percentage regarding forest fires, when compared with that recorded in the FORIS Food and Agriculture Organization database collection on-line, is at least three times higher and the grazing figure is also believed to be much higher. The low values appearing in the report of the First National Forest Inventory are due to several specific reasons. Firstly, only the burned trees left standing during the 5-year period prior to the inventory were measured; the remainder of the burnt land was recorded as either unstocked or regeneration land. Secondly, only trees with a diameter at breast height (dbh) greater than 5 cm were measured. However, damage by grazing occurs only to trees with dbh less than 5 cm. Unfortunately, it was not possible to obtain other information at this stage regarding a more realistic value of the degraded forest and other wooded land due to grazing.

Functions

According to the data supplied by the National Forest Inventory (1992) and the data reported by Albanis *et al.* (2000), over half of the forest and other wooded land in Greece (51.58%) is managed for production purposes, 5.18% for tourism-recreation, 14.4% for hunting and 28.84% for grazing and fuelwood production. The tourism-recreation land includes national parks, aesthetic forests, natural monuments, wetlands, recreational sites, urban forests, coastal forests, shelterbelts along the highways and forest areas surrounding lakes. It is worth noting that soil and water protection is one of the main management goals for the whole forest and other wooded land in Greece, due to the extent and intensity of erosion and torrential phenomena (Albanis *et al.*, 2000).

According to the United Nations Economic Commission for Europe/Food and Agriculture Organization (2000) classification, 1.03% of the Greek forest and other wooded land is placed under the International Union for Conservation of Nature (IUCN) categories I and II, i.e. virgin forests and National Parks. The same source also records 17.67% of the forest and other wooded land as IUCN land categories III and IV, i.e. aesthetic forests and other specially protected areas. However, this classification cannot be used to derive a good estimate of the productive land area as, in the National Forest Inventory of

1992, no distinction of land was made on the basis of the protection function of the land according to the IUCN categories.

Institutional Aspects

Ownership and size of properties

Forest and other wooded land in Greece, to a great extent (74.1%), is publicly owned. Private forestland accounts for only 6.5% of the total forest area. Also, the common forestland forms 9% and the ecclesiastical forestland, i.e. the forestland belonging to monasteries, charitable institutions and joint property, forms 10.4% of the total forest area. The distribution of this type of ownership structure in Greece reflects the special historic, socio-economic and political conditions of the country. The forest law of the country prohibits fragmentation of forestland and thus it has contributed to the maintenance of relatively large forest holdings. In addition, the maintenance of the high percentage of public forest and other wooded land in the country has many strong advocates, who believe that the State can be a better manager of this land, since it can afford funding, trained personnel and the required infrastructure for managing the land.

Forests were not classified by ownership size in the National Forest Inventory of 1992. Information referring to the number of properties and their area by size class as divided by ownership type was derived from the Forestry Development Programme 1976–1980 (Centre of Planning and Economic Research, 1976). According to this plan, 99.9% of the public and the community forests are more than 100 ha in size. Of the private forest properties, only 3.2% fall into the 2–50 ha-size class, which makes it one of the smallest percentages in Europe. The majority of private forest properties exceed 1500 ha in size.

Administration and policies

Forests in Greece are the responsibility of the Ministry of Agriculture. The Forest Service, which operates under the General Secretariat of Forests and Natural Environment (GSF&NE) in the Ministry of Agriculture, is the main body responsible for the protection and management of the state forests, as well as for the supervision and control of the private forests. GSF&NE consists of the Central Service, the Regional Services, the Forest Policy Council, the Revision Council for the property of forests, the Forest Technical Council and the Regional Councils and Committees.

The Central Service has a supervisory role and consists of six Directorates responsible for forest policy formulation, drawing up long-term programmes for forest development, forest fire protection, research supervision and promotion of the country's cooperation with European Union (EU) countries, other countries and international organizations.

The Regional Services have an operational role for implementing the forest policy formulated by the Central Service, local programmes and studies. They operate at two administrative levels, Intra-prefectural and Prefectural. The Intra-prefectural Services include seven Specific Regional Forest Inspectorates and 13 Forest Inspectorates for each of the 13 regions of the country. The Prefectural Services include 31 Forest Directorates with 80 Forest District Offices, 24 Directorates without Forest District Offices and two Directorates of Reforestation in Attiki and Thessaloniki Prefectures.

In addition to the GFS&NE, other bodies that contribute to the development and protection of the country's forest resources are the Panhellenic Confederation Union of Agricultural Cooperation (PASEGES), the Forest Owners Association of Greece (FOA), the Geotechnical Chamber of Greece (GEOCG), the Hellenic Forestry Society (HFS) and non-governmental organizations (NGOs). The main NGOs, which have an important influence on various issues related to forest resources in the country, are the Hellenic Society for the Protection of Nature, the Hellenic Ornithological Society, the World Wide Fund for Nature Hellas, the Greek Biotope/ Wetland Centre and the Arctouros Society.

In Greece, forest policy is based on the 1975 Constitution (Art. 25 and 117) and Laws No. 86/1969, No. 998/1979 and No. 1650/1986, and aims to protect and manage the forests and other wooded land through the implementation of the sustained yield principle. Law No. 86/1969 codified almost all the laws that had been issued since 1928 and that had been amended and completed by Law No. 4173/1929. Law No. 86/1969 and its associated following amendments by Laws No. 886/1971, No. 996/1971, No. 248/1976 and No. 998/1979 is the Forest Code of the country and regulates matters concerning protection, management, real property rights on forestland, taxation, exploitation of State and private forests and improvement operations (Albanis *et al.*, 2000).

Forestland use planning aims at applying multiple-use forestry on the sustained yield principle and it is implemented by the Forest District Offices. Forestland use changes are prohibited by Article 24 of the country's constitution, unless they are enforced for public interest. State and private forests and other wooded lands destroyed by fires and other causes should be reforested and their use for other purposes is prohibited. Expropriation of forest and other wooded lands, which belong to natural or public legal entities, is possible only in favour of the State, without, however, causing any change in the forestry land use.

General land use planning at the national and regional level is implemented by the Ministry of Environment, Physical Planning and Public Works. Law No. 998/1979 determines specific protection measures for maintaining, developing and improving forest and other wooded land of the country. Law No. 1650/1986 concerns the protection of the environment and it has introduced five new categories of protected areas as well as changes in their administration and management. These categories include areas of strict protection of nature, areas of nature protection, national parks, protected natural formations, protected landscapes and elements of landscape and areas of ecodevelopment. This law also determines the content of the specific environmental studies which should be carried out in order to formulate proposals for measures required for protection and management of the areas designated under any of the protected status categories.

It is worth noting that, in terms of protection and management of the natural environment during the past decade, the Ministry of Agriculture has gradually lost administrative power, as the corresponding responsibilities were taken over by the Ministry of the Environment, Physical Planning and Public Works. An important influence regarding natural environment protection and management is also exerted by NGOs. Participation of forest landowners' associations, individuals and other interested parties in the land use planning processes is extremely limited.

Contribution of the Forest to the National Economy

Gross domestic product and employment

Forestry is important in the economy of the mountainous and semi-mountainous communities of Greece. However, the contribution of the forestry sector to the Greek economy as a whole is very low, barely reaching 0.15% of the country's gross domestic product (GDP). This is due mainly to the fact that the majority of the country's forests are of low productivity. Also, the non-market goods and services derived from the forest resources are not registered in the national accounts. It is also worth noting that over the last decade, the contribution of forestry to the country's total GDP has decreased, and this can be attributed mainly to the higher productivity achieved by the other sectors of the economy (Albanis *et al.*, 2000).

The contribution of the timber-based industries, including the furniture and paper industries, is higher than that of forestry, reaching 1.17% of the GDP (National Statistical Service of Greece, 1995), mainly due to the higher value added by the timber-based industries.

Employment in forestry is rather difficult to estimate, although a decreasing trend in the number of people employed has been observed over the past decade. Generally, forestry offers supplementary employment to the population of the mountainous and semi-mountainous areas of the country. In addition, many forest operations are carried out by contractors – and no longer through the direct labour system of the Forest Service – and the number of workers employed is not registered.

About 26,000 people are employed in forestry, accounting for 0.61% of the national employment, of which 4000 are permanent staff, with the remainder being seasonal staff. In terms of employment in other forest-related activities, statistical information is even less

accurate. About 500 people, mainly foresters, are employed by Local Authorities as permanent personnel with responsibilities relevant to development and management of forest resources, parks and protected areas within their administrative boundaries. Also, seasonal personnel are employed for firefighting, which is now the responsibility of the Fire Service under the jurisdiction of the Ministry of Public Order. Over the past couple of years, about 4500 people were employed for firefighting on 5-month contracts.

Employment in timber-based industries forms 1.4% of the country's total employment. It is believed, however, that this figure is lower than the real one, as the National Statistical Service of Greece provides information only for timber-based enterprises which employ more than ten people, while the majority of the country's sawmills usually employ less than ten people.

International trade

The total amount of timber consumed and processed by the timber-based industries in Greece is derived mainly from imports, almost 66%, and from national production, 34% (Ministry of Agriculture, 1996). Greece imports unprocessed and processed wood products, such as wood pulp from the Nordic countries, France, Austria and Eastern European countries. It exports mainly paper, resin, briar roots, laurel leaves, carobs and acorns.

Forest product imports represent about 3.8% of the country's total import value, while forest product exports account for only around 1.3% of the total export value. The total value of forest product exports forms only 15% of the total value of forest product inputs, which makes Greece a net importing country of wood and wood products (Albanis *et al.*, 2000). The country is almost self-sufficient in terms of firewood production. The national firewood production covers 99.5% of the total amount consumed, supplemented by only around 0.5% of imports.

Other forest-based industries: tourism and green belts

It is well acknowledged that the importance of forest resources is increasing when all the other benefits are considered, except production of marketed goods, such as wood and non-wood forest products (NWFPs). Benefits, such as provision of recreational opportunities, climate regulation, scenic beauty, and soil and water conservation, are among the most important constituents for improving the quality of life. Therefore, when these non-market benefits are taken into account, the contribution of forestry to the national GDP is increasing considerably.

Many studies have confirmed the role that 'green belts' play when based on tourism, especially in areas designated as parks or other aesthetic areas, or areas where timber-processing industries are based. Studies based on the 'hedonic price method' typically show an increase in property value when located near woods or parks or other natural areas. Unfortunately, studies of this kind are lacking in Greece and, therefore, no specific information on the contribution of these effects can be reported at this stage.

The Values of Greek Forests

Table 15.1 presents a picture of the estimated values of Greek forests, according to the total economic value (TEV) categories of direct and indirect use values, option, bequest and existence values and other negative externalities. Though incomplete, due to insufficient information, it shows that the most outstanding values are given by the watershed protection function, grazing and timber, and negative externalities linked to erosion, floods and landslides due to poor or no forest management.

Direct use values

The most economically important forest products and their direct use values are shown in Table 15.1. Timber traded on the market is valued as roundwood at 'forest gate price', i.e. roadside price, while the 'saved increment' is valued at half of the stumpage price (Ministry of Agriculture, 1996). The value of honey is based on the average annual production during the previous 15 years and the forest contribution of 80% to the total production (Eleftheriadis and Efthimiou, 1988). Other goods including Christmas trees, resin and briar

roots are valued at the quantity traded on the market and the average market price (Albanis *et al.*, 2000).

With regard to estimates of forest grazing values in Greece, several methods have been proposed by different analysts. Eleftheriadis (1998) provides four estimates of the country's forest forage value based on different calculation approaches. One estimate is based on the 'substitute goods' method and assumes forest forage equivalent to any kind of animal food (hay or clover) over the feeding period. The total cost of feeding one animal is estimated to be equal to €40/year and the contribution of free range to animal feeding equal to 62%, i.e. approximately 7.5 months/year. Considering that there are about 14 million animals in total in the country, this estimate results in a total grazing value of €350 million (updated 2001 prices).

A second estimate provided by the same study is based on 'similar product' prices, assuming that forest forage is equivalent to the hay value. Considering that the forage production allowed for grazing is 5.59 Mt and the hay price is €130/t, the value of forest grazing is estimated at €727.5 million (updated 2001 prices).

The third estimate of the Eleftheriadis study (1998) is based on grazing fees. Grazing fees or rangeland rent are not determined in free market conditions in Greece. Grazing rights are granted for unlimited resource use, i.e. land and time, to interested individuals only by Local Authorities at very low annual prices, of €0.25 and €0.4 for small and large animals, respectively. Meanwhile, grazing in private lands is charged at about €5.5/animal/year (updated 2001 prices). Based on the private grazing fees and a number of 14 million grazing animals, it gives a forage value of €77 million.

The fourth estimate in the same study, based on the 'alternative cost' method, assumes that forest forage substitutes animal nutritional requirements for 7–8 months. The monthly animal nutritional requirements were estimated at 0.48 kg SE (starch equivalent)/day (free range) or 0.34 kg SE (domestic) for a 35 kg animal. The cost of 0.34 kg SE is assumed to be equivalent to the forage day value per animal for 180–240 days per year. The value of 1 SE is calculated to be equal to €0.13/day (updated 2001 prices). It results in an annual forest forage value ranging between €327.6 and €436.8 million.

Table 15.1.   Values of Greek forests.

Valuation method/output	Quantity	Value[b] (000 €)
Direct use values		
Market price valuation[a]		
Timber (m³)	700,000	54,600
Firewood (m³)	1,580,000	33,180
Net growth of standing timber stock (m³)	1,476,180	14,762
Honey (t)	10,000	45,000
Christmas trees (pieces)	95,050	713
Resin (t)	5,830	991
Briar roots (t)	3,600	720
Bay leaves (t)	3.7	252
Plant soil (t)	153.3	11
Decorative plants (t)	179	9
Other: resinous wood, pine cones, seeds	—	1
Surrogate market pricing		
Grazing (million FU)	1,749	227,400
Permit price		
Hunting (no. of hunters)	272,000	26,656
Consumer surplus (TCM, CVM)		
Recreation (day-visits/year)		7,815
National parks and protected areas	271,000	4,065
Other forest areas	1,500,000	3,750
Total direct use values		412,110
Indirect use values		
Cost avoided method		
Watershed protection		293,600
Flood prevention		5,600
Soil moisture conservation		288,000
Shadow pricing		
Carbon sequestration (tC)	463,000	9,260
Total indirect use values		302,860
Option, bequest and existence values		
Cost-based method: biodiversity conservation		12,680
Total option, bequest and existence value		12,680
Negative externalities		
Cost-based method		
Erosion, floods and landslides/avalanches (m³/ha)	6.47	−232,920
Losses due to forest fires (ha)	33,700	−32,800
Damages caused by illegal acts		−3,000
Total negative externalities		−268,720
TEV		458,930

[a]Quantities refer to 1996 (Albanis *et al.*, 2000); [b]Updated to 2001 prices.

Another estimate, provided by Albanis *et al.* (2000), is based on grazing rental prices. The annual rent per 0.1 ha of grazing land is considered equivalent to the value of 2.5 kg of meat of living animal and 3 kg of milk, i.e. about €6.9. The mean annual production of a rangeland is estimated to be 52 kg per 0.1 ha and the total forest grazed area of the country 5.3 Mha. Therefore, the total forest forage value equates to €365.7 million.

As can be observed, there is a substantial difference among the forest forage value estimates, due to the variety of the approaches used. Some approaches, such as the one based on grazing fees, clearly underestimate the true value of grazing in Greece, being referred to a nominal price, therefore not capturing the consumer surplus. For this study, an estimate is elaborated and reported in Table 15.1 based on the forage units (FU) consumed and the value of 1 FU. It is considered that 1 ha of forest grazed land is equivalent to 1 t of forest production allowed for grazing or 1 t of hay, or approximately 330 FU. Taking into account that the market price of hay is about €130/t, the price of 1 FU is equivalent to €0.13. Therefore, the grazing value estimate for the total forest grazed area of 5.3 Mha is €227.4 million.

Game resource management is carried out by the Ministry of Agriculture and the Hunting Confederation of Greece. Specific management measures involve the determination of the hunting period, prohibitions of hunting rare and threatened species, restrictions on the number of animals and birds that can be hunted per outing, the creation of game refuges, game breeding stations and controlled shooting areas, as well as the enrichment of forest with native game. The value of hunting reported in Table 15.1 was based on the hunting permit price (J. Paterakis, Hunting Club of Thessaloniki, 2001, personal communication) and the total number of hunters (Albanis *et al.*, 2000). There are three different types of permits allowing for hunting at local, regional and national level, and the larger the area of the permit validity, the higher the price. However, it was not possible to obtain information on the exact number of hunters in relation to the type of permit. Therefore, an average value is calculated by taking the three different prices corresponding to the three different spatial scales.

Recreation value is based on the number of day-visits/year and consumer benefit/visit in the forest. Two separate value estimates, one for the protected natural areas and the other for the forest recreation areas, are calculated and summed up, because the consumer surplus of protected natural areas (especially national parks) is different from that of other forest recreational areas. The consumer surplus of the use value of national parks (extrapolation from the Olympos National Park case study) is found to be equivalent to €15/day (updated 2001 price), based on a travel cost method (TCM) investigation (Eleftheriadis and Kazana, 1996). The consumer

surplus of other recreational areas is taken to be equal to about €2.5/day-visit, based on a contingent valuation method (CVM) survey (Kazana and Eleftheriadis, 1998). Estimates on the day-visits are elaborated from information extracted from different studies (Eleftheriadis and Efthimiou, 1988; Karameris, 1988; Malamidis *et al.*, 1992; Papageorgiou, 1996; Albanis *et al.*, 2000). To calculate recreation values, other approaches were also attempted. One such approach was based on information regarding the number of visits to the national parks and the entrance fee payable for the National Park of Samaria in Crete. The other approach used a forest recreation value/ha/year provided by a case study carried out at the 'Kedrinos Lofos' forest of Thessaloniki (Karameris, 1988). It is worth noting that both approaches gave similar results to the consumer surplus estimations as cited above.

Indirect use values

Monetary valuation of watershed protection services performed by forest cover is based on the cost avoided method. As a surrogate measure for flood prevention, it is assumed that 1% of agricultural production could be lost in the proximity of forests, if forest cover disappeared. Considering that the agricultural production is about €7 billion in Greece (2001 prices) and the agricultural areas situated in the proximity of forests comprise about 8%, the value of flood prevention is estimated at €5.6 million.

In addition, water supply in rural Greece depends to a considerable extent on springs. The loss of such springs due to poor or no watershed management could require a replacement cost of approximately €0.04/l of water, due to required drillings. Considering a water supply of 18 billion l/year (estimate elaborated from information published by Eleftheriadis, 1985), and assuming that at least 40% of it would require a replacement cost for drillings, the value of soil moisture conservation reaches €288 million. The total value of watershed protection is calculated by summing up the two value estimates.

The carbon that could be stored annually in Greece, based on the annual net increment of woody biomass, is estimated to be about 1,079,000 tC. However, the real carbon quantity stored annually calculated after subtracting the annual harvests from the net increment of woody biomass is about 463,000 tC (National Observatory of Athens, 1997). The monetary valuation of carbon storage is based on a shadow price of about €20/tC (Fankhauser, 1995).

Option, bequest and existence values

It is not possible to estimate the option, bequest and existence values of the country's forests, as relevant studies are completely lacking. A few CVM/TCM applications at local level (Kazana and Eleftheriadis, 1998) or for some of the country's national parks (Eleftheriadis and Kazana, 1996; Papageorgiou, 1996) have provided only estimates of mainly recreation direct use values.

However, an estimate of the forest biodiversity value may be attempted based on the cost method. A surrogate measure for estimating the existence value of forest biodiversity in Greece can be derived by assuming the value of forest biodiversity is equal to the value of the investment allocated for protection and management of forest biodiversity through various EU and national funds. The funds allocated for conservation and protection of the natural environment in Greece over the period 1993–2000 is equal to €88.7 million (Organization for Economic Cooperation and Development, 1999; Albanis et al., 2000; Ministry of Environment, 2001), resulting in an annual average investment value for forest biodiversity of €12.6 million. It should be noted that this figure only partly constitutes a surrogate of the forest existence values, as it refers to protected areas or areas under a special protection status and not to the total national forest area. In addition, this figure represents a stock and not a flow value.

Negative externalities linked to forests

Valuation of erosion and floods due to poor forest management is based on the replacement cost method. Soil loss due to erosion and floods

in the whole country equals 86 Mm³/year or 6.47 m³/ha/year according to the Ministry of Agriculture (1996). There are 2 Mha of degraded or poorly managed forest area, which requires protection from water hazards through watershed management (Albanis et al., 2000). The cost for returning soil to its original situation, i.e. reducing dam sedimentation by dredging accumulation of silt downstream, is estimated at about €18/m³ (own elaboration).

The total damage caused by forest fires is referred to an annual burnt area of 33,700 ha and is estimated on the basis of the restoration cost technique, i.e. it included the standing volume expectation value (€12.3 million), the watershed protection value lost (€3 million), the reforestation cost (€12.8 million) and the fire combating cost per year (€4.7 million).

Finally, the damage due to illicit activities such as illegal cutting and illegal land occupation is estimated by using as a surrogate measure the annual expenses for forest protection in relation to those aspects by the GSF&NE on average.

Towards the Total Economic Value of Greek Forests

The estimated total economic value (TEV) of the Greek forest outputs and its categories is shown in Fig. 15.2. The values represent the total minimum estimates for the different categories, i.e. the direct and indirect use values, the option, bequest and existence values and the negative externalities.

The overall TEV is about €0.5 billion, when some non-use values are not included due to

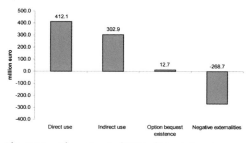

Fig. 15.2.   The estimated TEV of Greek forests.

insufficient information. The direct use values appear as the most important sector of the overall TEV of about €412 million, followed by the indirect use values of about €303 million. The negative externalities amount to €269 million, while the option value appears quite low (€12.7 million) due to limited data availability. It should be noted, however, that all these values are indicative estimates of the importance of forest non-use values, and detailed studies should be carried out to improve their accuracy.

The breakdown of TEV into the various forest outputs is shown in Fig. 15.3.

Water management is the most important function of the Greek forests, accounting for about 64% of the TEV. The high percentage reflects the importance of the water resource protection and management issues in the context of forest resource management. The remainder is distributed among the other outputs: grazing (50%), timber (12%), honey (10%), firewood (7%), hunting (6%), net growth increment (3%), biodiversity conservation (2%), recreation (2%) and carbon sequestration (2%), followed by the negative externalities (–59%). Negative externalities include mainly erosion, floods and landslides due to poor or no forest management, and damage caused by forest fires, illegal cutting, illegal land occupation and insects.

Conclusions and Perspectives

This analysis provides a holistic view of the information already known and of the missing information required to obtain a comprehensive estimate of the TEV of the Greek forests. However, some observations are worth noting. Forest public goods and externalities have a particularly important contribution to the TEV configuration, the most important being the watershed protection function. Traditional forest outputs are also significant, such as timber, grazing and firewood, among which grazing appears most relevant. Other goods, such as honey, hunting, recreation, carbon sequestration, and option and bequest values, also have a weight in the TEV, although their estimates are lower than expected, mainly due to lack of detailed studies and other information. Also worth noting is the crucial role played by negative externalities linked to the damage and defensive expenditures caused to society by floods and fires.

There is an apparent need for further studies to supply all necessary information for achieving a more accurate and comprehensive TEV estimate of the Greek forests. This type of knowledge will provide a useful framework for policy formulation in terms of protection and management of the country's forest resources in the wider context of sustainable development.

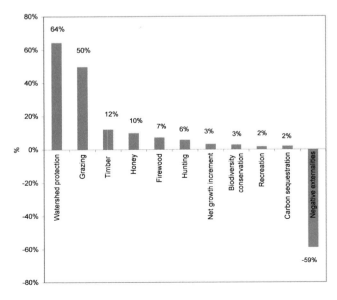

Fig. 15.3. The main components of the TEV of Greek forests.

References

Albanis, K., Galanos, F. and Boskos, L. (2000) *Criteria and Indicators for the Sustainable Forest Management in Greece*. General Secretariat of Forests and Natural Environment, Ministry of Agriculture, Athens.

Centre of Planning and Economic Research (1976) *Development Programme 1976–1980*. Section of Forests, Centre of Planning and Economic Research (KEPE), Athens.

Eleftheriadis, N. (1985) Contribution of forests to the economy and quality of life. *Forest Chronicle* Nos. 5–6/7–8, 7–23 (in Greek).

Eleftheriadis, N. (1998) Valuation of the forage production of the national ecosystems in Greece. *Geotechnical Scientific Issues* 9(2), 4–19.

Eleftheriadis, N. and Efthimiou, P.N. (1988) An integrated development programme for Greek forestry. *Journal of Environmental Management* 26, 179–190.

Eleftheriadis, N. and Kazana, V. (1996) Economic evaluation of environmental goods: the Olympos National Park case study. In: *Proceedings of the 7th Panhellenic Forestry Conference*. Hellenic Forestry Society, Thessaloniki, Greece, pp. 283–293.

Fankhauser, S. (1995) *Valuing Climate Change. The Economics of the Greenhouse*. Earthscan, London.

Food and Agriculture Organization (2000) *Forestry Information System (FORIS) – Country Profiles – Greece* (http://www.fao.org/countryprofiles/index.asp?lan=ed&iso3=GRC).

Karameris, A. (1988) The value of recreation in suburban forests. *Scientific Bulletin Geotechnica* 1, 75–79.

Kazana, V. and Eleftheriadis, N. (1998) A CVM economic valuation of recreation use value: the Mylopotamos Drama case study. In: *Proceedings of the 8th Panhellenic Forestry Conference*.

Hellenic Forestry Society, Thessaloniki, Greece, pp. 575–581.

Malamidis, G., Eleftheriadis, N. and Aidinidis, E. (1992) *Olympos National Park: Research on its Social Dimension*. TEI Kavala, Special Edition No. 3.

M'Hirit, O. (1999) Mediterranean forests: ecological space and economic and community wealth, *Unasylva* 197, 3–15.

Ministry of Agriculture (1992) *Results of the First National Forest Inventory*. General Secretariat of Forests and Natural Environment, Ministry of Agriculture, Athens.

Ministry of Agriculture (1996) http://www.minagric.gr/greek/2.5.2.2.html Access date: 2001.

Ministry of Environment, Physical Planning and Public Works (2001) http://www.minenv.gr/frame.html?2&1&2&/4/41/e100.html Access date: 2001.

National Observatory of Athens (1997) *Final Report CORINAIR 94*. Ministry of Environment, Physical Planning and Public Works.

National Statistical Service of Greece (1995) *National Accounts*. Directorate of National Accounts, NSSG, Athens.

National Statistical Service of Greece (1999) *Greece in Figures*. NSSG, Athens.

Organization for Economic Cooperation and Development (1999) *Economic Instruments for Pollution Control and Natural Resources Management in OECD Countries: a Survey*. Working Party on Economic and Environmental Policy Integration.

Papageorgiou, K. (1996) An Evaluation of the National Park System of Greece with Particular Reference to Recreational Management in the Vikos-Aoos National Park. PhD Thesis, University of Sheffield, UK.

United Nations Economic Commission for Europe/Food and Agriculture Organization (2000) *Global Forest Resources Assessment 2000*. Main Report, United Nations Publications, Geneva.

16   Albania

Kostandin Dano

General Directorate of Forests and Pastures, Tirana, Albania

Introduction

Albania covers 28,748 km², of which 36% are forests. The total population is around 3.3 million, 42% of which is urban (Directorate General for Forests and Pastures, 2004). The climate is of a Mediterranean type in the western lowland coastal area and continental in the rest of the country. Four climatic subdivisions can be distinguished:

- The Mediterranean flat zone, mostly represented by the coastal area, with relatively dry, hot summers and cool, wet winters. The area from Vlore to Saranda in the south is best suited for citrus cultivation, due to the very low incidence of frost. Average annual rainfall is between 900 and 1400 mm, but in the central and northern part of the coastal area it can reach 2000 mm.
- The hilly Mediterranean area, stretching from north to south, is located on the eastern side of the coastal zone. Hills reach an altitude of 300–500 m and have a similar climate to that of the flat zone.
- The submountainous Mediterranean zone is comprised of the highland plains in the south (plain of Korca) and the river valleys in Peshkopi, Kukes, Tropoje and Puke. It is has an altitude of about 800 m, lower temperatures than the coastal plain and frequent frost.

- The mountainous Mediterranean zone covers all lands above 800 m and has an almost continental climate. Annual precipitation is 1000–1800 mm and in the northern Albanian alps it exceeds 2000 mm. This is an area covered mostly by forests and natural pastures.

In the national economy, agriculture makes the most significant contribution: 34% of the gross domestic product (GDP) in 2001 (World Bank, 2002). However, this sector is still challenged by unfavourable climatic conditions due to frequent droughts and economic constraints caused by improper equipment modernization.

Forest Resources

Forest area and typologies

Albanian forests cover 1 Mha, of which 94.7% are natural forests and the remainder are plantations (Directorate General for Forests and Pastures, 1999). A map of forest area is shown in Fig. 16.1. The main forest types are: high forests (45%), coppices (30%), and *maquis* and shrubs (25%), which present a growing stock of 153, 47 and 30 m³/ha, respectively (Table 16.1). Degradation affects 34% of the forest area and is due mainly to overgrazing (14%), timber exploitation (11%), forest fires (3%) and other factors (6%) (Directorate General for Forests and Pastures, 1999).

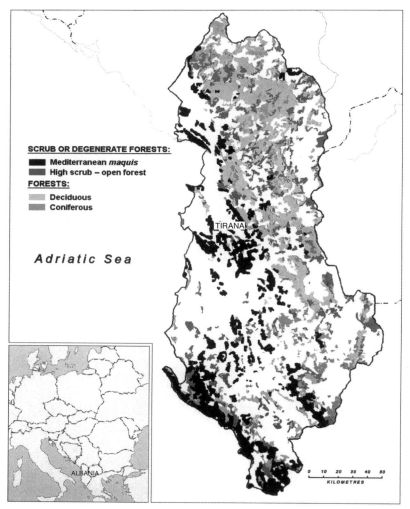

Fig. 16.1. Forest map of Albania. Source: adapted from Food and Agriculture Organization/General Directorate Forests and Pastures/World Bank (1995).

Functions

According to the main functions, forests are classified into production (85%) and protection stands (15%) (General Directorate for Forests and Pastures, 1999; Directorate for Protected Areas, 2000). The ecological role performed by forests is now being reconsidered in terms of various functions such as biodiversity and wildlife habitats (Table 16.2).

Until the mid-1990s, protected areas in Albania were divided into three categories: National Forest Parks, Game Hunting Reserves

(A and B) and Natural Monuments (World Commission on Protected Areas, 2000). In 1994–1995, an ecological survey of virgin forests was undertaken (World Bank/Forest and Pastures Research Institute, 1997), resulting in a recommendation to increase the number of protected areas and to enlarge those already existing. In 1999, the Prespa National Park and Lake Ohrid Landscape Protected Area were established. Subsequently, a national study on the current network of protected areas in Albania was carried out, and in 2000 an 'Action Plan' was initiated. In 2002, a new network of

Table 16.1. Classification of forests according to type and management regime.

Forest type	Area		Growing stock (m³/ha)	Mean annual increment (m³/ha)
	ha	%		
High forests	464,120	45	153.0	2.06
Coppices	307,430	30	47.0	1.15
Maquis, shrubs (if the case)	256,450	25	30.0	0.95
Total forest and other wooded lands	1,028,000	100	82.3	1.40

Source: Directorate General for Forests and Pastures (1985).

Table 16.2. IUCN protected areas in Albania.

	IUCN category	Size (ha)
National Parks	II	53,940
Strictly Nature Reserves/ Scientific Reserves	I	14,500
Nature Monuments	III	4,360
Managed Nature Reserves	IV	42,898
Protected Landscapes/ Seascapes	V	29,873
Managed Resource Reserves	VI	18,245
Total		163,816

protected areas in Albania (designated according to the IUCN categories) was set up by the Directorate of Protected Area Management in Albania (World Commission on Protected Areas (2000)). To date, this network includes four Strictly Protected Areas (IUCN category I), 12 National Parks (IUCN category II), four Nature Monuments (IUCN category III), 26 Nature Reserves (IUCN category IV), three Protected Landscapes (IUCN category V) and four Resource Reserves (IUCN category VI). In addition, 15 Important Bird Areas have been identified in Albania (Regional Environmental Centre for Central and Eastern Europe, 2000). The areas under protection make up 5.8% of Albanian territory.

Institutional Aspects

Ownership and size of properties

According to the Directorate General for Forests and Pastures (DGFP), Forest Cadastre Sector, 96.8% of forest area is public, 2.8% is common

land and the remaining 0.4% is private (Directorate General for Forests and Pastures, 1999). The objective for the near future is to increase the communal property so that it comprises 40% of the total forest area. The average size of private properties is 20 ha and that of common property is 1000 ha.

Administration and policies

In recent decades, Law No. 6607/1968 was the principal forestry law in Albania. Under the communist government in power at that time, it provided for a centrally planned system of forest management. The primary focus of this law was forest productivity. It did not specifically address other relevant aspects such as forest ownership and user rights, the role of the State in forestry activities or public participation in the forestry sector.

Nowadays, the Law on Forestry and Forest Police Service (Law No. 7223/1992) is the main law concerning Albanian forests. It replaces the former emphasis on production, with a greater accent on environmental issues and market-oriented, sustainable forest management. This law defines Albania's forest estate, whereas the State, communal and private forests are expressly considered part of it.

The DGFP has full responsibility for policy implementation as well as planning based on forest management plans. The management of State forests is the responsibility of the government's forest administration.

Communal forests, a new legally introduced category, are located on State land, but are handed over for the 'common use' to one or several villages or communes. Their management may be undertaken through agreements between forest and local authorities. Private forests are those that are already existing or planted on private land. Within the forest administration, a Forest Police Service has been created; officers have military police status and can carry arms.

Management planning requires the establishment of an annual allowable cut based on sustainability criteria. Forest national parks and natural reserves may be declared and designated by the Council of Ministers. Meanwhile, scientific reserves, natural monuments, landscapes and protected industrial woods may be declared and designated at a lower level by the DGFP. The DGFP is empowered to issue regulations on protected areas for which entry fees may be established.

The Law No. 7817/1995 on pastures and meadows was adopted and also deals with issues related to forestry. The Law regarding environmental protection (Law No. 7664/1993) is the principal law concerning environmental protection issues. It is especially relevant for forestry as it requires 'environmental licences' to be issued by competent authorities for certain activities that may have an impact on the environment.

The forestry administration is also responsible for wildlife management, which gained attention with heightened interest in hunting, following the political changes of 1991. The Ministry of Agriculture and Food (MAF) is mainly responsible for forestry. It includes the DGFP, which has administrative and financial autonomy. The DGFP is responsible for the production and protection functions of Albanian forests and pastures. Forest operations in the field are managed and carried out at the district and local section levels.

The Agency of Environmental Protection is the control authority for environmental matters. Based on the Law for the Revenue Account, the DGFP can now invest 60% of the total budget for the development of forests and pastures (natural regeneration, reforestation, watershed management, fire protection and non-commercial cutting). Also, different projects financed by international institutions and the Albanian government provide the possibility of increasing investment in this sector during the transition period of the national economy.

Contribution of the Forest to the National Economy

The contribution of forestry to the Albanian economy is very low in terms of market monetary value (0.03% of the GDP). Nevertheless, timber-based industry, including furniture and timber constructions, represents one of the main sectors of the national economy, accounting for some 5.11% of the GDP (Directorate General for Forests and Pastures, 1999; Statistical Institute, 2000).

During the last decade, the labour market in Albania was affected by a number of changes. The privatization process, followed by the closing down of many enterprises, has significantly reduced the employment level. By the end of 1992, it decreased by more than 26% compared with 1991. During 1989–1993, at least 34% of women and 23% of men already in employment lost their jobs. In 1994, the employment level rose slightly, but later on it decreased again, particularly after the crisis caused by the collapse of the fraudulent pyramid schemes. The share of the total labour force working in forestry (2.36%) is more significant than that of the timber-based industries (1.62%) and other forest-based industries, such as those processing non-wood forest products (NWFPs) (2.27%) (Directorate General for Forests and Pastures, 1999; Statistical Institute, 2000).

The trend of wood removals declined from 1.5 Mm^3 in 1991 to less than 0.25 Mm^3 in the following years. According to the Forest Cadastre (Directorate General for Forests and Pastures, 1999; Statistical Institute, 1999), national production of firewood obtained from the forest area is 200,000 m^3, while for timber production it is around 60,000 m^3.

The imports of firewood are not significant, while timber imports represent 25,000 m^3, or 31.2%, of the total timber consumption. In the overall wood market, imports make up only 9.3% of the total wood consumption.

The Values of Albanian Forests

Although the forests' contribution to the GDP appears of low importance, the role of forests in improving welfare is significant, especially due to the non-market benefits provided by forests. Table 16.3 presents the valuations of the different total economic value (TEV) components. It should be noted that there is a large gap in the knowledge of the different value types, both market and non-market.

Direct use values

Direct use values refer mainly to wood forest products (WFPs), represented by timber and firewood, traditional NWFPs such as cork and medicinal plants, as well as grazing and hunting. Valuation of WFPs, cork and medicinal plants is undertaken with reference to the quantities traded in the market and the average market price (Table 16.1).

Grazing is estimated on the basis of the number of forage units (FU) annually consumed and the imputed price of 1 FU. In Albania, there are 1,200,000 goats grazing on 600,000 ha of forests. The amount of fodder grazed yearly is around 300 FU/ha. Based on a price of €0.1/FU, derived from the market price of hay, the total value of grazing is about €18 million.

Valuation of hunting is based on the prices of hunting permits. In Albania, the State sells hunting permits and identifies the hunting areas. Enforcement and control are delegated to District Forest Services. Hunting benefits for foreign hunters is based on the permit price (€100/day), number of foreign hunters (550/year) and permitted days (four/year). The value of local hunting is estimated by means of the permit price (€10/hunter) and number of Albanian hunters (10,000). Consequently, the overall hunting value reaches €0.3 million.

Table 16.3. Values of Albanian forests.

Valuation method/output	Quantity	Value (000 €)
Direct use values		
Market price valuation[a]		
Timber (m³)	60,000	2,700
Firewood (m³)	200,000	3,000
Cork (t)	755	151
Medicinal plants (t)	20,000	360
Substitute goods pricing		
Grazing[b] (million FU)	180	18,000
Permit price		
Hunting[c]	10,550	320
No. of foreign hunters	550	220
No. of local hunters	10,000	100
Total direct use values		24,531
Indirect use values		
Cost avoided method		
Watershed protection[c] (ha)	500,000	3,000
Shadow pricing		
Carbon sequestration (tC)	120,000	2,400
Total indirect use values		5,400
Negative externalities		
Cost-based method		
Erosion, floods and landslides due to poor forest management (m³/ha)	33.7	−10,000
Losses due to forest fires (ha)	9,000	−9,000
Losses of forest natural quality due to illegal acts (no.)	4,700	−1,400
Total negative externalities		−20,400
Total economic value		9,531

[a]Source: Directorate General for Forests and Pastures (1999); INSTAT (1999).
[b]Source: Directorate General for Forests and Pastures (1999).
[c]Source: Directorate General for Forests and Pastures (2000).

Official statistics do not provide data related to other activities generating direct use values, such as beekeeping, as during the last 10 years small private business was practically non-existent. In addition, the value of recreation is difficult to estimate due to the lack of reliable data.

Indirect use values

From the wide range of ecological services provided by Albanian forests, only the values of watershed protection and carbon sequestration are estimated.

According to the Directorate General for Forests and Pastures (2000), there are 500,000 ha of land that would be exposed to a high risk of soil erosion if the present land use was changed. The cost-based method referring to the defensive expenditures employed at national level is used. This gives a monetary value of about €3 million, which can be assumed as the minimum social value attributed to this function.

The value of carbon sequestered by forests is based on the annual quantity stored by forest biomass of 120,000 tC (United Nations Economic Commission for Europe/Food and Agriculture Organization, 2000) and a shadow price of €20/tC (Fankhauser, 1995). This leads to a total value of carbon sequestration of about €0.2 million.

Negative externalities linked to forests

Several negative externalities linked to Albanian forests can be, to a certain extent, estimated in monetary terms: for example, erosion due to poor forest management, damage caused by forest fires and loss of the forest's natural quality due to illegal acts.

Erosion and other watershed disservices due to poor forest management are valued by using the replacement cost method. There are 200,000 ha of degraded/poorly managed forest area affected by an annual soil erosion of about 3 mm/ha (Directorate General for Forests and Pastures, 2000). The cost of dredging the eroded soil is assumed to be around €50/ha. This figure is based on different minimal

measures forecast for soil restoration. On this basis, a rough estimate for the average value of soil erosion can be calculated as €10 million. It should be added that erosion is an important issue in Albania. A recent risk evaluation study (Lako and Dano, 2004) shows that 70% of forest area is at high, or extremely high, risk of erosion, caused by human and natural factors. Table 16.4 shows the distribution of area according to the erosion risk level.

The damage caused by forest fires is estimated on the basis of the restoration cost method. The average forest area affected by fires during 1990–2000 was 9000 ha/year. The replacement cost necessary for restoring the burnt area is about €1000/ha. This gives a total value of €9 million.

The losses of forest biodiversity and landscape value due to illegal actions are estimated by using the amount of fines paid for the violation of Forest Law No. 7623/1992. A total of 4700 illegal actions were reported for which the payments amount to €1.4 million (Directorate General for Forests and Pastures, 2000).

Towards the Total Economic Value of Albanian Forests

This study highlights several important forest values in Albania. This picture is, however, incomplete due to the scarcity of data for all TEV categories. The estimated values are represented mainly by direct use values (€24.5 million), followed by indirect use values (€5.4 million) and negative externalities (€20.4 million). Option, bequest and existence values, although not estimated due to the lack of data, are increasingly important for society, and this is confirmed by the wide concern for establishing protected areas, which nowadays represent 5.8% of the total area of the country.

Figure 16.2 shows the estimated components of the TEV of Albanian forests. Grazing in forests is one of the most significant activities in rural areas, estimated at €18 million. Wood production accounts for €5.7 million; however it does not consider illegal wood cutting, which is a common practice in Albania. Watershed protection, estimated at €3 million, is based on defensive expenditure which might greatly

Table 16.4. Area of main land use categories according to the erosion risk level.

Land use category	Area according to the erosion risk level (%)			
	Slight	Moderate	High	Extremely high
Forest	2.5	29.9	51.7	15.8
Coniferous forest	2.7	28.5	58.3	10.5
Broadleaved forest	2.1	26.0	54.4	17.5
Shrubs	1.6	21.5	57.9	19.0
Open forest	3.7	42.2	41.8	12.3
Pasture	3.8	30.9	53.0	12.2
Agricultural land	11.9	57.8	24.2	6.0
Other lands	4.9	20.1	58.4	16.5

Source: Lako and Dano (2004).

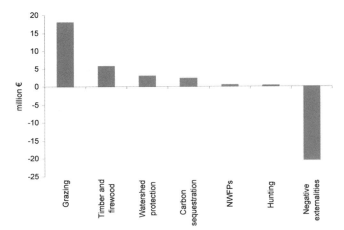

Fig. 16.2. Estimated values of the Albanian forests.

underestimate the true value of forest protection functions. Carbon sequestration in forest biomass is valued at €2.4 million. Negative externalities are represented mainly by the damage caused by erosion due to poor forest management, roughly estimated at €10 million. It should be noted that erosion is an important issue in Albania, caused by a number of factors such as excessive wood cutting and grazing in forests.

Conclusions

There are several important issues characterizing the situation of Albanian forests:

- Resources have been degraded significantly over the last 10 years. This has been caused mainly by important, but unrecorded, human pressure on forest resources (uncontrolled wood cutting and overgrazing).
- Investments in forest management have dropped considerably since the mid-1980s.
- Forests have various protection problems.

In general, Albanian forests have been used mainly for the production of WFPs and traditional NWFPs, principally oriented to the domestic market. The difficulties faced during the transition period to a free market economy were also felt through the poor management of the forest private sector.

A new strategy for the development of the forestry and pastures sector has been developed recently according to the Council of Ministers' decision (VKM) of 23 April 2004. It aims at re-establishing and restoring the natural balance between forestry and pasture resources, improving the sustainable management of the forestry and pasture production systems, ensuring an optimal contribution of the forestry and pasture sector to the economy, reduction of poverty, and sustainable development in the country.

References

Directorate for Protected Areas (2000) *Protected Zones*. DPA, Tirana.

Directorate General for Forests and Pastures (1985) *National Forest Inventory*. Forests Management Sector, DGFP, Tirana.

Directorate General for Forests and Pastures (1999) *Forest Cadastre Sector*. DGFP, Tirana.

Directorate General for Forests and Pastures (2000) *Forest Statistics*. DGFP, Tirana.

Directorate General for Forests and Pastures (2004) *National Forest Inventory*. Forests Management Sector, DGFP, Tirana.

Fankhauser, S. (1995) *Valuing Climate Change. The Economics of the Greenhouse*. Earthscan, London.

Food and Agrriculture Organization/Directorate General for Forests and Pastures/World Bank (1995) *Albanian Forestry Project*. FAO/DGFP/WB, Tirana.

Lako, T. and Dano, K. (2004) *Evaluation of Erosion Risk in Albania*. Albania National Forest Inventory (ANFI) Project, Tirana.

Regional Environmental Centre for Central and Eastern Europe (2000) *Annex 1. Country Report Albania*. REC Country Office.

Statistical Institute (1999) *Buletini Statistikor Vjetor/ Statistical Bulletin No. 4*. INSTAT, Tirana.

Statistical Institute (2000) *Statistical Annual Year Book*. INSTAT, Tirana.

United Nations Economic Commission for Europe/ Food and Agriculture Organization (2000) Global Forest Resources of Europe, CIS, North America, Australia, Japan and New Zealand (Industrialized temperate/boreal countries). *Global Forest Resources Assessment 2000*. Main Report, United Nations Publications, Geneva.

World Bank (2002) *Albania at a Glance*. World Bank, Washington, DC.

World Bank/Forest and Pastures Research Institute (1997) *Ecological Survey of High Forests of Albania*. DGFP, Tirana.

World Commission on Protected Areas (2000) http://www.iucn.org/themes/wcpa Access date: January 2004.

17 Croatia

Rudolf Sabadi[†], Dijana Vuletić and Joso Gračan

Forest Research Institute, Jastrebarsko, Cvjetno naselje 41, 10450 Jastrebarsko, Croatia

Introduction

Croatia has a heterogenous territory, as a result of its particular location at the junction of the Panonian, alpine–Dinaroides and Adriatic–Mediterranean spaces, within the northern temperate zone. It occupies 56,540 km^2 and is divided into the central part (34.8% of total area), the eastern part (19.6%), the mountainous region (14.0%), the northern littoral (10.8%) and the southern littoral (20.8%). The Mediterranean karst area stretches over 21,429 km^2, i.e. 38% of the total country area. Croatia has 4.38 million inhabitants, about 32% of which live in the Mediterranean-littoral area (Central Bureau of Statistics, 2001).

The country is shaped like an upturned 'V', with the apex pointing north-westerly. The southern branch, the longer and narrower, extends along the Adriatic coast between Kotor Bay – now in Montenegro, and Savudrija, at the Slovenian border, north of the Istria peninsula (Castellan and Vidan, 1998). Three regions can be distinguished:

1. The *Adriatic* or *Mediterranean region* comprises the Adriatic coast, with its 1185 islands, islets and rocks. The small coastal fields are cultivated with legumes and fruits, particularly vines and olives. The autochthonous vegetation is formed by Mediterranean plants. The climate is represented by mild winters, and warm, dry summers. Precipitations are rare, however, reaching 800 mm in the southern islands and up to 1500 mm in the Rijeka region. Due to the mild climate, the coastal region (Primorje) is densely populated; the regions around Rijeka, Zadar and Split exceed 200 inhabitants/km^2.

2. The *Panonian region*, located in the north, covers half of the country's area, with 66% of the population. It includes the huge plain between the rivers Sava and Drava, divided by hills and plateaus up to 500 m altitude. To the west and north of Zagreb, Croatian territory reaches the foot of the Slovenian mountains, barely higher than 1000 m. This region has a continental climate, with sharp differences between temperatures in winter and summer.

3. The *mountainous chain* of Dinara, situated between the Adriatic and Panonian regions, is well endowed with forests (beech and fir) in its western part. Climate is characterized by abundant precipitations exceeding 3000 mm. To the south-east, between the coast and inland, extends the beautiful mountain of Velebit. The whole region, except for the littoral part, is thinly populated, especially after the war of independence (1991–1995).

[†] Retired.

Forest Resources

Area

Croatian forests cover 2.5 Mha including productive and unproductive, covered and bare, fertile and unfertile areas. They form 43% of the country's area, according to the national forest management plan (Hrvatske Šume, 1998). This places the country among those in Europe relatively richly endowed with forests, with more than 0.5 ha of forest/inhabitant.

The country's plant biodiversity is rich and heterogeneous, containing about 4500 species and subspecies of vascular plants, almost 50% of which belong to the forest ecosystems. There are 260 autochthonous woody plants, among which more than 50 are economically significant. In addition, the vegetation comprises numerous species of bush, shrub, herbaceous plants, mosses and mushrooms, as well as a multitude of microorganisms within the soil. The forest communities, zoocenoses and habitats include more or less natural and stable ecosystems.

From the phyto-geographical viewpoint, Croatian forest vegetation can be divided into two regions: the Mediterranean region – with two belts and five zones – and the Euro-Siberian-North American region – with six belts and 15 zones. There are 68 forest associations and subassociations, from thermophytic and mesophytic to the cryophilic.

Forest soils are classified into auto-morphous and hydromorphous, and their formation was influenced by extremely different factors. The most productive soils are the auto-morphous luvisols, dystric cambisols, calceo-cambisols, brunipodsols and rankers, while less significant are podsolic colluvial, regosols and lithosols. Hydromorphous forest soils are dominated by eugley, semi-gley, pseudogley and fluvisol. The increasing water and air pollution caused by anthropogenic and technological factors has negative consequences on soil stability, with increasing negative influences on forest associations.

Croatian karst, or the Mediterranean-littoral region, including the Adriatic islands, is among the richest regions for endemic European flora. For example, the Velebit mountain littoral side contains 93 local endemic plants, Biokovo littoral slopes contain about 31 and the Konavli archipelago contains around 22.

Typologies

Of the total forest area, 84% is formed by productive forests and other wooded land, 14% is without tree cover – but most of it suitable

Table 17.1. Croatian forests according to ownership and structure.

Forest ownership	Structure of forests	Area (000 ha)	Growing stock	Annual growth
			million m³	
Public Corporation	Even-aged forests	890.6	190.3	6.0
Hrvatske Šume Ltd.	Uneven-aged forests	394.9	88.1	2.1
	Without management plan	330.0	—	—
	Subtotal 1	1,615.5	278.3	8.1
Other public institutions	Even-aged forests	10.1	1.4	34.8
	Uneven-aged forests	18.3	6.5	133.0
	Subtotal 2	28.4	7.9	167.8
Private ownership	Even-aged forests	393.9	34.8	1.3
	Uneven-aged forests	40.5	3.6	0.1
	Subtotal 3	434.4	38.4	1.4
Total	Even-aged forests	1,294.6	226.5	7.3
	Uneven-aged forests	453.7	98.2	2.3
	Without management plan	330.0	—	—
	Total	2,078.3	324.6	9.7

Totals might not add up exactly due to rounding.
Source: Hrvatske Šume (1998).

for afforestation – and the remaining 2% is unfertile.

Forest classification according to ownership and structure is given in Table 17.1. Total growing stock is 324.6 Mm^3, of which 88% belongs to public forests. The total annual increment is 9.7 Mm^3, i.e. 2.9% of the growing stock. Even-aged forests account for 74% of total forest area, 70% of the growing stock and 76% of the total annual increment. According to the *Rules of Forest Management*, during the next 10 years, some uneven-aged forests will be converted into even-aged stands.

Around 85% of the total growing stock and about 87% of the total annual increment is comprised of broadleaved stands and the remainder of conifers. In total growing stock, the largest category is beech (36.3%) and various oaks (27.4%) among the broadleaved; and fir (9.4%) and spruce (1.9%) among the conifers.

The actual annual cut in public forests for 2003 was planned[1] to be 3.6 Mm^3. The official data on annual cut in private forests are highly unreliable, since the owners of small forest holdings do not always possess cutting permits, to avoid tax payment. However, it is estimated that about 1.2 Mm^3 is cut annually in private forests, of which only a quarter is covered by permits. Therefore, the total annual cut is about 4.8 Mm^3, i.e. 50% of the annual increment.

Croatian forests can be divided into continental and Mediterranean forests. Of the *continental forests*, the even-aged stands

represent the most valuable part from the economic viewpoint. They occupy 730,600 ha and have a growing stock of 177 Mm^3 (Hrvatske Šume, 1998). Due to forest diseases, their growing stock has diminished below the normal value, causing further drops in the annual growth and allowable cut. The uneven-aged continental forests extend mainly in the western mountainous part of Croatia and are mostly exploitable forests. They cover 294,400 ha and have a growing stock of about 76 Mm^3 (Hrvatske Šume, 1998).

In addition, there are treeless areas, mostly including former pastures, covered by weeds or bush and envisaged for reforestation, as well as small areas of meadows within forests, envisaged as grazing for game and wild animals.

Mediterranean forests are located in the karst and littoral Croatian area, covering 1 Mha (Table 17.2). Their growing stock is 36.5 Mm^3 (28% of the total), of which 20% is provided by conifers and the remaining 80% by broadleaved. Even- and uneven-aged stands comprise 53 and 47%, respectively. The average annual increment of Croatian Mediterranean forests is 724,126 m^3 (1.35 m^3/ha), which is about 7.5% of the total Croatian annual growth.

The stocked area of Mediterranean forests (741,338 ha) is made up of high forests (11%), coppices (52%), *maquis* (4%), *garrigue* (2%), scrub (25%) and plantations (6%) (Hrvatske Šume, 1998). They play a multiple role, through their recreational, aesthetic, touristic, climatic,

Table 17.2. Area of karst and littoral forest region (Mediterranean forests).

Region	Forest land	Ownership			
		Hrvatske Šume Ltd.	Other public forests	Private forests	Total area
		000 ha			
Istrian-littoral	Stocked	219.1	7.7	91.1	318.0
Karst	Unstocked	87.8	0.0	0.0	87.8
Karst of Dalmatia	Stocked	330.2	4.5	88.7	423.4
	Unstocked	220.4	1.5	0.0	221.9
Total	Stocked	549.4	12.2	179.8	741.3
	Unstocked	308.2	1.5	0.0	309.8
Total area of Mediterranean forests		857.6	13.7	179.8	1051.1

Data elaborated from management plans of earlier administrative distribution of the Istrian–Croatian littoral, Lika region and Dalmatian karst region. As most parts of the plans were designed without preliminary forest management plans, many data will be most probably subject to later changes, which could cause slight differences in totals.
Source: Hrvatske Šume (1998).

erosion control and watershed protection functions, of which the most important is that of protecting the region's hydroelectric plants from sedimentation. Due to extensive and frequent forest fires[2] over large areas, Mediterranean forests have gradually become degraded. Of these forest types, *maquis* and *garrigue* are under high risk of degradation, while high forests and plantations perform their productive and protective roles well.

Figure 17.1 shows the map of forest vegetation according to Quézel categories (Quézel, 1981) and adapted to the Croatian peculiarities as follows:

- *Thermophilic wild olive and pistachio scrub* are scattered only in a few places on the islands of Vis, Lastovo, Mljet and Korčula, near Dubrovnik.

- *Mediterranean conifer forests of Aleppo pine, brutia pine and maritime pine* are present on almost all the islands along the coast.

- *Sclerophyllous evergreen oak forests of holm oak* occupy the greatest part of the area as a dense, high *maquis*, located only in places such as high forests. It is a typical transitional community of the northern part of the Mediterranean and it always develops in the border zone between evergreen and deciduous vegetation types.

- *Deciduous forests of zeen oak, afares oak, hornbeam, ash and occasionally beech* in Croatia are firstly a climax community in the sub-Mediterranean vegetational zone, growing at altitudes from 0 to 600 m. Through some river valleys, it penetrates

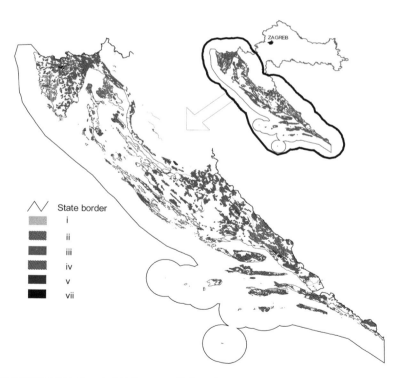

i) thermophilic wild olive and pistachio scrub   ii) Mediterranean conifer forests of Aleppo pine, Brutia pine and maritime pine   iii) sclerophyllous evergreen oak forests of holm oak   iv) deciduous forests of zeen oak, afares oak, hornbeam, ash and occasionally beech   v) mountain or high-altitude forests of cedar, black pine and firs   vii) non-Mediterranean forest type of dislocated forests of English oak (*Genisto elatae – Quercetum roboris*)

Fig. 17.1. Vegetation types of Croatian forests according to the Quézel classification. Prepared by V. Lindic, in cooperation with Mirta Co. Ltd and Oikon Co. Ltd.

deep into the continental background and, as enclaves and extrazonal forms, it extends abundantly into the interior of the country, too. Secondly, it is a clima-regional community of the epi-Mediteranean zone of the Mediterranean-montane vegetation belt, which is vertically connected with the *Querco-Carpinetum orientalis* forests. It grows at altitudes from 400 to 950 m.

- *Mountain or high-altitude forests of cedar, black pine and firs* can be found only on Biokovo mountain near Split and the Pelješac peninsula near Dubrovnik.
- *Non-Mediterranean forest type of dislocated forests of English oak (Genisto elatae–Quercetum roboris)* is common in Croatian lowlands and can be found only along the river Mirna in the Istrian peninsula, located on the north-western coast.

Institutional Aspects

Ownership and size of properties

About 81% of Croatian forest area is public property, while the remaining 19% is private. Most public forests are managed by *Hrvatske Šume* Ltd. (Croatian forests), a large State company, and the remaining public land is managed by other governmental institutions or authorities (Table 17.3).

Private forests are dispersed in numerous smallholdings which hardly exceed 2 ha/holding. Except for the stands in Gorski Kotar region, private forests are degraded and some of them are devastated. There are three reasons

for such a situation: the fragmentation of holdings, totalling 599,056 with an average size of plots equal to 0.76 ha; the overextended woodcut with the aim of widening agricultural areas and providing investment means for mechanization in agricultural holdings[3]; and the uncertainty of private ownership dominating for the 50 years of communist rule, which pushed owners to wood overcutting activities, regardless of the consequences for the biological state of their forests.

Administration

Croatia is divided into 12 administrative regions and 16 forest regional units, totalling 171 forest offices (Hrvatske Šume, 1998). The total forest area managed by these forest offices covers 2 Mha, of which 0.9 Mha is karst forest area and 1.1 Mha are continental productive forests. The area administered by each forest office is divided into several management units. These are formed by numerous forest departments and sub-departments and represent the basic management and planning areas. This division is done according to the forest type, condition, stage of development, soil configuration and natural borders such as waters, hill peaks and roads.

For each management unit, there is a general management plan that applies for 20 years, must be revised mid-term and provides general directions for the next 20 years. For the first 10 years, the management plan is very detailed; however, operative management plans are made on an annual basis. The revision of a management plan has to be approved by the Ministry of Agriculture and Forestry.

Table 17.3. Forest area ownership.

Ownership type	Stocked forest land	Unstocked forest land		Non-fertile forest land	Total forest land	
		Productive	Unproductive			
	000 ha					%
Hrvatske Šume Ltd.	1592.9	323.1	14.5	61.0	1991.5	80
Other public institutions	31.3	1.2	0.1	0.3	32.9	1
Private ownership	454.1	7.0	0.0	0.1	461.1	19
Total	2078.3	331.3	14.6	61.4	2485.6	100

Source: Hrvatske Šume (1998).

Contribution of the Forest to the National Economy

Gross domestic product and employment

The contribution of forestry is around 0.84% of the gross domestic product (GDP) (Central Bureau of Statistics, 2000). During the last decade, forestry employment registered a decreasing trend, from 12,000 before the war to 9850 permanent employees at present, i.e. 0.6% of total national employment. It is foreseen that it will be reduced further to 5000 permanent staff in the next 2–5 years. Employment in timber-based industries is around 13,000, i.e. around 0.8% of total employment. The situation of employment and production of timber-based industries is presented in Box 17.1.

International trade

Croatian foreign trade of wood is based on continental forestry and its products. There is no market for wood products of Mediterranean tree species (*Pinus nigra* and *P. halepensis*), except for firewood (*Quercus ilex*) which is sold to local people. Actually, the whole Mediterranean part of the forest area is characterized by low wood productivity and, therefore, by negligible commercial importance. Only 4.5% of the total annual cut is realized in the Mediterranean part of the forest area.

In 2000, Croatian foreign wood trade was based mostly on exports (Sabadi, 1994, 2001) of 0.9 Mt, at a value of €188 million, while imports

accounted for 0.3 Mt, or €81 million. In the same year, exports of wooden furniture were about €42 million, while imports of wooden furniture accounted for around €50 million.

Forest-based industries: tourism and green belts

Croatia has numerous national and nature parks, strict and special reserves, natural and horticultural monuments, protected landscapes and forest parks. In total, there are 280 continental and 76 Mediterranean forest sites with particular aesthetic value and/or historical significance, covering 589,200 ha, i.e. 7% of the country's land (Ministry for Environmental Protection and Physical Planning, 2002). Of this, national parks cover an area of 99,800 ha, 53% of which is occupied by the Mediterranean national parks of Paklenica, Mljet, Kornati, Brijuni and Krka. The national parks of Croatia are visited frequently by thousands of people, more than half of whom are foreigners. It is planned to establish eight new nature parks, of which four will be in the Mediterranean part of the country.

The Values of Croatian Forests

Table 17.4 presents the estimates of forest values grouped into direct and indirect use values, option, bequest and existence values as well as negative externalities linked to Croatian forest and its Mediterranean part. A more detailed

---

Box 17.1. Employment and production of timber-based industries during the war period.

Before the war, forest industries in Croatia utilized 86 large public sawmills, veneer and board mills, employing about 14,300 staff, and some 200 public companies with about 36,400 employees. In production and manufacturing of pulp, paper and other forest products, there were about 50 public companies employing about 10,000 persons.

In the same period, annual timber production consisted of about 0.25 Mm³ of conifer, 0.3 Mm³ of oak, 0.4 Mm³ of beech and about 0.17 Mm³ of other species. Intermediate processed timber was represented annually by about 48,000 m³ of veneers, 7000 m³ of plywood, 1000 m³ of blackboards and about 75,000–120,000 m³ of particle board. To these should be added an annual production of about 3.7 million pieces of furniture, 45,000 m³ of parquetry, 200,000–250,000 pieces of banded furniture, 150,000 m³ of joinery and about 22,000 m³ crates and palettes (without barrels). During and after the war, numerous primitive sawmills were erected, overall exceeding 400, of which only about 70 are worth mentioning.

Table 17.4. Values of Croatian forests.

Valuation method/output	Total quantity	Total value (000 €)	Value in the Mediterranean part (000 €)
Direct use values			
Market price valuation			
Timber for trade (Mm³)	2.3	188,260	3,000
Firewood (Mm³)	2.4	31,500	6,700
Net growth of standing timber stock (Mm³)	4.8	97,044	19,400
Forest fruits: chestnuts, walnuts, berries, other	—	7,583	3,000
Mushrooms (t)	1,000	2,000	1,000
Truffles (t)	20	1,600	1,600
Medicinal herbs and plants (t)	2,000	1,000	900
Fibrous materials (t)	2,500	1,000	1,000
Honey (t)	1,000	2,000	800
Surrogate market pricing			
Fodder and forage (t of fodder)	261	13,066	11,760
Consumer surplus			
Recreation (000 day-visits)	2,010	27,854	11,152
In forest reserves	10	83	43
In national parks	2,000	27,771	11,109
Hunting (no. of hunters)	47,000	10,105	2,500
Local hunting	42,000	9,030	2,258
Foreign hunting	5,000	1,075	242
Total direct use values		383,012	62,812
Indirect use values			
Cost-based method			
Water protection and erosion control (Mha)	2	20,780	12,470
Carbon sequestration (MtC)	8.9	178,000	24,920
In forest soils	2.7	54,000	7,560
In biomass	6.2	124,000	17,360
Total indirect use values		198,780	37,390
Option, bequest and existence values			
Tourists' potential use of recreational services		18,000	6,600
Locals' potential use of environmental services (forest reconstruction and protection)		131,400	43,800
Total option, bequest and existence values		149,400	50,400
Negative externalities			
Damage caused by forest fires (ha)	4,600	−2,683	−2,247
Total negative externalities		−2,683	−2,247
TEV		728,509	148,355

description of the methods and indicators employed is given in the following sections.

Direct use values

Several direct use values of Croatian forests, such as timber, firewood and forest fruits, are estimated on the basis of market price. Fodder and forage for grazing are valued by means of surrogate market pricing, while hunting and recreation are estimated on the basis of the contingent valuation method (CVM) and travel cost method (TCM).

Timber and firewood

Timber from Croatian Mediterranean forests has little importance for the wood industry, primarily because of the inadequate tree species for exploitation. Significant value is attributed to firewood, most of which comes from continental forests. Quantitative valuation of timber and firewood considers the annual cut with and without permits. Timber removals are about 2.3 Mm^3, of which around 5.4% come from the Mediterranean forests. Firewood production is about 2.4 Mm^3, to which Mediterranean forests contribute 80%. Based on the roadside prices[4], the monetary value of timber and firewood is €188.2 million and €31.5 million, respectively. The value of wood forest products (WFPs) in the Mediterranean part of Croatia is only €9.7 million.

Net growth of standing timber stock

Quantitative valuation is the difference between the annual increment and total annual cut (4.8 Mm^3). Estimated at half of the stumpage price, it gives a total value of €97 million. Of that, the estimate for the Mediterranean part corresponds to €19.4 million.

Forest fruits

In Croatia, there is a variety of forest fruits, firstly thanks to the close-to-nature silviculture, and secondly, due to the spread of berries – particularly blackberries – among other shrub vegetation in places where degradation progressed as a result of forest fires, clearcutting, overcutting and grazing (Šatalić and Štambuk, 1997). Monetary valuation is undertaken for chestnuts, walnuts, berries and others (miscellaneous).

On average, the harvest of chestnuts reaches 11,400 t/year which, valued at a market price of around €0.6/kg, gives a total value of €7.13 million. About 1.5 t of walnuts are picked from forests yearly, which leads to a total monetary value of €3000, considering an average price of €2/kg. It is supposed that 80% of the total quantity of chestnuts and about 30% of walnuts are sold on the market, the rest being self-consumed.

The most common forest fruits in Croatia are a wide variety of berries as well as apples, pears, cherries and, less often, medlars, which are mostly collected for self-consumption. However, it is estimated that only insignificant quantities of these products (100 t) are sold on the local market, with a total annual value of €150,000. Other forest fruits collected from the forests for marketing and self-consumption are estimated at about 500 t, having a total market value of €300,000. Adding up these figures, the total monetary value of forest fruits reaches €7.6 million, of which slightly less than half can be attributed to the Mediterranean forests.

Mushrooms and truffles

Mushrooms are generally harvested by individuals for subsistence or recreational purposes. Nevertheless, significant quantities – most of them growing in the forest, except for *champignons* – are sold seasonally on the green market throughout the country. There are no official data on the quantities of mushrooms harvested from the forests. However, it can be assumed that there are around 1000 t, out of which one-third is sold on the market and the rest self-consumed. The total value estimated for both uses at market prices reaches about €2 million, half of which is collected from the Mediterranean part of the country. In the Istrian peninsula, about 20 t of truffles are produced annually, of which around 90% is exported, mainly to Italy, and the remainder are consumed domestically. At a price of about €80/kg, it gives a total value of around €1.6 million.

Medicinal plants

Forest and forest soil in Croatia are inexhaustible sources of medicinal plants, which for centuries have been collected for local use, and in later times for industrial purposes.[5] It is assumed that about 2000 t of various medicinal herbs are collected annually from the forests. At a market price of approximately €0.5/kg, this gives a total annual value of €1 million, of which 90% belongs to the Mediterranean part.

Fibrous materials

A Mediterranean bush of Spanish broom, or gorse (*Spartium junceum L.*), was in earlier times in demand as raw material for production of cotton-like fibrous material[6]. It is assumed that about 2500 t of textile fibrous plants are collected annually from forests. At an average price of €0.4/kg, this gives a total value of €1 million, all of which is generated in the Mediterranean part of the country.

Honey

Of the value of honey produced in about 100,000 beehives in Croatia, 50% might be considered as being generated from forest and forestland. Taking 10 kg/year as the capacity of an average beehive and the price of €4/kg, the value of honey produced annually from forest and forestland is approximately €2 million, of which 40% is produced in the Mediterranean part.

Tourism and recreation

Access to forests is generally free, except for national parks, some natural parks and special nature reserves where entrance fees are required. Results of TCM surveys show an average consumer surplus of €2.85/ha for various forest sites with recreational value (Benc, 1997; Vuletić, 2002). Applying this benefit to the area of forest reserves and special reserves of 29,000 ha, it gives a value of about €82,650 (corresponding to €8.3/day-visit). National parks extending on 61,700 ha have the highest recreational value[7] in the country, of approximately €450/ha (€14/day-visit) (Benc, 1997; Vuletić, 2002), or €27.8 million overall. Based on these results, the total recreational value amounts to €27.9 million. Results of CVM applications show similar results. The recreational value of the Mediterranean forests is assumed to be equal to 40% of the total, calculated on the basis of the ratio of protected areas and the number of visitors.

Hunting

Income from hunting is a very important part of the national revenue. There is plenty of game in almost every part of Croatia, thanks to its relatively small population, to the extended forest cover and reserve areas and of course to the long-term care of the right number of game per specific area. This includes high ethical hunting standards followed by domestic and foreign hunters.

Valuation of hunting is undertaken by several methods, among which the CVM seems the most reliable.[8] The results of a CVM survey in Italy showed an average consumer benefit of €250/hunter (De Battisti *et al.*, 2000). This value is probably too high for Croatia, despite the large number of foreign hunters willing to pay for shooting trophies, in addition to other taxes and associated costs (accommodation). The benefit of €215/hunter seems closer to the Croatian reality.

The number of national and foreign hunters decreased during the recent war and will probably remain low due to the numerous landmines scattered almost everywhere over the area formerly held by Serb rebels. Considering that annually there are about 5000 foreign hunters and around 42,000 national hunters, the value of hunting can be assessed within the probability limits of ±20% at approximately €10.1 million. Based on the 25% ratio of game found in the Mediterranean part, the value of hunting in this area is €2.5 million. It should be noted that, in numerous scientific papers, game hunting is often overvalued. An objective estimate of this value is needed.

Fodder and forage

The use of forest and other wooded land for fodder and forage to feed domestic livestock was once a very important activity in Croatia, particularly in semi-arid areas of the littoral hinterland and in the southern and south-eastern mountainous regions. Grazing in the forests is forbidden by law, a rule very often ignored. Recently, the trend of raising goats has been revived, on the grounds that they prevent forest fires by clearing the lower layers of inflammable vegetation. Anyway, grazing in woodlands is generally tending to decline, as a result of more intensive animal husbandry. Besides, after the recent war, the marginal, poor areas of Croatia have practically been emptied: their population, previously relying on half-nomadic animal husbandry, has migrated to more fertile areas of the country. Thus the number and size of flocks of sheep have also decreased.

It is estimated that there are 40,000 cattle, 1500 horses, 50,000 sheep and 20,000 goats annually grazing in the forests. Grazing by cattle, horses and sheep takes place on 250 days/year, and by goats on 270 days/year. The daily consumption of forage (grass and leaves) is estimated to be as follows: 15 kg for cattle and horses, 5 kg for sheep and 8 kg for goats. The average market price of fodder is about €50/t. Based on these data, the value of grazing is €7.8 million for cattle and horses, €3.1 million for sheep and €2.2 million for goats. Adding up these figures gives a total value of €13 million, of which around 90% accrue to the Mediterranean part.

Indirect use values

Erosion control and watershed protection

About 55% of the total forest area, or 1.15 Mha, is located on soils which would be exposed to heavy erosion if no forest cover was available. The absence of the rest of forest cover would provoke only a moderate erosion level. Annual average precipitation in the country amounts to 1.13 billion m^3 water, one-third of which is retained by forests. This diminishes the water impact by about one-tenth, which is sufficient to control floods most of the year (Common, 1988). The monetary value of such retention power could be calculated by using the cost–benefit method in a 'with–without forest' situation, i.e. how much damage would cause a water impact of an additional 11 Mm^3 of rainwater throughout a year. Since there are no reliable studies to prove or reject the above assumptions, they should be accepted with the utmost care.

Furthermore, the existing forest cover protects the soil against erosion, sometimes up to 100%, and this should enter the same cost–benefit analysis, under the same assumptions as above. The forest has an enormous influence on water balance. Croatia has no great problems with the drinking water supply, even close to the relatively great urban agglomerations. In the European industrially developed countries, such as Germany, water protection costs amount to DM72/ha/year (Brill and Burness, 1994). In Croatia, such a cost may be regarded more moderately, probably at no more than about €10/ha/year (Prpić *et al.*, 1989; Prpić, 1992a,b). Thus, the water protection and erosion control function of forests may be estimated at €20.7 million. Of

this value, 60% is assumed to accrue to the Mediterranean part, based on the proportion of area with high risk of erosion occurring in this zone.

Carbon sequestration

Valuation of carbon sequestration is based on the quantity of carbon annually stored in forest biomass and soils of 8.9 MtC (Box 17.2) and a shadow price of €20/tC (Fankhauser, 1995). It gives a total value of €178 million, of which around 14% can be attributed to Mediterranean forests.

Option, bequest and existence values

Data from two CVM surveys provide information on the willingness to pay of tourists and local populations for the recreational benefits of Croatian forests. The first was conducted among tourists in Korèula forest during two tourist seasons (Sabadi, 1992a,b). Using spatial models to relate total visits to forest area (Krznar and Lindić, 1999; Krznar *et al.*, 2000; Vuletić *et al.*, 2000; Vuletić, 2000, 2001, 2002), it found a willingness to pay for recreational opportunities of about €9/ha of forests. Extrapolated to the total forest area, this gives an estimated value of €18 million (of which €6.6 million is for the Mediterranean forests). The second survey estimated the annual benefits to the local population of the Coastal Forest Reconstruction and Protection Project (Horak and Tadej, 1995; Horak and Weber, 1997; Horak *et al.*, 2001). It found a willingness to pay for protection from forest fires and reforestation of burned areas of €30/person, or €131.4 million, for the total forest area (of which €43.8 million is for coastal forests). Because forest use was impossible at the time of the surveys, we interpret these willingness to pay results as referring to option value.

Negative externalities linked to forests

Forest fires are the main negative influence on forests in the Mediterranean region of Croatia, affecting 70–90% of the country's burnt area (Hrvatske Šume, 2002). Valuation is based on the average burnt area (4600 ha) and the

---

Box 17.2. Value of carbon sequestration in forest biomass and soils.

The total amount of carbon (C) sequestered in Croatian forest biomass and soils is calculated as follows:

Total C content = C content in wood + C content in the above-ground vegetation + C content in dead trees and roots + C content in forest soil
C content in wood = growing stock × expansion factor to include branches, thicket, leaves and roots (1.45) × coefficient of transformation from wood into dry biomass × conversion factor of biomass into C (0.5)

The growing stock of Croatian forests is 324.3 Mm³ and refers to the above stump wood with a diameter larger than 7 cm. Adding up the volume of trees outside forests, i.e. 6.2 Mm³, one arrives at a total growing stock of 330.5 Mm³. Of this, the growing stock is 284.2 Mm³ for broadleaved, and 46.3 Mm³ for conifers, within and outwith the forests. The coefficient of transformation from wood into dry biomass is 0.37 for broadleaved and 0.55 for conifers. Based on these figures:

C content in wood = (46.3 × 1.45 × 0.37 ×0.5) + (284.2 × 1.45 × 0.55 × 0.5) = 125.74 MtC   (1)
C contents in the above-ground vegetation, dead trees and roots, and forest soil are estimated based on the productive forest cover (2,078,300 ha) and the following average indicators:
C content in above-ground vegetation = 1 tC/ha
C content in dead trees and roots = 5 tC/ha
C content in the forest soil = 150 tC/ha
C content in above-ground vegetation + in dead wood and
roots = 2.0783 Mha × 6 tC/ha = 12.5 MtC   (2)
C content in the forest soil = 2.0783 Mha × 150 tC/ha C = 311.7 MtC   (3)

Adding up (1), (2) and (3), the total C content in wood biomass and forest soil is 449.94 MtC. Applying an average growth rate of 2%, the annual quantity of carbon sequestered is 8.9 MtC.

Source: Willis and Benson (1988).

---

restoration costs, which in total amount to €2.6 million. Of that, €2.2 million are attributed to the Mediterranean forests.

Towards the Total Economic Value of Croatian Forests

At the present time, it is possible to evaluate and express in monetary terms only a part of the well-known values of forests. According to the accepted concept of total economic value (TEV), they are grouped into direct and indirect use values, option, bequest and existence values, and are presented in Table 17.4 together with the associated negative externalities.

Total use and non-use values of forests, market and non-market, conservatively estimated reach €728.5 million, of which only 20% accrue to the Mediterranean part (Fig. 17.2). Direct use values are estimated at about €383 million. They are mainly represented by WFPs (30% of the TEV), followed by net growth of timber stock (13%), recreation (4%), hunting (3%) and other NWFPs of minor importance.

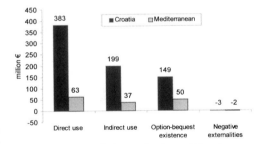

Fig. 17.2. Towards the total values of Croatian forests.

The indirect use values of Croatian forests are estimated at about €199 million and are provided primarily by the carbon sequestration function (24% of the TEV). The value of watershed protection is highly underestimated, referring to only one particular aspect for which data are available: protection of water supply. It is believed that afforestation and reforestation, in particular of the karst region, as well as of the agricultural land at and below marginal productivity, followed by forest fire control, will have significant positive environmental impacts. Not only will it increase the raw material supply for domestic industry and exports, but it will raise the value of water protection enormously, and also increase erosion control and carbon storage value. In turn, this will create prospects for extension of the existing hydroenergetic plants.

Option values for tourists and local populations, estimated at €149 million, account for some 20% of the TEV. Among the negative externalities, only the damage caused by forest fires is valued. This reaches €2.6 million and reduces the estimated TEV by 0.3%.

Conclusions

The research results give a comprehensive picture of the various benefits provided by Croatian forests. The picture is incomplete, however, due to the scarcity of data for several forest values. The results show the prevalence of WFPs in the estimated TEV. The value of carbon sequestration comes in second place, much of which is provided by carbon stores in forest soils. It is followed by recreation and tourism-related activities, which are very significant activities for this region today and represent major opportunities for the future development of the Mediterranean part of the country (Walsh et al., 1990). It should be noted that the protection functions of forests (water and watershed), though highly underestimated, remain very important for society. NWFPs, such as mushrooms and forest fruits, have only a minor importance compared with the above-mentioned values.

All in all, it should be stressed that there is a clear need for further research to supply the missing information necessary to achieve a more comprehensive TEV estimate of the Croatian forests. This information will be helpful if incorporated into the decision-making process aimed at improving the sustainable development of the country's forests.

Notes

[1]   It was 3.3 Mm^3 during 1986–1995, and 3.5 Mm^3 during 1996–2000.

[2]   According to the United Nations Economic Commission for Europe/Food and Agriculture Organization (2000), the annual damage to Croatian forests covers 27,000 ha and is due mainly to insects and diseases (56%), forest fires (41%) and, to a lesser extent, wildlife grazing.

[3]   A survey of private forests shows, besides devastation, an essential disturbed structure. The worst situation is in coppice forests of lowland pedunculate oak, where the clearcut has degraded the natural site, pushing the vegetation towards primitive species. At higher altitudes and on foothills, the dominating coppices of autochthonous tree species constantly show the increasing presence of the black locust (Robinia pseudoacacia).

[4]   By a government decision, the prices of wood products in Croatia have not been changed since 1996. Domestic wood industry and private sawmills buy the major part of production. A smaller amount goes on auction sale to domestic or foreign buyers. In that case, prices can be higher or sometimes lower than the conclusive one.

[5]   Among the most demanded medicine plants on the market, the following should be remembered: Digitalis purpurea, Atropa belladona, Aconitum napellus, Colchicum autumnale, Equisetum arvense, Potentilla formentilla, Alchemilla vulgaris and Matricaria chamomilla. Aditionally, there are more than 80 species of herbs and plants collected for medicinal use, from the arid and semi-arid littoral and mountainous part behind the Adriatic coast (typical Mediterranean), up to the humid, rainy northern and north-eastern part of the country (continental). Before the Second World War, the Dalmatian insect powder made from the flower of Chrysanthemum cinerariafolium was in demand as a high-quality and efficient insecticide. Later developments in chemistry pushed it out of the market while, recently, new ecological movements seem to want to bring it back. Consequently, on the whole, the quantity of medicinal herbs in Croatia could increase considerably in the future.

[6]   It is hoped that gorse will again be an interesting product for industrial use, provided that the abandoned technology the Italians (Callonia) developed at the end of the Second World War could be modernized. In the littoral and karstic part of Croatia, cultivation of gorse could be economically profitable. Reedmace (Typha latifolia and Typha angustifolia) is used locally as weaving material for baskets and mats. Common reed (Phragmites communis), rich in cellulose, is abundant in marshy terrains and serves only locally for thatched roofs or similar uses. Sedge (Carex caespitosa) is found throughout the world as a component of marshy habitats. A limited number of about 1000 species are used locally as food or bedding for animals, while a few are cultivated as

ornamental plants. Basket willow (*Salix viminalis*), almondleaf willow (*Salix amygdalina*) and purple willow (*Salix purpurea*) are used in large quantities as a weaving material for baskets, an important Croatian export item.

[7] In times of peace, each of the national parks of Plitvice Lakes, Paklenica and Krka waterfalls were visited annually by more than 500,000 people.

[8] Another valuation attempt is based on the prices of hunting permits sold to hunters by public authorities. Accordingly, the value of hunting, including various costs – wages of game keepers and other personnel, expenditures on armaments, ammunition, transport and insurance – is much lower, about €4.2 million. To a certain extent, this value underestimates the total value of hunting by not incorporating the benefits that accrue to illegal hunters.

References

Benc, S. (1997) Metodološka Primjerenost i Praktièna Iskoristivost Metode Cost–Benefit Analize u Ekošumskom Gospodarstvu Hrvatske. Methodological Suitability and Practical Use of a Cost–Benefit Method in Eco-forestry Economy in Croatia. Doctoral Thesis, Faculty of Economy, Zagreb.

Brill, T.C. and Burness, H.S. (1994) Planning versus competitive rates of groundwater pumping. *Water Resources Research* 30, 1873–1880.

Castellan, G. and Vidan, G. (1998) *La Croatie*. Presses Universitaires de France (PUF), Paris.

Central Bureau of Statistics (2000) *Statistical Yearbook*. Central Bureau of Statistics, Zagreb.

Central Bureau of Statistics (2001) *Statistical Report No. 1137*. Central Bureau of Statistics, Zagreb.

Common, M. (1988) *Environmental and Resource Economics: an Introduction*. Longman, London.

De Battisti, R., Val, A. and Rosato, P. (2000) *Il Valore Economico della Caccia nella Montagna Veneta*. Habitat, January.

Horak, S. and Tadej, P. (1995) *Evaluation of Forests in Croatian Coastal Tourism*. Institute for Tourism, Zagreb [in Croat].

Horak, S. and Weber, S. (1997) *Forests as Destination Attractiveness Factors: Case Study of Dalmatia (Croatia)*.Turizam Vol. 45, br. 11–12/97., str. 275–288, Zagreb [in Croat].

Horak, S., Marušić, Z. and Weber, S. (2001) *Coastal Forest Reconstruction Project in Croatia: the Aesthetic and Recreational Value of Croatian Coastal Forests to the Local Population*. Final Report, Institute for Tourism, Zagreb, pp. 3–47.

Hrvatske Šume (1998) *Long-term Management Plan of Croatian Forestry*. Zagreb.

Hrvatske Šume (2002) *Long Term Report of Croatian Forestry*. Zagreb.

Krznar, A. and Lindić, V. (1999) Methodology for evaluating the usefulness of health and landscape benefits of forests. *Radovi* 34, 103–118 [in Croat].

Krznar, A., Lindić, V. and Vuletić, D. (2000) Methodology for evaluating the usefulness of tourist-recreational benefits of forest, *Radovi* 35, 65–81 [in Croat].

Ministry for Environmental Protection and Physical Planning (2002) *National Parks and Nature Parks in the Republic of Croatia*. MEPPP [in Croat].

Prpić, B. (1992a) *O vrijednostima opæekorinih funkcija šume*. About other forest functions values. ŠL 6–8, pp. 301–312 [in Croat].

Prpić, B. (1992b) *Ekološka i Gospodarska Vrijednost Šuma u Hrvatskoj*. Šume u Hrvatskoj. Ecological and Economical Value of Forests in Croatia. Forestry Faculty in Zagreb, Hrvatske Šume, Zagreb, pp. 237.

Prpić, B., Rauš, Đ., Matić, S., Pranjić, A., Meštrović, Š., Seletković, Z., Lukić, N., Vukelić, J., Prebježić, P., Sklenderović, J. and Žnidarić, I. (1989) *Detaljna Studija za Hidrološku Sanaciju Šume Repaš – Šumarska Komponenta*. Detailed study for Hydrological Reclamation of Repaš Forest – Forestry Component. Institute for Research in Forestry, Faculty of Forestry, Zagreb, p. 84 [in Croat].

Quézel, P. (1981) Floristic composition and phytosociological structure of schlerophyllous mattoral around the Mediterranean region. In: Di Castri, F., Goodall, D.W. and Specht, R.L. (eds) *Ecosystem of the World. Mediterranean-type Shrublands*, Elsevier Scientific Publishing Company, Amsterdam.

Sabadi, R. (1992a) *Ekonomika šumarstva*. Forest economy, Školska Knjiga, Zagreb [in Croat].

Sabadi, R. (1992b) *Šumarska politika*. Forest politics, Hrvatske Šume, Zagreb [in Croat].

Sabadi, R. (ed.) (1994) *Review of Forestry and Forest Industries Sector in Republic of Croatia*. Ministry of Agriculture and Forest of the Republic of Croatia and Public Corporation 'Hrvatske Šume', Zagreb, pp. 1–120.

Sabadi, R (2001) Economic development in forestry and forest industries in Croatia from the establishment of the new Croatian state until the end of 2000. *Rad. Šumar Institute* 36, 61–89, Jastrebarsko [in Croat].

Šatalić, S. and Štambuk, S. (1997) *Šumsko Drveæe i Grmlje Jestivih Plodova*. State Service for Environmental Protection, Zagreb [in Croat].

United Nations Economic Commission for Europe/ Food and Agriculture Organisation (2000)

Global Forest Resources Assessment 2000. Main Report. United Nations Publications, Geneva.

Vuletić, D. (2000) *Ecological and Biological Parameters in Integral Forest Evaluation.* Proceedings of VIIth Croatian Biological Congress, Hvar 24–29 September 2000, Croatian Biological Society, Zagreb [in Croat].

Vuletić, D. (2001) Results of evaluating the social forests functions of island Mljet. In: *Science in Sustainable Management of Croatian Forests.* Forest Faculty in Zagreb, Forest Research Institute, Jastebarsko, pp. 579–586, Zagreb [in Croat].

Vuletić, D. (2002) Metode Cjelokupnog Vrednovanja Turistièkih i Rekreacijskih Usluga Šuma za Otok Korèulu kao Pilot Object. Methods of Complete Evaluation of Touristic and Recreational Forest Services for the Island of Korèula as a Pilot Object. Doctoral thesis. Forestry Faculty [in Croat].

Vuletić, D., Vrbek, B. and Novotny, V. (2000) The evaluation results of the benefits of the health and landscape forests functions. In: *XXI IUFRO World Congress, Forest and Society: the Role of Research,* 7–12 August 2000. Kuala Lumpur, Poster Abstracts Vol. 3., pp. 325, Malaysia.

Walsh, R.G., Bjonback, R.D., Aiken, R.A. and Rosenthal, D.H. (1990) Estimating the public benefits of protecting forest quality. *Journal of Environmental Management* 30, 175–189.

Willis, K.G. and Benson, J.F. (1988) A comparison of user-benefits and costs of nature conservation at three nature conservation reserves, *Regional Studies* 22, 417–428.

18   Slovenia

Robert Mavsar, Lado Kutnar and Marko Kovač
Slovenian Forestry Institute, Vecna pot 2, 1000 Ljubljana, Slovenia

Introduction

With a forest cover of 57% of the total land area, Slovenia is the third most forested country in Europe, after Finland and Sweden. Forests are therefore an essential feature and constituent part of the environment, and awareness of their protective and social importance is steadily increasing.

Slovenia is mainly a mountainous country. The occurrence and the structure of the forests are, therefore, to a great extent, shaped by the climate, parent rocks, soils and relief. In the north, Slovenia is enclosed by the calcareous Julian and Kamnik Alps, by the siliceous Karavanke Mountains and by the crystalline Pohorje and Kozjak Mountains. The southern and south-western parts belong to the Dinaric ridge. Since limestone of the Jurassic and Cretaceous period is the prevalent parent material, the area is strongly shaped by karstic processes. The eastern part descends over the tertiary hills of Slovenske gorice and Haloze to the Pannonian basin. A similar, hilly-like landscape can also be found in the Slovenian littoral area. Yet, these hills are not of the same geological origin, but are formed of Lias layers. A hilly central part of Posavje – formed of permo-carbonic schist and sandstone – is considered to be a part of the extreme south-eastern alpine boundary. Regions of plains are rather sparse and occur along the river network. The largest is the basin of Ljubljana. While its northern and central parts are formed of sedimentary materials, its southern part (Ljubljana's moorland) is predominantly loamy. Like the basin of Ljubljana, the basin of Celje is also formed of calcareous sedimentary matter. On the other hand, the plains, belonging to the Drava and Mura watershed, are formed of sand of magmatic origin (from the Central Alps) (Gregorcic *et al.*, 1975).

The macroclimatic types are influenced by the Mediterranean Sea, alpine mountain chain and Pannonian plain. The warm sea climate, penetrating from the coast, generally reaches the western Slovenian border where it is stopped by the first slopes of the hilly sub-Mediterranean interior and modified into the cool littoral climate. Highlands, such as the Julian Alps, the Karavanke and the Kamnik Alps, are characterized by a temperate subpolar climate. A transition from the sub-Mediterranean climatic zone to the continental zone is interrupted by the Dinaric mountain massif, which is characterized by an interferential climate. A large part of central Slovenia has a humid continental climate. However, when approaching the Pannonian plains, the climate changes rapidly, and becomes semi-arid continental (Gregorcic *et al.*, 1975).

According to Zupančič *et al.* (1987), Slovenia is divided into three large phytogeographic regions: the alpine–high-Nordic region; the Euro-Siberian-North American region and the Mediterranean region. The Mediterranean region occupies the south-western part of Slovenia, classified as the northern coast sector of the Adriatic province. From the floristic and vegetational criteria, the margin of Mediterranean climate influences is not well delimited, as only some sub-Mediterranean plant species grow and these are rarely true types. Only a sparse and fragmentary development of Mediterranean vegetation can be observed, while the sub-Mediterranean vegetation includes numerous elements from the Euro-Siberian-North American region.

Forest Resources

Area

According to the United Nations Economic Commission for Europe/Food and Agriculture Organization (2000), the forest area extends over 1,166,000 ha and occupies most of the littoral and continental regions. Despite rather favourable conditions, the forest cover has not always been so high. It began to increase around 130 years ago when the forest cover started to grow – from 737,000 ha (36.4%) in 1875 to 1.16 Mha (57.5%) in 1996 (Republic of Slovenia, 1996). The major reason for this was the abandonment of agricultural lands, which have been gradually overgrown by forest vegetation. Presently, efforts are being made to keep the extension of forest cover constant (Fig. 18.1). Nevertheless, as it is impossible to

halt the increasing trends in just a few years, a stable forest cover remains more of a desire than a reality.

Growing stock and annual increment have been increasing along with the rising share of forest area. Total growing stock increased by 87% during the last 50 years, while per hectare growing stock was augmented by 54%, reaching 266.4 m³/ha in 1996 (United Nations Economic Commission for Europe/Food and Agriculture Organization, 2000). During the same period, the current annual increment almost doubled, reaching 5.4 m³/ha (Ministry of Agriculture, Forestry and Food, 1997).

Typologies

Most of Slovenia's forests are located within a natural environment of beech, which – together with the beech–oak and fir–beech forests – covers about 70% of the nation's forestland (Table 18.1). As it is a mountainous country with a large rugged karst area, Slovenia also has a high proportion of difficult to access forests. This could be the main reason why forests have not undergone such rigorous human intervention as in most other Central European countries.

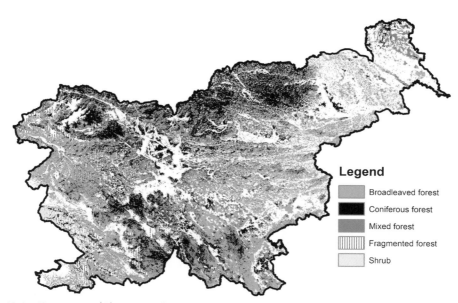

Legend
- Broadleaved forest
- Coniferous forest
- Mixed forest
- Fragmented forest
- Shrub

Fig. 18.1.   Forest map of Slovenia. ©Slovenian Forestry Institute.

According to various points of view, pertaining to matters such as tree species preservation, and vertical and horizontal stand structure, the forest situation can be said to be satisfactory. There are a total of 71 different tree species, ten of which are coniferous and 61 deciduous. The predominance of the broadleaved species in the natural forests can be seen in Fig. 18.2. The actual tree species composition is still far from ideal. While the ideal ratio in terms of growing stock between conifers and deciduous trees should be 20:80, the actual ratio is 49:51 (Republic of Slovenia, 1996).

The heterogeneous climate results in diverse vegetation patterns, which may change rapidly from place to place. Based on studies of the phyto-geographic regions (Wraber,

1969; Košir, 1979, 1994; Zupančič et al., 1987; Zupančič and Žagar, 1995), and on different geographic features such as geology, soil, relief, precipitation, temperature, potential vegetation, climate and phenology, the country was divided (Kutnar et al., 2002) into ecoregions and subregions as shown in Fig. 18.3 and described in Box 18.1. The syntaxonomy of forest vegetation is in accordance with Robic and Accetto (2001), while the characteristics of forest vegetation types are based upon Marinček and Čarni (2002).

Based on the classification of Quézel (1976), Slovenian forests under the influence of Mediterranean climate can be divided into the following vegetation groups:

- Sclerophyllous evergreen forests of oaks, including the holm oak forests (*Ostryo-Quercetum ilicis*), can be found only on a few scattered places such as Osp, Dragonja valley, Sabotin, and above the Hubelj spring. They cover the warmest places of south-oriented limestone rock walls.

- Hilly deciduous forests of oaks, hornbeam, ash and, occasionally, beech, including many different (sub-)Mediterranean deciduous forests in the south-western part of Slovenia, for example eastern hornbeam forests (*Q. pubescentis–C. orientalis*), hop hornbeam forests (*S. autumnalis–O. carpinifoliae*), different oak forests (*S. autumnalis–Q. pubescentis, Ostryo–Q. pubescentis, M. litoralis–Q. pubescentis, S. autumnalis–Q. petraeae, Carici umbrosae–Q. petraeae*), hornbeam forests (*O. pyrenaici–C. betuli, S. autumnalis– C. betuli*) and beech forest (*O. pyrenaici– Fagetum*).

Table 18.1. Distribution of forest types in Slovenia.

Forest species	Area (000 ha)	Proportion (%)
Willow and alder forests	8.1	1
Oak and hornbeam forests	94.6	8
Oak forests	36.6	3
Thermophilic deciduous forests	62.8	5
Pine forests	42.7	4
Beech–oak forests	124.7	11
Beech forests on carbonate parent rock	309.9	27
Acidophilic beech forests	194.4	17
Silver fir forests	53.3	4
Dinaric fir–beech forests	177.2	15
Norway spruce forests	16.8	1
Alpine forests	45.0	4
Total	1,166.0	100

Source: Republic of Slovenia (1996).

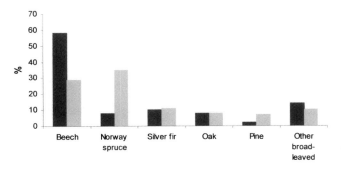

Fig. 18.2. The proportions of the most important tree species (percentage of growing stock). Source: Republic of Slovenia (1996).

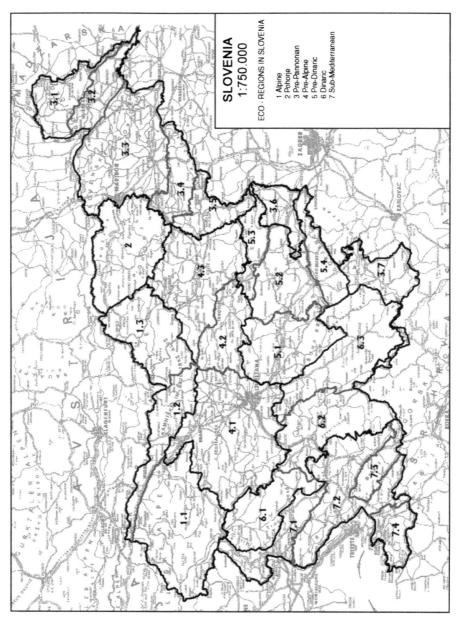

Fig. 18.3. Map of ecoregions and subregions in Slovenia. Reprinted from Kutnar *et al.* (2002), with permission.

Box 18.1.   Slovenian ecoregions and forest vegetation.

Sub-Mediterranean ecoregion (7)

In this region, limestone and flysch are the characteristic parent materials. Due to different geological and other ecological conditions, it is divided into four subregions: 7.1 Goriška Brda-Vipavska dolina; 7.2. Kras-Vremsko gričevje; 7.3. Brkini; and 7.4. Šavrinsko gričevje.

Goriška Brda-Vipavska dolina (7.1)

In this subregion, where the carbonate-rich flysch is the most diffuse parent material, *Ornithogalo pyrenaici–Carpinetum betuli* is a fairly common forest type. It covers the lower part of the hilly areas at altitudes from 90 to almost 300 m. Several forests of this type were felled for agricultural and urban purposes, while the more or less degraded coppice forests remained.

At higher altitudes, the alternation of main tree species results in the occurrence of submontane beech forest: *Ornithogalo pyrenaici–Fagetum*. Due to the warm climate, they do not make a continuous vegetation belt but retreat to shady sites providing more favourable moist conditions for the growth of beech trees. The parent material of this forest type consists of different flysch layers containing mainly marl and limestone covered by the deep eutric Cambisol.

Upon flysch parent material and in isolated instances of chromic cambisol (*terra rossa*), the sessile oak forest *Seslerio autumnalis–Quercetum petraeae* grows. It is found at altitudes of 100–400 m, on cold sites influenced by the Mediterranean climate, which is modified by a higher humidity and more constant temperature conditions. The trees are among the highest in this region, reaching on average around 15 m, while individual trees can exceed 20 m. These forests are among the highest quality forests in this subregion.

On hills with smooth slopes, on medium Eocene marl and tertiary flysch, *S. autumnalis–Quercetum pubescentis* is found. The slopes are interrupted sporadically by torrents and landslides along the watercourses. Generally, these are not well-managed forests, with shorter trees, well below 10 m in height. However, they assume an important protective role for soils against erosion and for landscape against the rigours of weather.

In the warm parts of the hilly and montane belt, on steep to very steep slopes, forests of *S. autumnalis–Ostryetum carpinifoliae* occur. The parent material is calcareous with rendzic Leptosols. These are secondary forests on primary sites of *Ostryo–Quercetum pubescentis* and of the *Seslerio–Fagetum* forests. Frequently, in communities of the *Seslerio–Ostryetum*, the tree layer is missing or is in the form of a low forest with the trees not exceeding 10 m. Their main role is the protection of the shallow karst soil, retention of rainwater and creation of barriers against the wind.

Kras-Vremsko gričevje (7.2)

This subregion is characterized by the laterite soils (*terra rossa*), large stony areas with shallow soils, and funnel-shaped craters, or Karst-dolinas, rich with deep soils. The most common forest type is *Ostryo carpinifoliae–Q. pubescentis*. It is a zonal forest community found in various positions and slopes on calcareous parent material. The forest stands take various forms of primarily mixed stands of coppice and seed source trees. Their main importance is for protection rather than for timber production, by safe-guarding the area against the rigours of weather and the soil against erosion. In the past, many sites of this type were used as pastureland. However, during the last two centuries, many of these sites have been planted with Austrian pine trees (*Pinus nigra*). After the Second World War, many of the pastures were abandoned, and have been gradually overgrown by the association of *O. carpinifoliae–Q. pubescentis*.

The Lias base and the plains, rich with deep and humid brown soils, are overgrown with well-preserved oak and hornbeam stands, while the montane parts are covered with beech forests – *S. autumnalis–Fagetum*. This is a zonal forest association that occurs in the montane belt of the littoral region up to 1100 m, in the coldest parts of the sub-Mediterranean region. It grows particularly on limestone and occasionally on dolomite, with shallow chromic cambisol and patches of rendzic leptosols. In certain areas, these forests also thrive on flysch parent material. The periphery of many forests of the *S. autumnalis–Fagetum* association has changed into secondary *S. autumnalis–O. carpinifolia* associations. On abandoned meadows of potentially natural *S. autumnalis–Fagetum*, the Austrian pine grows very well. On cold sites at low altitudes, the sessile oak forest *S. autumnalis–Quercetum petraeae* is present.

continued

Box 18.1. *Continued.*

Brkini (7.3)
In this subregion, the silicate-rich flysch is the most diffuse parent material, on which the azonal *Castaneo sativae–Fagetum* association is very common, known also as 'moderately acidophilous beech forest'. The soil is mainly medium deep to deep dystric cambisol. In exploited forests and on abandoned meadows, different *stadia* can be found, the most frequent being *Melampyro vulgati–Q. petraeae*. It grows in dry and warm places. These are mainly rural forests that were exploited excessively in the past for litter gathering and grazing purposes.

Šavrinsko gričevje (7.4)
In this subregion, the carbonate-rich flysch is the characteristic parent material. On hills of altitudes ranging from 300 to 500 m, with gradients of 10 to 25°, the oak forest *S. autumnalis–Q. pubescentis* is found. The slopes on this parent material are affected by torrents and landslides along the watercourses. Therefore, this forest type plays an important role in soil erosion protection.

A significant portion of the primeval forests of this subregion has been converted to a secondary association of *Molinio litoralis–Q. pubescentis*. However, it presents an important succession stage in the reforestation process of abandoned agricultural land in the coastal region. The primordial forest of *C. orientalis* can also be sporadically found in this subregion.

Dinaric ecoregion (6)
The Dinaric ridge separates the littoral and continental regions. It is characterized by a mixed fir–beech forest (*Omphalodo–Fagetum*, previous name *Abieti–Fagetum dinaricum*). It covers parts of high karst plateaus from Trnovski gozd to Kočevski Rog and Mount Snežnik, extending to the Gorski Kotar region in Croatia. It grows at altitudes from 700 to 1200 m. The parent material is mostly limestone with sporadic dolomite. The Dinaric fir–beech forests are economically the most significant forests in Slovenia.

In the lower part of the mountain belt, between 600 and 900 m above sea level, forests of *Lamio orvalae–Fagetum* occur. The mountain beech forests are also among the highest quality. From lowlands up to altitudes of around 600 m, the submontane beech forests of *Hacquetio epipactis–Fagetum* can be found. They are under the influence of intensive agricultural activities.

Pre-Dinaric ecoregion (5)
This zone extends from the central to the eastern part of the country. Here, *H. epipactis–Fagetum* and *Lamio orvalae–Fagetum* forests are very common, along with a variety of other forest associations. On gently configured hills and flatlands, at altitudes from 150 to 350 m, there are some remnants of *Abio albae–C. betuli* forests. The area was mostly deforested to create agricultural and urban areas, while the remnants are degraded and are mainly coppices.

On non-calcareous parent material, *C. sativae–Fagetum* forests grow. The majority of these moderately acidophilous beech forests have been changed either to monoculture spruce or to coppiced forest. In exploited forests and on abandoned meadows, different *stadia* can be found, the most frequent being *Melampyro vulgati–Q. petraeae*.

On dystric cambisol, acidophilous hornbeam forests of *Vaccinio myrtilli–C. betuli* occur. Like the majority of the forests in this hilly region, they have been intensively utilized due to their location close to settlements, to their easy access, as well as to favourable land configuration.

Alpine ecoregion (1)
This region consists of the Julian Alps, Kamnik-Savinja Alps and Karavanke. In the western part of this region, at altitudes of 900 m to the upper forest line of approximately 1500 m, beech forests of *Anemone trifoliae–Fagetum* association grow. In similar conditions, the mixed fir–beech forests *Homogyno sylvestris–Fagetum* can also be found. According to the vegetation pattern, they are similar to the Dinaric fir–beech forests, but not as rich in terms of floristic composition. Many beech and fir–beech forests have been converted to essentially pure monoculture spruce forests belonging to different associations, such as *Aposerido–Piceetum*, *Prenantho purpureae–Piceetum* and *Avenello flexuosae–Piceetum*. Some potentially natural sites of primary spruce forest, such as *Adenostylo glabrae–Piceetum* and *Rhytidiadelpho lorei–Piceetum*, can be found sporadically.

More towards the eastern part of the alpine region, at altitudes ranging between 900 and 1400 m, the so-called 'alti-mountain' beech forests of *Ranunculo platanifoliae–Fagetum* grow on carbonate parent materials. In the east, above 1000 m, secondary spruce forests of *Avenello flexuosae–Piceetum* extend over a large area, on non-calcareous parent material of potentially natural sites of acidophilous beech forests. At lower altitudes, on these parent materials, acidophilous beech forests of *C. sativae–Fagetum* association grow.

In the Alps and Karavanke, the upper tree line reaches 1700 m above sea level. The woody vegetation that can be found is dwarf mountain pine shrubbery (*Rhodothamno–Rhododendretum hirsuti* association), mixed with single larch (*Larix decidua*) trees.

Pre-alpine ecoregion (4)

This zone extends over the southern foothills of the Alps, Karavanke and the Pohorje Mountains. It is located between the alpine region (1) in the north, and the Dinaric (6) and Pre-Dinaric regions (5) in the south. It incorporates the Kranj-Ljubljana basin and the Celje basin. On these plain areas and gentle slopes, the azonal forest association of *Vaccinio–Pinetum sylvestris* grows. It is a pioneer association occurring mostly on poor sites. Much more common in this area are acidophilous beech forests of the *Blechno–Fagetum* association and the *C. sativae–Fagetum* association, which are growing on dystric cambisol on various non-calcareous geological formations. The first one is mainly found in the central part of Slovenia. Due to intensive exploitation of the *Blechno–Fagetum* forests, many have changed into secondary forests of *Leucobryo–Q. petraeae*, *Calluno–Q. petraeae* and *Vaccinio–Pinetum*. The majority of *C. sativae–Fagetum* forests are in the eastern and in the north-eastern parts of Slovenia. There are only a few well-preserved pure beech forests. In many cases, they have been changed either to spruce (*Picea abies*) monoculture forests, or to coppiced forests with a mix of sweet chestnut (*C. sativa*), Scots pine (*Pinus sylvestris*) and sessile oak (*Q. petraea*).

In this region, the azonal beech forests with *O. carpinifolia* cover large areas of the thermophilous sites with a relatively high amount of precipitation. *Ostryo–Fagetum* forests are found from the lowlands up to an altitude of around 1000 m.

In the so-called mountain belt, on dolomite and limestone of various geological formations, the *H. epipactis–Fagetum* and the *Lamio orvalae–Fagetum* forests are also very common. The former grows at lower altitudes than the latter, which, in turn, are well preserved and of higher quality than the beech forests closer to the lowlands.

Pohorje ecoregion (2)

This includes the crystalline Pohorje and Kozjak Mountains. In the lowland and hilly areas, mixed forests of beech, sweet chestnut and sessile oak (*C. sativae–Fagetum*) grow.

In the northern part of this region, on shady and very steep slopes at altitudes from 400 to 1000 m, *Galio rotundifolii–Abietetum* forests predominate. The parent materials consist of non-carbonate tonalite, gneiss, mica schist, amphibolite and others. The selectively managed forests prevail over the even-aged forests of this association. The *G. rotundifolii–Abietetum* forests are economically very important because they have a high growing stock.

Predominantly at altitudes between 900 and 1300 m, the so-called 'acidophilous alti-mountain' beech of the *Luzulo albidae–Fagetum* association occurs. The geological material consists of moderately acidic non-calcareous stones. These forests have always been under heavy human impact, such as clearing for meadows and livestock grazing. In the Pohorje mountain range, beech was exterminated locally, in the period of the glassworks. All these activities resulted in the complete predominance of Norway spruce (*P. abies*), which has evident influence on the sites. The secondary spruce forests were classified as the *Luzulo sylvaticae–Piceetum* association.

The *Cardamini savensi–Fagetum* is a zonal beech association of the upper parts of the Pohorje mountain range. It grows on silicate metamorphic and igneous rocks between 800 and 1300 m above sea level. In these forests, European silver fir (*A. alba*) and Norway spruce are mixed with beech.

Pre-Pannonian ecoregion (3)

This extends from Bela Krajina in the south-east, to Goričko and Prekmurje in the north-east of Slovenia. Forests cover only a small part of this region. Forests of hornbeam (*Carpinus betulus*) and oaks (*Quercus*

continued

Box 18.1. *Continued.*

sp.), which would be characteristic for lowlands and the hilly parts of Slovenia, were mainly cleared and the land has been used since then for agriculture.

On small areas along watercourses, the association of white willow (*Salicetum albae*), forests of black alder (*Alnetum glutinosae*) and pedunculate oak–ash forests (*Querco roboris–Ulmetum laevis*) occur. In the plains, which are influenced by high underground water and temporary flooding, oak–hornbeam forests (*Q. roboris-Carpinetum*) can also be found. These riparian forests are threatened since marshes and wetlands were intensively drained in the past. In many places, they have been converted to wet meadows. Where improvements took place, there are now agricultural areas.

In the northern part of this region, on the lower altitudes of hills, *Pruno padi-C. betuli* occur, while on higher altitudes, the zonal beech forests (*Vicio oroboidi-Fagetum*) grow. On the sunny side of the hills, at the medium-steep to steep slopes from 300 to 700 m above see level, moderately acidophilous beech forests with admixed sweet chestnut and sessile oak grow. They thrive on very different non-calcareous parent material. *G. rotundifoli-P. sylvestris* forests grow on luvisol and degraded soils.

In the southern part of this region, in Bela Krajina, the secondary association *Pteridio-Betuletum pendulae* is a result of the degradation of potentially natural forests of the association *Abio albae–C. betuli*.

- Mountain beech (*Fagus sylvatica*) forests of the *S. autumnalis–Fagetum* association, growing at 1100 m above sea level, extending from Banjščice to the Snežnik mountain chain. They grow on the coldest sites of the sub-Mediterranean region of Slovenia.

This vegetational distribution is often mixed with atypical forest types, which may cover considerably large areas. Reasons for this interchanging include the result of human actions as well as the highly diverse bedrock, relief and local climate. Due to this variation, silver fir, for instance, can be found in the acidic silicate rocks all over Slovenia from 200 to around 1200 m above sea level, as well as Scots pine forests in the plains of similar, and even poorer, edaphic conditions. On the alluvial soils – the most common in Prekmurje – large tracts of black alder may be found, as well as indigenous oak stands. Also, there are also some other *bio-coenoses*, which occasionally interrupt the landscape unity and contribute to the diversity of forest and non-forestlands.

Degraded areas

In spite of rather stable forests, which are due to well-preserved forest structures, there are many natural and anthropogenic disturbances affecting their health. It is reported that about 37,600 ha of forests are damaged or degraded annually, mainly due to abiotic factors, among other reasons (United Nations Economic Commission for Europe/Food and Agriculture Organization, 2000). Unfavourable weather conditions (such as drought, extreme winds and sleet) are the most frequent reasons for damage. Besides direct impacts, these factors can stimulate excessive development of insects or diseases.

Another factor having a detrimental effect on the forest's ecological stability is the impaired balance between vegetation and wildlife. Since an excessive population of herbivores seriously affects the forests' regenerative capacity, natural regeneration – being a prime condition of stable forests – may in some places become impossible (Republic of Slovenia, 1996).

Forest fires cause significant damage in the karst and sub-Mediterranean region. In the period 1991–1996, they damaged 4221 ha of forestlands. The main reasons were due to human activities, while only 2% of all fires were caused by natural forces (mainly lightning) (Jakša, 1996). Air pollution also has a negative effect on forest condition and stand stability. Severe damage is, in most cases, confined to particular locations close to thermal power plants and other industrial zones. However, despite such situations, forest health inventories show that degrading has slowed down within the last 5 years and that the health status has stabilized (Hočevar *et al.*, 2002).

Functions

Evaluation of factors other than timber production in the regional forest plans – which are produced throughout Slovenia – shows that at least one social or ecological function is significant in almost half of the nation's forests (Table 18.2).

As the country is poor with regards to other natural resources, the production of high-quality wood is one of the main objectives of forest management. Considering a highly variable relief with steep slopes[1], intensive precipitation and large karstic terrain, protecting against climatic extremes, landslides, torrents, floods and the like is one of the most important forest roles. Forests are also home to indigenous plant and animal species. Fourteen per cent of all endangered vegetation and 50% of animal species live in forests, among them large carnivores, which have been wiped out in most European countries. Furthermore, forests protect the supply of drinking water.

Additionally, forests provide a wide range of so-called social functions such as tourism, recreation, protection of cultural and natural heritage, education and research, and others. In cases where social and/or ecological functions are recognized as being important, forests with a special status can be declared. In these forests, the main objective is to ensure the protection and development of these functions.

Table 18.2. Distribution of the forest area according to the main forest functions.

Forest function	Forest area (%)
Productive	
Wood	55.6
Game	0.7
Other forest products	2.3
Ecological	
Protection	15.0
Habitat	5.6
Hydrological (water purification)	2.7
Climate regulation	2.1
Social	
Tourism and recreation	4.1
Natural and cultural heritage protection	3.8
Aesthetic	2.6
Education and research	2.0
Others	3.5

Source: Veselic (2001).

Institutional Aspects

Ownership and the size of properties

At the outset, it should be mentioned that the process of denationalization – which began in the early 1990s – has not yet been completed; therefore, further changes can be expected with regard to the structure of ownership and the size of forest property.

According to official statistics, in the early 1990s about 62.4% of forests were privately owned and 37.6% were publicly owned (Republic of Slovenia, 1996). In the past 10 years, the share of private forests has increased, due to the restitution of 7% of the forestland[2] to the former owners. Consequently, in 1996, 70.2% of forests were privately owned and 29.8% publicly owned (United Nations Economic Commission for Europe/Food and Agriculture Organization, 2000). It is estimated that at the completion of the denationalization process, about 80% of forests will be privately owned and about 15% will be State-owned (Ministry of Agriculture, Forestry and Food, 1998). Other forest owners will most probably be municipal authorities, the Roman Catholic Church or cooperatives.

Forest property is fragmented, having an average size of 2.3 ha, among the smallest in Europe. Moreover, forest properties are divided further into separate parcels with shapes – very long and narrow are common – which are often inappropriate for efficient management.

Privately owned forests are generally very fragmented, particularly in the Mediterranean region. One estimate is that the country has between 250,000 and 300,000 private forest owners. In 30% of the total forest area, forest properties are smaller than 3 ha and are of very limited economic importance to their owners (Fig. 18.4). Another shortcoming of highly dispersed ownership is that local foresters are obliged to cooperate with forest owners and, therefore, a great number of owners may mean that their efforts are less efficient. Conversely, such circumstances are favourable to the diversity of species composition and forest structure. Larger forest properties are found in mountainous regions, where forests represent an important source of income for the forest owners.

Administration and policies

Forests and forestry are the responsibility of the Forest Department of the Ministry of Agriculture, Forestry and Food. The Ministry is responsible for the forest policy (preparing the national legislation), adopting forest management plans, financing the public forest service and paying subsidies. In addition, it is also in charge of international relations.

The Slovenian Forest Service is a public forestry service. It is organized at the State level (central unit), regional level (14 units) and local level (94 local units and 430 forest districts). It employs about 700 forestry experts. According to the 1993 Forest Act, the Forest Service is responsible for forest protection (preparing forest fire protection plans), monitoring forest conditions and their development, providing guidelines for forest management, preparing forest management plans (at three levels), constructing and maintaining forest roads, advising forest owners and providing them with knowledge and training (courses).

The Chamber of Forest Owners was established in 1999 and is intended to represent forest owners and their interests. Membership is obligatory for almost all forest owners. The chamber will also take over some tasks that presently are under the responsibility of the Forest Service and are directly related to forest owners.

Forest policy

Forest policy is based on the 1993 Forest Act and it regulates forest protection, silviculture, exploitation and its use. In turn, the Act considers forests as natural resources in order to ensure their close-to-nature and multi-purpose management in accordance with the principles of protection of the environment and natural values, long-term and optimal development, so they can fulfil their functions. The Act also regulates the conditions of forestland and individual forest trees and groups of forest trees outside settlements in order to preserve and enhance their role in the environment (Republic of Slovenia, 1993).

Besides the Forest Act, in 1994 the Slovenian Parliament also adopted the National

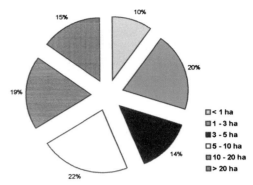

Fig. 18.4. The structure of private forests. Source: Republic of Slovenia (1996).

Forest Development Programme, which was prepared by the Ministry of Agriculture, Forestry and Food. Taking into account the natural principles of forest ecosystems, public interest, available State budget, as well as the needs and interests of forest owners, the Programme provides the basis for the conservation and development of all forests and their functions. It sets out strategies for the development of individual spheres of forest management and provides professional guidelines for cooperation with fields that interact with forestry in the landscape (Republic of Slovenia, 1996).

Forest management planning

Forest management plans define conditions for harmonizing the use of forests and interventions in the forests and forestlands, the necessary extent of silviculture and protection of forests, the highest allowable degree of their exploitation and conditions for managing wildlife (Republic of Slovenia, 1993). The plans are public and are conducted at three levels: regional and management unit; forest management plans; and silviculture plans. Regional and management unit plans are prepared by the regional branches of the Forest Service. Regional plans are prepared for forest regions and expire after a decade, for example 2000–2009. Plans designed for forest management units cover considerably smaller areas (between 2000 and 6000 ha), they are prepared successively (10% of the region's area annually) and they also expire after 10 years.

Once the plans have been prepared, they must be subject to a public hearing, the location and timing of which must be made known promptly to forest owners and the public. After the public hearing, the regional unit of the Forest Service must deal with the expressed comments and proposals and must do its best to include them within their proposals. On this basis, a new proposal of the forest management plan should be prepared. Regional forest management plans are adopted by the Government of the Republic of Slovenia, while the plans for forest management units are adopted by the minister responsible for forestry (Republic of Slovenia, 1998).

Contribution of the Forest to the National Economy

Gross domestic product and employment

Despite a high share of forests, Slovenian forestry is not an important economic sector. In 2001, the share of forestry in the gross domestic product (GDP) was only 0.36% (Statistical Office, 2002). Such a low economic contribution can be explained partly by the structure of forest ownership and partly by the GDP accounting methods. The fragmentation of private forests is one reason that causes forest estates to be of little economic interest for the owners. Moreover, with the increasing number of forest owners, private forest estates are becoming even smaller. Such a situation represents a serious obstacle to professional intervention in private forests, to the production of high-quality timber and to potential forest utilization.

The accounting system is the second reason for such a low share of forestry in the GDP. The official statistics evaluate only a part of the market products: timber and firewood. Non-wood forest products (NWFPs), such as honey, mushrooms and game meat, are assigned to other sectors, for example agriculture or food production, whereas non-market public goods and externalities are not accounted at all.

The change in forestry organization – initiated in the early 1990s – brought about a continuous decrease in the number of employees, which in 2000 fell to about 32% of that in

1991 (Winkler, 1998). According to national statistics (Statistical Office, 2001), this corresponds to only 1970 people. However, it should be mentioned that national employment statistics take into account only those employees who work in forest enterprises, of which the main activities are forest utilization and the timber trade. Individual private entrepreneurs, who often work as contractors, are also included in this figure. However, neither the farmers nor other forest owners who also work in forestry (e.g. for silvicultural work and logging), nor the people working at the Slovenian Forest Service (755 people in 2000) are included in this figure (Slovenian Forest Service, 2000). Accordingly, the total number of persons working within the forestry sector is significantly higher.

International trade: timber export/import

As shown by the Statistical Office (2001), the import of wood is greater than the export. In 2000, Slovenia imported 495,000 m^3 of timber and 2000 m^3 of firewood. Within the same year, the country exported 251,000 m^3 of timber and 62,000 m^3 of firewood. National wood production in 2000 was 2,253,000 m^3, of which 24% was for firewood and 76% for timber. The overall wood market was dominated by national wood production, while imported wood represented only 18%.

The Values of Slovenian Forests

The estimates of Slovenian forest values are presented in Table 18.3, divided into direct and indirect use values, option values and negative externalities. A more detailed description of the methods and indicators used for valuation is given in the following sections. While the physical estimates refer to the most recent year available, the monetary valuations are updated to 2001.

Direct use values

Slovenian forests provide wood forest products (WFPs) and a wide range of NWFPs. WFPs are

Table 18.3.   Values of Slovenian forests (2001).

Valuation method/output	Quantity	Value (000 €)
Direct use values		
Market price valuation		
Timber (m³)	1,603,000	88,607
Firewood for sale (m³)	539,000	12,050
Net growth of timber stock (m³)	3,958,000	87,996
Honey (t)	1,980	5,771
Decorative plants (Christmas trees) (pieces)	300,000	5,648
Mushrooms – commercial use (t)	839	3,948
Berries (t)	600	2,294
Chestnuts (t)	1,500	3,088
Game meat (t)	896	3,733
Game trophies (pieces)	13,000	6,060
Permit price		
Hunting	13,200	5,733
Total direct use values		224,921
Indirect use values		
Cost avoided method		
Watershed protection and erosion prevention (ha)	269,505	—
Shadow pricing		
Carbon sequestration (tC)	1,305,300	26,106
Total indirect use values		26,106
Negative externalities		
Cost-based method		
Loss of agricultural crops and cattle due to forest wildlife	—	−247
Total negative externalities		−247
TEV		250,787

considered as being timber and firewood. The value of annual production is assessed on the basis of the annual timber production in 1998 (Statistical Office, 1999) and average roadside prices. The net growth of timber stock is valued based on the quantity of wood increment net of removals and half of the average stumpage price, weighted by the actual cut structure (Krajčič, 2000).

NWFPs are considered to be honey, mushrooms, Christmas trees, berries, chestnuts, hunting trophies and game meat. Values for most of these products are assessed on the basis of average market prices. However, the value of some products such as Christmas trees, chestnuts and hunting trophies is taken from other other research (Hočevar et al., 1999).

Honey

The value of honey is based on the average quantity of honey produced by bees in the forests and the honey market price. It is estimated that the annual production of honey in 2000 was 2450 t (Statistical Office, 2001), of which about 80% was produced by beehives kept on forestlands. The market price valuation results in a total value of €5.7 million.

Mushrooms

In Slovenia, mushroom collecting is regulated by law. Though a permit for mushroom collecting has not yet become a legally required document, the quantity of mushrooms that may be picked in 1 day is limited to 2 kg/person. The physical valuation includes the quantity of mushrooms purchased by companies for further processing and the quantity commercialized on the local markets. It does not include the quantities collected for free. The monetary valuation, based on the market price, is about €3.9 million.

Game meat

An estimate is derived from the available data on wild animal species (such as red deer, roe deer, chamois and wild boar) shot in 1998 (Slovenian Hunting Association, 2000), average weights of species and the average prices of 1 kg of meat of particular species. The estimate also includes the game that was shot in hunting reserves[3].

Hunting

Wildlife and granting the rights for hunting is the domain of the State and is not bound to forestland ownership. The members of local hunting societies are allowed to hunt within hunting reserves, the area of which averages 4000 ha. Every member of the society is requested to pay an annual subscription of €100 and is obliged to volunteer at least 30 h a year to work within the reserve (for game feeding, mowing, game management planning and estate maintenance). Physical valuation of hunting considers the number of hunting societies' members to be about 23,000. As hunting is not bound exclusively to forestland, but may occur in much wider areas (such as grasslands and arable lands), the above figure is weighted by the share of forestlands (57.5%). The monetary value of hunting, based on the subscription rate and the value of volunteer work, totals €5.7 million.

Indirect use values

Considering the climatic and relief characteristics of Slovenia, the forests play an important role in providing a set of indirect benefits. Undoubtedly, the most important are water protection (flood prevention and water flow regulation, watershed protection, water purification and moisture conservation) and soil erosion prevention. It is estimated that around 269,505 ha of forests perform the watershed protection and erosion prevention functions. However, estimating this function in monetary terms is not possible.

Furthermore, forests represent a major carbon sink in Slovenia. Forest trees and soils sequester annually 1,305,000 tC (Box 18.2; Simončič *et al.*, 1999). Based on a shadow price of about €20/tC, the monetary value of carbon sequestration is €26.1 million.

Negative externalities related to forests

To obtain a complete picture of the total value of Slovenian forests, negative externalities should also be considered. As with the option, bequest and existence values, valuation studies are also scarce for regarding the negative externalities. Only the loss of agricultural crops and cattle due to forest wildlife can be estimated on the basis of the annual compensation payments for the damage caused by protected species, of €247,200 (Ministry of Agriculture, Forestry and Food, 1998).

Towards Total Economic Value of Slovenian Forests

On the basis of the available data, the TEV is estimated to be around €250.7 million. The largest share is ascribed to timber and net growth of timber stock, followed by indirect values (carbon sequestration), NWFPs and others (Fig. 18.5). Data scarcity allows only a partial estimation of the indirect use values and negative externalities, while option, bequest and existence values could not be estimated at all. A more comprehensive estimation of these values would probably significantly change the structure of the TEV of Slovenian forests.

Conclusions and Perspectives

In Slovenia, forests are considered to be the most valuable natural resource – a description that can be found in almost all documents

Box 18.2.   Calculation of the carbon sequestered by Slovenian forests and soils.

Verified IPCC methodology was used in 1998 to calculate emissions and absorptions of carbon dioxide (CO_2) by forestry and other land use changes in Slovenia. Calculating the quantity of carbon sequestered in forest biomass and soils involved the following steps:

1. Accumulation of growing stock
Step 1: quantity of carbon sequestered in the annual increment.

Year	A1 Increment – conifers (000 m³)	A2 Increment – broadleaves (000 m³)	B1 Conversion m³ ⇒ dry mass (dm) for conifers (kt)	B2 Conversion m³ ⇒ dry mass (dm) for broad-leaves (kt)	C1 Annual increment – conifers	C2 Annual increment – broadleaves	D Share of carbon in dry mass	E (C1 + C2) × D Carbon sequestration (kt C)
1996	3366	3744	0.4	0.6	1346	2246	0.5	1796

Step 2: quantity of dry matter from fellings.

		F Cutting (000 m³)	G Conversion m³ ⇒ dry mass (dm)	H (F × G) Biomass cut	H1 Total biomass cut	I Deforestation	J Forest fires (kt dm)	J1 (0.1 × H1) Unrecorded fellings (kt dm)
1996	Conifers	1512	0.40	605		247 ha	288 ha	
	Broadleaves	818	0.60	491	1096	32	17	110

Step 3: quantity of carbon sequestered in biomass.

	K (H1 + I + J + J1) Total biomass used (kt dm)	N Share of carbon	O (K × N) Released carbon/year (kt C)	P (E–O) C sequestration/ year (kt C)
1996	1255	0.5	628	1168

2. Land use change
Step 1: spontaneous afforestation of abandoned agricultural land.

	A Area under afforestation (000 ha)	B Biomass increment (t dm)	D Share of C in the above-ground biomass	E1 C sequestration/ year
1996	67	117	0.5	59

Adding up the results (P + E1) gives an estimate of 1,227,000 tC stored by forest biomass. In addition, the quantity of CO_2 sequestered in the soil due to the land use changes is 290 Gg CO_2, corresponding to 78,300 tC. Overall, the total quantity of carbon stored by forest biomass and soils is 1305 kt C.

Source: Simončič et al. (1999).

regarding forests and forestry. But is it true? Forestry in Slovenia can be considered as very progressive and ecologically oriented. For many decades, the concepts of sustainability and multi-functionality have been implemented in management practice and forest policy. However, public awareness regarding the environmental and social values of forests has increased only to a limited extent. For the majority of the general public, forests remain only as a provider of wood and some other by-products. How can we change this perception? As shown by this chapter, the TEV of the Slovenian forests has been only partially calculated. Data have been given in some detail regarding forest composition, roles and functions; however, estimates concerning the values of the goods and services originating from

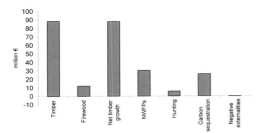

Fig. 18.5. Main components of the TEV of Slovenian forests.

forests are scarce. This should be taken as a warning for us (those working in forestry) to put more effort into deriving the necessary data and to make the results of such studies public. By public, this means not only the intended professional and scientific community, but also the general public. The possibility of presenting the total value of Slovenian forests could be a strong argument when trying to convince the decision makers how to plan future land uses and for how limited resources should be spent for investment projects. It may also help the forestry sector to improve its position in society and to improve the sustainability of forestlands.

Notes

[1] Over 50% of forest sites are highly unstable and vulnerable.

[2] Nationalized in the 1940s and 1950s.

[3] The value of game meat is not included in the price of hunting permits or volunteer work, so it has to be categorized separately.

References

Gregorcič, V., Kalan, J. and Kosir, Z. (1975) The geological and forest-vegetative image. In: *The Forests of Slovenia.* Borec, Ljubljana, pp. 26–62.

Hočevar, M., Behin, L., Jurc, M., Kovač, M., Ferlin, F., Kutnar, L., Cas, M., Bozic, G., Medved, M. and Ogulin, A. (1999) *Final Report: UN-ECE/FAO Temperate and Boreal Forest Resources Assessment 2000 for Slovenia.* Slovenian Forestry Institute, Ljubljana.

Hočevar, M., Mavsar, R. and Kovač, M. (2002) Forest conditions in the year 2000. *Zbornik gozdarstva in lesarstva* 67, 119–157.

Jakša, J. (1996) Size and consequences of forest fires in Slovenia between 1991 and 1996 and the place of forestry in fire suppression. *Slovenian Journal of Forestry* 55, 386–395.

Košir, Z. (1979) Ecological, phyto-sociological and forest management characteristics of the Gorjanci region in Slovenia. Research Reports. *Forestry Wood Science and Technology* 17, 1–242.

Košir, Z. (1994) *Ecological and Phyto-sociological Conditions of the Mountainous and Hilly SW Outlying Regions of Pannonia.* Zveza gozdarskih društev Slovenije, Ljubljana, Ministry of Agriculture, Forestry and Food.

Krajčič, D. (2000) Ratio between potential and yields in Slovenian forests. Research Reports. *Forestry and Wood Science and Technology* 61, 95–119.

Kutnar, L., Zupančič, M., Robič, D., Zupančič, N., Zitnik, S., Kralj, T., Tavčar, I., Dolinar, M., Zrnec, C. and Kraigher, H. (2002) The delimitation of the regions of provenance of forest tree species in Slovenia based on ecological regions, Research Reports. *Forestry Wood Science and Technology* 67, 73–117.

Marinček, L. and Čarni, A. (2002) *Commentary on the Vegetation Map of Forest Communities of Slovenia on a Scale 1:400.000.* Zalozba ZRC, ZRC SAZU, Biološki inštitut Jovana Hadzija, p. 79.

Ministry of Agriculture, Forestry and Food (1997) *Slovene Agriculture, Forestry and Food Industry in Figures.* Ministry of Agriculture, Forestry and Food, Ljubljana.

Ministry of Agriculture, Forestry and Food (1998) *Report on the State of Agriculture, Forestry and Food Industry for the Year 1998.* Ministry of Agriculture, Forestry and Food, Ljubljana.

Quézel, P. (1976) Les forêts du pourtour méditerranéen. In: *Forêts et Maquis Méditerranéens: Écologie, Conservation et Aménagemants.* Note technique MAB, 2, 9–33, UNESCO, Paris.

Republic of Slovenia (1993) Forest Act of the Republic of Slovenia. *Official Journal of the Republic of Slovenia* No. 30.

Republic of Slovenia (1996) The Forest Development Programme of Slovenia. *Official Journal of the Republic of Slovenia* No. 14.

Republic of Slovenia (1998) Regulation on forest management and silviculture plans. *Official Journal of the Republic of Slovenia* No. 5.

Robic, D. and Accetto, M. (2001) Overview of the syntaxonomical system of forest vegetation of Slovenia. Department for Forestry and Renewable Resources, University of Ljubljana.

Simončič, P., Kobler, A., Robek, R. and Zgajnar, L. (1999) *Assessment of GHG Sink for Forestry and*

Changes in Land Use – Project Report. Slovenian Forestry Institute, Ljubljana.

Slovenian Forest Service (2000) *Work Plan 2000*. Slovenian Forest Service, Ljubljana.

Slovenian Hunting Association (2000) *Statistical Data of Slovenian Hunting Organisations*. Slovenian Forest Service, Ljubljana.

Statistical Office (1999) *Statistical Yearbook of the Republic of Slovenia*. Statistical Office, Ljubljana.

Statistical Office (2001) *Statistical Yearbook of the Republic of Slovenia*. Statistical Office, Ljubljana.

Statistical Office (2002) *Statistical Yearbook of the Republic of Slovenia*. Statistical Office, Ljubljana.

United Nations Economic Commission for Europe/ Food and Agriculture Organization (2000) *Global Forest Resources Assessment 2000*. Main Report. United Nations Publications, Geneva.

Veselic, Z. (ed.) (2001) *Analysis of Development Possibilities of Forests and Forestry*. Slovenian Forest Service.

Winkler, I. (1998) Efficiency of forestry under the new social and economic conditions. *Slovenian Journal of Forestry* 56, 3–11.

Wraber, M. (1969) Pflanzengeographische Stellung und Gliederung Sloweniens. *Vegetatio* 17, 176–199.

Zupančič, M. and Žagar, V. (1995) New views about the phytogeographic division of Slovenia – I. *Razprave IV razreda* 26, 3–30.

Zupančič, M., Marinček, L., Seliškar, A. and Puncer, I. (1987) Considerations on the phytogeographic division of Slovenia. *Biogeographia – Biogeografia delle Alpi Sud-Orientali* 13, 89–98.

19 Italy

Lelia Croitoru[1], Paola Gatto[2], Maurizio Merlo[1,†] and Paolo Paiero[2]
[1]University of Padova, Centre for Environmental Accounting and Management in
Agriculture and Forestry (CONTAGRAF), Via Roma 34, Corte Benedettina,
35020 Legnaro (PD), Italy; [2]Department of Land and Agro-Forestry Systems, Agripolis,
Via Roma, 35020 Legnaro (PD), Italy

Introduction

Italy is a peninsula jutting out into the central
Mediterranean Sea covering 30.1 Mha, of
which 29.5% are forests. Around 60% of them
are mountain forests; the remaining 35% are
located in hilly areas and only 5% are situated
in the plains (Istituto Nazionale di Statistica,
1997a). From north to south, four broad geographical regions can be identified: the Alps; the
Po valley; the Apennines; and the two main
islands of Sicily and Sardinia.

- The Alps are mainly crystalline massifs in
  the central-western part and limestone
  massifs in the eastern part, which can reach
  elevations of over 4000 m. At high altitudes, they are covered predominantly by
  alpine conifer forests (spruce, Scots pine,
  larch and arolla pine) and, at a lower level,
  by beech mountain woods. The pre-alpine
  belt stretches along the southern edge of
  the Alps and is covered by deciduous
  forests of oaks, hornbeam, ash and
  chestnut.
- The Po valley extends to the south of the
  Alps, bordering the northern Apennines
  and the Adriatic Sea. The Po river flows
  across the centre of the plain and pours
  into the sea through a large delta. The

whole area is densely inhabited and intensively cultivated. A small discontinuous belt
of woodland runs along the river, which is
composed of poplar, holm oak and other
riparian species.
- The Apennines extend for over 1200 km,
  stretching from the Ligurian Alps down to
  the extreme south of Calabria, also including the northern Sicilian mountains. They
  are covered by forests of beech, black
  pines and firs, as well as deciduous mixed
  stands of oak, hornbeam and ash, while, in
  the south, forests of Mediterranean oaks
  can be found.
- Sicily and Sardinia are the largest islands in
  the Mediterranean. In this region, typical
  Mediterranean vegetation can be found:
  holm oak, cork oak and other thermophilic
  oaks prevail in the centres of the islands,
  whereas the coasts are dominated by
  maquis and garrigue vegetation types.

The diversity and distribution of forest species are influenced by the climatic variability
throughout the country. In the Alps, the long
cold winters and cool rainy summers favour
the growth of conifer forests. In the Po valley,
climatic conditions are more continental, with
cold, foggy winters, and warm, sultry summers,
becoming milder around the pre-alpine lakes.

† Deceased.

In the Apennine region, the sub-Mediterranean climate prevails, with cold snowy winters, dry summers and heavy rainfall concentrated on the Thyrrenian side. In Sicily and Sardinia, the climate is typically Mediterranean, with hot dry summers and low precipitation, which mainly occurs in the winter.

Italy has 56.3 million inhabitants, 45% of whom are concentrated in the north of the country, with the highest densities in urban areas (Istituto Nazionale di Statistica, 2001a). Recently, however, a gradual population decrease was felt in the large conurbations, mostly in favour of small and medium-sized towns. At the same time, the rural population is now registering a slight growth, especially in the centre and north-east, where modern industrial community life coexists with traditional forms of rural living.

Forest Resources

Area

According to the National Forest Inventory (NFI), forests cover 8.6 Mha, of which 97% are natural or semi-natural, and the remainder are plantations (Istituto Sperimentale per l'Assestamento Forestale e per l'Apicoltura e Ministero dell'Agricoltura e delle Foreste, 1985). This is based on a definition that a forest, to be classified as such, must extend over a minimum area of 2000 m^2, have a minimum width of 20 m and a crown cover of at least 20% of the area.

Differences from other sources need to be highlighted for a better understanding of these figures. The Istituto Nazionale di Economia Agraria (1998), quoting ISTAT (Istituto Nazionale di Statistica – National Statistical Office), reports a total forest area of 6.8 Mha for 1997, of which 98% are natural forests. These data are based on a more restrictive definition compared with that of the NFI: forest should extend over a minimum area of 5000 m^2, with a crown cover of at least 50% of the forestland. Besides, the area covered by Mediterranean vegetation types is not always included within the forest area. ISTAT data are derived from an annual estimation performed by the local forest services, originally based on the information provided by the Cadastre and subsequently updated according to individual valuation of Forest Service officers. At the same time, the NFI is based on standard statistical methods, as is the case with forest inventories of other countries. The difference of 2 Mha between the ISTAT and the NFI data is composed principally of mountainous and hilly farmland, mainly pastures and meadows, that have been abandoned in recent decades, leaving room for spontaneous reforestation.

The United Nations Economic Commission for Europe/Food and Agriculture Organization (2000) reports for 1995 an area of 10.8 Mha for total forest and other wooded land, of which 89% are natural forests. The difference between United Nations Economic Commission for Europe/Food and Agriculture Organization (UNECE/FAO) and NFI data might be due, at least to some extent, to further expansion of forests on abandoned land during 1985 and 1995. In addition, it includes the 'other wooded land' category, comprising large areas covered by Mediterranean vegetation types. On the other hand, the source of UNECE-FAO data is based on national data adjustments to the FAO definitions of forest area, which sometimes reflect reduced accuracy. Certainly, the above information and discrepancies should make it clear that the 1985 NFI needs to be updated, being a basic tool for any forest and environmental policy.

The main rationale for choosing NFI estimates is that, compared with the other sources, the NFI definition is most suitable for the situation of forest fragmentation of privately owned forests throughout Italy. It also provides consistency, with the whole range of data and information needed to outline the state of Italian forests.

Typologies

Italian natural forests are formed by high forests (25%), coppices (42%) and other formations including shrubs and *maquis* (25%), with an average growing stock of 211, 115 and 90 m^3/ha, respectively. Actively managed forests (generally high forest stands, but sometimes also coppices) provide the greatest part of marketable production, as well as a large share of public goods and positive externalities. Also, the ecological role of a poorly managed and abandoned forest – once almost unaccounted for – is now being reconsidered in terms of various functions, such as biodiversity and wildlife habitat. Plantations are composed mainly of poplars managed for industrial wood production: 20 m^3/ha average annual growth and almost 20% of total wood production.

Degraded forest area makes up 2.1 Mha (25% of total forestland), and is due mainly to forest fires (42%) and overgrazing (28%). This is the poorly/unmanaged forest area unable to provide the entire set of functions and often generating negative externalities. Strong degradation levels accrue to 29% of degraded area, 38 and 33% accounting for medium and slight levels of damage, respectively. The United Nations Economic Commission for Europe/Food and Agriculture Organization (2000) estimates a damaged forest area of 136,100 ha, quite clearly reflecting the annual damage, and not the overall status, of forest degradation.

An accurate representation of forest vegetation in Italy is given by Tomaselli (1973), based on the corresponding vegetational climax types proposed by Giacomini and Fenaroli (1958). Tomaselli's classification has been adapted to that proposed by Quézel (1976),[1] adding two more categories: the 'reforestations' and the 'other productive plantations'. The starting point of this representation is the 'plant association' (or groups of associations) representing a plant community whose flora is defined around an average value (Pignatti, 1994). Examples of associations include an oak forest, an alpine meadow and a riparian forest, all being recognizable through some characteristic species existing only in those phyto-coenosis. For example, the holm oak (*Quercus ilex*) gives its name to various coenoses of holm oak forest climax (*Quercion ilicis*), being the guiding species of the *Quercetum ilicis* association. The concept of 'association' is therefore essential for classifying the vegetation in a certain area: it allows the identification of the forest types which, in turn, identify specific ecological conditions, i.e. the environmental factors and their effects on living organisms.

Eight socio-ecological categories have been drawn in Fig. 19.1 and described in Box 19.1.

Functions

According to the NFI, 60% of the total forest area is managed mainly for production, 34% for protection and 5% for other environmental objectives. The concept of forest functions refers particularly to the main forest management objectives undertaken in Italy. However, the production function should be understood within the limits set by the 1985 Landscape Act, on the basis of which removals must respect the existing natural capital. Felling is allowed only to the extent of its compatibility with the conservation of the forest ecosystem: only management or, better, 'care' felling (*taglio colturale*) is allowed. For instance, removals account for one-third of the growth (8 Mm^3 compared with 25 Mm^3 growth), confirming that, in practice, the general objective of Italian forest management implies various forms of protection.[2] Another significant fact is that the low level of removals is due not only to conservation, but also to extensive forest abandonment, particularly in privately owned forests, as a result of the high costs of management and exploitation, and therefore low profitability of wood.

Institutional Aspects

First of all, it should be mentioned that forest property structures are rather static, as no evident changes took place during the second half of the 20th century except for a certain enlargement of public properties. Private owners abandon the forest, but rarely sell it; meanwhile public forests (State, local authorities and common properties) in general cannot be sold due to the law and traditional custom.

Ownership and size of properties

According to both the Istituto Nazionale di Statistica (1997a) and NFI (Istituto Sperimentale per l'Assestamento Forestale e per l'Apicoltura e Ministero dell'Agricoltura e delle Foreste, 1985), 66% of the forest area is private and 34% is public. More than 90% of the private forest area is owned by individuals. There are around 700,000 properties under private

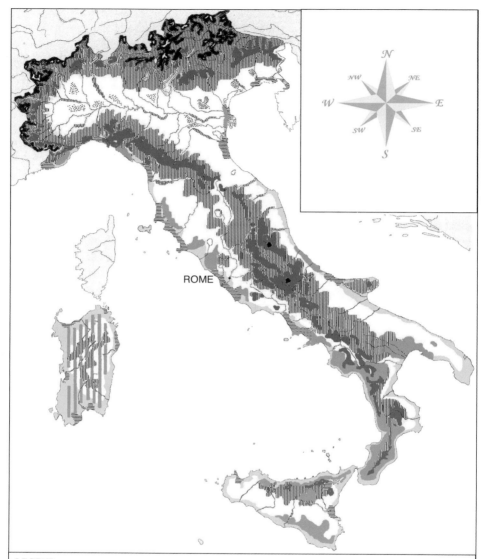

LEGEND

Mediterranean coenosis of *maquis* and *garrigue* including rock-adapted and psammophilous vegetation – corresponding to thermophilic wild olive and pistachio scrubs. Mediterranean conifer forests of pines and juniper *maquis* – corresponding to Mediterranean conifer forests of Aleppo pine, brutia pine, stone pine, barbary thuya and Phoenician juniper. Mediterranean sclerophyllous evergreen forest of holm oak and other thermophilic oaks – corresponding to sclerophyllous evergreen oak forest of holm oak, cork oak and kermes oak. Deciduous forests of oak, hornbeam, ash, riparian poplars and chestnut – corresponding to deciduous forests of zeen oak, afares oak, Lebanese oak, tauzin oak, hornbeam, ash and occasionally beech. Mountain forests of beech, black pine and occasionally fir – corresponding to mountain and high-altitude forests of cedar, black pine and fir. Other non-Mediterranean forest types (alpine conifer forest of spruce, Scots pine, larch, arolla pine). Oro-Mediterranean forest of mountain bush with juniper, heather, green alder, dwarf pine and occasionally willow, birch, oak – corresponding to oro-Mediterranean stage stands of arborescent juniper and thorny xerophytes. Productive plantations (poplar, gum, pine, hazel, locust)

Fig. 19.1. Map of Italian vegetation according to the adapted Quézel classification. Prepared by Paolo Paiero.

Box 19.1. Italian forest types according to the Quézel classification (Paolo Paiero).

According to the Quézel classification, Italian forests can be divided into eight categories:

(i) Mediterranean coenosis of *maquis* and *garrigue*, including rock-adapted and psammophilous vegetation (estimated area 1.1 Mha)
Starting from southern Italy and the islands, this category occupies a wide land strip, stretching, though at a diminishing rate, up to the northern Adriatic where it meets similar environments of the Istrian and Dalmatian coasts. It includes the coasts and the inland of Sicily and Sardinia up to 300–500 m; the Ligurian riviera and the Tyrrhenian and Adriatic coasts from Terracina to Gargano up to 200–250 m on the well-exposed mountain-sides.

This category is mostly represented by true *maquis* vegetation, made up of evergreen xerophyllous woody plants and the *garrigue* of thorny shrubs and therophytes. Where soils are not too degraded, the most important *maquis* components are oleander and carob (*Oleo-Ceratonion*), together with the most xerophilous evergreen oaks (*Q. ilex* and *Q. coccifera*), which are, however, less frequent. Two types of *maquis* can be found in Italy, the high *maquis* dominated by holm oak and sclerophyllous shrubs, such as the strawberry and mastic trees, and the low *maquis* with various *coenosis* characterized by cistus, euphorbia, dwarf palm and broom.

The *garrigue* occupies the most degraded areas of the southern Italian coasts. Colonization occurs discontinuously by means of thorny oak shrubs (*Q. coccifera*), prickly broom (*Ulex europaeus*) or other xerophyllous Mediterranean components that sometimes mix with graminaceous tufts (*Bromus* sp.pl.) or cushions of *Satureja*, *Helichrysum* or other composites. These are the only species, together with the aromatic rosemary (*Rosmarinus officinalis*), sage (*Salvia officinalis*), lavenders (*Lavanda stoechas, L. spica* and *L. officinalis*) and thymes (*Thymus capitatus* and *T. vulgaris*), able to colonize the rocky, arid and sunny limestones. Only the helichrysum (*Helichrysum italicum*) also spreads on to the siliceous rocky strips, anticipating the steppe with asphodel (*Asphodelus microcarpus*) and cistus (*Cistus incanus, C. creticus, C. monspeliensis* and/or *C. salvifolius*).

Along the coasts of Sardinia (the area around Sinis), in south-western Sicily, in the swamps of Migliarino in Tuscany, in the Po delta, near the Venetian lagoon, Marano and Grado, a halophilous vegetation colonizes the salty back-dune areas or, more rarely, the inner zones separated from the sea.

(ii) Mediterranean forests of pines and juniper *maquis* (estimated area 570,000 ha)
Within the same coastal strip where *maquis* grows, pine forests can also be found. Though often of artificial origin, these pine woodlands have a high landscape value, extending along the Thyrrenian up to the northern Adriatic coasts. Among the various pines, the most important are the stone pine (*P. pinea*) and the maritime pine (*P. pinaster*), followed by the Aleppo pine (*P. halepensis*) and the Calabrian pine (*P. brutia*) in the south and the Austrian black pine (*P. nigra* var. *austriaca*) on the northern Adriatic coasts. Their artificial origin is historically recorded, it being said that the stone pine was found in the ancient Roman gardens and farms where it was introduced from Portugal. It seems, however, that its origins might go back to Etruscan times.

These pines sometimes colonize the lowest slopes of the Apennines, such as in Liguria along the Riviera of Ponente, in Tuscany (stone and maritime pine), in Sardinia on Mount Limbara (maritime pine), in Puglia on the Gargano (Aleppo pine), in Sicily on the Peloritani Mountains (stone pine) and near Trieste (black pine). The most thermophilic species is the Aleppo pine, located in strictly Mediterranean areas, followed by the maritime pine, which can be found in both mountain and Mediterranean areas, e.g. Corte or Hamilton pine (a variety of maritime pine), which, in Corsica and Sardinia, can also be found above 1500 m, and the stone pine which finds its optimum in areas not too far from the sea.

Moving from the Thyrrenian to the Adriatic coast, the maritime pine can be found in some areas of the Ligurian Apennines, connected with the pine forests of the neighbouring province. In Tuscany – Viareggio, Migliarino, San Rossore and Cecina – the stands of maritime and Aleppo pine are close to the sea, while the stone pine is found more inland. In Lazio – Fregene, Fiumicino, Castelfusano and Castel Porziano – similar forest stands are notable. Meanwhile, in Campania, from the Vesuvio layers, the pine extends to Pompei and the Sorrentina peninsula. Ascending the Calabrian and Puglia coasts, stone and maritime pine forests can be found along the Lesina and Varano lakes (north of Gargano). Other noteworthy pine woodlands in northern Italy are: the stone pine woods of Cervia and Classe (Ravenna);

continued

Box 19.1.   *Continued.*

the stone and black pine woods of Aquileia and Grado; and the Aleppo pine woods on the Trieste and Istrian coasts.

Some coastal strips are diversified by the presence of juniper *maquis* (*Juniperus oxycedrus* and *J. phoenicea*), as frequently occurs along the Thyrrenian coast, Calabria and particularly in Sardinia, where the red juniper dominates.

(iii) Mediterranean sclerophyllous evergreen forests of holm oak and other thermophile oaks (estimated area 1.5 Mha)

The guiding species of the evergreen oak forests is the holm oak (*Q. ilex*), which is the most robust woody plant colonizing the rocky slopes of all the Mediterranean area. The holm oak can be found all along the coasts, either as a tree or as a shrub, forming, by itself or together with other sclerophyllous species, *maquis* and sometimes forests. Due to its coriaceous leaves, the holm oak can adapt to the warmest and most arid environments of the peninsula, also at a certain height (600–1000 m; 1400 m in the Aspromonte) both in the Apennines and in the central-eastern pre-Alps (Adige Valley, Colli Euganei, Lago di Garda, Carso and Friuli). The holm oak is the typical species of the phyto-climatic area of *Lauretum* (De Philippis, 1937). Meanwhile, in certain regions (Sicily and Calabria), it also has great economic importance, managed both as high forest and as coppice. Other oak species with similar ecological needs can be found in holm oak woodlands: the cork oak (*Q. suber*), more demanding in terms of soil moisture and also more thermophilic, and the kermes oak (*Q. coccifera*), definitively the most thermophilic of all Mediterranean oak species, often in a bushy form. In the most favourable areas, other oak species may also appear, such as the valonia oak (*Q. aëgilops*), the Macedonian oak (*Q. trojana*) and the European Turkey oak (*Q. cerris*).

The structure of these woodlands, only rarely actively managed, is rather complex. The tree layer, dominated by holm oak, sometimes associated with other deciduous species, such as the pubescent oak (*Q. pubescens*) and the Montpellier maple (*Acer monspessulanum*), can reach 15–18 m. The lower layer is made up of a very thick undergrowth of *maquis* shrubs, amongst which feature the strawberry tree (*Arbutus unedo*), the phillyrea tree (*Phyllirea media*), Mediterranean buckthorn (*Rhamnus alaternus*) and laurustinus (*Viburnum tinus*), in addition to other typical components of shady environments, such as black bryony (*Tamus communis*), clematis (*Clematis flammula*), honeysuckle (*Lonicera implexa*), smilax (*Smilax aspera*) and the evergreen rose (*Rosa sempervirens*).

In some areas – such as Gallura, northern Sardinia – the holm oak is replaced by the cork oak in the upper layer, while an intricate sublayer of holm oak, associated with various light-seeking graminaceus species taking the place of lianas, can be found.

(iv) Deciduous forests of oak, hornbeam, ash, poplars and chestnut (estimated area 3.7 Mha)

This is the most widespread forest category in Italy, with greatly variable aspects and composition. It ranges from the temperate-fresh stands, in the pre-Alps and in the Po plain (with mixed oak and hornbeam stands and xerophilous variants such as riparian forests of poplars and willows), to the temperate-warm stands in the well-exposed sites of the Apennines and islands, such as mixed oak forests, chestnut and European Turkey oak stands, including European ash and the mixed stands of plane and willow of Sicily.

This varied and wide forest cover is found in the flat area of the Po plain, in the foothills of the northern pre-Alps and Apennines up to 800–1000 m, in the central-southern Apennines up to 1000–1200 m, and in the Sicilian mountains from 200 to 1400 m.

The widest areas are covered by the meso-thermophilic oak stands and chestnut forests in the north and by the pure/mixed Turkey oak stands in the centre-south. Smaller areas are covered by riparian poplars – generally along the water courses – alder stands (river banks) and plane stands (river/seashores and sheets of water in eastern Sicily). Given the width of the area of mixed deciduous stands – and the fact that they are displaced within very diversified climatic areas – their composition changes substantially according to their geographical location. The forests of durmast and/or pubescent oak represent, on the initial pre-Alps, the subthermophilic variant of the mesophyllous bay oak and/or durmast oak forest. All these three types are mixed forests, rich in species composition, and are in a climax situation. Other shrub/tree species of the oak and hornbeam stands of the pre-Alps are: European hornbeam (*Carpinus betulus*), field maple (*Acer campestre*), smooth-leaved elm (*Ulmus minor*), narrow-leaved ash (*Fraxinus oxyphylla*), the limes (*Tilia platyphyllos* and *T. cordata*), the service and rowan trees (*Sorbus domestica* and *S. aucuparia*), hazel (*Corylus avellana*), the poplars (*Populus alba* and *P. nigra*) and white willow (*Salix alba*). Meanwhile, the stands of pubescent oak, being more xerophilous, have more thermophilic components, such as European hornbeam (*Ostrya carpinifolia*), manna ash (*Fraxinus ornus*), the cornel tree (*Cornus mas* and *C. sanguinea*), golden rain (*Laburnum anagyroides*), broom (*Genista tinctoria*) and common juniper (*Juniperus communis*).

In the Apennines and the pre-Alps, chestnut forests often take the place of pubescent oak mixed forests. The composition is almost the same, but it is sometimes differentiated by specific components, relics of former vegetation, superseded by chestnut. The natural poplar stands are a type of riparian forest developing along the large Italian rivers, such as the Po, Adige, Piave, Tevere and Arno, forming a very thick wood with willows, alders, ash and walnut, together with participation of thriving, recently introduced (naturalized) trees, such as black locust (*Robinia pseudacacia*) and false indigo (*Amorpha fruticosa*).

(v) Mountain forests of beech, black pine and occasionally fir (estimated area 1.7 Mha)

Above the oak forest, i.e. above 1000–1200 m, the typical mountainous habitat starts with the first beech forests. The beech, though demanding in terms of humidity and soil coolness, is clearly of an oceanic nature. Consequently, its distribution is limited to those alpine areas and other south-European reliefs reached by oceanic breezes coming from the nearby sea. This explains the lack of beech in the Valle d'Aosta and in the inner valleys of the central-western Italian Alps, where this species is substituted by more continental ones, particularly the Scots pine. There are two types of beech forests growing in the southern Alps: an Eastern one, with 'Illyrian' characteristics, wherein the beech mixes with the white fir; and a western one, similar to the Apennenic type, composed of beech and other broadleaves. The eastern variant, characterized by a scarce participation of other tree species other than the main one, grows well on carbonate soils, is very thick and has a rich undergrowth of shade-seeking plants: anemone (*Anemone nemorosa*), coral root (*Cardamine bulbifera*), sweet woodruff (*Galium odoratum*), violet (*Viola* sp. pl.), daphne (*Daphne mezereum*), prenanthes (*Prenanthes purpurea*) and paris (*Paris quadrifolia*). The western variant, generally distributed on arenaceous–marly soils, is characterized by the presence of mesophylous broadleaves: lime tree (*T. platyphillos*), maple (*Acer campestre*), manna ash (*Fraxinus ornus*), European hornbeam (*Carpinus betulus*) and, naturally, white fir (*Abies alba*) and sometimes yew (*Taxus baccata*). Only in the most southern areas of the peninsula and in Sicily do they diversify through the presence of tree and shrub species of clearly a Mediterranean type, such as the oak of Dalchamp (*Quercus dalechampii*), the maple of Lobelii and Naples (*Acer lobelii* and *A. neapolitanum*), the English holly (*Ilex aquifolium*), the yew tree (*Taxus baccata*) and the alder of Naples (*Alnus cordata*).

Within this category the forests of black pines of the western Alps and the Corsican pine (*Pinus nigra* var. *calabrica*) are also found. The black pine, found to the east of the Piave river, is similar to the black pine of Austria and Slovenia. This rustic conifer species, with ecological characteristics of the sub-Mediterranean area, when associated with the Scots pine forms typical mixed forests with some xerophilous broadleaves such as European hornbeam and manna ash, which can be considered as part of the Austro-alpine pine forest association.

(vi) Alpine conifer forests of spruce, Scots pine, larch and arolla pine (estimated area 720,000 ha)

Except for pastures, obtained in the past at the expense of forests, the superior mountain level of the Alps – ranging between 1500 and 2000 m – presents an almost continuous cover of conifers. At this altitude, the climax is represented by spruce (*Picea abies*) forests, a micro-thermal conifer which, in monospecific structures, occupies the most inner alpine valleys and is similar to other north European stands. This species is, however, characterized by a certain ecological flexibility, also being able to grow at lower altitudes – with beech and fir– or at higher subalpine altitudes – where it grows together with larch and arolla pine. The degree of mixture among these species and the rich undergrowth allows at least two types of spruce forest to be distinguished: the sub-alpine and the mountainous. The former develops at almost the border between the mountain level and the sub-alpine, where spruce mixes with 15–20% of larch and/or arolla pine in the clearings. The mountainous spruce forest can start from the inferior mountain level (1200–1500 m), it is mono-specific and dense. It is characterized by a rich undergrowth of moss and other acidophile broadleaves, while the sub-alpine spruce forest is dominated by more heath-like (*Calluna vulgaris*) undergrowth associated with blackberry, bilberry, rhododendron and other small mountain shrubs. Lichens are also frequent, encrusting the branches and the trunks of these

continued

Box 19.1. *Continued.*

thin stands, situated very close to the tree line. The two forest types also have a different origin: the mountainous spruce forests are strongly affected by man, who has largely favoured the conifers to the detriment of other components originally diffused in this climatic layer, such as beech; in contrast, the sub-alpine spruce forests are of a natural origin and border the highest pastures.

Together with larch, the arolla pine must also be considered in terms of the strict mixture of these two species when colonizing the rocky slopes of the central-western Alps. There, the arolla pine forest is located in an area usually pertaining to a characteristic vegetation of contorted shrubs. The rhododendrons and berries, in a compact shrub formation, form the undergrowth of this thin forest. A similar situation also occurs with the Swiss mountain pine (*Pinus montana* var. *uncinata*).

If compared with spruce, the Scots pine forest occupies only the most arid areas of the mountain level, where spruce does not grow (eastern Alps) or where beech cannot survive due to the lack of oceanic conditions (western Alps). The areas where this pine grows, the so-called 'Scots pine valleys', have a continental climate: Val di Susa, Valle d'Aosta, Valtellina, Val Venosta and Val Canale are the most significant examples.

(vii) Oro-Mediterranean forests of mountain bush with juniper, heather, green alder, dwarf pine and occasionally willows and birch (estimated area 450,000 ha)
Above the forest line – from the final trees to the start of alpine meadows – other timber species grow, which are very significant for soil protection and avoiding superficial runoff: rhododendrons, green alder and dwarf mountain pine (*P. montana* var. *mughus*). The latter represents probably the most characteristic type of alpine woodlands in the eastern Alps, also being spread on carbonated rocks in some high areas of the central Apennines. According to the variety of the main species, two structural types can be found:

- the forest type, with Swiss mountain pine (*P. uncinata*), growing on the crystalline soils of the central-western Alps, having a thin structure, similar to that of the Scots pine;
- the shrub type, thick and impenetrable, formed by dwarf pine on calcareous–dolomitic soils. In the past, the dwarf pine provided the raw material for a pharmaceutical distillate (mugho oil). Nowadays, this use has been abandoned, fortunately for the conservation of these important alpine ecosystems.

Another type of 'miniature forest' in the Alps is represented by the woods of green alder (*Alnus viridis*), which, in the siliceous–marly soils, takes the place and the colonizing function of Swiss mountain pine and dwarf pine woodlands.

The undergrowth is represented by rhododendrons (*Rhododendron hirsutum, R. ferrugineum* and *Rhodothamnus chamaecistus*), dwarf juniper (*Juniperus alpina*), common heather (*Calluna vulgaris*), bearberry (*Arctostaphylos uva-ursi* and *A. alpina*), mantoin heath (*Erica carnea*) and the berries (*Vaccinium myrtillus, V. vitis-idaea, V. uliginosum*). These shrubs anticipate the discontinuous clods of the higher alpine heather (*Calluna vulgaris*) moorland and alpine dryas (*Dryas octopetala*).

(viii) Productive plantations of poplars, gums, pines, hazels, walnuts, locust and ash (estimated area 130,000 ha)
Apart from the natural poplar stands along the shores of water courses, the cultivation of poplars has been practised for centuries in Italy, as in the rest of Europe. The most ancient form is the so-called riparian poplar growing, i.e. along rivers, canals, roads and edges of cultivated fields, still practised in the Po plain. These cultivations, producing wood and fodder, also serve as windbreaks. They also provide shade and have an ornamental value. Autochthonous species were used, i.e. the black poplar (*Populus nigra*, in Lombardy with its cypress variety *P. nigra* var. *italica*) or the white poplar (*Populus alba*) in different varieties and forms.

The specialized cultivation of poplars, spread at present over the flat areas of various Italian regions, has more recent origins and is carried out in plots of several hectares, using clones of hybrid fast-growing varieties for production of timber, pulp and wood chips. The most used poplar clones in Italy are some selections of *Populus × euroamericana* – a hybrid between the Euro-Asiatic black poplar and the American black poplar.

Apart from poplars, very few exotic deciduous species are used in Italy for fast-growing wood production. One of these, nowadays enjoying a certain success, is the black locust (*Robinia pseudacacia*), due, in part, to its extraordinary capacity for spontaneous diffusion. Originating from the Atlantic area of the USA, it has expanded spontaneously in Italy as well as in other European countries.

Of the exotic species introduced into Italy, the gums (*Eucalyptus* sp.pl.) are probably the most widespread, occupying at least 50,000 ha, concentrated in Sicily and Ionic Calabria. They are important for ornamental use, windbreaks (Lazio, Tuscany and Sardinia) and pulp for paper production (Sicily and Calabria). Gums should be cultivated exclusively within the Mediterranean zone with temperatures above 0°C, and on flat areas with abundant groundwater.

Plantations of walnut trees (*Juglans regia* and *J. nigra*) have also been grown recently for timber production (high-quality furniture). These cultivations require, however, intense management and are normally undertaken on agricultural land, where fertilizing and localized weeding can be easily carried out. Cultivations of hazels (*Corylus avellana*) can be found in Piemonte and Campania.

In around 1930–1950, experimentation with fast-growing exotic species was attempted in Italy, the major part of which failed to produce the expected outcome. Nowadays, there are very few arboreous species still in use, apart from Monterey pine (*Pinus insignis*), originating from the Pacific coast of North America. In Italy, this species gave positive results only in the south (Calabria and Sardinia) due to its peculiar ecological characteristics linked to a warm humid climate.

ownership, which are fragmented into small-sized plots and are often abandoned. The average size of a private holding is around 3–4 ha, with a large majority of very small-sized plots of less than 1 ha (Consiglio Nazionale dell'Economia e del Legno CNEL, 1999). These data, provided by agricultural censuses, however, hide the real world development, i.e. a certain informal aggregation by non-official leasing, as well as a clear trend, promoted by specific policies towards Forest Consortia. Up until now, these aggregations, even though not always successful, show the more open attitude of new generations of landowners towards cooperation and establishing consortia.

Common properties, under private law and, to a large extent, a national heritage, account for some 300,000–400,000 ha. They still prove to be an efficient institution for achieving the private and public objectives of forestry. Their role and value have been important in different stages of development: agricultural society, rural outmigration and industrial development, and post-industrial society. This role has been underlined by Pareto (1896) and demonstrated by research undertaken in recent years. According to Merlo (1995), where common properties are properly managed by local institutions – mainly groups of families – neither 'tragedy of the commons' (Hardin, 1968), nor free-riding takes place.

More than 70% of the public forests are owned and managed by communes and 20% by the State and regions. There are around 7000

public properties with an average size of about 400 ha (Istituto Nazionale di Statistica, 1982), often allowing a relatively effective forest management compared with private forests. During recent decades, the average size of Italian forest plots registered a slight increase, from 6.95 ha (1982) to 7.51 ha (2000). This is due mainly to forest natural regeneration as a result of private land abandonment (Istituto Nazionale di Statistica, 2001b). These figures hide, however, the dualism between the relatively large public properties, i.e. around 400 ha, and the small private properties, i.e. around 3–4 ha.

Administration and policies

The responsibility of forestry belonged to the Ministry of Agriculture and Forests[3] until its shift to the regions in the 1970s. At present, the State, and its Forest Administration (Ministry of Agricultural and Forest Policies), is responsible for forest police, fires (when aircraft are involved), general directives on forest policy implementation, statistics and international relations, in particular those related to the European Union (EU) and International Conventions (Navone and Shepherd, 1998).

The Central Forest Administration acts through the State Forest Corps (Corpo Forestale dello Stato, CFS), formed by around 8000 men/women, with the legal status of a police force. The CFS used to be the institution in charge of planning, monitoring, policy implementation, providing technical assistance and undertaking reforestation on private and public lands (Ciancio, 1996). Since the 1970s, most of the responsibilities of the CFS have been passed to the regions. At present, the CFS acts as a police force and mainly works through agreements signed with the regions, parks and local authorities. Debates and rumours about the Central Forest Administration's assignment to regions and/or to the Ministry of the Environment have always been the *leitmotiv* of forest policy in Italy during the last three decades.

Nevertheless, the regions have been given full responsibility for policy implementation, forest planning and soil conservation issues. Certain regions also passed further devolution to local authorities of lower levels, for example mountain communities formed by associations of communes (municipalities) in mountainous areas. These can be responsible for forest management and even administration of forest incentives in some cases.

Regions and the State cooperate in instances of problems of paramount importance (forest fire plans) or in the application of EU regulations (financial resources). In other cases, distinction between responsibilities and competencies between the Ministry and regions is far from clear, as proved by the failure of the 1988 National Forest Programme (NFP). Only eight regions expressly applied its directives, three were strongly against it in certain parts and the remaining nine regions did not even consider it (Corrado and Merlo, 1998).

Forest policy is based on the 1923 Forest Law, aiming at watershed protection and forest conservation. The rationale of this policy was the containment of agricultural and other land uses at the expense of forests on fragile soils, where watershed protection is badly needed. Through this law, 89% of forests became subject to a 'hydro-geological or soil conservation bond', prohibiting changes in land use and imposing specific management practices: continuous coverage, selective felling, uneven-aged and multi-specific stands, and natural regeneration, all primarily aimed at watershed protection.

The 1923 Forest Law also embraces issues such as reforestation, consortia amongst forest owners, forest management and exploitation rules, the social role of forestry in the rural economy and forest industries. Law enforcement through forest policing was also addressed. There are no doubts that it has been a good law, as unanimously agreed by the forestry profession, the administrators and those responsible for legal implementation (Carrozza, 1988). Certainly updating is needed, but this has been left to the regions, given the current institutional context. In fact, after the regional reform of the 1970s, the main principles of the 1923 Law were incorporated, often without substantial changes, into the different regional forest laws; eight regions with 'ordinary' autonomy (devolution) out of 15 have produced their own Forest Law (Corrado and Merlo, 1998). The same has been done for three out of five regions with 'special' autonomy.

Some of these developments certainly needed nationwide coordination, or even a national forest Law, as demanded by various circles since the 1970s. One answer has been the 1985 Landscape Act under the responsibility of the Ministry of the Environment, concerning forestry whenever environmental values are at stake, i.e. almost always given the nature-based character of most Italian forests. The result is that all forests are completely under the regime of this Act, and its few lines/indications devoted to forests. It is certainly a deterioration compared with the comprehensive nature of the existing 1923 Forest Law.[4] Now, the debate on a new forest law is still on the table, particularly in relation to society's increasing demand for multi-functional forests (recreation, biodiversity and conservation) as well as for compliance with the international agreements signed by the country.

The 1988 NFP was produced by the Ministry of Agriculture and Forestry as the main policy document for forestry. It aimed at a better coordination of interventions on forestland, forestry and forest-based industries. Though focused on forest conservation issues, its implementation failed due to uncertain institutional State–region relationships, as well as poor financial resources. Consequently, the 1923 Forest Law and 1985 Landscape Act remain the pillars supporting forest policy in Italy nowadays. Nevertheless, an NFP is demanded by certain international agreements and also more or less vociferously asked for by various stakeholders.

The role of forest stakeholders in Italy can be depicted as follows: within the public stakeholders, the Ministry for Agricultural and Forest

Policies has certainly lost ground, while the regions, park authorities and Ministry of Environment have gained importance. Within the category of private stakeholders, the role of forest owners has declined, while amenity societies, environmental trusts and timber-based industries have strengthened their position. Participation of farmers/forest owners, unions and lobbies in the land use planning by local authorities and parks is quite limited. Alliances between environmentalists and timber-based industry could be seen in supporting forest conservation, as 80% of the timber used by the industry comes from imports. Not surprisingly, the 'hot' issue of forest certification has seen the furniture and other timber-based industries in the front line, needing to improve their image and marketing perspectives. In addition, alliances between conservationists and land forest owners are often developed for the prevention of hunting access rights.

Contribution of the Forest to the National Economy

Gross domestic product and employment

The weight of forestry in the Italian economy is negligible in market monetary value – 0.05% of the gross domestic product (GDP). In contrast, timber-based industries including furniture, construction timber and pulp/paper represent one of the main sectors of the Italian economy, accounting for some 4.5% of the GDP. Table 19.1 shows that the value of forest products now amounts to 3.4% of the total output value of the timber industry (Istituto Nazionale di Statistica, 1991), while it is known that the share of forest products in the furniture industry is even less significant (1.04%). 'Intermediate

Table 19.1. Components of timber industries production value.

Timber industries production value	100.00%
Intermediate timber products	34.18%
Labour	17.06%
Forest products	3.4% in 1985; 7.7% in 1978; 9.5% in 1968

Source: Istituto Nazionale di Statistica (1991b).

timber products' (Food and Agriculture Organization, various years) include imported timber – around 80% of Italian consumption. The timber-based industries are more closely dependent upon forests of other countries (imports), except for the poplar–plywood industry and, to a minor extent, timber for construction in north-eastern Italy. This is, however, a basic feature of the national economy, which is almost completely based on the transformation of imported raw materials, ranging from metals to energy.

Statistics regarding forestry employment (0.05% of the total) mainly account for the workforce employed in the exploitation enterprises and as forest wardens, the latter acting mainly within public properties – communes and local authorities. The real employment figure in forestry is, however, higher as some of the silvicultural operations are performed by working farmers – often part-time and hobby farmers. Employment in other forest-related activities (services) stands at more than 10,000: the CFS alone employs some 8000 people. A further 10,000 jobs can be attributed to the regions and local forest services, while in certain Italian regions, substantial forest employment is due to various local authorities aiming at alleviating unemployment. Some 30,000–40,000 jobs can be counted mainly in Southern Italian regions, particularly in Calabria.

In contrast, employment in timber-based industries accounts for 4.7% of the total national employment and 13.2% of the industrial employment (Istituto Nazionale di Statistica, various years), i.e. around 500,000–700,000 jobs. Clearly, this is due to industries such as the construction of furniture and house fittings, which mainly use imported rather than home-grown timber.

International trade

Elaboration from data of the Food and Agriculture Organization (1999) shows that the total wood consumption accounts for 60 Mm^3 (almost 1 m^3 *per capita* of roundwood equivalent), made up of mainly imported timber or wood-based intermediate products, i.e. 83% of the total wood market. National production for the same year accounts for 10 Mm^3, equally represented by firewood and timber. The overall wood market is therefore dominated by timber imports (50 Mm^3), while wood exports are practically nil, except for plywood and particle panels in certain years. Timber is the third largest Italian import after oil and foodstuffs. Nevertheless, taking into account the export of furniture, house fittings and other timber-based products, the trade balance during the last 10 years has always been positive.

Other forest-based industries: tourism and green belts

When welfare and the tourist industry are taken into account, the situation is completely reversed. Forests become very important to the GDP and the quality of life, and people are well aware of this. Indirect market effects are well acknowledged, as previously demonstrated using 'hedonic pricing' for the location of woodland near houses. Input–output tables built at a local level confirm the role played by 'green belts' based on tourism, particularly where parks and areas of 'outstanding natural beauty' have been designated (Casini, 1993). It is interesting to note that forests and timber-processing industries can also be part of these 'green belts', when solid wood furniture and crafts, such as timber sculptures, are concerned. Local production in these areas is stimulated by the large number of visitors, as shown by Fodde (1995).

The Values of Italian Forests

The estimates of Italian forest values are presented in Table 19.2, grouped into direct and indirect use values and negative externalities linked to forests. Few attempts of valuing option, bequest and existence values have been made. A more detailed description of the methods and indicators employed is given in the following sections.

Direct use values

Italian forests provide wood forest products (WFPs) that are mainly timber and firewood, together with a wide range of non-wood forest products (NWFPs) such as mushrooms, truffles, chestnuts, pine kernels and other forest fruits. Most of these products are estimated based on the quantities traded on the market and their market price (Istituto Nazionale di Statistica, 1997a). Timber and firewood for trade are the exception; they are valued using the roadside prices. It can be seen that the market value of firewood is just slightly higher than that of timber, while, overall, the WFPs are central to the market forest values. The timber increment not yet exploited is valued based on the quantity of wood increment net of removals and half of the average stumpage price (~€20/m^3).

Mushroom valuation refers to the collected quantities and the average market price. In fact, a substantial part of the mushroom's value could be referred to the price of the collecting permits paid by professional or amateur collectors as foreseen by Italian legislation.[5] In addition, a considerable amount is collected for free, either because local authorities do not request permits, or because people do not comply with the rules. Since the average permit price is not available at a national level, and studies evaluating consumer surplus were not found, reference is made to the average market price for all mushrooms.

The value of honey is based on the average quantity produced by bees in and around forests. It is roughly estimated that 1.25 kg of honey can be produced from 1 ha of forest, with an average market price of €6/kg (Mezzalira, 1987). Apart from these NWFPs, a significant weight in the market forest values is represented by chestnuts and hazelnuts, followed by truffles, pine kernels and other forest fruits.

Valuation of grazing is based on the number of forage units (FU) consumed and the value of 1 FU. There are 2 Mha of grazed forest, with a minimum forage consumption of around 200 FU/ha (Talamucci, 1991). The market price of 1 FU grazed by the animals is derived from the market price of hay: 1 t of hay corresponds to about 330 FU. The market price of hay is around €150/t and includes the value of fresh forage as well as the costs of forage cutting, transportation, drying and other necessary operations for transformation. Based on the residual value method, the fresh forage price is calculated at around €0.15/FU. A similar value can be derived by using the surrogate price method, considering that the market price of 1 kg of barley

Table 19.2. Values of Italian forests.

Valuation method/output	Quantity	Value[b] (000 €)
Direct use values		
Market price valuation[a]		
Timber for trade (m³)	3,276,069	209,475
Firewood (m³)	5,076,013	227,183
Net growth of standing timber stock (m³)	16,000,000	320,000
Mushrooms for trade (kg)	2,613,947	25,471
Mushrooms for self-consumption (kg)	12,000,000	60,000
Honey (kg)	3,750,000	22,500
Cork – first harvest (q)	5,958	131
Cork – subsequent harvests (q)	92,035	4,050
Truffles (kg)	73,868	8,628
Chestnuts (q)	698,524	55,648
Pine kernels (q)	20,715	3,040
Hazelnuts (q)	154,719	16,266
Bilberries (kg)	381,423	1,200
Strawberries (kg)	72,338	389
Raspberries (kg)	68,234	288
Acorns (kg)	50,403	1,811
Surrogate market pricing		
Grazing (million FU)	400	60,000
Consumer surplus (CVM, TCM)		
Hunting (no. of hunters)	285,000	71,250
Recreation (million day-visits)	68–168	170,000–420,000
Total direct use values		1,257,330–1,507,330
Indirect use values		
Cost avoided method		
Watershed protection (ha)	3,000,000	1,321,000–2,000,000
Shadow pricing		
Carbon sequestration (million tC)	3.1–6.9	60,000–140,000
Total indirect use values		1,381,000–2,140,000
Negative externalities		
Cost-based method		
Erosion due to poor or no forest management (m³/ha)	2.18	−65,400
Damage by floods, landslides, avalanches		−367,000
Damage by forest fires (ha)	41,019	−60,700
Losses of natural quality due to illegal actions (no.)	15,068	−1,680
Losses of landscape quality due to illegal actions (no.)	3,594	−128
Total negative externalities		−494,908
TEV		2,143,422–3,152,422

[a]Quantities refer to 1994 (Istituto Nazionale di Statistica, 1997).
[b]Updated to 2001 prices.

corresponds to 1 FU. Accordingly, the total grazing value is around €60 million.

Other forest outputs originating from direct use values are hunting and recreation, calculated on the basis of people benefits, i.e. consumer surplus, deduced from applications of the travel cost method (TCM) and the contingent valuation method (CVM) at a local level. Valuation of hunting[6] was undertaken with reference to CVM, which was preferred to the permit price-based method for various reasons. Firstly, despite its shortcomings, the former is able to reflect 'real' values, while the latter expresses nominal values, as reflected by the fixed price of hunting permits.

Secondly, there is a wide variability of permit prices throughout the country, and using one of them would make it impossible for valuation to be generalized at a national level. The CVM survey, applied in two hunting reserves of northern Italy, estimated an average consumer surplus of €250/ hunter (De Battisti *et al.*, 2000). Extrapolated to 285,000 forest hunters (Istituto Nazionale di Statistica, 1994–1995), it results in an overall value of €71.2 million.

Recreation value is based on the number of day-visits/year and consumer benefit/visit in the forest. The first factor was rather difficult to estimate as an overall figure; however, according to Istituto Nazionale di Statistica (1998), 68 million visits in mountainous and hilly areas were recorded. At the same time, a specific survey (Scrinzi *et al.*, 1996) provided a figure of 168 million day-visits/year. Accordingly, a range of 68–168 million day-visits is considered a reliable range of estimation. The average consumer benefit is around €2.5/visit, obtained from several TCM and CVM applications undertaken in Italy during the last two decades. The evidence of these values has been corroborated in a few cases by the real world experience of visitors being obliged/stimulated to pay a price for access to the forests or, generally speaking, to services accompanying recreation in the forests, such as footpaths, guides and car parks. This is shown by a survey on marketing of public goods and externalities linked to forestry (Merlo *et al.*, 2000). Overall, the recreational value is estimated at €170–€420 million.

It should be noted that the recreational value can include various items, ranging from traditional hunting and shooting to fruit picking and access for nature-related interests, such as bird watching. The enjoyment of these benefits depends to a large extent upon the property rights regime. The traditional rights of the local people concerning the collection of firewood and other secondary products are also important. Seen in a modern post-industrial context, this regime pinpoints the multi-faceted public connotations of forest property rights in Italy, and, indeed, in other Mediterranean countries, which are less defined, assigned and enforced when compared with the central/northern European countries. There, an earlier economic–industrial development has made the enforcement of a stronger property rights regime possible, which today is impossible to change in a modern society.

Indirect use values

Italian forests provide a diversified array of environmental benefits *latu sensu*, such as water-related services, ranging from water flow regulation, watershed management and flood/landslide/avalanche prevention, to soil protection, moisture conservation and water purification. Other indirect use values provided by forests include landscape quality and carbon sequestration. Assigning monetary values to these functions is difficult, and the results are sometimes dubious.

Water-related services are valued on the basis of the 'costs avoided' method. It is assumed that the magnitude of this function is proportional to the costs of floods, erosion, landslides and other dis-services avoided by the existence of forest cover (*with* and *without* the forest situation). The total area of the country under high risk of water-related hazards (floods, landslides and erosion) is around 3 Mha (Ministry of Environment, 1997). This risk can be attributed either to poor forest management or to lack of appropriate forest coverage. In contrast, the other part of the forest area protects another 3 Mha of land rather well, mainly lowland with intensive land uses (cities, communication channels and industrial estates).[7] It is reported that recently the average national public expense for soil conservation and hydraulic works for maintenance/renovation amounted to €1321 million (Ministry of Environment, 1997). These costs can be attributed to the defence and protection by various means of the 3 Mha characterized as being of high risk from water-related hazards, where forests are not operating properly or there is lack of forest cover. On the basis of the 'costs avoided' method, it can be roughly assumed that the well-managed forests protecting over 3 Mha of intensive land uses perform a service that can be valued at around €1321 million. Another survey (Codemo, 1986, updating Patrone's 1970 estimates) calculated for the same service a figure of €2000 million. It can therefore be assumed, as a rather rough first estimate, that the value of water management function ranges between €1321 and €2000 million.

Carbon sequestration in forest biomass is valued by using the shadow pricing method. The net quantity of carbon fixed yearly in the Italian forest biomass was estimated as 3.1 MtC for 1991 (Cesaro and Pettenella, 1994) and as 6.9 MtC for 1995 (United Nations Economic Commission for Europe/Food and Agriculture Organization, 2000). As a first approximation, they are considered as the lower and upper limits of the interval wherein the annual carbon sequestration ranges. A shadow price of about €20/tC is applied (Cesaro and Pettenella, 1994; Fankhauser, 1995), resulting in an annual value ranging between €60 and €140 million.

Calculation of other indirect use values – such as landscape and natural environment including biodiversity services – is even more difficult. One reason is because some forest values are greatly inter-related with others and the valuation methods do not allow for a clear separation between values. This is the case for the change in landscape quality, whose value is partly incorporated into the recreational value (with a positive sign) and partly into the damage produced by forest fires (with a negative sign).

Option, bequest and existence values

Concerning option, bequest and existence values, no precise data can be reported. In fact, the available Italian surveys aiming at estimating the value of forests through CVM do not make a clear distinction between direct use/option/bequest/existence values. It can be assumed that the value of at least €5/day-visit, resulting from several surveys, certainly captures to some extent the option, bequest and existence values of those visiting the forests. At the same time, this clearly misses a substantial amount of these values and ignores the non-visitors' willingness to pay for protecting forests. More research in this field is certainly needed, taking into account the complex and different functions performed by Mediterranean forests. The research results undertaken in northern European countries and in North America cannot be simply transferred to the Mediterranean forests, given their peculiarities. It can be argued that the values for Mediterranean forests should

be higher, given the important tourist industry, the rich biodiversity and, in general, their multiple functions.

Existence value is estimated based on a CVM survey[8] (close-ended version) undertaken in a nature reserve in Sicily, showing an average benefit of €15/person (Signorello, 1990). Peculiarities of the area – wetland characteristics are not similar to those of the national forest area – do not allow wide extrapolation. Even so, the valuation is seriously considered as the reserve is characterized by Mediterranean vegetation of '*maquis*, shrubs'. Assuming an individual benefit ranging between €10 and €20, total existence value would be about €600–1200 million. This estimate has just an indicative significance and only partially captures the forest existence value.

The values reported so far certainly provide a limited view and quantification of the true value of forests. A wide range of public goods and positive externalities cannot be easily defined and even less quantified and captured within the total economic value (TEV) framework, due to various shortcomings including poor knowledge of the forest ecosystem and the reliability of valuation methods.[9]

Negative externalities linked to forests

In order to complete the picture of the forest TEV, additional aspects should be considered, more particularly the public bads and negative externalities linked to poorly managed or unmanaged forests. Examples range from erosion, floods, landslides, avalanches and damage produced by forest fires, to loss of landscape and natural quality due to illegal actions that negatively affect the TEV (Table 19.1).

Valuation of erosion due to poor/no forest management is based on the average erosion rate and its replacement cost. The quantitative indicator, obtained from the universal loss equation, ranges from 2 to 4 m^3/ha/year (corresponding to soil losses of 0.2–0.4 mm/year) in Italian alpine environments (southern slopes) up to 13–21 m^3/ha/year (namely 1.3–2.1 mm/year) in the southern Mediterranean areas (Benini, 1990). Applying a minimum soil erosion rate of 2.18 m^3/ha, a degraded forest area of 2 Mha and a replacement cost (costs of dredging soil accumulated downstream of the watershed) of about €15/m^3, it results in a total value of €65 million.

Damage arising from floods and landslides due to poor forest management refers to catastrophic events which occurred in Italy as reported by official statistics. During the last 45 years, they affected 4568 communes (56% of the total number), which underwent 2477 landslides. The value of this damage amounted to €16.5 million (Ministry of Environment, 1997), leading to an approximate annual value of €367 million.

Valuation of damage caused by forest fires refers to the events generated by human-related activities, leaving aside natural fires, and is based on the restoration cost method. On average, 41,019 ha were affected by forest fires during 1985–1994 (Istituto Nazionale di Statistica, 1994–1995). The cost, including standing trees expectation value, is estimated at €26.7 million and the extra cost for restoration and plantation exceeding the ordinary cost of these forest operations can amount to €34 million (Istituto Nazionale di Statistica, 1996). This leads to a total of €60.7 million.

The losses due to illegal actions are identified with environmental damage caused by all illegal actions violating Forest Laws No. 3267/1923, 431/1985 and the Regional Laws on water and forest protection, landscape and environmental safeguards (Istituto Nazionale di Statistica, 1994–1995). Valuation assumes the amount of fines as a proxy for these types of damage. This leads to a partial estimate, as neither all illegal actions nor the true value of damage can be entirely reflected through the value of fines paid. Bearing in mind these limitations, there were 5068 illegal actions on forest sites, 7595 of which were contrary to laws protecting flora and 7473 to the law protecting flora and fauna. The total amount of fines paid equalled €1.7 million. In addition, there were 3594 illegal actions against the landscape law, for which the annual payments averaged €128,000. Valuation of the negative impacts of pollen and other allergens on human health is not possible as the only available information related to the number of people suffering from breathing problems (663,581), which also includes other connected illnesses (Istituto Nazionale di Statistica, 1997b).

Towards the Total Economic Value of Italian Forests

The estimation of the TEV of Italian forests and its categories, far from being a purely academic exercise, highlights the way in which various forest outputs and the related values are perceived. This perception can help management, what is now called sustainable forest management, and inform forest policies.

Figure 19.2 reflects the minimum estimated values of different forest outputs, corresponding to direct/indirect use values and negative externalities. Overall, the estimated TEV ranges between €2 and €3 billion, if non-use values are ignored due to information scarcity. The indirect use values reflected by forest ecological functions have the greatest importance, ranging between €1381 and €2140 million, followed by direct use values (€1257–1507 million). Negative externalities account for €495 million. The existence value is only partially estimated as a flow value, ranging between €600 and €1200 million. However, these are only indicative estimates, intended to give insights into the significant importance of non-use values. The result thus obtained can also be seen as 'provocative', aimed at stimulating further research towards an improved forest valuation in terms of annual flow as well as total stock pertaining to the whole country.

The results show that the most important forest function in Italy is related to water issues, the value of which, as roughly estimated, falls between €1321 and €2000 million and accounts for 62% of the minimum estimated TEV (Fig. 19.3). Even though debatable to a certain extent, due to limited availability of data, the magnitude of this function confirms the central

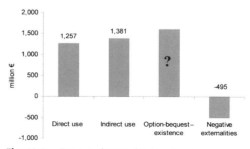

Fig. 19.2. Estimated TEV of Italian forests.

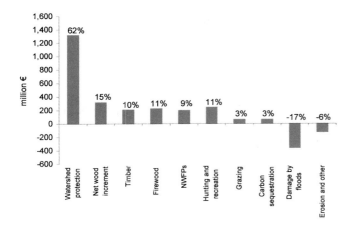

Fig. 19.3. Main components of the TEV of Italian forests.

aim of forest policies, based on the 1923 Forest Law and 1985 Landscape Act, that address forest management, aiming at conservation and protection against floods, landslides, avalanches and erosion.

The remaining 38% of the TEV is distributed mainly between the value of 'saved increment' (15%), hunting and recreation (11%), firewood (11%), timber (10%), NWFPs (9%), carbon sequestration (3%), grazing (3%) and negative externalities (−23%). It should be remarked that the greatest part of the negative externalities is imputable, and accounted, in terms of damage caused by floods and landslides, due to poor forest management, especially produced by the abandoned privately owned areas.

The indicative significance of the TEV components is also given in Table 19.3, presenting the estimated forest values per hectare of forests, in relation to the lower bound of the valuations undertaken. To a certain extent, they bear the limit of overestimation, as total forest area has increased since the last forest inventory.

Conclusions and Perspectives

The main results of the Italian survey highlight the growing importance of public goods and positive externalities, mainly represented by water-related issues. The role of forests in water and soil protection is crucial, especially in steep-sloped mountainous areas (Alps and Apennines) characterized by Mediterranean climate and a high risk of floods, landslides and sheet erosion. In fact, the protective function is better performed by public forests, where management is undertaken efficiently by public bodies (regions or communes), with proper consideration of many environmental issues and social concerns of local communities. In contrast, private forests are poorly managed due to fragmentation and high exploitation costs in recent decades. This results in a high risk of erosion and landslides on downstream agricultural lands, the incidence of forest fires and, overall, it leads to the deterioration and marginalization of the rural community structure.

The results show the high level of importance of forest recreation, which is sometimes associated with NWFPs, such as mushroom collecting, berry picking and hunting. This reflects the increasing social demand for outdoor activities as a result of the increased income of the Italian population. In many cases, market-based mechanisms (for recreation and connected goods) and institutional measures (regarding hunting) have been established and implemented; however, those who usually benefit from these services are visitors and the tourism industry, while those supporting the costs of their provision are a limited farming community (aside from public bodies). Therefore, in many of these areas, the need for sustaining rural development should be seriously considered. Addressing this issue could involve implementation of compensation schemes – through negotiated agreements, supported by institutional

Table 19.3.   Average estimates of forest values.

Average value	Watershed protection	Wood	Net increment	NWFPs	Hunting, recreation	Grazing	Carbon sequestration	Negative externalities	TEV
€/ha	154	50	37	23	28	7	7	−59	249

frameworks and public support (*State pays approach*) – between various forest stakeholders so as to allocate benefits and costs in a socially ethical way.

The role of timber and firewood in the total forest value is not very significant, as forest-based industries are supported mainly by imports rather than by national production. It is thought that organizing consortia among private forest owners could help improve the national wood supply and its competitiveness as an input for timber-based industries. This issue has been frequently discussed since the early Italian legislation (1923). The scarce estimates of option, bequest and existence values tend to mask, rather than to reveal, their true social value. In fact, the importance of environmental concerns is increasing for the general public as well as for many environmental groups and public authorities, as seen for example by alliances between environmentalists and timber-based industries in supporting forest conservation.

All in all, Italian forests, and especially their Mediterranean part, are a source of public goods and externalities increasingly important for local communities, visitors and general society. Their enhancement can be achieved through their full consideration in adequate institutional/ policy means and financial–economic measures suitable for improving forest management and rural development.

Notes

[1]   The original Quézel classification (Quézel, 1976) refers to the Mediterranean forest vegetation distributed according to the following categories: (i) thermophilic wild olive and pistachio scrub; (ii) Mediterranean conifer forests of Aleppo pine, brutia pine, maritime pine, stone pine, barbary thuya and Phoenician juniper; (iii) sclerophyllous evergreen oak forests of holm oak, cork oak and kermes oak; (iv) deciduous forests of zeen oak, afares oak, Lebanese oak, tauzin oak, hornbeam, ash and occasionally beech; (v)

mountain or high-altitude forests of cedar, black pine and firs; and (vi) oro-Mediterranean stage stands of arborescent juniper or thorny xerophytes.

[2]   Except for poplar plantations that are located in floodplains, where protection – at least from soil erosion – is not necessary.

[3]   Recently renamed the Ministry of Agricultural and Forest Policies.

[4]   Misunderstandings, sometimes ridiculous, are widespread, e.g. the environmentally oriented selection cuts have been seen by certain courts as equally applicable to poplars and to other short-rotation plantations aimed solely at timber production. This has often prevented new forest plantations that are much needed, due to the huge timber deficit of the country. It is the case of those considered by the 2080/1992 European Union Forest Regulation, applied to a very small extent in certain regions, because landowners were afraid to make future land use changes impossible.

[5]   The collection of mushrooms and small fruits growing spontaneously in woodlands is usually a free right. However, regulations have been issued recently due to increasing human pressure on forest products. Maximum quantities of small fruits and mushrooms to be picked per day and per person are established by national and regional legislation. Mushroom picking is now also regulated by National Law 352/1993, which has assigned mushroom property rights to the Local Authorities. Provinces, Mountain Communities and Municipalities can, and do, sell mushroom picking permits on their land, usually with different prices for residents and non-residents.

[6]   Game is traditionally public (State) property in Italy, as are hunting/shooting rights. The State sells shooting/hunting permits and generally identifies the hunting areas at a regional/local level. Enforcement and control is under the regions and Provinces. Hunters have free access to private properties, except when these are fenced or enclosed: *fondi chiusi*. Past and recent legislation (National Law 157/1992) has foreseen a certain limited direct remuneration of landowners in selling access and shooting rights to private estates, so-called game-farms: Aziende Faunistico-Venatorie and Aziende Agri-Turistico Venatorie.

[7]   In addition, some 61,200 ha of forests protect buildings and other infrastructures from landslides and avalanches (Istituto Sperimentale per

l'Assestamento Forestale e per l'Apicoltura e Ministero dell'Agricoltura e delle Foreste, 1985, p. 73).

[8] Another study (Ministry of Environment, 1998) shows the *in situ* conservation investments (conservation of protected areas, unprotected areas, environmental recovery) as a result of the Convention of Biological Diversity (1992). By means of the second 'Triennial Plan for Protected Areas', the Ministry of Environment allocated €77.3 million for environmental resource protection, which was distributed among: terrestrial protected areas (€2 million); protected areas of regional interest (€24.4 million); protected areas of national interest (€43.3 million); establishment of new national parks (€5 million); and marine protected areas (€2.5 million). Roughly assuming that from the total investment, only the first four categories can be included within the forest area, then the total investment for the biodiversity conservation would be €74.8 million for a 3-year period. That leads to an annual investment of €25 million. However, it is not clear which share of this value corresponds to the annual flow of biodiversity benefits.

[9] Indeed, public goods and the associated externalities are not always valued together. For instance, water protection function and erosion were estimated separately using different methods.

References

Benini, G. (1990) *Sistemazioni Idraulico Forestali.* UTET, Turin, Italy.

Carrozza, A. (1988) Linee e tendenze della legislazione forestale/Forest legislation trends. *Economia montana e linea ecologica* 20, 4.

Casini, L. (1993) La valutazione economica degli effetti dell'istituzione di un parco: l'analisi d'impatto sull'economia locale/Economic valuation of a park: impact on the local economy. *Rivista di Economia Agraria* 58, 95–129.

Cesaro, L. and Pettenella, D. (1994) Un'analisi degli effetti delle politiche forestali nella prevenzione dei cambiamenti climatici. *Rivista di Economia Agraria* 11, 37–61.

Ciancio, O. (1996) *Il Bosco e l'Uomo.* Accademia Italiana di Scienze Forestali, Florence, Italy.

Codemo, L. (1986) *Economia Forestale. Annuario di Economia Montana.* INEMO, Rome.

Consiglio Nazionale dell'Economia e del Legno (1999) *Il Sistema Foresta-legno Italiano: Problemi e Prospettive per il 2000 di una Politica dell'Offerta Interna di Legname.* Rapporto di ricerca, May 1998, CNEL.

Corrado, G. and Merlo, M. (1998) *The State of the Art of the National Forest Programmes in Italy.* Freiburg, Germany.

De Battisti, R., Val, A. and Rosato, P. (2000) Il valore economico della caccia nella montagna Veneta. *Habitat,* January.

De Phillipis, A. (1937) *Classificazioni ed Indici del Clima in Rapporto alla Vegetazione Forestale Italiana.* CNR, Rome.

Fankhauser, S. (1995) *Valuing Climate Change. The Economics of the Greenhouse.* Earthscan, London.

Fodde, F. (1995) Analisi dei Sistemi Foresta-legno nella Regione. Tesi Dottorato, Università di Padova, Italy.

Food and Agriculture Organization (1999) FAOSTAT Database collection, www.fao.org. Accessed October 2000.

Gellini, R. and Grossoni, P. (1996) *Botanica Forestale.* CEDAM, Padua, Italy.

Giacomini, V. and Fenaroli, L. (1958) *La Flora. Conosci l'Italia,* II. TCI, Milan, Italy.

Hardin, G. (1968) The tragedy of the commons. *Science* 162, 1243–1248.

Istituto Nazionale di Statistica (1982) *3° Censimento dell'Agricoltura: Caratteristiche Strutturali delle Aziende Agricoli.* Vol. 2, Tome 3, ISTAT, Rome.

Istituto Nazionale di Statistica (1991) *7° Censimento Generale dell'Industria e dei Servizi.* ISTAT, Rome.

Istituto Nazionale di Statistica (1994–1995) *Statistiche della Pesca e Caccia.* ISTAT, Rome.

Istituto Nazionale di Statistica (1996) *Statistiche dell'Ambiente.* ISTAT, Rome.

Istituto Nazionale di Statistica (1997a) *Statistiche Forestali.* ISTAT, Rome.

Istituto Nazionale di Statistica (1997b) *Statistiche della Sanità.* ISTAT, Rome.

Istituto Nazionale di Statistica (1998) *Annuario delle Statistiche del Turismo.* ISTAT, Rome.

Istituto Nazionale di Statistica (2001a) *14° Censimento Generale della Popolazione e delle Abitazioni – Primi Risultati – Sintesi.* ISTAT, Rome, 21 Ottobre, 2001.

Istituto Nazionale di Statistica (2001b) *5° Censimento Generale dell'Agricoltura – Caratteristiche Strutturali delle Aziende Agricole.* ISTAT, Rome.

Istituto Nazionale di Economia Agraria (1998) *Annuario dell'Agricoltura Italiana.* Vol. 52, Il Mulino, Bologna, Italy.

Istituto Sperimentale per l'Assestamento Forestale e per l'Apicoltura e Ministero dell'Agricoltura e delle Foreste (1985) *Inventario Forestale Nazionale.* ISAFA-MAF, Rome.

Merlo, M. (1995) Common property forest management in northern Italy: a historical and socio-economic profile. *Unasylva* 46 58–63.

Merlo, M., Milocco, E., Panting, R. and Virgilietti, P. (2000) Transformation of environmental recreational goods and services provided by forestry into recreational environmental products. *Forest Policy and Economics* 1, 127–138.

Mezzalira, G. (1987) An attempt to evaluate the multiple functions of shelterbelts. multipurpose agriculture and forestry, *Proceedings of the 11th Seminar of the European Association of Agricultural Economists*, IUFRO Symposium Wissenschaftsverlag, Vauk, Kiel, Germany.

M'Hirit, O. (1999) Mediterranean forests: ecological space and economic and community wealth. *Unasylva* 50, 3–15.

Ministry of Environment (1997) *Relazione sullo Stato dell'Ambiente/Report on the Environment*. Ministry of Environment, Rome (in Italian).

Ministry of Environment (1998) Italian National Report on the Implementation Status of the Convention on Biological Diversity. ENEA – Environment Department, Rome.

Navone, P. and Shepherd, G. (1998) Italy. In: Shepherd, G., Brown, M., Richards, M. and Schrecken berg, K. (eds) *The EU Tropical Forestry Sourcebook*. European Commission, pp. 247–260.

Pareto, L (1896, Italian edition, 1949) *Corso di Economia Politica/Course of Political Economy*. Einaudi, Turin, Italy.

Pignatti, S. (1982) *Flora d'Italia/Italian Flora*. Ed agricole, Bologna, Italy.

Pignatti, S. (1994) *Ecologia del Paesaggio/Landscape Ecology*. UTET. Turin, Italy.

Quézel, P. (1976) Les forets du pourtour mèditerranèen. In: *Forets et Maquis Méditerranéens: Écologie, Conservation et Amenagements*. Note técnique MAB, 2, UNESCO, Paris.

Scrinzi, G., Tosi, V., Agatea, P. and Flamminj, T. (1996) Coordinate quali-quantitative dell'utenza turistico ricreativa in Italia/Quali-quantitative features of touristic-recreative use. *Genio Rurale* 3, 53–76.

Sereni, E. (1961) *Storia del Paesaggio Agrario Italiano/History of Italian Agricultural Landscape*. Laterza, Bari, Italy.

Signorello, G. (1990) La stima dei benefici di tutela di un'area naturale: un'applicazione della contingent valuation/Estimate of protection benefits of a natural area: application of a contingent valuation. *Genio Rurale* 9, 59–66.

Talamucci, P. (1991) Pascolo e Bosco/Forest and pasture. Prolusione tenuta all'inaugurazione del 40° anno accademico dell'Accademia Italiana di Scienze Forestali 2/3/91, *L'Italia Forestale e Montana* 2.

Tomaselli, R. (1973) *La Vegetazione Forestale d'Italia/Italian Forest Vegatation*. Ministero Agricoltura e Foreste, Collana Verde, 33, Rome.

United Nations Economic Commission for Europe/Food and Agriculture Organization (2000) *Global Forest Resources Assessment 2000*. Main Report. United Nations Publications, Geneva.

20 France

Claire Montagné, Jean-Luc Peyron, Alexandra Niedzwiedz
and Odile Colnard
*Laboratoire d'Economie Forestiere (UMR ENGREF/INRA), 14 Rue Girardet,
CS 4216, F-54042 Nancy Cedex, France*

Introduction

With almost 60 million inhabitants and a total area of about 550,000 km^2 (55 Mha), France is an important country that borders the Mediterranean Sea. It constitutes a link between the south and north and between the west and centre of Europe. This particular situation has an effect as much on the exchange of people and goods as it does on species migration or on environmental conditions. Another major feature of France is the large range in altitude, from sea level to almost 5000 m (Mont Blanc), although the average altitude is around 350 m. The climate is also varied, even though the country can globally be considered as temperate with oceanic influences. Four main types of climates and biogeographical areas are encountered: the temperate oceanic climate in the west; the temperate climate with continental influences in the east; the mountainous climate in the Alps and the Pyrenees; and, finally, the Mediterranean climate characterizing the southeastern part, including Corsica. These factors help to explain the wide ecological diversity that is encountered in France.

The Mediterranean region is rather restricted in area when compared with the whole of France. Its precise definition is not easy because Mediterranean and other influences overlap, for example near the sea in high mountains or, on the contrary, far from the coasts in long valleys such as the Rhône corridor

(Quézel and Medail, 2003). The French natural Mediterranean region spreads over the low altitude plains and valleys around the Mediterranean Sea and over hills and mountains mainly created by the Alpine and Pyrenean tectonic plates. According to Illy and Pinatel (1997), the geographical Mediterranean area includes:

- The four 'departments' (counties) of the Languedoc-Roussillon region (Aude, Gard, Hérault and Pyrénées-Orientales) and the eastern slope of the Cévennes mountains in the Lozère department;
- All of the Provence-Alpes-Côte d'Azur region (departments of Alpes-de-Haute-Provence, Alpes-Maritimes, Bouches-du-Rhône, Var, Vaucluse and Hautes-Alpes), with the exception of the Drac and Séveraisse valleys in the Hautes-Alpes department;
- All of Corsica;
- The south of Ardèche and Drôme departments of the Rhône-Alpes region, mainly comprised of four valleys: the Ardèche and Drôme valleys and the Etrieux and Aigues valleys.

As can be seen, the boundaries of the French natural Mediterranean area do not coincide with the French administrative ones, neither at the departmental nor at the regional level. However, in practice, administrative boundaries are useful, for example when using statistical data. Thus, two different definitions will be used here for the French Mediterranean area, one based on regional and the other on departmental boundaries:

- The French Mediterranean regions will include Corsica, Languedoc-Roussillon and Provence-Alpes-Côte-d'Azur (Fig. 20.1a);
- The French Mediterranean departments will include those of these three regions, except Lozère, which is more mountainous than Mediterranean, plus the Drôme and Ardèche departments that are situated in the Rhône-Alpes region (Fig. 20.1b); this definition is consistent with that used by the National Forest Inventory (Ministère de l'Agriculture, de l'Alimentation, de la Pêche et des Affaires Rurales, 2001; Peyron and Colnard, 2002).

The Mediterranean area of France occupies 12% (regional definition) to 14% (departmental definition) of the total land area. It is located on the southern edge of France and at the northern edge of the Mediterranean region, with a northern limit just above the 45° latitude and a southern limit a little above 41°. The whole of the Mediterranean region lies between 27 and 45° latitude.

To obtain an estimate of the total economic value (TEV) of the French Mediterranean forests is not an easy task for many different reasons. Firstly, these forests possess very specific characteristics (such as ecosystems and functions), which can also be present elsewhere in France but not with such great importance as in the Mediterranean region. Secondly, forest economics is a relatively new field of study in France and the

number of references regarding forest environmental values is low. Even though some analyses have been carried out (Tessier and Peyron, 2000; Ministère de l'Agriculture, de l'Alimentation, de la Pêche et des Affaires Rurales, 2001; Peyron, 2001; Cazaly, 2002; Peyron and Colnard, 2002; Peyron et al., 2003; Tabourel and Peyron, 2004), forest accounting is only just developing. Moreover, in spite of the substantial weight of environmental and social concerns in the Mediterranean area, these aspects have not received a greater attention than in other regions. Thus, in this study, data are often derived from national analyses. Finally, it is not obvious to obtain, for the French Mediterranean forests, a rigorous description and a good assessment of their TEV. Therefore, the data and results must be considered as rough preliminary estimates.

Forest Resources

Much of the data presented in this section have been obtained from the indicators for the sustainable management of French forests (Ministère de l'Agriculture, de l'Alimentation, de la Pêche et des Affaires Rurales, 2001) and the preliminary works undertaken on French forest accounts in the Forest Economics Laboratory (LEF ENGREF/INRA, Nancy) on behalf of the French Institute for the Environment (IFEN) and the Statistical Office of the European Communities (EUROSTAT). The data given in the following developments can be considered as representative of the year 2001. However, due

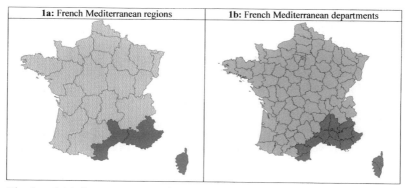

Fig. 20.1. The French Mediterranean area: administrative regions and departments. Source: Laboratoire d'Economie Forestière (2004), for the purpose of this chapter.

to some lack of annual availability, this date has to be taken approximately. Moreover, because of the windfalls of late 1999, 2000 and 2001, data for these periods are not representative of a 'normal' year. The data of year 1999 often provide a better estimate of the current situation of the French forests. Therefore, data given in this chapter are considered to describe France and its Mediterranean area 'around 2001'. The fact that the Mediterranean area is only a small part of France has been used here as an advantage: the same figures are given both for France as a reference and for the French Mediterranean area (regions or departments). Forestlands are successively considered in their territorial and human context, as forest stands and for the various functions they perform.

Area and people

It is as difficult to define the French Mediterranean forests as it is to define the French Mediterranean area. Even though several precise definitions apply to forests, the boundaries between forests and other natural lands are not so clear in the Mediterranean area. Results can be different according to the data source. However, the two main data sources, the TERUTI survey on land use, undertaken annually by the statistical service of the French ministry in charge of Agriculture (SCEES) and the National Forest Inventory (IFN), use more or less the same definitions. One main difference between these sources is the time lag between the two surveys: the TERUTI survey is made on an annual basis while IFN inventories are undertaken every 12 years on average in a given department. However, IFN inventories not only concern the wooded area, as is the case for the forest part of TERUTI, but also the wood volume and increment.

These two data sources are based on similar definitions of a 'forest' when groves (*bosquets*, area covering between 0.05 and 0.5 ha) are included:

- The trees and shrubs must belong to the forest species featuring on a closed list.
- They must have a tree-like shape: individualized stem, relatively straight, branching only above a certain level (~1.5 m), unless

the different shape is the result of treatment to obtain a specific type of product (pollarding) or natural deformation (wind or snow), which does not inhibit the normal use of the trees.
- The apparent forest tree cover eligible for inventory (diameter > 7.5 cm) must be at least 10% of the ground area, or, in the case of young forest trees not eligible for inventory, the density must be at least 500 well-spaced saplings per hectare.
- The stand must have a minimum area of 0.05 ha with a crown width of over 15 m.
- Poplar plantations are also considered.

According to this definition, the French Mediterranean departments contain up to 18% of all French forests. Almost all forests considered as Mediterranean lie within the Mediterranean departments. However, more than half of the wooded areas in these departments are lowland, hill and, above all, mountain forests, thus highlighting the debatable character of the Mediterranean boundaries (Table 20.1).

Around 13% of the French population live in the Mediterranean departments. The Mediterranean population density (104 inhabitants/km^2) is a little lower than the national average (108 inhabitants/km^2). However, important disparities exist between the departments: from 20 inhabitants/km^2 (Alpes-de-Haute-Provence) to 361 inhabitants/km^2 (Bouches-du-Rhône).

The Mediterranean population is mainly urban. Large cities are located along the coastline (such as Marseille, Nice and Montpellier) or in the Rhône river valley (such as Avignon, Nîmes and Orange). The hinterland is rural with a very low population density. This particular situation can be explained by several factors: the mountainous character of many rural areas; a significant decrease in agricultural activities; the importance of the sea and the Rhône corridor for transportation facilities; and the fact that the Mediterranean area is in the south of France and thus attractive for people initially living in other regions.

This attractiveness of the south and the sea also explains the fact that the Mediterranean people appear to be older than other people in France (Table 20.1): the mild climate and the living conditions make the area attractive for many retired people.

Table 20.1. Major land, forest and population characteristics for France and the French Mediterranean departments or regions.

Around 2001	France	Mediterranean	Mediterranean/ France
Land			
Surface area (Mha and %) (MAAPAR/SCEES, 2003)	54.9	7.4	15%
Biogeographical types (Ministère de l'Agriculture, de l'Alimentation, de la Pêche et des Affaires Rurales, 2001)			
Lowland and hill	62%	7%	
Mountain	29%	46%	
Mediterranean	9%	47%	
Total	100%	100%	
Forests			
Forest area (Mha and %) (IFN, 2002)	15.0	2.7	18%
Forest percentage (%)	27	38	
Forest area increase (%/year) (Ministère de l'Agriculture, de l'Alimentation, de la Pêche et des Affaires Rurales, 2001)	+0.5	+0.9	
Forest fires (source: Ministère de l'Agriculture, de l'Alimentation, de la Pêche et des Affaires Rurales, 2001)			
Number (thousand and %)	5.6	3.3	60%
Area (1000 ha and %)	28.6	20.1	70%
People			
Population (million inhabitants and %) (INSEE, 2002)	58.5	7.7	13%
Population distribution (INSEE, 2002)			
Urban (%)	85	76	
Rural (%)	15	24	
Population age structure (Mediterranean regions, %) (INSEE, 2002)			
−20 years old	25	23	
20–59 years old	54	52	
+60 years old	21	25	
Forest area per inhabitant (ha/inhabitant)	0.25	0.41	

Due to the high proportion of urban population, the harsh topography and the decline of agricultural activities, the Mediterranean area has a high forest cover (38%), much more than the rest of France (27%). As a result, the forest area per inhabitant is also higher (0.41 ha/inhabitant) than in the rest of France (0.25 ha/inhabitant). Despite the high forest percentage and the high fire risk (each year, a large proportion of the burned forest is situated in the Mediterranean area, see Table 20.1), the forest area increases annually at a rate of +0.9%, much higher than in the rest of France (+0.5%).

Typologies

Forests of the Mediterranean area are situated on hilly ground, with more than a quarter located above 1000 m in altitude (one-eighth for France) and more than half with a slope greater than 30% (a quarter for France). Consequently, for many, there is poor accessibility, depending on the gradient and distance to a road (Table 20.2).

Regarding the share of high forest: the stand structure in the Mediterranean area is quite similar to that of the rest of France. However, the Mediterranean area is characterized by its large share of coppice (twice that of France). Moreover, although stands inherited from the old broadleaved system of coppice with standards are not very numerous, the mixture of coppice and conifers is relatively well spread over the Mediterranean area. Finally, compartments that are temporarily deforested are encountered less often in the Mediterranean area than in the rest of France, mainly because the management is less intensive and the natural conditions are more severe (Table 20.2).

Table 20.2. Forest characteristics in France in general and in the French Mediterranean departments in particular.

Around 2001	France (%)		Mediterranean (%)		
1. Altitude classes					
0–250 m	41		18		
250–500 m	25		25		
500–750 m	13		18		
750–1000 m	9		14		
1000–1500 m	9		17		
> 1500 m	3		8		
Total	100		100		
2. Slope classes					
0–15%	59		26		
16–30%	16		20		
31–70%	23		48		
> 70%	2		6		
Total	100		100		
3. Accessibility classes					
Easy	73		58		
Average	21		19		
Difficult	6		23		
Total	100		100		

4. Forest structure			% of Mediterranean forest	% of structure total area in France
Poplar plantations	2		~0	~0
Other even-aged forests	47		48	17
Uneven-aged forests	5		4	14
Coppice	15		33	36
Coppice with broadleaved high forest	25		4	3
Coppice with coniferous high forest	5		11	33
Temporarily deforested	1		~0	5.8
Total	100		100	
Broadleaved trees (stand area)	64		54	
Coniferous trees (stand area)	36		46	

5. Forest species area	%	Rank	% of Mediterranean forests	Rank	% of species total area in France
Pubescent oak (*Quercus pubescens*)	6.7	6	17.8	1	48
Holm oak (*Quercus ilex*)	3.1	10	17.3	2	99
Scots pine (*Pinus sylvestris*)	8.2	5	16.0	3	37
Aleppo pine (*Pinus halepensis*)	1.7	> 10	9.8	4	100
Beech (*Fagus silvatica*)	9.4	4	5.9	5	9
Chestnut tree (*Castanea sativa*)	3.6	9	4.9	6	19
Maritime pine (*Pinus pinaster*)	10.0	3	4.2	7	8
European larch (*Larix decidua*)	0.7	> 10	3.2	8	81
Cork oak (*Quercus suber*)	0.6	> 10	3.2	9	99
Austrian pine (*Pinus nigra* var. *austriaca*)	1.3	> 10	3.0	10	42
Silver fir (*Abies alba*)	4.1	8	2.7	> 10	11
Pedunculate oak (*Quercus robur*)	16.9	1	0.1	> 10	9
Sessile oak (*Quercus petraea*)	13.6	2	1.0	> 10	8
Norway spruce (*Picea abies*)	5.4	7	1.2	> 10	4

Source: Ministère de l'Agriculture et de la Pêche/Institut Français de l'Environnement (2001).

Considering the covered area:[1] the Mediterranean forest species are mainly broadleaved trees, but their predominance (54%) is less important than in the whole of France (64%). The characteristics of the Mediterranean vegetation could be underlined through the main forest trees (Quézel, 1979; Table 20.2): for example, whereas the pedunculate oak is the main French forest species (it covers 17% of the French wooded area), it is virtually absent from the Mediterranean forests (Fig. 20.2).

Functions

Since the new forest law No. 2001–602 (9 July 2001), French forests are legally considered as multi-functional and their management is based on productive, ecological and social objectives. Multi-functionality has long been a major objective for French forestry, but its implementation is evolving and its nature has become more formal. Of course, this principle has to be adapted to local situations, owners and social needs. In the particular case of the Mediterranean area, where productive functions are less developed than in other French forests, multi-functionality is undoubtedly much more than a wish, it is an indisputable reality (M'Hirit, 1999; Benoit de Coignac, 2001). However, until recent years, wood revenues were supposed to finance the greatest part of the whole management, thus including actions towards ecological and social functions. Nowadays, it is recognized more and more that voluntary policies are required in order to develop non-wood benefits in an efficient way, in agreement with the wishes of society. In this context, economic evaluation is intended to play an increasing role.

IFN data allow 'productive forests' to be distinguished from 'other forests'. The 'other forests' category includes all forests unavailable

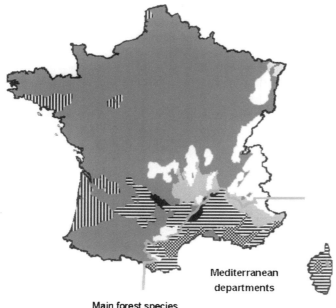

Mediterranean
departments

Main forest species
Castanea sativa
Quercus ilex, Quercus pubescens
Other broadleaved trees
Pinus pinaster
Pinus silvestris
Other pines
Other coniferous trees

Fig. 20.2. French Mediterranean forest types. Source: IFN as published in Institut Français de l'Environnement (2002).

for wood production (due to, for example, inaccessibility and steep slopes) or strictly protected forests (biological reserves), wooded areas with landscape, tourist, recreational or cultural objectives, and the wooded parts of military ground. According to this classification, the Mediterranean forests appear to be potentially less productive than those of the rest of France: 90% of the wooded area could be utilized for wood production versus 95% in the whole of France. Overall, Mediterranean forests appear

actually to be less productive than the rest of France. They only contribute to about 9% of French removals and mainly in low-quality roundwood and non-marketed firewood (Table 20.3).

The multi-functional nature of the French forest contributes to the difficulty of classifying forest areas into different possible uses: they are generally not excludable. However, several indicators[2] are useful in order to approximate these different uses (Table 20.3).

Table 20.3. Several indicators of forest multi-functionality and comparison between the French and Mediterranean forest functions.

Around 2001	France	Mediterranean departments
Productive forests		
Area (%)	95	90
Total roundwood removals (Mm³/year)	64.0	5.8
Hardwood logs	8.0	0.1
Softwood logs	15.2	0.8
Pulpwood	11.4	0.7
Other marketed	3.2	0.4
Non-marketed (firewood)	26.2	3.8
Roundwood losses (Mm³/year)	7.3	1.4
Harvesting	5.8	0.6
Fires	1.2	0.8
Storms	0.3	
Truffles (t/year)	60	48
Cork removals (000 t/year)	10	10
Tourism/recreation		
Global capacity (forest ha/inhabitant)	0.25	0.41
Urban fringe forest (urban fringe forest ha/urban fringe inhabitant)	0.13	0.18
National parks		
Buffer zone (Mha)	0.91	0.15
Million visits/year	6.2	2.0
Regional parks		
Total area (Mha)	4.9	1.0
Forest area (Mha)	1.3	0.4
Protective forests		
Specific habitats (area of old regular stands)	2.4%	5.8%
Mixed stands (≥ 3 species within a 25-m radius)	44%	23%
Erosion (gradient > 30%)	2.9%	7.8%
Avalanche (altitude > 1500 m)	25%	53%
Biodiversity in national parks		
Core zone (Mha)	0.35	0.07
Of which forest	0.08	0.02
Biodiversity in regional parks		
Total area (Mha)	4.9	1.0
Of which forest	1.3	0.4

Sources: IFN (data bought by LEF); FFT (personal communication); Ministère de l'Agriculture, de l'Alimentation, de la Pêches et des Affaires Rurales (2001); Ministère de l'Agriculture et de la Pêche/SCEES (2000); Parcs nationaux (website).

The recreational function of forests can be assessed through the available area per inhabitant. The recreational capacity of Mediterranean forests is almost twice that of the whole of France. However, this Mediterranean advantage diminishes when this recreational capacity is estimated within a radius of 50 km around the large cities (but 100 km around Paris). Although forests are very much present in the Mediterranean area, many of them are far from the main cities and are not easily accessible for most of the population.

Among the six metropolitan national parks, four are located partly or completely in the Mediterranean area, but only two are really characterized by Mediterranean vegetation and climate: Port-Cros and Mercantour. They represent 18% of the total area of French national parks, and their buffer zones are visited each year by around 2 million visitors (32% of national park annual visitors, Table 20.3). Eight of the 42 regional parks are also located in the Mediterranean area and are very important for the development of tourism.

If the productive function of Mediterranean forests is less important than the French average, their protective function is highly significant. A large proportion of the Mediterranean forests perform a protection function against avalanches (altitude > 1500 m). Considering soil protection: more than half of the Mediterranean forests play a significant role against soil erosion (gradient > 30%), whereas only a quarter of the French forests can be included in this category. This role can be very important at lower elevations, notably regarding the silting up of reservoirs, as studied by EDF (Électricité de France) in the context of the Durance dam in the Mediterranean area. The participation of forests in water protection is difficult to quantify: few data are available at the national level (Tabourel and Peyron, 2004), which is not representative of the Mediterranean area.

Wildlife is plentiful and plant biodiversity is rather high in the Mediterranean area (Quézel, 1999). Its conservation can be assessed through several indicators, for example national and regional parks. The cores of national parks are devoted to wildlife and habitat protection. About 20% are situated in the Mediterranean area. Moreover, the eight regional parks located in the Mediterranean area also perform an ecological function.

The area of very old regular stands forming specific habitats is larger in the Mediterranean area than in the whole of France (Table 20.3), but the proportion of mixed stands is much lower in the Mediterranean area than in the rest of France (Table 20.3).

Institutional Aspects

The special nature of French Mediterranean forests, as shown through the physical and ecological features described above, is also remarkable at the institutional level.

Ownership and size of properties

Almost three-quarters of the French forests are private property. Public forests make up the remaining 26%. They are divided into State-owned forests (10%) and forests that belong to communes and other public bodies (16%), mainly located in the eastern part of the country or in mountainous regions. The ownership structure is almost the same in the Mediterranean part (Table 20.4).

Public forests are managed by the French Forest Agency (ONF, Office National des Forêts). Private forests can be managed by forest consultants, cooperative bodies, marginally by ONF, or by the owner himself/herself. All forest estates larger than 25 ha must be managed according to a management plan approved by the regional forest ownership agency (CRPF, Centre Régional de la Propriété Forestière).

The average size of private forest properties is a little higher in the Mediterranean area (3.59 ha) than in the whole of France (3.05 ha). However, the size distribution is quite similar between France and the Mediterranean area, even though very large estates are less numerous in the south-east of France than in the rest of the country.

Administration and policies

The forest policy is under the jurisdiction of the State, which ensures its national coherence.

Table 20.4. Forest ownership structure in France and Mediterranean departments or regions.

Around 2001	France		Mediterranean departments	
Public forest	26%		28%	
State-owned forest	10%		11%	
Local community-owned forests	16%		17%	
Private forest	74%		72%	
Total area (Mha)	10.6		1.4	
Number of owners (million)	3.5		0.4	
Average size (ha/estate)	3.0		3.6	

	No. of owners (1000, %)	Total area (1000 ha, %)	No. of owners (1000, %) (Regions)	Total area (1000 ha, %) (Regions)
Less than 1 ha	2361	745	250	85
	(68)	(7)	(63)	(6)
More than 1 ha	1122	9875	148	1344
	(32)	(93)	(37)	(94)
Of which 1–10 ha	83	30	82	31
Of which 10–25 ha	11	18	11	18
Of which 25–100 ha	5	27	6	33
Of which 100 ha +	1	25	1	19

Sources: IFN (data bought by LEF); MAAPAR/SCEES, 2002.

These national orientations can be adapted and completed at the regional level through the Regional Forest Orientations that define local priorities. Moreover, this forest policy is also the result of other policies such as the Common Agricultural Policy, the environmental policy, the land use policy and others.

The legal status of forests concerns the administrative, civil and penal laws. Since 1827, the French forest code constitutes the core of the French forest policy, gathering together the main legislative provisions relating to forests. This first version of the forest code was in keeping with the general pattern at the beginning of the industrial era: it exactly reflected the human–forest relationship of the time, assigning to the forest a productive main objective. However, this first version also reflected the premises of the modern notion of sustainable management of forests, gathering together several measures regulating cuts and deforestation. This text was revised and modified many times until the most recent forest law, No. 2001–602.

This forest law of July 2001 has three main objectives:

- To insert the management of French forests into an international and European framework for sustainable forest management.

- To respond to new explicit and implicit social demands towards forest biodiversity and forests in general (improvement of living conditions, promotion of recreation in forests, preservation of the environment, air and water protection, production and utilization of wood-based ecomaterials).

- To foster the economic valuation of the increasing wood production potential in French forests.

Following these modern bases, the French forest policy is now explicitly characterized by sustainable management objectives, in order to ensure forest conservation, productivity and global multi-functionality for present and future generations.

Nowadays, the forest code manages: multi-functionality; interfaces between forests and other natural lands; relationships between public authorities and other actors; and protective, productive and recreational measures. It is divided into five parts. The first part constitutes the core of the forest policy relating to public forests, and defines the ONF assignments. Private forest management dispositions are the subject of the second part. The third part is devoted to general law and order maintenance measures (such as clearing control). The fourth part records all the measures aiming to avoid human or climatic forest damage (such as forest fire protection or mountainous area maintenance). Finally, the fifth part deals with qualitative and quantitative forest evaluation tools (National Forest Inventory).

As mentioned above, the forest policy also depends on other policies such as regional planning and sustainable land development. Several dispositions related to forest management and protection are written down in the rural code (afforestation rules and coastal zone law), in the general tax code (death tax modalities) and in the environmental code (nature protection and biodiversity conservation).

A set of incentive measures is available for forest owners in order to act in the spirit of the multi-functionality objectives. Direct subsidies are, for example, provided in order to promote development and competitiveness in the forest and wood chain. With this aim, public authorities encourage forest owners to maintain wooded areas and to improve wood supply in both quantity and quality, notably through subsidies devoted to afforestation and reforestation, coppice conversion into high forests, stand quality and stability improvement, forest road maintenance and creation, material and non-material forest and cork investment, start-up and development of forest companies, wood promotion, and equipment against disasters and forest fire. The government can also promote the development and competitiveness of forests and the wood chain through the training and grouping of forest owners. This education is encouraged through national assistance to the constitution of official management documents (*Plans Simples de Gestion*), a genuine tool for piloting a forest and ensuring its 'good' management. Public commitment in favour of forest owners' associations and management pooling is the responsibility of the private forest property division. This situation induces an organizational problem in the forest and wood chain (difficulty connected with the huge number of suppliers for the processing industry).

Public policy also encourages natural ecosystem protection and management. In that sense, two incentive measures are especially devoted to forest biodiversity protection. Both of them aim to favour the silvicultural processes that principally ensure biodiversity protection or restoration. One applies to Natura 2000 areas and the other to other sites. Other measures apply to a wide range of forest facilities and environmental concerns such as defence against forest fires, maintenance of mountainous areas, fixation of coastal dunes, restoration of acidified soils, creation and maintenance of hedgerows and scattered trees, and public recreation facilities.

Finally, forest owners benefit from several tax exemptions and tax allowances. In order to avoid detrimental cuts, forest areas are partly exempt from death duties. With the intention of compensating non-profitable silvicultural operations (planting and restocking), young-forested areas are temporarily exempt from land tax and the owner's taxable profit is reduced. With the aim of motivating investors and developing chain competitiveness, forest purchase carries a tax allowance. Public authorities allow a VAT discount on forest works with the goal of improving forest productivity. Finally, forest ownership integrity is encouraged by partial exemption from inheritance tax in cases where unity of ownership is maintained (non-division, grouping and others).

The financial funds required for this programme are provided by the general State budget, and for the management and maintenance of public forests, from the proceeds of wood sales and hunting rents. For 1999, public funds allocated to forest policy by the French ministry in charge of agriculture were around €265 million (Table 20.5).

Since 2000, the European Union (FEOGA-G) has contributed to the French forest policy up to 40% of forest measures included in the PDRN.[3]

In order to adapt this national structure, the French forest code provides for Regional Forest Orientations (ORF, *Orientations Régionales Forestières*), which is updated periodically. For the three Mediterranean regions, the main priorities appear to be related to the sustainable and multi-functional management of forests according to their special nature. The emphasis is put on biodiversity protection, recreational function, defence against forest fires, regional resource improvement, cut and use of timber products, and research and extension efforts (Ministère de l'Agriculture et de la Pêche, 1988, 1998, 1999).

Table 20.5. National funds allocated to forest policy.

Around 2001	France (million €)	Mediterranean regions (million €)
Research	3.1	—
Forest extension programme	17.7	—
National Forest Inventory	6.5	—
Compensatory payment (community forests)	133.9	Not available
Protection (fire, dunes, mountain, etc.)	48.2	12.5
Forest works (afforestation, roads, etc.)	36.3	21.8
Forest industries	16.8	3.3
Specific functioning	1.5	—
Acquisitions	0.8	—
Total	264.8	37.6

Sources: Institut Français de l'Environnement (2002); Office National des Forêts (2002).

Mediterranean forest authorities are noticeably interested in forest fire protection and in mountainous and littoral soil protection. Considering forest fire protection (DFCI, Défense des Forêts Contre l'Incendie), a large set of measures exists, from regulation decisions (such as obligation of clearing undergrowth, fire use and road traffic limitations) to action and information initiatives (equipment, planning and education). Mountainous soil protection is ensured (as in the whole of France) by the RTM programme (Restauration des Terrains en Montagne). In order to fight against soil erosion, landslide and avalanches, it consists of reforesting vulnerable lands or sowing them with grass seed or in correcting stream banks. Finally, littoral protection is covered by the adoption of the 'coast law' (Law 86–2 dated 3 January 1986) that concerns littoral planning, protection and land development. The main basis of this law is the idea that 'spaces close to the shore do not have the role of receiving a significant urbanization'. Free use by the public constitutes the fundamental role of beaches. This law, through such dispositions as the prohibition of isolated house construction or the necessity to preserve remarkable sites, landscapes and biological stability, concerns Mediterranean forests.

On a very small scale, the land forest charters (*chartes forestières de territoire,* or CFT), instituted by the new forest policy act, unquestionably provide a suitable framework for local appropriation of forest issues. They consist of a new multi-objective contractual tool at the local development stakeholders' disposal (such as public or private forest owners, local councillors and users) in order to ensure the best response to society's expectations with respect to forests. Over the first 21 charters, five sites are situated in the Mediterranean area. According to the Mediterranean forest priorities presented above, they focus mainly on defence against forest fire, conservation and development of non-timber forest products (cork, public recreation and firewood).

Finally, several very active organizations exist in order to promote Mediterranean forests and their products and to encourage dialogue between different stakeholders (e.g. 'Forêt Méditerranéenne', 'Observatoire de la Forêt Méditerranéenne', 'Institut pour la Forêt Méditerranéenne' and 'Conservatoire de la Forêt Méditerranéenne').

Contribution of the Forest to the National Economy

Gross domestic product and employment

The forest and wood chain industries provide almost 1.5% of the total French employment and 6% of the agricultural and industrial employment, structured in three job categories: forest management and primary processing industries, the timber industry and commercial activities (Daly-Hassen and Peyron, 1996). Only 3.5% of these jobs are situated in the Mediterranean area (Service des Etudes et Statistiques Industrielles, 2004).

In the Mediterranean area, more than in the whole of France, the forest and wood chain is characterized by its small-scale production: a significant proportion of firms have less than 20 workers. For example, whereas the average number of employees in a French sawmill is approximately nine, it is only four in the Languedoc-Roussillon region.

It can be assumed that the forest activities and industries represent approximately the same share of the total gross domestic product (GDP) as in the total employment. Under this assumption, the French forest activities and industries contribute about 1.5% of the GDP. As for the Mediterranean forest activities and industries, they represent about 3.5% times the 1.5% of the total French GDP. Since the Mediterranean area contributes up to 10.5% of the French GDP (Institut National des Statistiques et des Etudes Economiques, 2002, website), this corresponds to about 0.5% of the total GDP of this Mediterranean area. The forest contribution to the economy is much lower for the Mediterranean area than for the whole of France. However, non-market values are not taken into account in the GDP, nor in employment figures.

International trade – wood import/export

France is a forest country with a significant and expanding timber-based industry. It is the largest producer of sawn hardwood in Europe. However, it is a net importer of forest products and mainly of pulp and paper, sawn softwoods and pieces of furniture. Considering only logs and pulpwood production and processing, annual removals are 34.6 Mm^3 over bark/year (Table 20.3). Total net imports (imports minus exports) can be estimated for all forest products from roundwood to paper and pieces of

furniture in roundwood-equivalent cubic metres through appropriate coefficients (Peyron *et al.*, 1999), which is 16.9 million roundwood-equivalent m^3/year (Laboratoire d'Économie Forestière, unpublished). It means that, for logs and pulpwood, two-thirds of French needs are covered by French forest resources and one-third by imports. Considering not only logs and pulpwood but also all roundwood supplied to industries and households, the annual removals amount to 64 Mm^3/year (Table 20.3) and the imports are about one fifth of the total needs. The same figures are not available for the Mediterranean area but it is obvious that the shortage of wood is still greater than that nationally.

Mediterranean forest products are mainly intended to satisfy local needs and few products are exported (except cork products, sawnwood and veneer sheets, essentially to Spain and Italy). The share of the French Mediterranean area in the international trade of wood forest products is much higher in total imports (5.6%) than in total exports (1.4%) (Table 20.6).

Values of French Forests

In this attempt to assess the total economic value (TEV) of the French and Mediterranean forests, it must be made clear that: (i) due to the lack of information, some data are missing and some others are rough estimates; (ii) the classification between direct use, indirect use and other values gives an idea of the main nature of some global results that should be, but have not always been, divided into these categories; and (iii) the valuation methods are given in their broad principles but one must consider that the addition of the resulting values is debatable.

Table 20.6. International trade of French wood products.

Around 2001	France	Mediterranean area
Origins of main imported products	Germany	Italy, Congo
Destinations of main exported products	Germany, Belgium	Spain, Italy
Mediterranean area share in the French imported wood products		5.6%
Mediterranean area share in the French exported wood products		1.4%

Source: DGDDI (data bought by LEF for the years 1993–2002).

Direct use values

The roundwood value is based on the volumes removed from the forests and the stumpage prices. Since the latter are known only for whole trees, their estimates for parts of trees (such as logs and pulpwood) are obtained through a conversion matrix used for the French forest accounts (Tabourel and Peyron, 2004). Moreover, for most species and qualities except softwood logs, stumpage prices in the Mediterranean area are half of their price elsewhere in France. The results show clearly the importance of non-marketed roundwood (mainly firewood) in the Mediterranean area compared with industrial roundwood. Moreover, conifers are the main source of industrial roundwood.

About 80% of French truffles are produced in the Mediterranean area at a price that is about €400/kg on the wholesale market. Self-consumed production is supposed to represent 50% more than marketed production, which is valued at the same wholesale price. Cork is seldom harvested in France – only in the Mediterranean area. Resin is no longer collected. Medicine, aromatic and decorative plants, as well as mushrooms and berries, are used either by professional people or by households. In the latter case, they are a result of a recreational activity and have been considered with forest services (see below). When professional people buy them, they have been classified as non-wood products. On the supply side and for public forests, they are monitored by ONF as miscellaneous marketed products in a manner consistent with roundwood stumpage prices (communal forests represent an additional value of 16% relative to State forests). On the demand side and for both public and private forests, their quantities are also published for some of them, but the corresponding forest prices are difficult to assess. In order to avoid double counting and to estimate the correct prices, only ONF data have been used. Thus underestimation can occur because of the products harvested in private forests. Finally, marketed non-wood products are valued much less than roundwood, but their share is higher in the Mediterranean area (Table 20.7).

Forest services include hunting, picking and recreation. The value of these services is given by a national survey carried out on 2575 households at the national level and 352 in the Mediterranean area (Peyron et al., 2003). This approach constitutes a generalization of previous research undertaken in the French Lorraine region (Després and Normandin, 1998; Marchesi, 1999; Peyron, 2001). Following this study, it is considered to equal (at least) the travel costs (fuel and vehicle, €0.24/km on average) borne by households when they travel to forests by car. The opportunity costs of the time spent on travelling to the site have been ignored. Moreover, an attempt to derive the total consumer surplus has shown that this variable was certainly much higher (perhaps five times) than the total travel costs (Peyron et al., 2003), but the authors have considered that this result was not reliable enough to be published more precisely. Therefore, the value measure used here is the total travel costs, which are also quite easy to assess. The assumption is made that, since households support such costs, they assign at least the same value to their visits. This is a cost price method (and not a travel cost method which would estimate a consumer surplus).

In order to distinguish between hunting and picking (mainly mushrooms and berries), forest visits have been divided according to these activities and the total travel costs have been computed for each of them. Under this assumption, the results show that forest recreational activities play a major role. Despite the use of different estimation methods, they seem to have a higher value than forest products. The gap is even more important in the Mediterranean area, due mainly to the weakness of the roundwood value. In the Mediterranean area, picking is more frequent than in the rest of France but hunting is practised less. Considering all activities, the distance between home and forest is greater in the Mediterranean area (see the travel cost/visit).

Indirect use values

Indirect use values considered here are divided into two broad categories: carbon sequestration and other economic, social and environmental interests.

Carbon sequestration is an indirect forest use that contributes to a better air quality. The change in carbon inventory is a consequence of

Table 20.7.   Summary of annual direct use values for French and Mediterranean forest products and services.

Around 2001	France			Mediterranean departments			Mediterranean/France
	Quantity	Price	Value	Quantity	Price	Value	% of value
Roundwood (stumpage price method)	Mm^3	€/m^3	Million €	Mm^3	€/m^3	Million €	
Hardwood logs	8.0	62	496	0.1	31	2.5	0.5
Softwood logs	15.2	32	484	0.8	32	24.6	5
Pulpwood	11.4	7.2	82	0.7	3.6	2.4	3
Other marketed	3.2	9.4	30	0.4	4.7	1.6	5
Non-marketed	26.2	9.4	247	3.8	4.7	17.9	7
Total	64.0	20.9	1,339	5.8	8.5	49	4
Non-wood products (Wholesale or forest price)	t/year	€/t	Million €	t/year	€/t	Million €	
Truffles	60	400,000	24	48	400,000	19	80
Cork	10,000	300	3.0	10,000	300	3.0	100
Miscellaneous	–		9.2	–		1.4	15
Total	10,060		36.2	10,048		23.4	65
Forest services (cost price method)	Million visits	€/visit	Million €	Million visits	€/visit	Million €	
Hunting	10	10.0	96	0.3	20.1	5.8	6
Picking	20	4.2	85	3.0	8.9	27	31
Other recreation	394	4.4	1,718	46	7.0	325	19
Total	424	4.5	1,899	50	7.2	358	19

Sources: Ministère de l'Agriculture, de l'Alimentation, de la Pêches et des Affaires Rurales (2001); Peyron *et al.* (2003); Office National des Forêts (2004; unpublished).

two physical phenomena: an increase in forestlands and an increase in the standing volume (Table 20.8). The increase in forestlands induces changes in carbon stored in soils, ground vegetation and foliage. The results are based on average values of the carbon content of all soils (70 tC/ha) and, for ground vegetation and foliage, on average values related to broadleaved and coniferous areas. When the predominant species is unspecified by the National Forest Inventory, then the broadleaved (lower) value has been used. The increase of the standing volume is both above-ground (stems and branches, alive or dead) and underground (roots). It is the result of the annual growth, removals, mortality and other losses and is given simply by the change in inventory. The corresponding carbon content is computed from average figures related to poplars, other broadleaved trees and conifers (Peyron and Colnard, 2002; Table 20.8). All these changes are assumed to occur immediately and are considered independently of the previous land use and carbon content. Since the average carbon

content of forest soils is 70 tC/ha, each hectare reforested adds 70 tC to the carbon stock and each hectare deforested represents a loss of the same amount. The price per tonne of carbon has been considered to be €20/tC, which seems to correspond to the average price found in the literature (Fankhauser, 1995; Table 20.9).

Under these assumptions, the Mediterranean area is characterized by a lower annual sequestration in the woody biomass than the rest of France, but by a much higher one in the ground, the vegetation and the foliage. Despite a low productivity of Mediterranean forests, the rapid (and mainly natural) afforestation rate creates a new carbon sink. Consequently, the change in above-ground and underground woody biomass is responsible for only a small fraction, a little more than one-third of the increase of forest carbon storage in the Mediterranean area, but for about two-thirds in France as a whole. Nevertheless, these results are based on French average carbon contents and ought to be improved by the inclusion of regional data.

Table 20.8. Figures related to carbon sequestration in France and the Mediterranean area.

Around 2001	Annual increase	Carbon factor	Carbon sequestration (Mt C/year)
France			
Soil (1000 ha/year)[a]			
All forestlands	74.2	70	5.2
Vegetation and foliage (1000 ha/year)[a]			
Broadleaved area	46.6	4.4	0.2
Coniferous area	9.2	10.6	0.1
Unspecified area	18.4	4.4	0.1
Total area	74.2		0.4
Standing timber (Mm³/year)[b]			
Poplar	20.0	0.30	6.0
Other broadleaved	10.4	0.28	2.9
Coniferous	−0.2	0.28	−0.1
Total volume	30.2		8.9
Roots (Mm³/year)[b]			
Poplar	20.0	0.054	1.1
Other broadleaved	10.4	0.062	0.6
Coniferous	−0.2	0.062	0.0
Total volume	30.2		1.7
Mediterranean departments			
Soil (1000 ha/year)[a]			
Total area	27.4	70	1.9
Vegetation and foliage (1000 ha/year)[a]			
Broadleaved area	12.8	4.4	0.1
Coniferous area	2.6	10.6	0.0
Unspecified area	12.0	4.4	0.1
Total area	27.4		0.2
Standing timber (Mm³/year)[b]			
Poplar	2.0	0.30	0.6
Other broadleaved	1.4	0.28	0.4
Coniferous	0.0	0.28	0.0
Total volume	3.4		1.0
Roots (from standing timber) (Mm³/year)[b]			
Poplar	2.0	0.054	0.1
Other broadleaved	1.4	0.062	0.1
Coniferous	0.0	0.062	0.0
Total volume	3.4		0.2

Sources: Institut Français de l'Environnement (2002); Ministère de l'Agriculture, de l'Alimentation, de la Pêche et des Affaires Rurales (2001).
[a]Carbon factor = tC/ha.
[b]Carbon factor = tC/m³.

Economic, social and environmental interests that are maintained by the forest coverage give another indirect use value. An example of such interests is the protection of landscapes and soils against forest fires, avalanches, landslides, floods and erosion. Few valuations have been made and, in spite of their significance, they are usually related to a precise site, as is the case for the value of landscape in the Cévennes mountains (Noublanche, 1999).

However, defensive expenditure is necessary to maintain or improve the capacity of low-production forests to ensure their functions. It constitutes a measure of the minimum value attributable to the benefits provided by these forests, since these benefits are likely to be at least equal to the expenses incurred for avoiding damage. In the Mediterranean area, they concern mainly the fight against forest fires, with prevention and firefighting costs that have been roughly estimated here to be 75% of the total French costs.

Option, bequest and existence values

The total removals are much lower than the annual current increment, both in the Mediterranean area and in France in general. This fact contributes to an increase in the total standing timber that has been termed here as 'saved increment'. This saved increment is given in Table 20.10. It is on average 1.9 m³/ha/year in France and 1.2 m³/ha/year in the Mediterranean area. The so-called stumpage price method has been used to value this saved increment by using half of the stumpage price

in order to take into account the fact that this volume will probably not be totally harvested, or not harvested at the right time. Results show that this saved increment is low in the Mediterranean area, as are all variables related to wood production.

Biodiversity is also considered to have an option, bequest and existence value. The method used to estimate this value is the so-called contingent valuation method that has been implemented in 2001 at the national level (Peyron et al., 2003). Two main questions were asked in a survey of a sample of 2575 French households (among which 352 were Mediterranean): (i) 'In France, among animal species (vertebrates) living in forests, 2% are threatened with disappearance; moreover, 12% are vulnerable and 6% are rare; as for plants, about 2% are

Table 20.9. Summary of indirect use values for French and Mediterranean forests per year.

Around 2001	France	Mediterranean departments	Mediterranean/France
Carbon sequestration			
(shadow pricing: €20/tC)			
Soil	€104 million	€38 million	37%
Vegetation, foliage	€8 million	€3 million	36%
Standing timber	€178 million	€20 million	11%
Roots	€34 million	€4 million	11%
Total	€324 million	€65 million	20%
Conservation of some economic,			
social and environmental interests			
(defensive expenditure method)			
Fight against forest fires	€70 million	€50 million	75%
Prevention of forest fires	€31 million	€23 million	75%
Defence costs on mountains	€16 million	€6 million	40%
Defence costs along coasts	€1 million		
Total	€118 million	€79 million	67%

Sources: Institut Français de l'Environnement (2002); Office Nationale des Forêts (2002).
See also Table 20.8.

Table 20.10. Summary of option, bequest and existence values for French and Mediterranean forests.

| Around 2001 | France | | | Mediterranean departments | | | Mediterranean/ France |
	Quantity	Price	Value	Quantity	Price	Value	% of value
Saved increment	Mm³	€/m³	Million €	Mm³	€/m³	Million €	
(half stumpage price method)	29.2	10.45	305	3.3	4.25	14	5
Biodiversity	Households	€/hh	Million €	Households	€/hh	Million €	
(contingent valuation method)	23.8 million	15.2	362	3.0 million	21.1	63.3	17

Sources: Ministère de l'Agriculture, de l'Alimentation, de la Pêche et des Affaires Rurales (2001); Peyron et al. (2003).

threatened or vulnerable. Thus the biological diversity of forests appears as a patrimony to be preserved by various protection and maintenance measures that have direct and indirect costs. On behalf of your household, would you be ready to dedicate annually to biodiversity of the French forests an amount of [x, 0 < x < €90]?' (the precise amount was a multiple of €6 and 15 different amounts were randomly proposed); and (ii) 'What is the maximum amount you would pay?'

Table 20.9 gives the total number of households in France and in the Mediterranean area, and the average and the total willingness to pay of French and Mediterranean households. One must note that the Mediterranean result is the willingness to pay of Mediterranean people for the French forests and not the willingness of the French to pay for the Mediterranean forests. However, these two figures are probably not very different. Moreover, in that case, the result is a consumer surplus. The willingness to pay of Mediterranean people is slightly higher than that of the French in general.

Negative externalities linked to forests

Among negative externalities, agricultural damage caused by forest game can be estimated from the compensation that is paid to farmers (Table 20.11). As for allergic factors, only one part has been estimated from the chemical treatments organized against caterpillars and for health reasons. They are particularly important in the Mediterranean area in comparison with the whole of France. These externalities are certainly underestimated because pollen impacts on health, drawbacks linked with shade and non-forest risks due to forest fires have not been valued. However, it is obvious that these negative externalities are limited in comparison with all positive values.

Towards the Total Economic Value of French Forests

The previous results are summarized in Fig. 20.3.

Table 20.11. Negative externalities for French and Mediterranean forests per year.

Around 2001	France	Mediterranean departments	Mediterranean/France
Allergic factors (caterpillars)	2.3 M€	0.6 M€	25%
Agricultural damages by forest game	21 M€	2.6 M€	13%

Source: Ministère de l'Agriculture, de l'Alimentation, de la Pêche et des Affaires Rurales (2001); Office National des Forêts (2002); Office National de la Chasse et de la Faune Sauvage (unpublished)

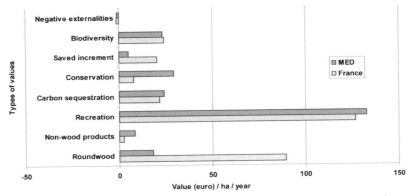

Fig. 20.3. Market and non-market estimated values for the whole of France and for the Mediterranean area (euros per forest hectare and per year).

The various values that have been estimated are not of the same nature. Some are
estimated on the basis of market prices and
others are consumer surpluses. Their addition is
therefore questionable. However, if one considers that the objective is to approximate the TEV
with the best available information and without
any double counting, these different values can
be compared or added with care. Under such a
strong assumption, the TEV of French forests
is approximately €4363 million (€291/ha), and
€648 million (€240/ha) for the Mediterranean
area.

Figure 20.3 shows some similarities, but
also significant differences, between France and
the Mediterranean area. In both cases, recreation is the main source of value. It represents
about half of the TEV (44% in France; 55% in
the Mediterranean area). Moreover, biodiversity
and carbon sequestration seem to have the
same importance (7–8% in France and 10% in
the Mediterranean area). The main difference
concerns roundwood values (including saved
increment), which are very low in the Mediterranean area (10%) compared with the whole of
France (38%). Conversely, non-wood products
and conservation (prevention of fires and
erosion) are more important for Mediterranean
forests (16% in total) than for the whole of
France (4% only).

The presented results need to be improved,
developed, detailed and repeated. However,
they constitute a first attempt to quantify the TEV
of Mediterranean forests. They also have the
advantage of allowing a comparison between
the whole of France and the Mediterranean area.
For the latter, they show that, even though the
economy of roundwood is very poor, forests
have a large national value that ought to be
taken into account in all decisions concerning
them.

Some policy implications can be drawn
from such results. Firstly, recreation certainly
appears much more important than is presently
considered in forest management. Secondly,
although roundwood is a large source of market
value and also an indirect source of other values
through maintaining forests in a healthy state, it
is not the main one. Of course, it must not be
neglected, but neither should it mask all other
values that could perhaps be enhanced through
an improved management. Finally, as this

phenomenon is still more acute in the Mediterranean area, Mediterranean regions could become
a reference for the management of a genuine
multi-functionality of forests.

Notes

[1]   Areas refer to stands in which the considered
species is the main one but does not cover the whole
land and may be mixed with one or several other less
represented species.
[2]   Data are generally available in an aggregate
(national) format, thus the choice of the
indicators presented in this section was guided
by the availability of 'departmental' or 'regional'
data.
[3]   PDRN (Plan de Développement Rural National;
National Rural Development Plan) is a public
document that defines, within the European
framework, the French rural policy for the period
2000–2006.

References

Benoit de Coignac, G. (2001) Fonctions de la forêt
    méditerranéenne et inventaire forestier. Revue
    Forestière Française 53, 304–309.
Cazaly, M. (2002) La forêt méditerranéenne française
    et son public; résultats d'une enquête par
    sondage. Forêt Méditerranénne 3, 173–184.
Daly-Hassen, H. and Peyron, J.-L. (1996) L'emploi
    salarié dans la filière bois française: situation et
    évolution récente. Revue Forestière Française
    48, 153–166.
Després, A. and Normandin, D. (1998) Evaluation des
    Services Écologiques et Récréatifs des Forêts, un
    Exemple en Lorraine. INRA-ESR Nancy, Unité
    Économie et Politique Agricoles et Forestières.
Fankhauser, S. (1995) Valuing Climate Change.
    The Economics of the Greenhouse. Earthscan,
    London.
Illy, G. and Pinatel, B. (1997) Histoire de la
    forêt méditerranéenne. Comptes Rendus de
    l'Académie d'Agriculture de France 83, 37–53.
INSEE (2002) Annuaire Statistique de la France (105°
    Edition). INSEE Paris.
Institut Français de l'Environnement (2002) Forêt,
    Économie et Environnement; Rapport de la
    Commission des Comptes et de l'Économie de
    l'Environnement. IFEN, Orléans, France.
Marchesi, S. (1999) Evaluation des Services
    Environnementaux des Forêts en Lorraine.
    ENGREF Thesis.

M'Hirit, O. (1999) La forêt méditerranéenne: espace écologique, richesse économique et bien social. *Unasylva* 50.

Ministère de l'Agriculture, de l'Alimentation, de la Pêche et des Affaires Rurales/Service Central des Enquêtes et Etudes Statistiques (2003) *Statistiques Forestières 2001*. Agreste, Chiffres et Données, Agriculture, no. 147.

Ministère de l'Agriculture, de l'Alimentation, de la Pêche et des Affaires Rurales (2001) *Indicators for the Sustainable Management of French Forests (2000 Edition)*. Ministère de l'Agriculture, de l'Alimentation, de la Pêche et des Affaires Rurales, Paris.

Ministère de l'Agriculture et de la Pêche/SCEES (2000) *Statistiques Forestières 1998 et 1999*. Agreste, Chiffres et Données, Agriculture, no. 129, 89 p.

Ministère de l'Agriculture et de la Pêche (1988) *Orientations Régionales Forestières de la Région Corse*. Ministère de l'Agriculture et de la Pêche, Ajaccio, France.

Ministère de l'Agriculture et de la Pêche (1998) *Orientations Régionales Forestières de la Région Languedoc-Roussillon, Tome 1*. La forêt et ses produits: descriptions et enjeux. Tome 3: Développement durable. Ministère de l'Agriculture et de la Pêche. Montpellier, France.

Ministère de l'Agriculture et de la Pêche. (1999) *Orientations Régionales Forestières de la Région Provence Alpes Côte d'Azur, Tomes 1 et 2*. Ministère de l'Agriculture et de la Pêche, Marseille, France.

Noublanche, C. (1999). Evaluation Économique du Paysage. Quelles Possibilités d'Identification des Composantes de la Demande pour l'Aide à la Décision Publique. Thèse pour le doctorat de sciences économiques et de l'Université Montpellier 1.

Office National des Forêts (2002) *Rapport Annuel 2001*. Office National des Forêts, Paris.

Peyron, J.-L. (2001) *Première Évaluation Économique Globale des Dégâts Forestiers dus aux Tempêtes de 1999*. Ministère de l'Aménagement du Territoire et de l'Environnement, Paris.

Peyron, J.-L. and Colnard O. (2002) Vers des comptes de la forêt? In: *Forêt, Economie et Environnement: Rapport de la Commission des Comptes et de l'Économie de l'Environnement*. Institut Français de l'Environnement (IFEN), Orléans, France, pp. 169–190.

Peyron, J.-L., Normandin, D. and Berthier, A. (1999) Production et consommation des produits du bois. *Chambre d'Agriculture* no. 892, dossier 'la forêt française: constats et perspectives' pp. 14–17.

Peyron, J.-L., Harou, P., Niedzwiedz, A. and Stenger, A. (2003) *National Survey on Demand for Recreation in French Forests*. LEF, ENGREF/INRA, Nancy, France.

Quézel, P. (1979) La région méditerranéenne française et ses essences forestières: Signification écologique dans le contexte circum-méditerranéen. *Forêt Méditerranéenne* 1, 7–18.

Quézel, P. (1999) Biodiversité végétale des forêts méditerranéennes, son évolution éventuelle d'ici à trente ans. *Forêt Méditerranéenne* 20, 3–8.

Quézel, P. and Medail, F. (2003) Que faut-il entendre par 'forêt méditerranéenne'? *Forêt Méditerranéenne* 24, 11–31.

Service des Etudes et Statistiques Industrielles (2004) *Régions. Edition 2003*. CD-ROM. SESSI, Caen, France.

Tabourel, S. and Peyron, J.-L. (2004) *Les Comptes de la Forêt: Enjeux et Méthodes*. LEF ENGREF/INRA, Nancy and IFEN, Orléans, France. Working paper.

Tessier, A. and Peyron, J.-L. (2000) *Comptabilité Économique de l'Environnement, les Comptes de la Forêt Française*. IFEN, Orléans and LEF ENGREF/INRA, Nancy, France. Working paper.

21 Spain

Pablo Campos[1], Alejandro Caparrós[1] and Enrique Sanjurjo[2]
[1]Instituto de Economía y Geografía, Consejo Superior de Investigaciones Científicas
(CSIC), Pinar 25, 28006 Madrid, Spain; [2]Instituto Nacional de Ecología,
Anillo Periférico 5000, 04530 Coyoacán, México, Distrito Federal, Mexico

Introduction[1]

The European Union (EU) Parliament proposed in 1995 to develop a green accounting approach for measuring the green total economic value (TEV) of forestlands. However, such a task is not easy. Lack of suitable commercial market data and scientific controversy regarding non-market valuation techniques for forest outputs are the main difficulties in the implementation of a new forest accounting system at the EU level. Dislocated market output, missing market output and environmental output are the additional elements that need to be aggregated to forestlands' conventional commercial net added value to calculate forestlands' green TEV.

However, the Spanish forestland conventional net added value as applied by the European Economic Accounts for Forestry (EAF) (Eurostat, 2000a) produces a poor measurement of the true market and non-market net added value that is accrued to society from Spanish forestlands.

Nevertheless, adding consumer surplus to the forestlands' conventional commercial net domestic product may be the wrong way to express the forests' economic contribution to the green TEV.[2] On the contrary, simulated exchange values for environmental outputs should be taken into account to produce homogenous aggregation conditions with market outputs (Caparrós et al., 2003).

As an illustrative example, the incomplete green TEV for Spanish forestland in 1999 is calculated.[3] However, figures have to be taken carefully, since the estimation is based on different ad hoc criteria and assumptions. Nevertheless, the criteria chosen are such that it is believed they give a conservative measurement of forestlands' dislocated market outputs, missing market outputs and non-market outputs (environmental).

Methodology

The definition of forestlands

The first difficulty encountered in the estimation of the income of Spanish forestlands arises due to the different available definitions of forests. However, the differences are more important for detailed classifications rather than for the overall area considered as forests.

According to the Spanish National Forest Inventory (Instituto de Conservación de la Naturaleza, 1996), forestlands[4] extend on 26,052,000 ha, or 51% of the country's land surface. Of these forestlands, 53.5% are areas with at least 5% of the land covered by trees. Box 21.1 presents the situation of the forest plantations and timber-based industry in Spain.

Box 21.1. Forest plantations and timber-based industry in Spain (M. Palahí, J. Maria Solano and A. Ottitsch).

1. Forest plantations in Spain: an opportunity and challenge for the Spanish forestry sector

In Spain, 14.7 Mha, more than half of the total forestland area, is covered in trees and considered as forest. Forest plantations make up approximately a quarter of the total forest surface (~3.7 Mha), and provide up to 81% of the total timber production in the country. The importance of forest plantations in Spain is the result of a long process of forestation that started in the second half of the last century and has been intensified during the 1990s by the afforestation of agricultural lands promoted through EU funds. Afforestation in the last century represented a unique process in the history of Spanish forestry, explained by a combination of factors: the implementation of afforestation policies by the public administrations; the active role of private investors and the paper industry; and the profound changes in the socio-economic situation of rural areas, resulting in the abandonment of agricultural land.

These factors led to the forestation of about 3.2 Mha during 1940–1980. The main species used were *Pinus sylvestris* (15.9% of the reforested area), *P. nigra* (11.0%), *P. pinea* (7.4%), *P. halepensis* (15.2%), *P. canariensis* (0.7%), *P. pinaster* (6.9% for protection and 15.4% for production), *P. radiata* (5.9%), *Populus* sp. (2.2%), *Eucalyptus* sp. (10.7%) and other species (8.8%) (Ministero de Medio Ambiente, 2002). It is remarkable that by 1985, already 1.3 Mha of fast-growing plantations had been concentrated in northern Spain, of which more than 80% were planted by private investors and the paper industry (Ministerio de Medio Ambiente, 2002). The main species used for this purpose were *P. radiata* (0.2 Mha), *P. pinaster* (0.5 Mha), *Eucalyptus* sp. (0.5 Mha) and *Populus* sp. (0.1 Mha). *P. radiata* can be found mostly in the Vask Country, while *P. pinaster* occupies vast areas in north-western Spain (Galicia). *Populus* sp. have been widely planted in riparian zones, while *Eucalyptus* sp. was planted in the north-western and south-western regions of Spain.

The role of foresting agricultural land during the last decade is also important. In 1993, a new EU forestation programme for agricultural lands was initiated through an Accompanying measure (2080/1992 EU Forest Regulation). Its implementation during 1994–1999 resulted in the forestation of 450,000 ha of Spanish agricultural land, i.e. an annual average of 75,000 ha. Although a greater extension of land (800,000 ha) was expected to be forested, it is remarkable that the actual forested land in Spain represents 50% of the total forested area in Europe through the implementation of this regulation.

The forestation programme implemented during the last decade enabled the planting of more hectares than in previous years and was one of the main tools for increasing the forest area in Spain. The new 1257/1999 EU Regulation, Art. 31, aimed at diversifying rural activities, increasing employment and income, improving prevention of soil erosion and water cycle regulation in the areas affected by these problems, will continue to promote the afforestation of agricultural lands. The forecasted forested area for the period 2000–2006 is of about 0.2 Mha; of which 0.12 Mha had already been forested by 2001 (Ministerio de Medio Ambiente, 2002).

2. Demand for new plantations

The average wood consumption has been increasing continuously during the last five decades throughout most developed countries. It is estimated that the demand for forest products in Spain will continue to increase in the short and medium term, calling for an increase in the wood supply. This will have to be met partially by products coming from new fast-growing plantations and productive reforested areas (Ministerio de Medio Ambiente, 2002). During recent years, already 81% of total timber production has come from plantation production in Spain. This share is around 60% in countries such as Argentina, Brazil and Japan and almost 100% in Chile and New Zealand (Pandey and Ball, 1998).

Spain traditionally has been an importer of timber and other wood-based products. However, the country still has extensive areas that could potentially be forested for production purposes (Ministerio de Medio Ambiente, 2002). In the north and north-west of Spain, at elevations below 400 m, climatic conditions are excellent for the introduction of fast-growing plantations of *Eucalyptus globulus*,[5] *P. radiata* and *P. pinaster*. Furthermore, in other regions at higher elevations, conditions can also be good for planting productive forests with species with longer rotations, mainly broadleaves and native pines. The Ministerio de Medio Ambiente (2002) suggests that new forestations could take place in areas with high productive potential (> 6 m³/ha/year) and which are currently lacking an adequate vegetation cover. Currently, an area of around 1 Mha could be classified as potentially adequate to be reforested by different types of productive plantations in Spain (Ministerio de Medio Ambiente, 2002).

3. Timber-based industry and international trade

The Spanish timber-based industry is characterized by a diversity of products and a large number of enterprises (20,000), employing 210,646 people in 1999. The value of timber and furniture production is €6.3 million and €6.4 million, respectively. In recent years, the Spanish trade in timber and furniture has experienced an increasing deficit (from €274 million to > €350 million between 1999 and 2001) due to the increased demand for wood products. However, during this period, the value of the timber-based industry has grown by 12%, while that of the furniture industry has increased by 9%. Currently, the timber-based industry is rather important within the EU, Spain being in third place behind Germany and France in timber-based products and in fifth place after Germany, Italy, France and the UK in furniture.

The Spanish sawmill industry has also experienced important changes in recent years and currently employs 12,440 people, having a total turnover of €907 million (Ministerio de Medio Ambiente, 2002). This industry is concentrated in the north of the country, with 40% of the total value of sawnwood produced in Galicia and the Vask country. The transformation affecting this subsector in the last 10 years caused the disappearance of 30% of the enterprises, leading to an increase of about 40% in the production of the remaining ones. The sawmill and plywood industries process around 7.5 Mm^3 of debarked roundwood, and only part of it is currently supplied by Spanish wood. In fact, imports of roundwood are around 4 Mm^3 while sawnwood imports amount to 3 Mm^3. These imports have a value of €992 million, which represents 34% of the total imports of the whole forest industry sector in 2000 (€2.89 billion). This production deficit calls for new plantations as well as new tools to promote a more production-oriented management of existing forests.

4. Final remarks

As the environmental movement continues to exert pressure for the protection and setting aside of more native and natural forest areas, less of this forest type will be available for logging, and the costs of obtaining wood from these sources will continue to rise. The forces discouraging logging in natural forests are unlikely to go away in the foreseeable future and the pressure to reduce harvesting in old-growth and some second-growth natural forests might add to the attractiveness of the movement to invest in planted forests.

The Economic Accounts for Forestry

The European Economic Accounts for Forestry (EAF) calculate the Spanish forestlands' TEV by incorporating a group of commercial final outputs (FOs) net of intermediate consumption (IC): the forestry and logging activities related to artificial plantations, timber felling, cork stripping, seeds and plants, firewood cutting and resin gathering. All these forestry EAF-estimated values are final outputs, i.e. the EAF does not measure the forestlands' grazing resources as intermediate output (IO) consumed by livestock. TEV measurement requires calculating total outputs (TOs) as the sum of the FO plus IO.

Spanish Atlantic and alpine forestlands are used mainly for timber and firewood production, so that the absence of IOs does not have a great influence on the results obtained. These types of forestlands are similar to those existing in northern Europe, explaining to some extent the focus of European accounts on timber products (nevertheless, most of them are also grazed by livestock and game). On the contrary, Mediterranean forestlands have multiple uses. Timber, pine nuts, firewood and cork are important products, but grazing resources are the main commercial goods in these systems. Therefore, one of the main improvements of the TEV approach presented in this chapter compared with the EAF consists of the incorporation of grazing resources produced by forestlands as an IO (to be added to the final production to obtain total production).

The EAF IC incorporates only raw materials and services used in silvicultural, logging and firewood activities. This forestry IC is maybe less important in Spanish forestlands than animal (livestock and game) and government ICs. The EAF *fixed capital consumption* (FCC) is also focused on silviculture activities, neglecting FCC for animal and government forestland activities. In other words, the EAF system does not calculate the operating income for all forestland uses, but only for the so-called silviculture activity.

Forestland total income should be the sum of operating income and capital gains/losses that accrue to society from all the outputs produced and used[6] (Campos, 1999, 2000). However, this chapter considers only a group of production outputs, and capital gains/losses are not considered (i.e. only part of the forestlands' green TEV is presented in the following sections). Lack of reliable data is the main reason for this incompleteness of this Spanish forestlands' total social income or TEV measurement exercise.

Forestland commercial outputs unaccounted by the European Economic Accounts for Forestry

Two groups of forestland commercial outputs unaccounted by the EAF system are considered. For classification purposes, they have been termed *dislocated* and *missing* outputs. The first group is included by the System of National Accounts, but in the European Economic Accounts for Agriculture (EAA) (Eurostat, 2000a) and not in the EAF, while the second group is not accounted at all.

Dislocated outputs

Dislocated *intermediate* outputs are *grazing resources*, *hay harvested* and *acorn grazing by pigs* (Rodríguez and Campos, 2001). Because these outputs are intermediate resources, they are not included either as livestock IC or as forestry IOs. Hunting is a dislocated *final* output. It occurs both in forestlands and in agricultural areas; nevertheless, commercial hunting in Spain takes place mainly in forestlands.

Missing outputs

The main commercial missing outputs are non-felled annual *timber natural growth*. Natural growth of timber has a commercial value even though it is not felled. Nevertheless, the EAF does not currently incorporate this concept. Other missing outputs are *truffles* and *forest nuts*.

Forestland environmental outputs not accounted by the European Economic Accounts for Forestry

Environmental values contribute to the welfare of society and can be added to the income figures estimated by national accounts. However, as the latter are estimated as exchange values linked to scarce goods and services, the former must be estimated in a similar way, i.e. in order to maintain the coherence with the income concept calculated in conventional national accounting, exchange values must be estimated, simulating exchange conditions where market data are not available (Caparrós et al., 2003).

This study incorporates: mushroom picking, public free access consumption of recreational services, owners' self-consumption of environmental services, carbon sequestration and conservation values. The latter are estimated as government expenditures on forest conservation: wildlife species, habitats and landscape conservation (protection and improvement). This is not an exhaustive list of the environmental outputs produced by Spanish forestlands, but they have the double characteristic of relevance and relative data availability.

Mushroom picking

Mushrooms are picked freely by the public in most Spanish forests. Market prices (at coniferous forest gates) are available and therefore used for their valuation (Martínez, 2003).

Free public access and consumption of recreational services

Market prices for the public consumption of environmental services are not available. Thus, prices are imputed from willingness to pay using data obtained from several contingent valuation method (CVM) studies. The chapter aims to estimate the prices that could occur if markets were established for these services. Consumer surplus values are not considered since they do not enable aggregation with market-based data. Instead, the median multiplied by 50% of the total number of visitors is used (Caparrós et al., 2003). This figure is approximately the maximum amount that could be obtained by establishing a single price for all the visitors (consumer surplus could only be fully internalized if *every* visitor paid the maximum amount they were willing to pay, an unrealistic assumption in a market situation).

Owners' self-consumption of environmental services

Owners' self-consumption of environmental[7] services has been recognized in scientific journals (Torell *et al.*, 2001). Market prices of forestlands in western industrialized countries discount the annual flow of owners' self-consumption, but these consumption flows have seldom been measured and published in scientific journals, except[8] for an example developed by Campos and Mariscal (2003) in the *dehesas* of Monfragüe using the CVM.

In theory, it would also be possible to distinguish within the self-consumption of private owners the part that corresponds to recreational services and the part that should be attributed to conservation values. Nevertheless, this classification is extremely difficult to undertake in practice. Therefore, these two values are shown together under the heading of *owners' self-consumption of environmental services*.

Carbon sequestration

Carbon sequestration is an example of an emerging value of forests that is making rapid progress towards its internalization by markets. At COP 7 in November 2001, the Marrakech Accords (United Nations Framework Convention on Climate Change, 2001) were signed by most countries of the world (with the sole significant exception of the USA). This agreement will probably enable the Kyoto Protocol to come into force. If this occurs, precise rules will enable the internalization of carbon sequestration, with the caps and limitations adopted in the Marrakech Accords (Caparrós and Jacquemont, 2003). Nevertheless, at present, uncertainties regarding the quantities sequestered and the applicable prices remain (Elzen and Moor, 2001). Therefore, the results with different assumptions are shown.

The limitations and caps imposed on this internalization imply that only a part of the carbon sequestered will be internalized by markets. However, the environmental benefits associated with the reduction in the carbon dioxide concentration of the atmosphere apply to all carbon sequestered in the forest biomass. Thus, the total carbon sequestered in Spanish forests will have a commercial part (the carbon sequestration internalized) and a noncommercial part (the portion left out of markets due to the limitations imposed on carbon sequestration in the Marrakech Accords). Even though the Kyoto Protocol and the related Marrakech Accords are not yet enforced, this classification is considered of interest and carbon sequestration is incorporated as an unaccounted commercial value for the part that will probably be internalized by markets and as unaccounted environmental values for the rest.

Conservation values

Differences between option and existence values are clear on a theoretical basis, but their empirical determination is full of difficulties. Therefore, applied studies do not usually separate them, but rather estimate a value for *free public access conservation willingness to pay in protected areas*. In addition, conservation values, which should apply to society as a whole, are commonly estimated only for visitors to selected areas. Hence, the authors consider that scientific evidence is, for the time being, not sufficient to provide an accurate estimation for the conservation value of Spanish forests by means of CVM.

An alternative approach is to estimate these conservation values by considering the *government expenditures dedicated to forest conservation values*. The idea behind this approach is that society values forest for at least as much as it is willing to pay for its protection. The difficulty arises while trying to isolate the portion of these expenditures that should be attributed to option and existence values and the portion that corresponds to commercial production. Nevertheless, this issue has only a classification interest, since the overall result remains invariant regardless of the alternative followed. The alternative taken has been to attribute all the expenditures to conservation, assuming that no margin appears for this activity (i.e. total output equalizes total costs). It is further assumed that 45% of the total forest conservation expenditure is due to commercial intermediate consumption, 5% to fixed capital consumption and 50% to labour cost. To facilitate direct interpretation of the results, government intermediate consumption and fixed capital consumption are classified as environmental costs.

Conservation values – option and existence values calculated for free public access to protected areas, have not been included in the estimation of the aggregated TEV of Spanish forestlands. This is to avoid possible problems of double accounting that may occur if both conservation values measurement alternatives – visitors' willingness to pay and government expenditures – described above to estimate conservation values, are used simultaneously. Visitors are also part of the total population, so that if government expenditure is supposed to equalize the forestland conservation value output for society, the part that corresponds to the visitors should also be included in the latter output. This way of estimating the conservation value probably yields an underestimated figure, but, as pointed out in the Introduction, this study aims at offering a conservative value of Spanish forestlands' TEV.

Results

Active use values

Active use values[9] are the most evident economic values of forests. Commercial values have been the first to enter official statistics, and most studies dedicated to the estimation of non-commercial values have focused on active use values.

Commercial direct active use values

Table 21.1 shows the commercial values generated by forests and measured by the EAF system. It also shows additional commercial values that are not incorporated by EAF, but which are accounted in the EAA system, and other values that are not measured at all. The value of timber felling represents €725 million and comprises 77% of the value of the EAF system commercial final output of forestlands in Spain. In addition, during the period 1990–1999, annual timber felling accounted for only 39% of the annual growth (Eurostat, 2000b).[10]

Table 21.1. Final commercial outputs and cost of forestry activities.

Class	Unit	Quantity[a]	EAF system final output[b] (million €) (A)	Final output not included in EAF system[b] (million €) (B)	Commercial final ouputs[b] (million €) (C = A + B)
Final outputs (FO)			943	44	987
Timber felling	(000 m³)	15,362	725		725
Firewood	(000 stere (st))	2,890	38		38
Reforestations[c]	(000 ha)	76	117		117
Seeds and plants	(000 plants)	112,425	20		20
Cork	(t)	62,361	41		41
Resin	(t)	4,175	2		2
Truffles and nuts				30	30
Other outputs				14	14
Intermediate consumption (IC)			131		131
Fixed capital consumption (FCC)			16		16
Gross value added at market prices (GVA = FO – IC)			812	44	856
Net value added at market prices (NVA = GVA – FCC)			796	44	840

[a]Refers to 1999.
[b]Updated to 2001 euros (International Monetary Fund, 2004).
[c]The reforestations final output is accounted by EAF, taking into account total silvicultural soil preparation and plantation costs. The reforestation final output equalizes the annual cost incurred in forest plantations.
Sources: authors' elaboration, Ministerio de Agricultura, Pesca y Alimentácion Agroalimentaria (2000, 2002) and Eurostat (2000b).

Timber is used to illustrate the relative importance of the different incomes measured. It is the most well-known income offered by forests, and data are relatively reliable for this figure. Nevertheless, since cost data are generally unreliable or even non-existent for some incomes (especially for detailed classifications), FO figures – or TO figures – are used for the purpose of comparison.

In Table 21.2, the FO of the game activity in Spanish forests can be found. Its value represents an FO of €180 million, i.e. 25% of that of timber felling output, and is therefore not a negligible value.[11] Table 21.2 shows the output value of forestry grazing resources as the IO generated by cattle breeding activity in forestlands. The relevance of grazing in Spanish forests is shown by the fact that its value represents an IO of €203 million, or 28% of that of the timber-felling output. This is a typical feature in Mediterranean forests, since in extensive areas (such as the *dehesas*), grazing is the most important commercial production (Campos *et al.*, 2001). Nevertheless, the EAF system does not incorporate this source of income in the forestry sector (since IO is not measured by the EAF system, the latter value is attributed entirely to the livestock sector).

The gathering of mushrooms in Spanish forestlands is estimated to reach 17,811 t/year and their average price is €3.6/kg (Martínez, 2003). Its final output has an estimated value of €64 million and represents 9% of that of the timber-felling output value. We assume that there are no costs of IC and no FCC when free-access visitors pick mushrooms. Therefore, the FO, gross value added (GVA) and net value of mushrooms added together have the same value.

Non-commercial direct active use values

Probably the most evident non-commercial active use of Spanish forests is related to their recreational use. In any case, this is the non-commercial active use value for which more empirical studies exist. Table 21.3 summarizes the recreational values enjoyed by visitors, based on CVM studies. Table 21.4 shows the results of the estimation of the final production of recreational services – using the data described in Table 21.3 – and the owners' self-consumption of environmental services from a Monfragüe CVM survey (Campos and Mariscal, 2003). The figures obtained imply that the value of the final output of recreational services

Table 21.2. Game and grazing resources final output[a].

Game	Big game	Small game	Game birds	Total
Number of head (000)	139	5,676	15,038	20,853
Final output (FO) from game[b] (million €)	50	60	71	180
Meat sales revenues (million €)	15	33	27	75
Complementary revenues (million €)	35	27	44	105

Grazing resources	Total MSU[c] (million)	Forestry area MSU (million)	Forestry grazing FU[d] (million)	Total forestry grazing output (million €)
Total intermediate ouput[e] (IO) and FU				
from forestry grazing	23,669	5,762	3,112	203
Cattle	11,580	2,299	1,242	81
Sheep	10,437	2,879	1,554	101
Goats	1,652	584	316	21

[a]1999 physical data at 2001 euros, updated using the index of consumer prices (International Monetary Fund, 2004).
[b]Final output value (FO) = gross value added (GVA) = net value added (NVA), we assume IC = 0 and FCC = 0.
[c]Spanish total extensive reproductive female measured in MSU (Merino sheep unit) (Martín Bellido *et al.*, 1986).
[d]FU = forage unit (1 kg of barley).
[e]Intermediate output value (IO) = gross value added (GVA) = net value added (NVA), we assume IC = 0 and FCC = 0.
Sources: authors' elaboration and MAPA (2002).

Table 21.3.   Free public access recreational services of Spanish forests: contingent valuation studies.

Source	Natural area	Sample size	Question type	Visit value (€/visit)
Riera and Delcalzi (1994)	Pallars Soibrá	300	Mixed	6.61
Rebolledo and Pérez (1996)	Dehesa del Moncayo	427	Mixed	5.65
Campos (1998)	Dehesa del Monfragüe	336	Mixed	8.17
Barreiro (1998)	Ordesa y Monte Perdido	405	Mixed	7.21
Pérez et al. (1998)	Posets Maladeta	382	Mixed	4.96
Caparrós and Campos (2002)	El Paular, Valsaín y Peñalara	741	Mixed (double)	4.38
Júdez et al. (1998)	Tablas de Daimiel	366	Dichotomous	12.24
González (1997)	Monte Aloia	402	Dichotomous	5.67

Source: authors' elaboration.

Table 21.4.   Final production of recreational services and of owners' self-consumption of environmental services in Spanish forests[a].

Class	Area (000 ha)	Visits			Price[b]	Final output[c] (FO) (million €)
		Visits/ha	Total (000)	Willing to pay the price (000)		
Free public access recreational services	2,579	15.02	38,730	9,365	5.21	101
National Parks	263	37.43	9,840	4,920	5.21	25
Other protected areas	2,316	12.48	28,890	14,445	5.21	75
Owners' self-consumption of environmental services from private forestands	21,876				87.43	1,913

[a]1999 physical data at 2001 euros, updated using the index of consumer prices (International Monetary Fund, 2004).
[b]Expressed as euros/visit for the recreational services linked to the free public access and as euros/ha for the owners' self-consumption of environmental services.
[c]Total final output (FO) value = gross value added (GVA) = net value added (NVA), we assume IC = 0 and FCC = 0.
Sources: authors' elaboration from Table 21.3, Campos and Mariscal (2003) and Ministerio de Medio Ambiente (1997).

accounts for €101 million and only reaches 14% of the value of timber felling at the forest gate. However, it is important to remark that this figure is estimated including only visitors to protected areas (since CVM studies have been done mainly in these areas and no estimations of the number of visitors to non-protected areas are available). The FO of the owners' self-consumption of *environmental services* amounts to €1913 million and is 164% higher than the timber-felling output value.

Indirect active use values

Table 21.5 shows the results for different amounts of carbon sequestered internalized and for different prices. As can be seen, most of the previous estimations were rather optimistic if the recent cap on forest management included

in the Marrakech Accords is taken into account. Using a price of €20/tC and the estimation based on Eurostat data, the total carbon sequestration output value generated is €173 million and represents 24% of the timber felling FO value. However, if the maximum amount permitted for forest management in the Marrakech Accords (United Nations Framework Convention on Climate Change, 2001) is valued at this price, carbon sequestration has an FO value of €13.4 million; the latter reaches only 2% of the timber-felling output value. Nevertheless, afforestation, reforestation, cropland management, grazing land management and revegetation are free of limitations in the Marrakech Accords (Caparrós and Jacquemont, 2003). Additional forest management credits can also be earned to compensate deforestation.

Table 21.5. Annual value of carbon sequestration.

| Source | Carbon sequestration 000 tC/year | Total final ouput[a] (FO) value for different prices (million €) | | |
		€5/tC	€20/tC	€100/tC
Rodríguez (1999)	4400–9800	35.5	142.0	710.0
Price Waterhouse Coopers (1999)	8067	40.3	161.3	806.7
Maximum for forest management (United Nations Framework Convention on Climate Change, 2001)	670	3.4	13.4	67.0
Eurostat (2000b)	8637	43.2	172.7[b]	863.7

[a]Total final output value (FO) = gross value added (GVA) = net value added (NVA), we assume IC = 0 and FCC = 0.
[b]We assume this value for TEV aggregation.
Sources: authors' elaboration and cited references.

Passive use values

The values collected under this heading are those which are most controversial, and criticism of environmental valuations is generally focused on the possibility of offering precise monetary figures for these values. In addition, the separation of option, bequest and existence values in practice is extremely difficult, if even possible at all. Therefore, most of the empirical studies available opted to estimate a value for *conservation*, which is in fact the sum of all the concepts discussed above.

As stated above, the value given by Spanish society to the conservation of Spanish forests is not measured by means of CVM studies. On the contrary, the estimation is based on the government's expenditures on forest conservation (Table 21.6). The final output for forest conservation has a value of €1039 million, and this latter value is 43% higher than the timber-felling output value.

Towards the Measurement of the Green Total Economic Value of Forestlands

In the previous section, the discussion was presented in terms of TOs, given the difficulty in estimating and attributing concrete cost values for the different goods and services generated in the Spanish forestlands. Nevertheless, the estimation of the TEV of forest implies the utilization of the concept of total social income, and more precisely the Hicksian income (Campos, 1999). This figure can be estimated by

Table 21.6. Forest conservation values measured by Government expenditure in Spanish forestlands[a].

Class	Economic variables (million €)
Final output value[b] (FO)	1039
Intermediate consumption (IC)	471
Fixed capital consumption (FCC)	42
Gross value added at market prices (GVA = FO – IC)	568
Net value added at market prices (NVA = GVA – FCC)	525

[a]1999 physical data at 2001 euros, updated using the index of consumer prices (International Monetary Fund, 2004).
[b]Total government expenditures (cost).
Sources: authors' elaboration and Ministerio di Medio Ambiente (1999).

summing the value added to the capital gains experimented during the accounting year. As stated, value added figures are difficult to estimate for detailed classifications, but for aggregated concepts their estimation is possible with sufficient accuracy. Thus, the discussion under this heading is carried out in terms of net value added. On the contrary, capital gains values for the whole of the Spanish forestry system are extremely difficult to obtain, and are therefore not included in the analysis. In what follows, we will use the concept of TEV as synonymous with forestry net value added or total social income.

Figure 21.1 shows the aggregated figure of €3999 million, representing the TEV or total net value added estimated for the Spanish

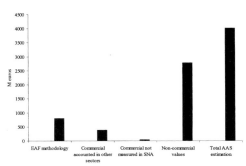

Fig. 21.1. Net value added or TEV at market prices generated in Spanish forestlands (1999 data at 2001 euros, updated using the index of consumer prices (International Monetary Fund, 2004)).

forestlands using the extended ASS methodology (Campos, 1999; Caparrós *et al.*, 2003). It includes: (i) €796 million of commercial net value added estimated following the EAF methodology; (ii) €383 million of commercial net value added generated by the forests from game and grazing resources but accounted or mislocated in the EAA sector; (iii) €44 million of commercial net value added from truffles, nuts and other outputs not measured or missing from official data; and (iv) €2776 million of environmental (non-commercial) net value added generated by mushroom gathering (€64 million), free public access to recreational services (€101 million), owners' self-consumption (€1913 million), carbon sequestration (€173 million) and conservation values (€525 million).

The Spanish forestland unaccounted commercial net value added (dislocated and missing) contributes €427 million and represents 11% of the forestlands' TEV at market prices. Environmental net value added at market prices contributes another 69% to the total green net value added or TEV. Therefore, the Spanish application of the EAF system has measured in the year 1999 only about 20% of the actual forestlands' green TEV at market prices. In other words, the tentative measurement of Spanish forestland TEV at market prices is five times higher than its conventional official figure given by the EAF system.

It is not surprising to find a relatively high value of forestlands' environmental net value added. Only the owners' self-consumption of environmental services

represents approximately 69% of the total output from the five environmental services considered (mushroom picking, free public access consumption of recreational services, owners' self-consumption of environmental services, conservation values and carbon sequestration). A pure definition of environmental services will exclude owners' self-consumption of environmental services from this category. Owners' self-consumption of environmental services has a private capital market value when land is traded and, following the economic income theory, it could be considered as an implicit commercial flow value. Taking into account four environmental services – and including owners' self-consumption of environmental services as an implicit commercial service – then the environmental net value added amounts to €863 million and represents approximately 22% of the Spanish forestland TEV.

The EAF gross value added (GVA) at market prices has a value of €812 million and represented 0.14% of the Spanish gross domestic product[12] (GDP) in 1999 (INE, 2002). The total AAS green gross valued added (GVA) at market prices totals €4057 million and accounted for 0.7% of the total Spanish GDP in 1999.

The main conclusion that can be drawn from the results obtained from this study is that there is a need for a systematic – and not only case study-based – estimation of Hicksian income generated by the Spanish forestlands. Available data for non-wood products and services are scarce and not well suited for extrapolation, and even timber income is not well accounted since commercial timber natural growth differs significantly from timber harvest, and capital gains are not considered.

Notes

[1]   This study is part of a collaboration between Eurostat (programme TAPAS 2001–2002), the Spanish Ministerio de Agricultura, Pesca y Alimentación (MAPA) and the Spanish Consejo Superior de Investigaciones Científicas. Any shortcomings of this study are the responsibility only of the authors.
[2]   The green TEV approach aims to consider all outputs and costs that accrue to society from any forestland economic activity. What is intended to be measured here is the green total social income (TSI),

one of the most relevant forestland economic indicators of the TEV approach.

[3] Data from 1999 are updated to 2001 euros using the index of consumer prices (International Monetary Fund, 2004), to ensure homogeneity with the other country chapters.

[4] In Spain, the forestlands (*monte*) include the grasslands: '*monte* or forestland is the land where there are growing trees, coppices, scrublands or grass species, irrespective of the fact that they were spontaneous or originated by seeded or planting, except when the species are characteristic of cropping [or they are] treeless meadows' (Spanish Monte Law, 1957). The Food and Agriculture Organization's definitions of forest and other woodedlands do not include grasslands, but Food and Agriculture Organization (2001) statistics for Spain take into account grasslands, giving a total area for 'forest' and 'other forestlands' of 26,267,000 ha.

[5] At higher elevations and in areas with risk of frost, *Eucalyptus nitens* is also being tested.

[6] Use can be at the site of production or off-site.

[7] They could be seen as commercial values, as they contribute to increasing the market price of forestlands.

[8] Three forthcoming new contingent valuation surveys have been applied to measure owners' self-consumption of environmental services in two Spanish regions of cork oak forests (Cádiz and Girona), Portuguese *montados* and Californian ranches. The last two have been done by Inocencio Seita Coelho and Lynn Huntsinger, respectively, in collaboration with the CSIC authors of this chapter.

[9] There is no consensus on the most appropriate TEV classification. We prefer to use 'active use values' instead of 'use values', and 'passive use values' instead of 'non-use values'.

[10] We assume that non-felled timber natural growth is non-commercial in a relevant portion. Thus, commercial timber natural growth is similar to commercial felling in physical terms. Hence, since the timber harvest stumpage value is not considered in the cost side, standing commercial natural growth is not taken into account in the final timber output side.

[11] As stated in the methodology section, the EAF system does not attribute this figure to forests.

[12] 1999 data at 2001 euros, updated using the index of consumer prices (International Monetary Fund, 2004).

References

Barreiro, J. (1998) El problema de los Sesgos en Valoración Contingente. Una Aplicación a la Estimación de los Valores Ambientales del Parque Nacional de Ordesa y Monte Perdido. PhD Thesis, Universidad de Zaragoza, Zaragoza, Spain.

Campos, P. (1998) Contribución de los visitantes a la conservación de Monfragüe, bienes públicos, mercado y gestión de los recursos naturales. In: Hernández Díaz-Ambrona, C.G. (ed.) *La Dehesa: Aprovechamiento Sostenible de los Recursos Naturales*. Editorial Agrícola Española, Madrid, pp. 241–263.

Campos, P. (1999) Hacia la medición de la renta de bienestar del uso múltiple de un bosque. *Investigación Agraria: Sistemas y Recursos Forestales* 2, 407–422.

Campos, P. (2000) An agroforestry economic accounting system: towards the measurement of social income. In: Jöbstl, H., Merlo, M. and Venzi, L. (eds) *Institutional Aspects of Managerial Economics and Accounting in Forestry*. IUFRO Division 4, Roma-Ostia, Italy, pp. 9–19.

Campos, P. and Mariscal, P. (2003) Preferencias de los propietarios e intervención pública: el caso de las dehesas de la comarca de Monfragüe. *Investigación Agraria: Sistemas y Recursos Forestales* 12, 87–102.

Campos, P., Rodriguez, Y. and Caparrós, A. (2001) Towards the dehesa total income accounting: theory and operative Monfragüe study cases. *Investigación Agraria: Sistemas y Recursos Forestales* Special Issue 1, 43–67.

Caparrós, A. and Campos, P. (2002) Valoración de los usos recreativo y paisajístico en los pinares de la sierra de Guadarrama. *Revista Española de Estudios Agrosociales y Pesqueros* 195, 121–146.

Caparrós, A. and Jacquemont, F. (2003) Carbon offset programs and biodiversity: an economic and legal analysis. *Ecological Economics* 46, 143–157.

Caparrós, A., Campos, P. and Montero, G. (2003) An operative framework for total Hicksian income measurement: application to a multiple use forest. *Environmental and Resource Economics* 26, 173–198.

Elzen, M.G.J. and Moor, A.P.G. (2001) *The Bonn Agreement and Marrakech Accords: an Updated Analysis*. RIVM report 728001017/2001. RIVM, Bilthoven, The Netherlands.

Eurostat (2000a) *Manual on Economic Accounts for Agriculture and Forestry – EAA/EAF 97 (Rev.1.1)*. Commission of the European Communities, Luxembourg.

Eurostat (2000b) *Sustainable Management of EU Forests: Partial but Significant Role in Limiting Greenhouse Gas (CO_2)*. Eurostat News Release No. 121/2000. Website: www.europa.eu.int

Food and Agriculture Organization (2001) *Global Forest Resources Assessment 2000*. Main Report, FAO, Rome.

González, M. (1997) Valoración Económica del Uso Recreativo-paisajístico de los Montes: Aplicación al Parque Natural del Monte de Aloia en Galicia. PhD Thesis, Universidad de Vigo, Vigo, Spain.

Instituto de Conservación de la Naturaleza (1996) *Segundo Inventario Forestal Nacional 1986–1995*. MMA, Madrid.

Instituto Nacional de Estadística (2002) *Anuario Estadístico 2001*. INE, Madrid.

International Monetary Fund (2004) *World Economic Outlook*. Website: www.imf.org. Access date: May 2004.

Judéz, L., De Andrés, R., Perez Hugalde, C. and Ibáñez, M. (1998) Évaluation contingente de l'usage récreatif d'une réserve naturelle humide. *Cahiers d'Économie et Sociologie Rurales* 48, 38–59.

Martín Bellido, M., Espejo, M., Plaza, J. and López Carrión, T. (1986) *Metodología para la Determinación de la Carga Ganadera de Pastos Extensivos*. INIA-MAPA, Madrid.

Martínez, F. (2003) *Producción y aprovechamiento de* Boletus edulis Bull.: Fr. *y* Boletus pinophillus *Pilát & Dermek en un bosque de* Pinus sylvestris L. Junta de Castilla y León, Soria, Spain.

Ministerio de Agricultura, Pesca y Alimentación Agroalimentaria (2000) Servicio de Cuentas y Balances Agroalimentarios (25/1/2000). In: *Boletín Mensual de Estadística de Enero de 2000*. MAPA, Madrid.

Ministerio de Agricultura, Pesca y Alimentación Agroalimentaria (2002) Distribución general del suelo según usos y aprovechamientos. In: *Anuario de Estadística Agroalimentaria 2001*. MAPA, Madrid.

Ministerio de Medio Ambiente (1997) *El Nedio Ambiente en España*. MMA, Madrid.

Ministerio de Medio Ambiente (1999) Gasto Público en Medio Ambiente 1996: Análisis Comparativo 1987–1996. MMA, Madrid.

Ministerio de Medio Ambiente (2002) *Plan Forestal Español*. http://www.mma.es/conserv_nat/planes/estrateg_forestal/pdfs/pfe.pdf.

Pandey, D. and Ball, J. (1998) The role of industrial plantations in future global fibre supplies. *Unasylva* 49, 37–43.

Pérez, L., Barreiro, J., Barberán, R. and Del Saz, S. (1998) *El Parque Posets-Maladeta, Aproximación Económica a su Valor Recreativo*. Publicaciones del Consejo de Protección de la Naturaleza de Aragón, Zaragoza, Spain.

Price Waterhouse Coopers (1999) *Evaluación del Sector Forestal en el Desarrollo Rural*. Madrid.

Rebolledo, D. and Pérez, L. (1996) *Valoración Contingente de Bienes Ambientales: Aplicación al Parque Natural de la Dehesa de Moncayo*. Gobierno de Aragón, Zaragoza, Spain.

Riera, P. and Delcalzi, R. (1994) El valor de los espacios de interés natural en España. Aplicación de los métodos de valoración contingente y coste de desplazamiento. *Revista de Economía Española* Special Issue, 207–229.

Rodríguez, J.C. (1999) El ciclo mundial del carbono. Método de cálculo por los cambios. In: Hernández, F. (ed.) *El Calentamiento Global en España: un Análisis de Sus Efectos Económicos y Ambientales*. Consejo Superior de Investigaciones Científicas, Madrid, pp. 97–139.

Rodríguez, Y. and Campos, P. (2001) Aporte energético del pastoreo en un rebaño de cabras trasterminante entre Monfragüe y la Sierra de Gredos. *Pastos* 29, 201–216.

Torrel, L.A., Rimbey, N.R., Tanaka, J.A. and Bailey, S.A. (2001) The lack of profit motive for ranching: implications for policy analysis. In: Torrel, L.A., Bartlett, E.T. and Larrañaga, R. (eds) *Current Issues in Rangeland Resource Economics*. New México State University, New Mexico, pp. 47–58.

United Nations Framework Convention on Climate Change (2001) *The Kyoto Protocol to the UNFCCC*. website: www.unfccc.int. Access date: March, 2003.

22 Portugal[1]

Américo M.S. Carvalho Mendes

*Faculty of Economics and Management of the Portuguese Catholic University,
Rua Diogo Botelho, 1327, 4169-005 Porto, Portugal*

Introduction

Continental Portugal is a rectangular-shaped territory with a total area of 8.9 Mha, the formation of which dates back to the first half of the 13th century. The northern and central regions are hilly, as is the southernmost region of the Algarve. The average altitude is about 400 m. The mountain tops often reach more than 1200 m, but never above 2000 m. The valley of Tagus and the region of Alentejo are flat, with an average altitude of about 250 m. Most of the major rivers flow from east to west. With the exception of the northern region, the principal mountain chains also have a similar orientation.

The total population in 2001 was about 9.9 million inhabitants (Instituto Nacional de Estatística, 2003a): 68.5% in the coastal area between Viana do Castelo and the Setúbal Peninsula. A total of 3.2 million inhabitants lives in the Lisbon and Oporto metropolitan areas. During 1991–2001, the population grew by 5.3%, mainly due to the positive migration balance; in the interior regions, the total (rural and urban) population decreased, with only the urban areas showing some resistance to this general trend.

The average forest area in Continental Portugal is 0.34 ha *per capita* – higher than the average for Europe and, indeed, higher than the averages for all the other regions within Europe except the nordic countries (Kuusela, 1994). According to the 1995 Forest Inventory, there were about 565,200 ha of forest in the coastal regions between Viana do Castelo and

the Setúbal Peninsula (16.9% of total forest area). This part of the country thus has 0.08 ha of forest *per capita*, compared with 0.89 ha *per capita* in the rest of the territory.

With some regional variations, the climate is essentially characterized by hot and dry summers and humid winters. This feature, combined with the fact that most vegetal species in forests and scrublands have a relatively high degree of inflammability, put forest resources under a high natural risk of damage by fire throughout the summer. The regional variations result in a series of climatic regions ranging from Atlantic to Mediterranean, with some intermediate types where either one of these two influences is dominating. Macedo and Sardinha (1993) distinguish the following climatic regions (average rainfall and average temperatures are in parentheses):

- Atlantic region, including the north-western part of the country (1000–2500 mm; 10–14°C).
- Atlantic region with a Mediterranean influence, including the western central part of the country (600–1000 mm; 15°C).
- Mediterranean region with an Atlantic influence, including the intermediate central part of the country, going southwards to the western part of Alentejo (450–800 mm; 16°C).
- Continental region, including the north-eastern part of the country, with the exception of the Douro Valley (500–1200 mm; 10–13°C).

- Mediterranean region with a continental influence, including most of the eastern part of the country (500–800 mm; 14–17°C).
- Mediterranean region, including the Douro Valley and the Algarve (400–600 mm; 15–18°C).

Forest Resources

Trends in land use and forest area

In 1995 (the date of the most recent forest inventory), forest area was about 3.3 Mha (Table 22.1), which is an increase by a factor of 2.7 from 1867 (the date of the first estimate of this land use). Up until the 1950s, there was simultaneous growth of forest and agricultural land, due to the large amount of uncultivated land made available after a secular process of deforestation. With the intense rural emigration in the 1960s and 1970s, the area of farmland started to fall, while forestland continued to expand. However, since the 1970s, not all abandoned farmland has been reforested, which recently resulted in an increase in scrubland.

The increase in forestland is the combined result of the following dynamics in the three main forest species:

- Continued expansion of maritime pine (*Pinus pinaster*) forests until the end of the 1960s, followed by a decline due to forest fires.
- Continued expansion of cork oak (*Quercus suber*) forests until the end of the 1930s, followed by a decline that has been reversed since the mid-1980s.
- Continued and rapid expansion of eucalyptus (*Eucalyptus globulus*) plantations since the mid-1950s as a substitute for farming or for other forest uses, namely burnt pine forests.

In spite of the increase in area, forests are currently threatened by several types of risks. According to the United Nations Economic Commission for Europe/Food and Agriculture Organization (2000), in 1995 there were 641,000 ha of degraded forest and other wooded land, i.e. 18.5% of the total forestland; 11.3% due to insect damage and diseases; and

2.5% due to forest fires. Fires are publicly perceived as the main threat to forests, not only because they are much more visible, but also because they have been increasing in severity since the mid-1960s.[2] During this period, on average, the forest area burnt was more than twice the area (re)afforested with support by public incentive schemes. The most affected areas were the central and north-western regions, where pine forests are concentrated.[3] More recently, the risk of forest fires has expanded to the southern regions of cork oak and holm oak forests, probably due to the decline in farming activities and to the increase in scrublands and other forms of accumulation of inflammable materials in or near the forests.

Forest functions and forest biodiversity

In 1995, the main function of 51.8% (22.4% of conifers, 17.8% of broadleaves and 11.6% of mixed stands) of the forestland was for wood supply (Leite and Martins, 2000a,b). The second main function, corresponding to 48.2% of the forestland, was for non-wood forest products (NWFPs), mostly cork production in the southern regions. The 11.3 million m^3 over bark (o.b.) of annual fellings for wood supply are almost of the same amount as the 12.9 million m^3 o.b. of net annual increment in the forests with this function (Table 22.2). Therefore, the derived demand by forest industries is in tight tandem with the wood supply.

There are 1.5 Mha (17.1% of the total land area of Continental Portugal) under some sort of special protection status, including Natura 2000 and the National Network of Protected Areas (such as national or natural parks) (Table 22.3). In the Natura 2000 sites, there are around 594,500 ha of forests, and in the National Network of Protected Areas there are 162,600 ha. Since almost all protected areas are Natura 2000 sites, it may be assumed that 39.1% of the area of these sites is covered with forests. As expected, the species of main commercial interest such as maritime pine, cork oak and eucalyptus have a lower incidence in these zones.

All tree species existing in the country, including all those endangered, are associated with forest ecosystems. These ecosystems are

Table 22.1. Land use in Continental Portugal since 1867 (000 ha).

Land type	1867	1902	1910	1920	1929	1939	1950/56	1963/66	1968/78	1980/85	1995/98
1. Forest and other wooded land[a]	1,240.0	1,736.9	1,956.5	2,022.5	2,332.0	2,467.0	2,832.3	2,825.7	2,969.1	3,108.2	3,349.3
A. Forest land by tree species dominance											3,201.1
Conifers	210.0	250	430.2	913.7	1,132.0	1,161.0	1,189.5	1,287.6	1,293.0	1,252.3	976.1
Maritime pine											
Other conifers											
'Montados'	370.0	712.9	782.7	868.9	940.0	1,050.0	1,274.5	1,215.4	1,192.5	1,128.7	1,174.4
Cork oak	121.0	325.5	366.0	413.7	560.0	690.0	651.4	636.8	656.6	664.0	712.8
Holm oak	249.0	387.5	416.7	455.1	380.0	360.0	623.1	578.6	535.9	464.7	461.6
Other oaks and chestnut	60.0	174.0	131.0	174.0	193.0	188.0	170.0		99.8	143.2	171.5
Other oaks	NA	78.2	47.0	78.2	108.0	108.0	94.0		70.6	112.1	130.9
Chestnut	NA	95.8	84.0	95.8	85.0	80.0	75.0		29.3	31.1	40.6
Eucalyptus	0.0	—	—	—	8.0	NA	113.3	98.9	213.7	385.8	672.1
Other	600.0	600.0	612.7	66.0	59.0	68.0	85.0		170.0	198.2	207.0
B. Other wooded land	NA	NA	NA	NA	NA	NA	NA	NA	NA	NA	148.2
2. Agricultural land	1,886.0	NA	3,111.9	3,229.0	3,282.0	3,380.0	4,762.0		4,205.9	3,902.4	2,972.9
Uncultivated land fit for cultivation	5,462.9		3,426.7	3,245.7	2,883.2	2,648.0	885.6		1,279.9	1,419.3	2,054.6
Productive, but uncultivated land (fallow, grazing, etc.)	2,116.0		1,926.0	1,639.0	1,565.0	1,484.0	395.6	NA	NA	NA	NA
Other uncultivated land fit for cultivation	3,346.9		1,503.8	1,606.7	1,318.2	1,164.9	490.0	NA	NA	NA	NA
3. Land unfit for cultivation	291.0	374.0	381.7	382.7	382.7	384.0	400.0	NA	425.0	450.0	503.1
4. Total land area	8,772.5	8,772.5	8,772.5	8,772.5	8,772.5	8,772.5	8,772.5	8,772.5	8,772.5	8,772.5	8,772.5
5. Inland waters	107.3	107.3	107.3	107.3	107.3	107.3	107.3	107.3	107.3	107.3	107.3
6. Total area	8,879.9	8,879.9	8,879.9	8,879.9	8,879.9	8,879.9	8,879.9	8,879.9	8,879.9	8,879.9	8,879.9
Forest coverage (ratio of forest land : total land area)	14.1%	19.8%	22.3%	23.1%	26.6%	28.1%	32.3%	32.2%	33.8%	35.4%	38.2%

[a]'Other wooded land' is defined here as being burnt forests, areas of clearcut and land with trees below the density needed to be classified as 'forests'.
Sources and methodology: Mendes (2002).
NA = not available.

Table 22.2. Area, growing stock, increment, fellings and removals in 1995.

	Area (000 ha)	Growing stock volume (000 m³ o.b.)	Annual net increment (000 m³ o.b.)	Fellings (000 m³ o.b.)	Annual removals	
					(000 m³ o.b.)	(000 m³ u.b.)
Trees in forest, total	3,383	275,760	14,312	11,500	11,300	9,400
Coniferous	1,179	147,782	8,323	6,200	6,100	4,900
Broadleaved	2,204	127,978	5,989	5,300	5,200	4,500
Trees in forest for wood supply[a]	1,897	188,020	12,900	11,200	11,000	9,100
Coniferous	1,021	140,871	7,890	6,200	6,100	4,900
Broadleaved	876	47,149	5,010	5,000	4,900	4,200
Trees in forest with other purposes		87,740	1,412	300	0	
Trees in other wooded land		16,246	213	0	0	
Trees outside forest and other wooded land			670	0	0	
Total		292,006	15,195	11,500	11,300	

[a]The 344,000 ha of mixed stands are split evenly between coniferous and broadleaved species.
Sources: Direcção Geral das Florestas (1998, 1999).

Table 22.3. Total area under special protection status in the year 2000 (000 ha).

Protection status			Total protected area	Forest land in protected areas
Natura 2000		Directive Birds	744.8	
		Directive Habitats	1094.3	
		Total[a]	NA	594.5
National Network of Protected Areas (NNPA)	Areas of national protection status	National parks	70.3	
		Natural parks	527.1	
		Natural reserves	63.2	
		Botanic reserves	0.0	
	Areas of regional protection status	Protected landscapes	12.8	
		Classified sites	2.3	
	Total (without double counting)		638.3	162.6
Total (without double counting)			1520.0	NA

Sources: Direcção Geral das Florestas (2001) and data collected from the DGF Internet site, on 19 November 2000.
NA = not available.
[a]This sum is not double-counted.

also important for animal species, especially mammals, birds and butterflies. Of the endangered animal species, 64% of mammals and 30% of birds are associated with forests.

Potential natural forest types and current forest cover[4]

Aguiar and Capelo (2004) distinguish six types of potential natural forests in Continental Portugal: deciduous oak forests; birch forests; oak forests adapted to calcareous soils; evergreen oak forests; other evergreen forests and other forests adapted to calcareous soils; and hygrophilic forests (Fig. 22.1).

Deciduous oak forests

Potential natural forests of deciduous oak trees are of two main subtypes according to the dominant species: *Quercus robur* or *Quercus pyrenaica*. Potential natural forests dominated by *Q. robur* (alone or mixed with *Q. pyrenaica*) correspond to the north-western part of

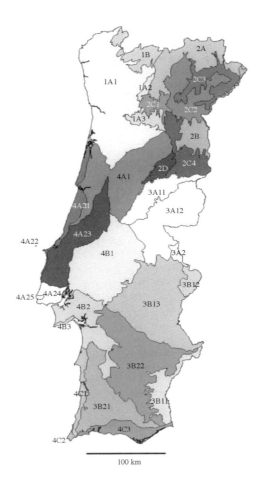

Fig. 22.1. Map of potential and actual forest vegetation[6] in Continental Portugal.

Legend:
1A1 and 4A1: potential natural forests of deciduous oak trees with *Quercus robur* as the dominant species, currently substituted by agricultural land uses, maritime pine and eucalyptus; 2A, 2B and 2C: potential natural forests of deciduous oak trees with *Quercus pyrenaica* as the dominant species, partially substituted nowadays by agricultural land uses and maritime pine; 1B and 2D: potential natural forests of birch (*Betula celtiberica*); 4A21, 4A22, 4A23, 4A24, 4A25 and 4B3: potential natural oak forests adapted to calcareous soils (*Quercus faginea*) which, nowadays, are residual after destruction by fires and pastoral land uses, or substituted by maritime pine in the sandy lands along the coast; 2C3, 3A11, 3A12, 4A23, 4B1, 4B2, 4B3, 4C1, 4C3 and 3B21: potential natural evergreen forests of cork oak, currently man-made, in most cases, and managed for cork harvesting; 2B, 3B11, 3B12, 3B13, and 3B22: potential natural evergreen forests of holm oak, currently legally protected by conservation laws; 4A24 and 4C3: potential natural forests of *Olea europaea* adapted to calcareous soils; 4C3: other potential natural forests adapted to calcareous soils composed of *Quercus faginea* and carob.

Continental Portugal with an Atlantic climate (1A1 and 4A1).[5] Potential natural forests dominated by *Q. pyrenaica* are typical of the highlands of the north-eastern and central eastern regions (2A, 2B and 2C), where the Atlantic climate is subject to Continental and Mediterranean influences.

Since Neolithic times, these forests have regressed due to agricultural and pastoral land uses. This regression process was concluded by the end of the Middle Ages, leaving only residual areas. These forests have been replaced with agricultural lands in the lower altitudes, and by single or mixed stands of maritime pine (*Pinus pinaster*) and *Eucalyptus globulus* in the lower or intermediate altitudes, and shrub lands of *Cytisus* sp.pl., *Genista florida* subsp., *Polygalaephylla, Ulex* sp.pl. or *Erica* sp.pl. in higher altitudes.

Birch forests

The typical species of potential natural birch forests in Continental Portugal is *Betula celtiberica*. This species can coexist with the deciduous oak forests where hydric conditions are appropriate, as is the case of the mountains of Gerês (1B) and Estrela (2D).

Oak forests adapted to calcareous soils

Potential natural forests of *Quercus faginea* subsp. *broteroi* (class *Quercetea ilicis*) are typical of the calcareous soils of the Estremadura and Arrábida regions (4A21, 4A22, 4A23, 4A24, 4A25 and 4B3), where there is a mixture of Atlantic and Mediterranean climatic influences. These forests are now very residual, after a long period of regression due to fires and pastoral land uses. In the sandy lands along the coast, there are forests of maritime pine planted by the Forest Services at the beginning of the 20th century to prevent erosion.

Evergreen oak forests

The two major species of potential natural forests of evergreen oak trees in Continental Portugal are cork oak (*Quercus suber*) and holm oak (*Quercus rotundifolia*, class *Quercetea ilicis*). Cork oak is present in the provinces of Trás-os-Montes (2C3), Beira Baixa (3A11 and 3A12) and Estremadura (4A23, 4B1), and in the western part of Alentejo and the Algarve (4B2, 4B3, 4C1, 4C3 and 3B21). Holm oak is typical of some of the more arid and interior areas, in the provinces of Beira Alta (2B) and Alentejo (3B11, 3B12, 3B13 and 3B22) and in the eastern part of the Algarve (4C3). Most of the stands of cork oak existing today are man-made and managed for the production of cork.

Other evergreen forests and other forests adapted to calcareous soils

Other potential natural forests of evergreen species and of species adapted to calcareous soils are typical of the region surrounding Lisbon (4A24) and the eastern part of the Algarve (4C3). The main species here is *Olea europaea* subsp. *sylvestris*. On the calcareous soils of the Algarve, *Quercus faginea* subsp. *alpestris* and carob (*Ceratonia siliqua*) can also be found. Nowadays, carob trees are managed for fruit production.

Hygrophilic forests

As far as riparian forests are concerned, the main species existing in Portugal are alders (*Alnus glutinosa*) and willows (*Salix atrocinerea* and *S. neotricha*). In other wetlands, swamps excluded, the main tree species is ash (*Fraxinus angustifolia*).

Institutional Aspects

Forest ownership

Most forests and other wooded land (93.4%) are under private ownership and management, the remainder being almost entirely communal forests managed by the Forest Services (Table 22.4). The main stakeholders involved in forest management in Continental Portugal are:

- The non-industrial private forest owners (NIPFO), managing more than four-fifths of the pine forests (typically with small holdings, in the northern and central regions) and almost all of the cork oak forests (often with large holdings, in the southern regions).
- The Forest Services, managing the public forests and most of the communal forests, often dominated by maritime pine.
- The pulp and paper industry, managing 28% of the eucalyptus forests, the rest being almost entirely with non-industrial private forest owners.

There are contrasting forestland ownership structures between the north and south of Continental Portugal: small-scale forestry (mostly < 10 ha) in the northern and central regions, and much larger holdings (mostly > 100 ha) in the south. The communal forests are located mostly in the northern and central regions.

Table 22.4. Area of forest and other wooded land by types of ownership and tree species in 1995 (000 ha).

Type of owner	Total		Maritime pine		Eucalyptus		Cork oak		Other forests and wooded land	
	Area	%	Area	%	Area	%	Area	%	Area	%
State forests	40	1.2	27	2.8	0	0.0	2	0.3	11	1.1
Communal forests	180	5.4	116	11.9	14	2.1	1	0.1	49	5.0
NIPF	2910	86.9	822	84.2	470	69.9	690	96.8	928	93.9
Forest industries	219	6.5	11	1.1	188	28.0	20	2.8	0	0.0
Total	3349	100.0	976	100.0	672	100.0	713	100.0	988	100.0

Sources: Mendes (2004) based on the authors' estimates and on data from Direcção Geral das Florestas (1992, 2001) and from Associação da Indústria Papeleira (2002).

Administration and policies

The Forest Services, with different nomenclature since their creation at the end of the 19th century, and, most of the time under the umbrella of the Ministry or Secretary of State for Agriculture, are the national forest authority. Given the large amount of uncultivated land existing at the time these services were created, afforestation was their main policy priority for a long time. Their initial mission in this area, accomplished during the first quarter of the 20th century, was the afforestation of the sandy lands along the coast to prevent erosion. This was an important and effective project. Another important project carried out between the 1930s and the start of the 1970s was the afforestation of the communal lands in the mountain areas of the northern and central regions. This was effectively accomplished, however, sometimes against the wishes of the local population and in contrast to traditional land uses.

In a country where a very large share of forestland is privately owned, the main aim of the Forest Services should have been to support improvements in the management of private forests. Here the track record of the Forest Services is not so good. Until the beginning of the 1980s, only a few attempts were undertaken: a programme started in the mid-1950s provided support in kind[7] and cheap credit for improving the cork oak stands; and, during the 1960s, eucalyptus plants were distributed free of charge.

From 1980 to 1988, a World Bank-funded programme was the first major attempt to intervene in the afforestation of private lands. The funds were allocated to a pulp and paper company and to the Forest Services who carried out all the afforestation works on the lands of those private owners willing to participate in this initiative. This strategy of public direct intervention in private forestry failed to meet some of the initial targets.

A major turning point in Portuguese forest policy towards private forestry happened with the first (re)afforestation programme co-funded by the European Union (EU), which started in 1987. Here the Forest Services switched from an intervention focused mostly on public and communal lands, complemented by assistance in kind, or direct intervention on private forestry, to a position of incentive regulator of private forestry which has since remained. The difference in this new role now compared with the mid-1980s is that the Forest Services gradually lost their initial role throughout the decision chain of designing the incentives schemes, and reviewing, approving and monitoring the applications of private owners for grants. Nowadays these responsibilities are with the public institute in charge of processing the agricultural structural funds (IFADAP).

The grant-driven (re)afforestation, which has been happening since the mid-1980s with the strong support of the EU-co-funded programmes, led to the organization of the private forest owners into associations capable of providing the technical assistance they need to prepare grant applications and to undertake the (re)afforestation works. The number of these types of associations grew rapidly in the late 1990s: 67 in 1998, 110 in 1999 and 130 in 2002 (Direcção Geral das Florestas (DGF) data).

Contribution of Forests to the National Economy

Contribution to the gross domestic product

In Portugal, in 1998, the forest sector represented 2.93% of the gross domestic product (GDP), which places the country in a top position within the EU 15, in terms of this indicator, being surpassed only by Finland and Sweden (Table 22.5).

Contribution to foreign trade balance

Exports, and not domestic consumption, have been the main driving force behind forest production in Portugal over the last one hundred years. In particular, exports account for the increase in the production of cork, eucalyptus pulpwood and pine wood, contributing positively to the trade balance. Until recently, these products were the second major export group before a big project in the car industry took off. Nowadays, forest products occupy fourth position, generating 11% of the total export value, a figure maintained since the 1880s, most of the time.

Contribution to employment

Official statistics underestimate the employment in the forest sector by excluding some of the small and medium-sized forest enterprises, as well as most of the services related to forestry and forest industries. The author's own estimate for 1995 (Mendes, 2004), taking into account these omissions, gives a total of around 227,800 persons employed in the sector, which

is 5.1% of the total workforce. This number is comprised of: 33,600 persons (0.8%) in silviculture, logging and hunting; 188,900 persons (4.2%) in forest industries and connected activities; and 5300 persons (0.1%) in non-marketed supporting services. This places the forest sector in seventh position, in terms of employment, after the following clusters of activities: agricultural and food industries, wholesale and retail, construction, public administration, textile and clothing industries, educational services and research.

Economic Value of Forest Production in Continental Portugal

Scope of the estimates

The scope of this estimation is the economic valuation of forestry *outputs* in Continental Portugal, including those that are not marketed. Some of these outputs contribute positively to society's well-being and are therefore referred to as *social benefits*, while others contribute negatively, being referred to as *social costs*. This study is mainly concerned with the 'resources' side of a forestry production account (in the national accounting sense of the word), extended to include some forest public goods and other non-marketed forest goods and services (Bergen, 2001). Estimates of some of the 'uses' in the forestry production account are given only for the depreciation in forestry capital due to fires. Therefore, a complete estimate of the net social added value for forestry is not obtained.

To make these estimates comparable with the other country chapters, timber and cork

Table 22.5. Gross value added of the forest sector at current market prices (million €).

		1995	1996	1997	1998
Forest sector	Forestry	644.2	595.5	559.5	606.4
	Forest industries	1,873.8	1,723.6	1,766.2	1,880.4
	(1) Total	2,517.9	2,319.1	2,325.7	2,486.8
(2) Economy		69,044.1	73,879.6	78,994.4	84,964.0
(1)/(2)		3.65%	3.14%	2.94%	2.93%

Sources: Instituto Nacional de Estatística (2001); Ministério da Agricultura, do Desenvolvimento Rural e das Pescas (2000).

production is evaluated at roadside prices. This implies that we are dealing with the 'resources' side of the consolidated production account of forestry, logging and cork extraction. Hunting and animal production based on acorns and grazing from forest areas are not included in this consolidation. What is estimated related to these two activities is the value of forest outputs that are their intermediate consumption.

Direct use values

Timber harvested

Data regarding the production of the different types of timber harvested are provided by the official agricultural statistics (Instituto Nacional de Estatística, 2003b). These data, published in cubic metres under bark, were converted[8] into cubic metres over bark. Monetary valuation is based on roadside prices for 2001 (Sistema de Informação de Cotações de Produtos Florestais na Produção, 2003a), considering that: the price for coniferous pulpwood, sawlogs and fuelwood refers to maritime pine; the price for broadleaved pulpwood refers to eucalyptus; the price for broadleaved sawlogs refers to oak sawlogs;[9] the price for other industrial wood refers to oak sawlogs; and the price for broadleaf fuelwood is a weighted average of the roadside prices for eucalyptus, chestnut and oak fuelwood.

Net growth in timber stock

Physical valuation considers the difference between the annual forest increment and timber harvested in 1998. Direcção Geral das Florestas (1999) reports a forest increment of 15 Mm^3 o.b., of which 54% accrues to conifers and 46% to broadleaves. Based on data of the Instituto Nacional de Estatística (2002a),[10] the quantity of timber harvested is 11.3 Mm^3 o.b., of which 55% is coniferous and 45% is broadleaved. Monetary valuation is based on half of the stumpage price for sawlogs, considering that not all of the net growth in timber stock has an exchange value. This valuation does not include the annual variation in the value of timber stock as a carbon sink, which is a public good. The latter is incorporated in the value

of the forest environmental services. The stumpage prices are the price of maritime pine (for coniferous growing stock) and of oak (for broadleaved growing stock) (Sistema de Informação de Cotações de Produtos Florestais na Produção, 2003b).

Cork

Data for production of virgin and reproduction cork in 2001 come from the official agricultural statistics (Instituto Nacional de Estatística, 2003b). The source for the roadside market price ('*preço de venda na pilha*') of reproduction cork is SICOP's leaflet (Sistema de Informação de Cotações de Produtos Florestais na Produção, 2003a). The price for the virgin cork is given by the SICOP website (Sistema de Informação de Cotações de Produtos Florestais na Produção, 2003b). It was assumed that the price reported for virgin cork is a roadside price.

Resin

Data for production come from the official agricultural statistics (Instituto Nacional de Estatística, 2003b). The producer market price per kg for 2001 was calculated considering the producer market price per incision for 2001, according to SICOP (Sistema de Informação de Cotações de Produtos Florestais na Produção, 2003a), and a production of 1.8 kg of resin per incision (Goes, 1991).

Honey

Valuation of honey distinguishes between origin-labelled production and other production. For the former, data regarding production and price in 2001 are provided from the answers to questionnaires sent by the Instituto de Desenvolvimento Rural e Hidráulica (Oliveira, 2004) to producer groups. The price refers to sales of those groups to wholesalers and other buyers.

Data for other production were obtained by subtracting the origin-labelled production from the total production of the country in 2001 (except 4 t of production in the Azores), as reported by official agricultural statistics (Instituto Nacional de Estatística, 2003b). The price is the average export price in 2001 according to these statistics (Instituto Nacional de Estatística, 2002a).

Pine nuts

There have been no official data regarding the production of pine nuts since 1972. The volume of production reported in Table 22.6 is an estimate made by Alpuim *et al.* (1998), and not the actual production for 2001. The price for 2001 is the producer market price, according to SICOP (Sistema de Informação de Cotações de Produtos Florestais na Produção, 2003a).

Chestnuts

The data for production and the market producer price in 2001 come from the official agricultural statistics (Instituto Nacional de Estatística, 2003b).

Carob

There have been no official data for carob production since 1977. According to the official agricultural statistics, the average annual production for 1968–1977 was 43,193 t. Current opinions of local experts give estimations ranging from 28,000 to 35,000 t. The valuation considers the average of the two estimates (31,350 t) and the producer market price for 2001 as reported by the official agricultural statistics (Instituto Nacional de Estatística, 2003b).

Arbutus berries

The most recent data for *Arbutus unedo* come from the first revision of the Forest Inventory (1969/1974), according to a Forest Services' publication (Direcção Geral do Ordenamento e Gestão Florestal, 1979). The data for production of berries per hectare come from Goes (1991). The price paid to pickers at the distillery gate is the author's own estimate based on a price of €15/l of arbutus brandy, a transformation ratio of 100 kg of berries per 15 l of brandy (Goes, 1991) and about 50% of the price of the brandy corresponding to the cost of berries at the distillery gate.

Elderberries

Data regarding quantity are the author's own estimate of the average annual production for Continental Portugal based on local informants from the area where this species is more frequent (Vale do Varosa) published in the CESE report (Conseho para a Cooperação Ensino Superior/Empresa, 1996; Mendes, 1997). The market price paid to pickers is the price for 1995 obtained from local informants in that area, inflated to 2001 prices, according to the producer price index for agricultural products (Instituto Nacional de Estatística, 2002a).

Mushrooms

Production is based on the author's own estimate for the average quantity of mushrooms picked and sold in the period 1997–1999, based on a report prepared by a working group under the coordination of the Instituto da Conservação da Natureza (2001). The price paid to pickers is based on information collected in October 2000, from local sources, in the border regions with Spain where this activity is more intense (Paulino, 2000). This price is less than half the export price.

Plants

The production is the author's own conservative estimate based on the quantities exported in the period 1988–1992, under positions 0910 and 1211 of the Foreign Trade Statistics. The averages for this period were 60.6 t for cooking plants (with a maximum of 75.3 t in 1992) and 822.6 t for the aromatic and medicinal plants (with a maximum of 1027.5 t in 1992). The market prices paid to pickers in 2001 are the author's own estimates.

Forest products for intermediate consumption in animal production

There are four types of forest goods to be considered as intermediate consumption for animal production: (i) acorns; (ii) grazing resources under forest cover; (iii) grazing resources in scrubland; and (iv) litter lying on the forest floor.

Table 22.6. Economic values of forest products in Continental Portugal (2001).

Outputs	Physical production (intermediate or final)	Valuation method	Unit value (euros per physical unit)	Value of production (000 €)
Direct use values				
Wood forest products				543,594
Timber harvested				430,604
Pulpwood				
Coniferous	2,153,000 m³ o.b.	Roadside market price	19.54/m³ o.b.	42,070
Broadleaved	6,684,000 m³ o.b.	Roadside market price	31.70/m³ o.b.	211,883
Saw logs				
Coniferous	4,733,000 m³ o.b.	Roadside market price	33.42/m³ o.b.	158,177
Broadleaved	221,000 m³ o.b.	Roadside market price	41.89/m³ o.b.	9,258
Other industrial wood	220,000 m³ o.b.	Roadside market price	41.89/m³ o.b.	9,216
Fuelwood				37,273
Coniferous	286,000 m³ o.b.	Roadside market price	38.22/m³ o.b.	10,931
Broadleaved	488,000 m³ o.b.	Roadside market price	53.98/m³ o.b.	26,342
Net growth in standing timber stock				75,717
Coniferous	2,060,000 m³ o.b.	50% of the stumpage price	19.53/m³ o.b.	40,232
Broadleaved	1,794,000 m³ o.b.	50% of the stumpage price	19.78/m³ o.b.	35,485
Non-wood forest goods				584,771
Cork harvested				390,726
Reproduction cork	128,000 t	Roadside market price	2,937/t	375,936
Virgin cork	30,000 t	Roadside market price	493/t	14,790
Resin	15,444 t	Roadside market price	200/t	3,089
Honey				7,619
Origin labelled honey production	172.5 t	Market price at producer group gate	3,970/t	684
Other honey production	4,361.5 t	Average export price	1,590/t	6,935
Fruits collected				53,310
Pine nuts	70,000,000 pine cones	Market price at farm gate	0.20/pine cone	14,000
Chestnuts	26,118 t	Market price at farm gate	9,976/t	26,055
Carob	31,500 t	Market price at farm gate	2,723/t	8,577
Arbutus berries (*Arbutus unedo*)	15,130 ha × 200 kg/ha	Market price paid to pickers at distillery gate	1,125/t	3,404
Elderberries (*Sambucus nicra*)	650 t	Market price paid to pickers	1,196/t	1,274
Edible wild mushrooms picked for sale	6,500 t	Market price paid to pickers	2,500/t	16,250

continued

Table 22.6. *Continued.*

Outputs	Physical production (intermediate or final)	Valuation method	Unit value (euros per physical unit)	Value of production (000 €)
Plants picked for sale				
Thyme, laurel and other cooking plants	80 t	Market price paid to pickers	3.75/kg	1,400
Aromatic and medicinal plants	1,100 t	Market price paid to pickers	1/kg	300
				1,100
Forest goods for intermediate consumption in animal production				112,377
Acorns grazed by pigs in extensive rearing	51,450,000 FU	Surrogate market price	130.3/1000 FU	6,704
Grazing resources under forest cover	673,900,000 FU	Surrogate market price	130.3/1000 FU	87,809
Grazing resources in scrub land (consumption by goats)	137,100,000 FU	Surrogate market price	130.3/1000 FU	17,864
Acorns and other products grazed by other animal species				No estimate
Net growth in the production capacity of non wood forest goods				No estimate, but probably positive
Recreational services				37,883
Hunting	219,005 hunters	Cost-based method		21,383
Informal forest recreation	6,000,000 day-visits	CVM	2.75/day-visit	16,500
Total direct use values				1,166,248
Indirect use values				
Carbon storage	1,450,000 tC	Shadow pricing	20/tC	29,000
Protection of agricultural soil				49,209
Protection of water resources	8,772,520 ha	Cost avoided method	3.30/ha	28,934
Forest landscape and biodiversity conservation	594,509 ha	Cost-based method	95.36/ha	56,695
Total indirect use values				163,838
Negative externalities				
Damage caused by forest fires		Cost-based method		136,850
Costs of fire prevention				17,350
Social costs of firefighting				35,853
Losses of forest products burnt				38,320
Reforestation costs				45,327
Other forest externalities				No estimate
Total negative externalities				136,850
Total economic value				1,193,236

ACORNS. The main sources of acorns currently grazed by animals are the cork oak and holm oak stands in the southern regions. The total and mean annual production of acorns of these stands are reported in Table 22.7, as given by the Forest Inventory of 1995 (Direcção Geral das Florestas, 2001). Not all this production is actually grazed by animals. For the farms surveyed in the project carried out by Moreira *et al.* (1995), the production of acorns grazed by pigs in an extensive regime ('*porco de montanheira*') is 37 kg/ha/year. This is about 5.5% of the mean production reported in Table 22.7. Applying this percentage to the total production reported in that table, a figure of 22,714 t for the cork oak stands and 16,903 t for the holm oak stands is obtained, which makes a total of 39,617 t. This is possibly a lower bound estimate of the amount of acorns grazed by pigs in extensive rearing. Another estimate can be made based on the number of pigs in this regime and their feeding needs. According to the same research project (Moreira *et al.*, 1995), in 1989 there were 6000 sows, each of these animals giving birth to ten suckling pigs per year. If eight out of these ten suckling pigs go on for fattening up to the age of 2 years, this gives 48,000 fattening pigs per year. If each of these pigs needs 1400 kg of acorns, a total of 67,200 t of acorns grazed by fattening pigs in an extensive regime is obtained. An estimate for this kind of use of acorn production is around 70,000 t/year.

To convert this quantity into forage units (FU), the coefficients proposed by Vieira da Natividade (1950, p. 317) are taken as a basis: 730 FU/t for acorns from cork oak and 743 FU/t for acorns from holm oak. Considering an intermediate value of 735 FU/t, 70,000 t/year of grazed acorns correspond to 51.5 million

FU/year. This quantity of grazed acorns is a lower bound estimate of the amount of acorns used in animal production because there are other animal species, besides pigs, in an extensive regime, fed with this type of forest good. No attempt is made to estimate this kind of intermediate consumption of acorns. To value this forest good, the price of barley for animal consumption in 2001 (Instituto Nacional de Estatística, 2002a) is used as a surrogate market price, assuming the equivalence 1 kg of barley = 1 FU.

GRAZING RESOURCES UNDER FOREST COVER. Based on information provided by the 1995 Forest Inventory (Direcção Geral das Florestas, 2001) on natural and artificial grasslands under forest cover, their total forage production is estimated as reported in Table 22.8. The mean annual production of forage in terms of dry matter (DM) is the author's own estimate, based on the information provided by Moreira (1980), as is the ratio of FU per kg of DM: 0.3 FU/kg DM for the natural grasslands and 0.45 FU/kg DM for the artificial grasslands.

With a total of 1.4 million t DM/year, most of which is from cork oak and holm oak stands, it is possible to raise livestock equivalent to 1.4 million head of sheep. According to Moreira *et al.* (1995), in 1989, the livestock in the southern regions of '*montados*' (forest stands dominated by cork oak and holm oak trees), pigs excluded, corresponding to autochthonous races usually in an extensive regime, amounted to a number of female adult animals equivalent to 1.5 million head of sheep. This is an indication that the estimate of forage production presented in Table 22.8 is probably of the same magnitude as the forage production actually used by livestock (pigs excluded) in an extensive regime, at least for the southern regions. To value this forest good, we use, as a surrogate market price, the price of barley for animal consumption in 2001 (Instituto Nacional de Estatística, 2003b), assuming the equivalence 1 kg of barley = 1 FU.

Table 22.7. Total and mean annual production of acorns in cork oak and holm oak stands in 1995.

Species	Type of stand	000 t	kg/ha
Sobreiro	Pure	343.0	579
	Mixed dominant	49.5	411
	Mixed dominated	20.4	177
Azinheira	Pure	266.4	688
	Mixed dominant	31.8	428
	Mixed dominated	9.1	130

Source: Direcção Geral das Florestas (2001)

GRAZING RESOURCES IN SCRUBLAND. Rego (1991) considers a mean forage production in scrublands to be 1.5 t DM/ha/year. According to the 1995 Forest Inventory, there were 2 Mha of scrublands. Applying that coefficient, a total of

Table 22.8.   Estimate of the forage production of grasslands under forest cover in Continental Portugal, in 1995.

Forest species	Natural grasslands				Artificial grasslands			
	ha	t DM/ha/year	t DM/year	000 FU/year	ha	t DM/ha/year	t DM/year	000 FU/year
Maritime pine	0	2	0.0	0	9,761	3.0	29,283.0	13,177
Cork oak	46,282	1	46,282.0	13,885	257,715	2.5	644,287.5	289,929
Holm oak	22,336	1	22,336.0	6,701	13,443	2.5	623,130.0	280,409
Eucalyptus	0	0	0.0	0	249,252	2.5	33,607.5	15,123
Other oaks	4,690	2	9,380.0	2,814	8,945	4.0	35,780.0	16,101
Stone pine	4,101	1.5	6151.5	1,845	6,956	3.0	20,868.0	9,391
Chestnut	0		0.0	0	6,670	4.0	26,680.0	12,006
Other broadleaves	0		0.0	0	6,955	4.0	27,820.0	12,519
Other conifers	0		0.0	0	0.0	3.0	0.0	0
Total	77,409		84,149.5	25,245	559,697		1,441,456	648,655

3 Mt DM/year is obtained. Considering a ratio of 0.5 FU/kg DM (1978), a total of 1540.9 million FU/year can be calculated. Most of this production is left without being used by animals, and therefore contributes to forest fires. The animals most likely to consume this type of vegetation are goats. In Continental Portugal, in 2001, there were 544,000 animals of this species (Instituto Nacional de Estatística, 2002a). Assuming that each of them consumes 300 FU/year from this kind of grasslands, a total of 137.1 million FU is obtained. This amount is assumed to have been consumed in animal production, in 2001.

LITTER LYING ON THE FOREST FLOOR.   Litter composed of leaves and fallen branches lying on the forest floor is a product that can be consumed by livestock, at least partially. Another part of these materials is needed to maintain the fertility of the forest soils. What is unused for these purposes contributes to the risk of forest fires.

Based on the coefficients proposed by Rego (1991) and the areas of forest in the 1995 Forest Inventory, the annual production of litter is 1.2 Mt DM in cork oak and holm stands (1.2 Mha ×1 t DM/ha) and 5.0 Mt DM in other forest stands (2.0 Mha × 2.5 t DM/ha). Adding up these estimates gives a total of 6.2 Mt DM/year. Based on a coefficient of 0.6 FU/kg DM (Vieira de Sá, 1978), this corresponds to 3744.7 million FU/year. It is assumed that all this production is left on the ground, or burns in forest fires.

Comparison between the value of forest goods used as intermediate consumption in animal production and the value of animal production

Since grazing resources are the most valuable non-wood forest goods after cork, it is important to verify the reliability of the estimate using a different method. In national accounts, the estimated value of €112.4 million of forest products used in animal production in 2001 is part of the value of animal production and not part of the value of forest production. That amount should be compared with the value of the following components of animal production: meat, milk and cheese from goats; origin labelled meat and cheese; origin labelled meat from cattle; and origin labelled meat from pigs.

According to the official agricultural statistics (Instituto Nacional de Estatística, 2003b), the value of meat production from sheep and goats in 2001 was about €163 million. According to the questionnaires sent by the Instituto de Desenvolvimento Rural e Hidráulica (Oliveira, 2004) to the producers' groups of origin labelled products in 2001, the value of origin labelled meat products from cattle and pigs was €117.2 million and the value of origin labelled cheese from sheep and goats was €12.8 million. Adding up these values, a total of €187.4 million is obtained for the animal production likely to be dependent on grazing products from forests and scrublands. Therefore, the previous estimate of €112.4 million for the value of these forest products can be considered as a reasonable approximation.

Net growth in the production capacity of non-wood forest goods

The net growth in the production capacity of non-wood forest goods is not estimated; instead, qualitative information regarding the trends in this forest resource is given. Cork harvesting is subject to regulations preventing removals beyond sustainable limits. It is believed that the industrial demand for cork induces harvesting of all sustainable production. Since the end of the 1930s, the cork oak area did not change substantially, but the stand's quality improved considerably during a programme carried out by the Forest Services in the late 1950s. Since the mid-1980s, the EU financial incentives have prompted a renewal and expansion of the cork oak stands. Thus, the future trends in the productive capacity of cork oak stands are likely to be positive.

The demand for pine nuts, chestnuts and carob is in tandem with the harvest, which is believed to be within sustainable limits. Since the mid-1980s, these species have also benefited from public financial incentives. So, the conclusion for this group of products is similar to the case of cork. In the case of mushrooms, there are situations of over-picking, but there are also areas of underpicking where there are no workers available and willing to do this job. Therefore, it is difficult to make a well-founded guess about the trend in the production capacity of this product. With respect to resin, honey, arbutus berries, elderberries, plants, acorns and grazing resources, there are reasons to believe that the trends in production harvested may not be following the trends in the production capacity. Starting with resin, the situation can be described as follows:

- A sharp decline in resin tapping since the mid-1980s: from 115,200 t on average per year in the period 1980–1986 to 21,300 t in the period 1996–2002.
- A decline in the area of maritime pine not as large as the decline in resin tapping: from 1.3 Mha in the second revision of the Forest Inventory (1980/85) to 976,000 ha in the third revision (1995/98), the decline continuing in more recent years because of forest fires.[11]

These trends led to a decline in the production capacity of resin. Other products (honey, berries, plants, acorns and grazing resources) are harvested below potential production; their production capacity is probably growing, not only because of no overuse, but also due to the growth in forest and other wooded land. The global conclusion is that the net change in production capacity of non-wood forest goods is probably positive.

FOREST HUNTING BENEFITS. The value of the hunting benefits of forests is estimated by using the costs paid by hunters, including hunting permits, fees for gaming services in hunting zones with excludable access, and membership fees to associative hunting areas.

- *Hunting permits.* In the 2001/02 hunting season, 219,000 hunters paid €5.5 million for their hunting permits[12].
- *Gaming services paid by hunters in hunting zones with excludable access.* According to Cipriano (1999), in the 1996/97 hunting season, average expenditure per hunter on gates, posts, game management and other gaming goods and services in hunting zones with excludable access was €674 in touristic zones, €311 in associative zones and €104 in social zones.[13] Assuming that the distribution of hunters across all types of zone in the 2001/02 hunting season was the same as in 1996/97, the total amount paid is €26.5 million[14].
- *Membership fees to associative hunting areas.* Membership fees to associative hunting areas averaged €207 (Cipriano, 1999, updated to 2001 euros). Given 96,000 members in 2001 (Bugalho and Carvalho, 2001), this amounts to €19.9 million.

Adding up these figures results in a total cost paid by hunters of €51.9 million. Not all of it can be attributed to forests, however. Although forests are very important for game feeding, other areas – agricultural areas and uncultivated lands – also play a role. A crude but simple criterion to impute the value of hunting to forests is to multiply it by the percentage of forests and other wooded lands in the total area with hunting capacity, which is 41% (Bugalho and Carvalho, 2001). Thus the value of hunting benefits attributable to forests is estimated at about €21.4 million.

INFORMAL FOREST RECREATION. No data are available regarding the number of visits to forests and other wooded lands for recreational purposes. Therefore, available data reporting the number of days spent in campsites are used as part of a proxy for that variable; as almost all camping grounds are under forest cover, it is reasonable to assume that enjoyment of forests may be one of the motivations of most campers.[15] This makes a total of 4.6 million days spent in campsites, in 2001 (Instituto Nacional de Estatística, 2002b, 2003c).

In addition, 0.4 million nights were spent by guests in rural tourist facilities. This figure does not include a large and increasing number of urban people who visit forest areas on weekends and holidays without staying overnight. The number of such visits is estimated very roughly by assuming that half the households in the two metropolitan areas of Porto and Lisbon (1.2 million households in 2001: Instituto Nacional de Estatística, 2003a) visit forest areas at least once a year, and count for just 1 day-visit per household, for a total of 0.6 million day-visits. This gives a total of about 6 million days a year for all types of visitors to forest areas.

The willingness to pay per day-visit is based on the only available empirical study of the recreational value of a Portuguese forest area (Loureiro and Albiac, 1996). Using a contingent valuation method (CVM), the authors found a mean willingness to pay for access to a forest reserve in the Terceira Island of the Azores of €2.75/day-visit (in 2001 euros). Given the estimated 6 million day-visits, the total value of informal recreation in forests is estimated at about €12.5 million.

Indirect use values

Carbon storage

The net annual increment of carbon storage in the woody biomass of Portuguese forests amounts to 1.45 MtC/year, based on the United Nations Economic Commission for Europe/Food and Agriculture Organization (2000). If this flow is evaluated at the mean social cost

of carbon emissions of €20/tC, as estimated by Fankhauser (1995, p. 64) for the decade 1991–2000, an estimate of €2.9 million is obtained.

Protection of agricultural soil

Estimating the protection of agricultural land begins with the regions facing a higher risk of desertification, such as Trás-os-Montes, Beira Interior and Alentejo, where the annual erosion of agricultural soil is 5–10 t/ha (Poeira *et al.*, 1990). Considering an apparent specific weight for sediments of 1.5 t/m^3 and a depth of 30 cm for agricultural soil, this erosion corresponds to an annual rate of soil loss of between 0.11 and 0.22%. The average of these rates (0.165%) is used, assuming that it corresponds to the rate of loss in agricultural production.

Based on Rocha *et al.* (1986), the ratio of erosion between land with forest cover to land without is 2:3. Assuming this is proportional to the forests' contribution in reducing erosion, the value of the crops preserved due to soil protection by forest cover is equal to $((1-\frac{2}{3})/\frac{2}{3})$ × 0.165% × gross value of crops.

If the (avoided) losses of crops were irreversible, for a 2% discount rate, the value of €1.068 million (Table 22.9) would correspond to a capital loss avoided of €53.4 million. If an amount of losses equal to v lasts for n years, the corresponding capital loss V_n is given by the following expression:

$$V_n = v \left[\frac{1-(1+r)^{-n}}{r} \right]$$

Table 22.9. The value of crops preserved due to the soil protection provided by forests.

	Gross value of crops in the year 2000[a] (000 €)	Gross value of crops preserved in the year 2000, due to the soil protection provided by forests (000 €)
Trás os Montes	526,260	434
Beira Interior	236,470	195
Alentejo	531,970	439
Total	1,294,700	1,068

[a]Source: Instituto Nacional de Estatística (2003e).

Considering a period of 50 years to recover from soil losses due to erosion and a 2% discount rate, the annual value of losses avoided in the three regions is €33.6 million.

To estimate the value of agricultural soil protection in other regions, an annual rate of soil erosion of 0.055% is assumed – one-third of the average for the three regions. Based on the same method, a gross value of crops of €1812 million is obtained, corresponding to an annual value of about €15.6 million. Adding up the two estimates (annual flows) gives a total value of €49.2 million.

Protection of water resources

The protection of water resources is estimated by using the public costs of watershed management avoided by the existence of forests. These costs are considered as a lower bound for the forests' benefits in water conservation. The Management Plans for the main watershed basins (Instituto Nacional da Água, 2000) provide data for the total public costs planned

for 2001–2020. They relate to the protection of ecosystems (PO3), flood prevention (PO4), fish and wildlife management (PO5) and water management (PO6) (Table 22.10).

To estimate the costs that would be borne in the absence of forest, it was assumed that the watershed management costs would increase in the same proportion as erosion would increase without forest cover. The increases in erosion were estimated for each watershed based on data from the 1995 Forest Inventory, as reported by the Direcção Geral das Florestas (DGF) software AreaStat, and data taken from the work of Rocha *et al.* (1986) on soil erosion. The sixth column in Table 22.11 is the coefficient by which we have to multiply the costs in order to obtain the amount of public costs annually avoided in watershed management due to existence of the current forest cover. The results of this estimation for each watershed are reported in the last columns of Table 22.11. Since the Watershed Management Plans on which this estimation is based are from 2000, the estimate is not corrected for inflation.

Table 22.10. Total public costs of watershed management for the Portuguese international rivers planned for the period 2001–2020 (million escudos).

Watershed	PO3	PO4	PO5	PO6	Total cost for 2001–2020	Annual cost
Minho	980	206	858	630	2,674	134
Lima	391	1,021	63	2,076	4,118	206
Douro	1,498	763	578	10,572	18,613	931
Tejo	11,739	822	450	15,910	28,921	1,446
Guadiana	1,460	7,840	2,915	1,250	13,465	673

Table 22.11. Rates of forest cover, forest cover correction factors for soil erosion rates and the annual public watershed management costs avoided by the existence of forest cover (thousand escudos).

Watershed	Total area (000 ha) (1)	Forest area (000 ha) (2)	(2)/(1) %	C	(1-C)/C	Annual costs with current forest cover for 2001–2020	Annual costs avoided for 2001–2020 due to the existence of the current forest cover — Total	Per ha
Minho	79.9	29.4	36.8%	1/3	2	133,675	267,350	3.3
Lima	117.2	34.7	29.6%	2/3	1/2	205,900	102,950	0.9
Douro	1,853.9	506.0	27.3%	2/3	1/2	930,650	465,325	0.3
Tejo	2,432.9	1,124.3	46.2%	1/3	2	1,446,054	2,892,108	1.2
Guadiana	1,146.0	344.2	30.0%	2/3	1/2	673,235	336,618	0.3
Rest of Continental Portugal	3,142.6	1,310.8	41.7%	1/3	2		3,736,534	1.2
Continental Portugal	8,772.5	3,349.3	38.2%				5,800,885	0.7

Converting into euros, a value of €28.9 million is obtained.

Forest landscape and biodiversity conservation

FOREST LANDSCAPE CONSERVATION IN PROTECTED AREAS.   The estimated value of forest landscape and biodiversity conservation is based on the only study available in Portugal (Santos, 1997). Using CVM, Santos estimated the willingness to pay of visitors to the Peneda-Gerês National Park for three different programmes of rural landscape conservation, one of which dealt with oak forest conservation. The best point estimate he obtained for the year 1996 amounted to 6634 escudos per household and per year (Santos, 1997, p. 587). Based on the total number of households visiting the park between September 1995 and August 1996, an aggregated willingness to pay of 397,377 million escudos per year was calculated (Santos, 1997, p. 590).

Data regarding the area of forests and other wooded land in Peneda-Gerês National Park are not available, but can be estimated at around 60,000 ha, natural grasslands included. Dividing the aggregated benefit by this surface gives an estimate of 6623 escudos/ha. In order to arrive at a national level estimate, it is assumed that all protected forests in Continental Portugal have the same characteristics (visitor numbers, visit frequency and site composition) as those in the Peneda-Gerês National Park. Extrapolating this estimate to the total forest and other wooded land existing in the Natura 2000 sites (Table 22.3) results in a total willingness to pay of €3937.4 million in 1996. Converting and updating[16] this value to 2001 prices, an aggregate willingness to pay of about €20.4 million is obtained.

PUBLIC EXPENDITURE FOR FOREST LANDSCAPE AND BIODIVERSITY CONSERVATION.   The official statistics regarding the environment (Instituto Nacional de Estatística, 2003d) report data on investment and operating expenditures for landscape and biodiversity conservation by the Public Administration (Central Administration, municipalities and public institutes) and the public non-profit organizations in the whole country, without specifying the share for Continental Portugal. Based on these data, it is estimated that, in

2001, the operating expenditure for this part of the country is about €145 million. It is assumed that 39.1% of this amount refers to forests and other wooded land, based on the share of forests in the total area under some protection status. This gives an estimate of €56.7 million. This value does not include the contribution of public investment expenditures in landscape and biodiversity conservation for the increase in the capacity of forest areas to provide these kinds of services. Therefore, this value is a lower bound for the cost-based estimate of these services.

Adding up the €56.7 million with the €20.4 million estimated above for forest landscape conservation in protected areas would be double counting. Therefore, the former value is considered as the estimate for these services.

Forest negative externalities

Costs of forest fires

In 2001, of the 866 forest fires for which the cause was discovered, 95.2% were started by human actions: negligence (such as the burning of grasslands, picnicking and cigarettes); accidental ignition (due to the operation in or near the forests of farm or forestry machinery, vehicles, trains and electric lines); conflicts regarding hunting; and arson.

This illustrates that forest owners are seldom among the initiators of forest fires; however, they bear part of the costs, together with other people in society (such as volunteer firefighters and taxpayers) not responsible for starting fires. Therefore, the costs of most of the forest fires in Portugal may be considered as negative externalities borne by the forest owners and other people in society who share those costs with them. Some of the components of these costs are estimated below.

COSTS OF FOREST FIRE PREVENTION.   There are five main stakeholders in the forest fire prevention system: the non-industrial private forest owners; the pulp and paper companies; the Ministry of the Interior; the Ministry of Agriculture; and the municipalities. In recent years, the pulp and paper companies spent more than €3 million per year on this kind of operation (Associação da Indústria Papeleira, 2003). In

2001, the Ministry of the Interior spent €8.1 million, most of it in transfers to forest owners' associations and municipalities for fire prevention actions (Ministério da Administração Interna-Gabinete do Ministro, 2003). Out of this funding, €3.1 million were allocated to the co-funding of brigades of fire sappers managed by forest owners' associations. This co-funding represents about 50% of the total operating costs of those brigades. Through the EU-co-funded programmes of the Ministry of Agriculture, €3 million were transferred to public and private beneficiaries in 2000 to support forest fire prevention (Ministério da Agricultura, do Desenvolvimento Rural e das Pescas-Gabinete de Planeamento e Política Agro-Alimentar, 2001). Although no data for 2001 are available, the same amount as in 2000 can be assumed. Data on how much the Ministry of Agriculture spent from its own funding in running its network of forest fire detection are not available.

Adding these four components we get a total of €17.4 million, which is a lower bound for the social costs of forest fire prevention in 2001.

SOCIAL COSTS OF FOREST FIREFIGHTING. There are three main stakeholders involved in firefighting: the Ministry of the Interior;[17] the local fire departments;[18] and the pulp and paper companies. In 2001, the Ministry of the Interior spent more than €21 million on forest fire prevention and firefighting (Ministério da Administração Interna, 2002), through its special agency in charge of supervising the fire departments (SNB, Serviço Nacional de Bombeiros). This money was spent directly by SNB and indirectly through transfers to the local fire departments. The source of this information does separate the amount allocated for fire prevention and firefighting. Subtracting the €8.1 million spent by the Ministry in fire prevention, a figure of €12.9 million spent on firefighting is obtained. The data source does not specify either the amount allocated to the local fire departments or the matching funding added by these departments. The pulp and paper companies contributed more than €1.5 million (Associação da Indústria Papeleira, 2003). The calculation of the opportunity cost of the time spent by voluntary firefighters is based on the number of fires – 26,942 acording to DGF – and the assumption of 20 volunteers per fire, each contributing 1 day

of work per fire, giving an equivalent total of 2700 full-time workers per year. The value added per full-time worker in agriculture and forestry, in 2001, was €8000. Assuming the same labour productivity for volunteer firefighters, the opportunity cost of their time spent in firefighting amounts to about €21.5 million.

COSTS OF LOSSES IN WOOD AND NON-WOOD FOREST PRODUCTION. For 2001, DGF estimates wood production losses at about €38.3 million (Direcção Geral das Florestas-Corpo Nacional da Guarda Florestal, 2003). Valuing the losses of NWFPs could be based on previous estimates (Table 22.6). However, as the burnt areas are not those where the more valuable NWFPs grow, such an attempt would overestimate these losses. Therefore, without further information, the estimate is limited to the losses of wood production.

COSTS OF THE RESTORATION OF BURNT FORESTS. DGF estimates the area of burnt forests at about 45,300 ha in 2001. Reforestation through new plantations would cost around €2250/ha. Reforestation through management of natural regeneration (in the case of pine forests) and stand improvement would cost up to €1000/ha. Using the least expensive option, a value of €45.3 million is obtained.

Other negative forest externalities

Other possible negative forest externalities not estimated here include: erosion, floods and landslides due to poor forest management; loss of landscape quality and recreational opportunities due to poor forest management; and loss of biodiversity and landscape quality and other losses due to intensive forestry and damage due to pest infections. It should be noted that the main consequence of poor forest management is the increase in the risk of forest fires. Therefore, some of the consequences of this kind of management are already covered by the estimation presented above.

Conclusions

Taken as an aggregate, the NWFPs turn out to be the main item in the TEV of forest production in Continental Portugal (€584.8 million). Cork stands out as the main contributor to this value (€390.7 million). Acorns and grazing form the second major element (€112.4 million) whose value is not imputed to forestry in national accounts as forest final production, since they provide intermediate consumption for livestock production. Wood forest products (WFPs) amount to €543.6 million, pulpwood (€253.9 million) being the main item in this group. Recreational services provided by forests are on the rise, but they are still a minor component of the total direct use value (€37.9 million). Also, a good part of this value is not yet internalized by forest owners.

The estimation of both indirect use values and negative externalities of forests is incomplete. However, the available estimates show that the costs of forest fires are about 83.5% of the value of forest public goods (carbon storage, protection of agricultural soils, water resources and landscape quality).

Notes

[1] This chapter is limited to the continental part of Portugal, hereafter called 'Continental Portugal'. Therefore, it does not include the islands of Madeira and Azores.

[2] They attained very tragic dimensions in 2003 when around 283,000 ha, or 8.5% of total forestland, were burnt.

[3] From 1982 to 1995, the area of pine was reduced by 41% and 21%, respectively, in the north-western and central regions.

[4] This section is the author's own summary of the contribution prepared by Aguiar and Capelo (2004) for this chapter, based on their earlier joint work regarding the biogeography of Continental Portugal (Costa *et al.*, 1998). The author takes full responsibility for all the possible shortcomings and errors in preparing this summary of their work.

[5] Here, and in the rest of the section, this kind of notation refers to the biogeographic regions shown in the map.

[6] This is the author's own adaptation of the biogeographical map of Continental Portugal made by Costa *et al.* (1998) where the names of the regions were changed in order to make them correspond to the forest types. The author takes full responsibility for all possible mistakes and shortcomings in this adaptation.

[7] Distribution of selected seeds free of charge, accompanied by technical assistance.

[8] Using the coefficients: 1 m^3 o.b. = 0.7 m^3 u.b. for conifers and 1 m^3 o.b. = 0.82 m^3 u.b. for broadleaves.

[9] Probably due to the small number of observations, the roadside price reported in SICOP's leaflet for oak sawlogs in 2001 is lower than the stumpage price. However, the information reported in SICOP's website gives a price lower than those two prices, but does not provide data on roadside prices. So the roadside price reported in SICOP's leaflet was retained.

[10] Converted into m^3 o.b. by using the same coefficients as for the timber harvested.

[11] 47,264 ha of maritime pine burnt from 1996 to 1999, according to the Forest Services.

[12] 134,000 national hunting permits issued for residents (€24.94); 85,000 regional hunting permits for residents (€12.47); 2000 hunting permits for non-residents (€44.89); and 33,000 special hunting permits for big game (€29.93) (DGF data).

[13] All amounts have been converted to 2001 euros using the consumer price index for leisure, recreation and culture.

[14] According to Cipriano (1999), 17% of hunters go only to zones with excludable access (touristic, associative, social or national); 44.4% go only to zones in the 'general' regime (free access); and 38.6% go to both types of zone. Within zones with excludable access, 16.7% go to touristic zones, 64.7% to associative zones, 2.5% to social zones and 16.1% to national zones. The distribution of hunters, as reported by Cipriano, is somewhat ambiguous because it may include some double counting; in the calculations, it is assumed that this is not the case.

[15] The number of stays in the campsites of the Algarve have been omitted since they are mainly located near beaches. Therefore, going to the beach is probably the basic motivation for camping in this region.

[16] By using the consumer price index for recreation, leisure and cultural services, as of December 2001, base 100 = 1997.

[17] From where originates most of the public funding for this purpose transferred to the local fire departments, or spent in the lease of aeroplanes and helicopters.

[18] The majority of which are based on volunteers.

References

Aguiar, C. and Capelo, J. (2004) *Vegetação Natural Potencial Arbórea de Portugal Continental* (unpublished manuscript).

Alpuim, M., Carneiro, M., Vacas de Carvalho, M.A. and Rocha, M.E. (1998) *Pinheiro Manso. 1ª Parte – O Pinheiro Manso e a Produção de Pinhão*. UTAD, Vila Real, Portugal, 28–29 September (unpublished manuscript).

Associação da Indústria Papeleira (2002) *Indústria Papeleira Portuguesa. Boletim Estatístico 2001*. CELPA, Lisbon.

Associação da Indústria Papeleira (2003) *Indústria Papeleira Portuguesa. Boletim Estatístico 2002*. CELPA, Lisbon.

Bergen, V. (2001) Forest public goods in national accounts. *Investigación Agraria – Sistemas y Recursos Forestales*, Fuera de Serie, No. 1, 27–42.

Bugalho, J. and Carvalho, J. (2001) *O Associativismo Cinegético em Portugal: a Importância do Associativismo Cinegético para a Conservação dos Recursos Naturais e o Seu Reflexo no Desenvolvimento Rural*. Ministério da Agricultura, do Desenvolvimento Rural e das Pescas, Direcção Geral do Desenvolvimento Rural, Série Estudos e Análises, Lisbon.

Cipriano, R.J.R. (1999) Análise do Panorama Cinegético em Portugal a Partir de Um Questionário aos Caçadores. Report submitted in fulfillment of the Bachelor's degree in Forest Engineering. Instituto Superior de Agronomia, Lisbon.

Conseho para a Cooperação Ensino Superior/ Empresa (1996) *O Sector Florestal Português*. [Lisbon and Porto]: CESE-Conselho Para a Cooperação Ensino Superior/Empresa (mimeo).

Costa, J.C., Aguiar, C., Capelo, J., Lousã, M. and Neto, C. (1998) *Biogeografia de Portugal Continental*. Quercetea 0, 5–56.

Direcção Geral das Florestas (1992) *Portugal, País de Florestas*. Direcção Geral das Florestas, Lisbon.

Direcção Geral das Florestas (1998) *The Portuguese Forest by Numbers*. Direcção Geral das Florestas, Lisbon.

Direcção Geral das Florestas (1999) *Anuário Florestal 1999*. Direcção Geral das Florestas, Lisbon.

Direcção Geral das Florestas (2001) *Inventário Florestal Nacional. Portugal Continental. 3ª Revisão, 1995–1998. Relatório Final*. Direcção Geral das Florestas, Lisbon.

Direcção Geral das Florestas-Corpo Nacional da Guarda Florestal (2003) *Determinação das Causas dos Incêndios Florestais em 2002 (Resumo Técnico)*. Direcção Geral das Florestas, Lisbon.

Direcção Geral do Ordenamento e Gestão Florestal (1979) *Distribuição da Floresta em Portugal Continental. Áreas florestais por concelhos 1978*. DGOGF – Estudos e Informação No. 284. Direcção Geral do Ordenamento e Gestão Florestal, Lisbon.

Fankhauser, S. (1995) *Valuing Climate Change. The Economics of the Greenhouse*. Earthscan, London.

Goes, E. (1991) *A Floresta Portuguesa, Sua Importância e Descrição das Espécies de Maior Interesse*. PORTUCEL, Lisbon.

Instituto da Conservação da Natureza; Centro de Micologia da Universidade de Lisboa; Direcção-Geral das Florestas; Direcção-Geral de Fiscalização e Qualidade Alimentar; Universidade de Trás-os-Montes e Alto Douro; Universidade de Évora; Forestis (2001) *Conservação, Valorização e Comercialização de Cogumelos Silvestres*. Relatório (mimeo).

Instituto Nacional da Água (2000) *Planos das Bacias Hidrográficas dos Rios Luso-Espanhóis – Síntese*. INAG. Lisbon (mimeo).

Instituto Nacional de Estatística (2001) *Agricultura em números 2000*. Instituto Nacional de Estatística, Lisbon.

Instituto Nacional de Estatística (2002a) *Estatísticas Agrícolas 2001*. Instituto Nacional de Estatística, Lisbon.

Instituto Nacional de Estatística (2002b) *Estatísticas do Turismo 2002*. Instituto Nacional de Estatística, Lisbon.

Instituto Nacional de Estatística (2003a) *Recenseamentos Gerais da População e da Habitação. Dados Comparativos 1991–2001*. (CD-ROM publication) Instituto Nacional de Estatística, Lisbon.

Instituto Nacional de Estatística (2003b) *Estatísticas Agrícolas 2002*. Instituto Nacional de Estatística, Lisbon.

Instituto Nacional de Estatística (2003c) *Estatísticas do Turismo 2002*. Instituto Nacional de Estatística, Lisbon.

Instituto Nacional de Estatística (2003d) *Estatísticas do Ambiente 2001*. Instituto Nacional de Estatística, Lisbon.

Instituto Nacional de Estatística (2003e) *Contas Económicas da Agricultura 2002*. Instituto Nacional de Estatística, Lisbon.

Kuusela, K. (1994) *Forest Resources in Europe 1950–1990*. Cambridge University Press, Cambridge, UK.

Leite, A. and Martins, L. (2000a) Recursos florestais. In: *Florestas de Portugal. Forests of Portugal*. Direcção-Geral das Florestas, Lisbon, pp. 139–143.

Leite, A. and Martins, L. (2000b) Funções produtivas. In: *Florestas de Portugal. Forests of Portugal*. Direcção-Geral das Florestas, Lisbon, pp. 157–159.

Loureiro, M.C. and Albiac, J. (1996) *Aplicação da Metodologia de Valorização Contingente para Determinação do Valor de Uso Recreativo Anual* da Reserva Florestal de Recreio do Monte Brasil (mimeo).

Macedo, F.W. and Sardinha, M. (1993) Fogos Florestais, Vol. 1, 2nd edn. Publicações Ciência e Vida, L.da., Lisbon.

Mendes, A. (1997) Estimativa do valor económico da floresta portuguesa. In: Puertas, F. and Martin Rivas, T. (eds) I Congreso Forestal Hispano Luso – II Congreso Forestal Español. 23–27 June 1997, Libro de Actas, Tomo VI. Gobierno de Navarra, Pamplona, Spain, pp. 135–140.

Mendes, A. (2002) A economia do sector da cortiça em Portugal. Evolução das actividades de produção e transformação ao longo dos séculos XIX e XX. Paper presented at the XXII Meeting of the Portuguese Association of Economic and Social History, University of Aveiro, Portugal, 15–16 November 2002 (http://www.egi.ua.pt/xxiiaphes/Artigos/Américo_Mendes.pdf).

Mendes, A. (2004) Cost Action E 30. Country Report for the Phase One of the Action. 'State of the Art'. Portugal. April 2004 (unpublished manuscript delivered to the action coordinators).

Ministério da Administração Interna (2002) Relatório Anual de Segurança Interna – 2001. Ministério da Administração Interna, Lisbon.

Ministério da Administração Interna-Gabinete do Ministro (2003) Livro Branco dos Incêndios Florestais Ocorridos no Verão de 2003.

Ministério da Agricultura, do Desenvolvimento Rural e das Pescas-Gabinete de Planeamento e Política Agro-Alimentar (2000) Panorama da Agricultura 1999. Ministério da Agricultura, do Desenvolvimento Rural e das Pescas, Lisbon.

Ministério da Agricultura, do Desenvolvimento Rural e das Pescas-Gabinete de Planeamento e Política Agro-Alimentar (2001) Apoios à Agricultura 2000. Ministério da Agricultura, do Desenvolvimento Rural e das Pescas, Lisbon.

Moreira, M.B., Coelho, I.S. and Reis, P.S. (1995) Análise Técnico Económica de Sistemas de Dehesa/Montado. Relatório Final (UE-DGVI-Programa CAMAR: CT90–0028). Instituto Superior de Agronomia, Lisbon (unpublished manuscript).

Moreira, N. (1980) Cultura de Forragens e Pastagens. Instituto Universitário de Trás-os-Montes e Alto Douro, Vila Real, Portugal, (mimeo).

Oliveira, V. (2004) Evolução dos Produtos Tradicionais com Nomes Protegidos (Produção, Valor da Produção, Índices de Quantidades, Preços e Valores). 1997 a 2001. Instituto de Desenvolvimento Rural e Hidráulica-Direcção de Serviços de Planeamento-Divisão de Estudos Planeamento e Prospectiva, Lisbon.

Paulino, A. (2000) Preservação de cogumelos corre perigo. Jornal Expressos October 30, 2000.

Poeira, M., Eliseu, J. and Ramalho, J. (1990) Erosão e gestão do solo. Reflexões sobre o caso português. DGF Informação 1, 28–33.

Rego, F. (1991) A silvopastorícia na sua relação com os incêndios florestais em Portugal. Ingenium 6, 67–72.

Rocha, J., Botelho, O., Maló, T., Nunes, N. and Monteiro, C. (1986) A floresta e a erosão do solo. Comunicações – 1° Congresso Florestal Nacional, Fundação Calouste Gulbenkian, Lisbon, 2–6 December 1986, pp. 374–378.

Santos, J. (1997) Valuation and Cost–Benefit Analysis of Multi-attribute Environmental Changes. Upland Agricultural Landscapes in England and Portugal. Thesis submitted in fulfillment of the degree of Doctor of Philosophy. Department of Town and Country Planning, University of Newcastle upon Tyne, UK.

Sistema de Informação de Cotações de Produtos Florestais na Produção (2003a) Indicadores de Mercado 2000–2003. Leaflet. Direcção Geral das Florestas-Divisão de Inventário e Estatísticas Florestais, Lisbon.

Sistema de Informação de Cotações de Produtos Florestais na Produção (2003b) Website: http://crytomeria.dgf.min-agricultura.pt Updated as of 12 April 2004.

United Nations Economic Commission for Europe/Food and Agriculture Organization (2000) Forest Resources of Europe, CIS, North America, Australia, Japan and New Zealand (Industrialized Temperate/Boreal Countries). UN-ECE/FAO Contribution for the Global Forest Resources Assessment 2000. Main Report. Geneva Timber and Forest Study Papers, No. 17. United Nations, Geneva.

Vieira da Natividade, J. (1950) Subericultura. Imprensa Nacional, Lisbon.

Vieira de Sá, F. (1978) A Cabra. Livraria Clássica Editora, Lisbon.

23 Institutional and Policy Implications in the Mediterranean Region[1]

Paola Gatto[1] and Maurizio Merlo[2,†]

[1]Department of Land and Agro-forestry Systems, University of Padova,
Viale dell'Università 16, 35020 Legnaro (PD), Italy; [2]University of Padova, Centre for
Environmental Accounting and Management in Agriculture and Forestry
(CONTAGRAF), Via Roma 34, Corte Benedettina, 35020 Legnaro (PD), Italy

Issues at Stake

As the previous chapters of this book have clearly highlighted, multi-functionality – the capacity for producing more than one piece of goods or services at the same time – is intimately embedded in the nature of Mediterranean forests. Many different factors, e.g. geographical location, climate, local economies and traditions, and cultural aspects, contribute to an all-round provision of wood and non-wood forest products: from timber to fuelwood, mushrooms and berries, from grazing to regulation of water flows, soil protection, carbon sequestration and conservation of biodiversity.

In economic terms, the essential feature of this multiple production is *jointness*: two or more outputs are jointly produced by the same set of production factors. In the specific case of forests, market and non-market goods are produced at the same time by the same forest unit. In other words, 'the possibilities of providing both timber harvests and the various amenity services of the forest are related *in common* to the conditions of the standing stock of timber' (Bowes and Krutilla, 1989, p. 56).

Of course, the nature of this jointness among the forest outputs is complex: economic relationships of complementarity, indifference or competition between the two sets of goods – market and non-market – may arise. Complementarity takes place when non-market goods and externalities are inputs of the production function and/or intentional products reducing marginal costs of producing market goods. Examples include some naturally oriented silvicultural practices aimed at obtaining natural regeneration in the forest – such as those carried out in the woodlands of the northern Mediterranean region, or low-level harvesting, which, by improving visibility and habitat, may reduce the cost of providing amenities (Bowes and Krutilla, 1989, p. 56). When the economic effects on the production function are nil, we speak of indifference. Competition – or substitution – occurs when non-market goods progressively take on the status of intentional products involving trade-offs with traditional market products. Therefore, an increase in the production of one results in higher marginal costs for the other. After a certain point, 'further harvesting of stocks might then begin to reduce wildlife habitat, add to stream siltation, and spoil scenic quality. At high levels of harvest, timber and amenity services will almost certainly be competitive' (Bowes and Krutilla, 1989, p. 56). The options

---

† Deceased.

aimed, for instance, at recreation and sports can result in being, to a large extent, not compatible with natural environment conservation, the latter being sometimes in conflict with watershed management. Again, a forest ecosystem reaches a high biodiversity when certain practices are applied, while biodiversity is reduced when other practices, or species, are employed.

'Problems of competing/complementary objectives/products have always been the rule for many Mediterranean forests' (Merlo and Rojas, 2000). For instance, recreation and landscape were certainly an unintentional pure externality in the past when forests were not scarce and people were not so crowded together. Then they developed into by/joint products when people started to become concerned with environment conservation. In more recent times, with the increase in available income accompanied by the positive income elasticity of demand of forest amenities, recreation and landscape have become intentional products competing with the long-established ones such as timber production and soil/water conservation.

These competitive relationships are responsible for generating conflicts in the use of the forest resource. Market failures – unpriced public goods and externalities, incomplete definition and assignment of property rights, lack of market transparency – make the identification of the optimal output mix difficult, if not even impossible. The structure of forest ownership together with institutional problems and some inertia of the public forest administration typical of Mediterranean areas makes the picture even more complex here than in temperate or boreal woodlands.

Hence, the attempt to reconcile these conflicts can be considered one of the greatest challenges facing forestry and the design of forest policy in the near future. This is even more badly needed in the Mediterranean areas, where the identification of an appropriate policy – or rather *policy mix* – is essential in order to guarantee the sustainable use of forest resources and the perpetuity of the flow of goods and services they provide.

The path through which this can be accomplished consists of two steps. The first step is of a descriptive positive nature, i.e. inventorying market and non-market goods and bads produced by Mediterranean forests, assigning

values and positioning them in the total economic value (TEV) framework, as investigated in Part II of this book. The second step, which is based on the evidence derived from the first, has a normative nature, i.e. how the problems of public goods/bads and externalities giving rise to market failures can be faced and solved through a political process. Exploring how this can be accomplished is the main focus of this chapter.

Firstly, the general set of policy tools employed in forests will be described, with regard to their development and the overall picture, pros and cons, successes and failures. As far as possible, an attempt will be made to discover boundaries and inter-relationships, synergies and contrasts among different tools. Then, reference will be made to the present situation of policies and institutions in the Mediterranean countries. Finally, some conclusions will be drawn with reference to implications and recommendations for a more efficient multi-purpose policy of Mediterranean forests.

The Set of Tools Employed in Forest Policy[2]

A possible framework for classification of instruments and measures for Mediterranean forest policies and mechanisms is presented in Fig. 23.1. A basic distinction is made between mandatory and voluntary instruments. The terms should be understood in a broad sense, bearing in mind that clear boundaries cannot be defined. Mandatory tools are those that must be complied with, without any possible escape. In contrast, voluntary tools have to be accepted by the involved parties or, at least, should leave the option of changing behaviour given implementation of a certain policy tool. Figure 23.1 also shows that the two sets of tools can, or better should, be integrated by complementary measures of information and persuasion.

The different instruments can be, and often are, employed simultaneously and have international, regional and local dimensions. Each of the major approaches, of juridical, financial, market-based and persuasion instruments, are now considered in turn.

Of course, institutions, including administration and services, as indicated in Fig. 23.1,

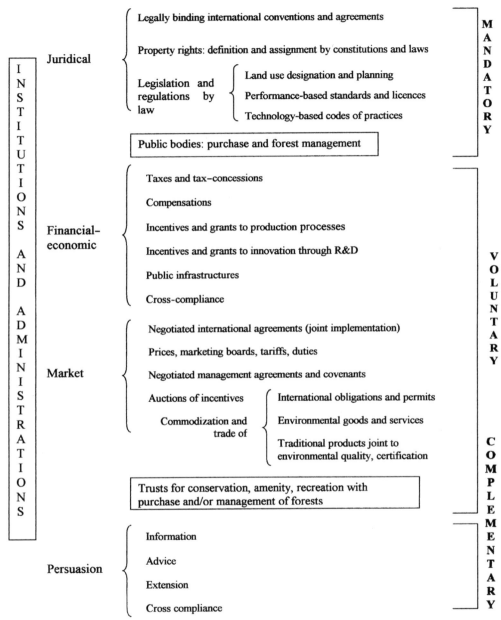

Fig. 23.1. The framework for classification of forest policy instruments and measures. Source: Adapted from Gatto and Merlo (1997).

constitute an all-embracing tool, a key concept in forest policy formation and implementation, essential not only for juridical and financial/economic tools but also for market and persuasion tools. To a very large extent, administration/ services are *the tool* through which the application of any other possible forest policy tool should go. At the same time, poor administration/services can also be the reason preventing the application of the various instruments, as is often the case in Mediterranean forestry. The interaction between administration/services and policy tools, though crucial to the formation/implementation of all types of forest policy, has up to now remained in the shadows. Certainly it seems to deserve further investigation, often being at the root of forest policy failures.

Juridical mandatory tools

Forest policies traditionally have been implemented through juridical mandatory tools, i.e. legally binding instruments included within constitutions and forest laws. The framework of property rights, legislation, regulations and planning, and, in some countries, purchasing and management of forestland, must be mentioned. The objectives of juridical mandatory tools have evolved from timber resources, game and forage conservation, to the welfare of local communities, including soil and watershed management, recreation, landscape and biodiversity maintenance.

1. *Legally binding international conventions and agreements*. These could represent important policy tools in the future. Their growing importance is a consequence of the acknowledgment of the existence of environmental problems at a global scale after the Conference of Rio in 1992. For instance, the Conventions on Biodiversity, Climate Change, Protected Areas and Endangered Species part of UNCED could be designated as mandatory. Conventions and agreements up to now are, however, far from being mandatory or legally binding for signatory countries: for example, the agreements signed in Rio, including Agenda 21 and the Principles of Sustainable Forest Management, remain 'non legally binding authoritative statements of principles' (Food and Agriculture Organization, 1994, p. 3). International conventions and agreements must in any case have a follow-up, including enforcement, at national, regional and local levels – and this is the weakest point of these instruments together with the difficult quantification of the agreed measures, such as forest area and forest management, carbon storage, biodiversity and fauna protection.

2. *Property rights: definitions and assignment by constitutions and laws*. Apart from the international context, juridical tools now rely mainly on forest property rights as stated by constitutions and laws.

3. *Legislation and regulation by law*. At a lower institutional level, legislation and regulation provide the technical–legal rules through which forest management options are defined. This framework is expressed by land use designation and planning, performance-based standards and licences supported by technology-based codes of practices, including local customs. Planning/programming procedures, able to differentiate local situations and needs, are the most advanced mandatory tools. Far from being indiscriminate, they should allow the most appropriate courses of action to be chosen. Of course, societies' ethical and cultural values, common understanding and consensus have always been at the root of juridical tools. An outstanding example of this is the uneven-aged and multi-purpose management of the alpine collective properties, dating back to medieval times.

4. *Purchase and forest management by public bodies*. This has also been much used to guarantee timber supply and the provision of environmental goods and services (EGS). The experience in general is positive whenever management is supported by effective administration/services and the pressure on forest resources remains acceptable. The general trend, according to 'subsidiarity', is, however, towards devolution from State to regional and local authorities. Neither purchase nor management by public authorities are now considered efficient tools. Costs are generally too high, as proved by analysis based on accounting data. Therefore, private forestry is often advocated, including the sale of public forests. Of course, generally speaking, the quality of management and the public services provided by public bodies are generally higher than in the case of private entrepreneurs.

Financial–economic tools

More recently, in industrialized societies, as the demand for timber and EGS has increased and the supply has decreased, voluntary economic instruments – such as compensation, incentives and free infrastructures – have emerged. This generation of instruments is based on the 'carrot' rather than on the 'stick', in other words, positive instruments aimed at convincing people to implement certain measures in exchange for various economic advantages. Quite clearly the economic–financial instruments follow a Pigouvian rationale (Pigou, 1920) aimed at internalizing forestry's positive externalities. According to the Organization for Economic Cooperation and Development (1996), a 'State pays' approach is applied. The Keynesian side of these instruments, often applied for creating jobs and activating depressed rural economies, must not be neglected. Examples are widespread around the world from developed to developing countries. Of course, economic–financial instruments can take various forms according to their economic meaning and their administration.

1. *Taxes and tax concessions* are aimed respectively at preventing undesirable land uses or maintaining traditional products and the new EGS. Tax concessions are perhaps the oldest financial instrument applied to direct land uses. For example, in France, tax relief is a well-established measure to secure afforestation. Now other land uses producing positive environmental impacts can enjoy tax concessions in several countries. Similar measures have been applied for a certain time in the UK and many other developed and developing countries. The opposite strategy is produced by taxes, perhaps the most commonly applied, or at least envisaged, instrument for controlling land use and therefore the forestry environment. They follow the much-advocated, and little applied, 'polluter pays' principle.

2. *Compensation* for higher costs and/or lower revenues met to maintain forests, to achieve afforestation or attain multi-purpose forestry and the related EGS, is perhaps the most common financial tool. In other words, lump sums or, more frequently, annual payments are made for maintaining/establishing forests – and not other more profitable land uses – and/or for adopting environmentally friendly forestry practices.

3. *Incentives and grants to production processes* – again lump sums or annual payments –

to encourage forestry practices and/or afforestation which would otherwise be neglected, are another possible financial tool applied in various countries. Unlike compensation, the payment is such as to include something more, i.e. a surplus to stimulate participation in a programme. Both compensation and incentives/grants are old-established policy tools applied to promote afforestation and to guarantee the provision of EGS. In certain countries, the incentive is provided by low interest rate loans aimed at promoting forest investments. Once the mechanism is started, specific self-sustained funds should make it work without the intervention of States, donors or other organizations. Also, sophisticated financial instruments have been designed to collect the required financial resources in the market. It is a field where economic–financial measures, far from being limited to the 'State pays' approach, are now moving towards fully market-based mechanisms.

4. *Incentives and grants for innovation through Research and Development* (R&D) in forestry and wood processing can be applied in a wider context than individual enterprises. One must look to public investment for R&D, education and extension. What is particularly important is the push towards new technologies, for example in the field of energy saving, management, maintenance, harvesting and development of long-life wood products.

5. *Public infrastructures* aimed at supporting forestry and the related EGS are also an important economic–financial tool. One must think of land arrangements, consolidation, drainage, roads, fire prevention and mechanization patterns. All these infrastructures are, to a certain extent, collectively provided at public expense and, therefore, represent an economic advantage for individual forest enterprises which consequently are oriented towards goals/objectives defined by a certain forest policy.

6. *Cross-compliance* is now very much an advocated policy tool, particularly within the European Union (EU). Cross-compliance is to a certain extent an indirect financial instrument. The concession of existing payments, such as income support for rural development, is taken as conditional on the adoption of environmentally friendly forestry techniques. In other words, cross-compliance could be considered a type of strong persuasion in line with economic/financial instruments.

It is sometimes difficult to frame the exact economic nature of the various instruments, in particular the difference between compensation aimed at re-establishing the *status quo* and the other payments, or various forms of economic–financial support, including something more than compensation. An example is the 1992 EU Common Agricultural Policy (CAP) reform of agroenvironmental and forestry measures providing payments to those who undertake certain practices or forestry investments. It is clearly stated that these payments have to be proportional to any increase in costs or loss of income. However, their application according to flat rates inevitably gives some forest entrepreneurs compensation greater than their forgone income. In other words, the total revenue for forestry and for providing EGS includes a certain amount of producers' surplus. This fact has been demonstrated by Whitby and Saunders (1996) by comparing standard payments with differentiated compensation agreed upon with individual entrepreneurs, i.e. the management agreements. Codification and standard rates are clearly the main shortcomings of economic–financial instruments (Colman *et al.*, 1992; Bishop and Phillips, 1993). They have also been criticized for undermining ethical commitment to stewardship and conservation (Colman, 1994). Therefore, an inherent risk in economic–financial instruments is paying for something that would be done anyway with lower payments or none at all. It has therefore been argued that economic–financial instruments should be conceived and applied only for a limited period of time in order to activate forestry, wood production and the provision of EGS. Once that forestry is consolidated, market forces should be sufficient to maintain existing forests and establish new tree plantations. However, experience in many countries shows that plantations are established only if subsidies are provided continuously, for instance in Argentina, Italy and other countries where self-sustained forestry funds are far from being consolidated financial structures. Only in a few countries (such as Brazil, Chile and New Zealand) have the subsidies created a self-sustained forest industry. The case is even more crucial for the provision of EGS where

the passage from the 'State pays' approach to the 'Beneficiary pays' or market approach (Organization for Economic Cooperation and Development, 1996) still appears wishful. It seems, however, unavoidable that a certain level of support for forestry should be provided in any case at least in terms of infrastructural and institutional measures, as is the case with many other industries where success is also based on a favourable institutional environment – such as given by public funding of research, education and physical infrastructures.

Market-led tools

At present, market-led tools are much discussed, and sometimes applied, as a viable alternative to economic–financial instruments. Often, however, they are part of wider policy packages including legal and economic–financial measures.

1. *Negotiated international agreements* are a sort of 'macro' management agreement negotiated at an international level amongst sovereign countries – the so-called *joint implementation*, now accomplished through the *Clean Development Mechanism* (see Box 23.1). Certain countries in fact have comparative advantages, i.e. lower costs for mitigation policies. This ratio could form the basis for international cooperation and the underlying compensation amongst countries and should cut the global costs of reducing emissions, improving the environment and safeguarding values such as biodiversity on a global scale.

2. *Price, marketing boards, tariffs and duties.* First of all, the traditional market and price policy adopted by individual countries or by blocks of countries must be considered. Support can be oriented to favour certain products while disadvantaging others. This has been the case in the last 30 years for cereals and livestock in the EC, now the EU. Price support of agricultural commodities has been a powerful instrument whose effects on forestry have never been fully understood. What would European forestry be if, say, a timber guaranteed price – higher than world prices – had been adopted? The question cannot be answered. It is clear, however, and well demonstrated, that cereal and livestock price support

Box 23.1. The Kyoto Protocol (by Roger A. Sedjo)[a].

The Kyoto Protocol, which is designed to provide for global reduced greenhouse gas emissions through country-targeted emission reductions, recognizes the role that forests can play in managing carbon. In concept, credit can be obtained through afforestation, reforestation and the prevention of deforestation. However, the implementation of policies to achieve these goals is often difficult. At one level (Article 3.3), the Protocol calls for countries to maintain and indeed increase their stock of forest. Indeed, credits toward achieving Kyoto country emission reduction targets can be accumulated through this process. For example, Japan and Canada can meet a substantial portion of their Kyoto targets via maintaining and expanding their forest stock (Amano and Sedjo 2003). Additionally, the Clean Development Mechanism (CDM) of the Kyoto Protocol offers arrangements whereby carbon-reducing and sequestering projects in developing countries can provide carbon credits to the developed countries that finance and sponsor carbon-mitigating activities. CDM projects include various energy-related activities as well as forestry.

For forestry, monitoring issues are substantial. Not only must it be confirmed periodically that there are trees on the ground, but the amount of carbon must be measured with some degree of accuracy. Furthermore, the sequestered carbon will change through time, increasing with growth as the trees mature, levelling off toward maturity as biomass stabilizes, and declining with forest damage, disease and forest destruction. Thus, unlike a carbon emission reduction whereby a particular unit of carbon is never released into the atmosphere, forest carbon may be sequestered for short, intermediate or long periods of time. However, forest carbon always has the potential to return to the atmosphere either by accident or by design, should a forest be converted or destroyed. Some have suggested that biological sequestration of carbon be treated as temporary, and a number of alternatives have been proposed as to how credits might be treated in this situation. One approach is that forest carbon offset credits could be treated as a permanent offset if the period is sufficiently long (over 50 or 100 years). Another approach would be that forest carbon offsets be treated as short-term assets with a limited life that would need to be renewed periodically by demonstrating either that the particular forest is still functioning as a sink or that the offset has been transferred to another acceptable location, either temporarily or permanently. Also, as with a permanent carbon emission reduction credit, the possibility of trading temporary forest carbon offsets has been raised. Thus, a portfolio of credits might include both permanent and temporary carbon credits, with the mix of credits amenable to management and change over time.

In the Marrakech Accords, agreed at the Seventh Conference of Parties (COP7), the parties agreed to allow afforestation and reforestation projects under the CDM, but did not agree on the detailed rules for such projects. At COP9 in Milan, the parties adopted a decision setting forth the modalities and procedures for sink projects in the first commitment period, with the treatment of sink projects under the CDM for the second period left to be decided. The main issue has been how to address the non-permanence of sink projects. In particular, if a sink project is destroyed, for example a forest burns down, who should be liable – the project developer, the host country or the holder of the Carbon Emission Reduction (CER) credit? The COP9 decision makes the holder liable by making CERs generated from sink projects of limited duration. The decision defines two types of CERs: temporary CERs (tCERs), which are valid for only one commitment period, and long-term CERs (lCERs) valid for the project's full crediting period. Sink projects can have a crediting period of either 20 years, with the possibility of two renewals up to 60 years total, or 30 years with no renewal. The CERs cannot be banked and thus both types must be used for the commitment period for which they were issued and both must be replaced by another credit prior to their expiration. Project participants can choose which of the two approaches to use. In practice, the two approaches are similar. The tCERs will be reissued if a sink project still exists. However, lCERs will need to be replaced before the end of the crediting period if monitoring indicates that the sequestration from a sink project has been reversed.

The COP9 decision also addressed the issues of additionality, leakage and uncertainties regarding socio-economic and environmental impacts. The latter was the most controversial due to concerns by some countries of genetically modified organisms (GMOs). The decision requires that GMOs be evaluated in accordance with the host countries' national laws and that information on the species be identified in the project design. Finally, the agreement also defines small-scale projects, which are eligible for fast-track approval, as those that result in net anthropogenic sequestration of less than 8 kt of carbon dioxide (CO_2) annually and are developed or implemented by low-income communities or individuals. Modalities for small-scale projects are to be considered in COP10.

continued

has displaced – and disincentivized – forestry, particularly plantation forestry such as poplars. It is remarkable how, notwithstanding cereal and livestock support, poplars in the Italian Po valley have been, in certain years, competitive with cereal and livestock (Favaretti, 1976; Prevosto, 1986). Again, marketing boards have been powerful instruments employed in agriculture and are almost unknown in forestry. What could their effects have been in regulating the supply of timber produced by short-rotation species often negatively affected by price cycles? Tariffs and duties are also traditional market instruments that can affect the supply of timber and therefore forest management. Their potential has also not been fully understood. It is, however, well proved by the experience of various countries (Repetto, 1988) how badly intended tariffs and duties have distorted the market. In some cases, low duties have provided an incentive for non-sustainable forest management. In other cases, tariffs on export have disincentivized forest investments (Hartwig Carte, 1994). Traditional market tools have now been suggested to augment, and to regulate, the supply of EGS linked to forestry. Widely discussed in recent years, they advocate a Coasian rationale (Coase, 1960). In particular, they should overcome the objection, often made against economic–financial instruments, of being indiscriminate and unethical. It has also been stressed (Steinlin, 1997) that forest conservation during the 18th and 19th centuries was stimulated by the price of timber and by the favourable market conditions. Therefore, the rationale of the supply of EGS should rely on market creation and prices of EGS.

3. *Negotiated management agreements and covenants* providing payments subject to negotiation between forest entrepreneurs and the responsible public authority are fast-growing policy tools, particularly within the EU. It is still the 'State pays' approach (Organization for Economic Cooperation and Development, 1996) but mitigated by the negotiation process. With respect to standard payments they should avoid possible excess payments, resulting in forest owners' rent. At least in theory the negotiation process should approximate the compensation/incentives to the marginal cost of forestry, plus the profit necessary to stimulate the agreement. Management agreements have been applied for some time in the UK (Colman *et al.*, 1992; Bishop and Phillips, 1993) and The Netherlands (Slangen, 1992). Various examples of *Vertragsnaturschutz* (contracts for nature protection) can be found in Austrian nature reserves. They are also applied in Italy, mainly within nature parks. High transaction costs seem to represent their major limitation (Whitby, 1995) although recent evidence shows that costs are substantially reduced once a policy is well established. It has also been argued that the initial assignment of environmental property rights to the community could improve the final results of management agreements (Bromley and Hodge, 1990). However, excessive assignment of rights over the forest to the public at large could disadvantage forest owners and lead to neglect/abandonment, a sort of 'tragedy of the commons', as is now the case with many forests in southern Europe and the Mediterranean basin (Rojas, 1995). A compromise solution between the compensation agreed through individual management agreements and the standard payments given to all forest enterprises applying for them is given by standard payments differentiated according to sites and specific practices. This option has been chosen by several EU countries and regions in applying the recent 1992 agroenvironmental and forestry measures included in the CAP reform. A more extended view of management agreements, requiring contract registration, is given by the so-called covenants – *servitutes praediorum* – legally attached to the land. From the community's point of view, they represent a stronger commitment to forestry.

4. *Auction of incentives* is a tool very rarely applied now, but which is nevertheless observed in certain countries such as the USA (Hamsvoort and Latacz-Lohman, 1996) and in Argentina's afforestation policy (Corradini *et al.*, 1993). It is based on competitive bidding directly submitted by landowners wishing to start forestry or to enter an EGS production programme. Bids represent the amount which forest entrepreneurs are willing to pay to start afforestation or to accept the restrictions imposed by the programme. Only bids lying below an unknown exclusion level (the bid cap) are accepted. Recent studies have demonstrated an increase in cost effectiveness by switching from fixed rate compensation to auction schemes (Hamsvoort and Latacz-Lohman, 1996). It is also argued that the larger the information gap between the entrepreneur and the government (in other words the less the government is aware of individual bidders' switch-over cost), the more the cost efficiency increases. Unlike management agreements, negotiating costs are not as important in programme implementation; however, other costs may become significant, for example the costs for the preparation of bids which may, in some cases, prevent landowners from participating in the auctions.

5. *Commoditization and trade of international obligations and permits* to adopt certain land uses and code of practices, with positive environmental impacts, emission permits and quotas limiting negative impacts, can also be seen as sophisticated market instruments able to regulate positive and negative externalities. This approach requires the creation of specific markets. For instance, permits and quotas for emission of greenhouse gases, or engagements to afforestation according to environmentally friendly codes of practice, could be the object of an international market. Of course, the market organization must be based on conventions defining and allocating quotas.

6. *Commoditization and trade of EGS* according to a pure Coasian approach (Coase, 1960) foresees specific markets where forestry-related EGS can be sold directly to consumers. Definition and assignment of property rights are certainly crucial issues. In order to create a market, assignment of a certain right to an entrepreneur (a landlord, a community or other economic entities) must take place. However, this assignment is not always easy. For instance, in many countries, forest access for berry and mushroom picking, far from being a private right, is virtually a public right of every citizen, guaranteed by specific laws and traditional customs. The same applies to many other forestry-related EGS; therefore, the role of a pure Coasian approach must not be exaggerated (Bromley, 1991; Zamagni, 1994). However, developments in the field of marketing EGS provided by forestry can be expected. In the past, forest entrepreneurs' associations (such as *groupements forestiers* and *consorzi forestali*) have been established in various countries to achieve economy of scale in timber production. Extension to forest-related EGS does not seem impossible. Economies of scale for EGS production are generally even more necessary than with timber production. This largely unexplored path could play a key role in defining environmental policies. Of course, practical solutions must be associated with local rights and customs. It has been suggested (Mantau, 1994) to by-pass the property rights issue, adopting modern marketing strategies in the framework of existing rights. What are now called 'structured' EGS are, in practice, additional market goods and services attached to forestry. 'Non-structured' EGS linked to forestry should be associated with – or preferably included within – complementary goods and services, i.e. 'structured' EGS, marketable under existing property rights. Such cases are expanding and are favoured by the growth of tourism and 'green consumerism'.

7. *Commoditization and trade of traditional products,* whose image is connected with environmental quality, is another option. This policy tool can be applied whenever direct selling of environmental goods and services is impossible or 'politically' unwise. Remuneration can be achieved by selling traditional quality products, with the price being influenced by environmentally friendly forestry and more generally by the environment in which they are produced. Long-established experiences in this field should not be overlooked, for example *appellation d'origine controlée* agricultural products (Organization for Economic Cooperation and Development, 1996). Regulations aimed at consumer protection were established, limiting the right to use the guaranteed origin label to local producers alone, providing certain protocols and sometimes including stewardship practices. Thus market rules have been modified, giving rise to differentiation and monopolistic competition on markets, allowing remuneration not only for the quality of the product, but also for the related practices and the environment. The now much-discussed timber and forest certification can be conceptually included within these marketing strategies as proposed by Ferro *et al.* (1995) and now accepted by the Organization for Economic Cooperation and Development (1996). Sustainable forest management could therefore be achieved through voluntary acceptance on the part of forest enterprises of certification schemes, which in perspective could create favourable market impacts as a result of differentiation and competitive advantage. Agrotourism and ecotourism are also very promising market activities that allow the conservation of unique landscapes and the associated forests (Ferro *et al.*, 1995). Given the growth of 'green consumerism', the sophistication and willingness to pay of consumers, these policies seem to have a future, and therefore an impact on the conservation of forests.

8. *Trusts for conservation, amenity and recreation.* Trusts for purchasing/managing forestland for conservation, amenity and recreation have recently assumed an important role in forestry. It is an option to be considered when important social values are at risk, and have to be protected. In certain cases, leases, management and the purchase of specific rights can also be considered. These various types of non-governmental bodies are now playing the same role once assigned to public organizations such as state forest enterprises. It is quite interesting to note that the flexibility of trusts and other similar non-public bodies favours, where possible, the marketing of environmental goods, thereby allowing direct remuneration of EGS provided by forestry.

Persuasion tools

Persuasion is, to a very large extent, complementary to the previous instruments. Information, advice and extension services represent various possible steps of persuasion to implement mandatory and voluntary forest policies. Cross-compliance is perhaps the most advanced step, also defined as a 'strong persuasion', if not blackmail, to adopt a certain policy. Tools of persuasion therefore represent important supports to juridical, economic–financial and market instruments. Their help is also essential in pursuing the social acceptability of forest policy through a 'bottom-up' approach based on participation. It is well known how the consensus and, even better, the convinced participation of local populations are essential for the success of a given forest policy in both developed and developing countries. Involvement of rural communities is even more important when forestry is under public or common property regimes. It is interesting to note that all international non-legally binding conventions can to a large extent be considered as tools of persuasion supported by the 'moral' authority of international organizations.

An Assessment of Present Forest Policies in the Mediterranean Countries

Based on the overview of possible tools to be employed in forest policy, this section deals with the assessment of forest policies in place in the 18 Mediterranean countries. In this context, the following concepts and elements are central and are integrated into the following discussion: (i) the availability of information on forest resources – in terms of both extension and consistency – in the Mediterranean area; (ii) the distribution of forestland ownership and the structure of the property, especially its size; (iii) the forest policy objectives; (iv) the existing policy tools; and (v) the current forest institutions operating.

Information constitutes the background for any collective action or public choice. It is therefore at the root of the governance of any economic system, being the basis on which decisions are taken. For example, carrying out a forest-related investment requires specific information: if the project necessitates choosing between competing uses, such as forest and agriculture, one should define both forest and non-forest scenarios, identify and quantify the related inputs and outputs, and refer the distribution of costs and benefits to the project stakeholders. In the case of afforestation or reforestation projects – community or social forestry projects offsetting deforestation and its negative externalities – one should identify and quantify the inputs related to land, labour and capital. Watershed management projects require the identification and quantification of the real impacts and their causes and the inputs and outputs involved in the activities on the upland use and management, based on a with–without project approach (Kengen, 1997).

It appears clear – and the recent developments of information theory have proved it very well – how sound information, essential in any economic decision, is costly to obtain – a hidden transaction cost often not considered in the analyses.

However, in a Mediterranean context, the contribution of the forest sector to national income is often so low as to make the collection of information a debatable – if not worthless – investment from a strictly financial point of view. This may partly explain the scarce and fragmented information available on the present situation of forest resources, as highlighted by the analysis of the institutional situation of regions and countries. Another reason could be that linked to the not always stable political and institutional situation of some areas, where forest authorities in charge of data collection have changed over time.

To give examples of this situation, the descriptive data on structural aspects (areas and types of ownership) are at times the outcome of estimates rather than of official statistical sources. When surveys and inventories on areas

and growing stocks are available, they are often not updated, or – when they are – the results of surveys carried out at different times or by different institutions are often not comparable. Of course, when the value of biodiversity and of other non-use values of forests is taken into account, the cost of collecting information would be totally justified by the high values involved, as the analysis carried out in this book has unmistakably shown.

Table 23.1 shows the most important institutional issues on which the present analysis is based.

Looking at the distribution of forestland ownership, a clear and pervasive dominance of public property is observed. In more detail, two main groups of countries can be recognized: the southern and eastern aggregates, where, with the exception of Egypt, public property is nearly absolute, and the northern group, where there are wider differences, with private property prevailing in Cyprus, Slovenia, Italy, France, Portugal and Spain.[3] In all the countries, the distribution of forestland ownership has very old historical roots, even if, in some cases (Morocco and Syria), it is still in the process of being defined, or in other countries (Slovenia) it is undergoing dramatic changes. It has a clearly defined status in the countries of the northern Mediterranean basin, where private property is stated as a constitutional right of any citizen. In any case, ownership distribution in the southern and eastern countries is always intimately interwoven with the assignment of property rights to local people and to nomadic tribes for collection of wood and other forest products and, usually, for grazing. Even in this case, the picture is not always clear and well defined, with many transformations still ongoing.

In principle, forest policy analysis looks at public property as one of the most secure tools to ensure conservation of the resource for itself and of society's interests for provision of goods and services over time. Of course this needs to be supported by clear public policy objectives and by an efficient forest administration, able to enforce legislation and rules. Conversely, uncertain property distribution or lack of control fuels free-riding and 'tragedy of the commons' types of problems. The high value of 'illegal logging' and of illegal actions in general, accounted for in the Moroccan and Turkish TEVs, gives an

Table 23.1. Key institutional issues in the Mediterranean countries.

Countries	Structural aspects					Institutional framework		
	Forestland ownership %			Size of properties (ha)		Policy objectives	Policy tools	Forest Institutions
	Public	Private	Other[a]	Public	Private			
Southern								
Morocco	99	1			0.5	✓ Conservation and management of forests ✓ Wildlife and hunting ✓ Afforestation ✓ Erosion control	National Forest Fund (for afforestation and research)	Directorate of Water and Forests
Algeria[b]	80	8	12			✓ Environment and conservation of forests ✓ Fighting desertification ✓ Socio-economic aspects	Strong enforcement through central and local Forest Authorities	Directorate for Forestry
Tunisia	84	15	1		mostly < 5	✓ Soil protection from erosion ✓ Social role (job creation)		General Directorate for Forestry
Egypt	53	47				✓ Protection of existing forests ✓ Sustainable wood supply ✓ Public awareness and commitment	National Forestry Action Plan	Ministry of Agriculture and Land Reclamation Ministry of Irrigation
Eastern								
Palestine	91	9				✓ Protection of natural forests ✓ Strengthening of the institutional structure ✓ Inventorying ✓ Public awareness	Palestinian Forestry Policy, complying with the National Biodiversity Strategy and Action Plan and the National Policies for Physical Development	Ministry of Agriculture (Forestry Department)
Israel	100					✓ Traditional ones: land development and afforestation ✓ More recent ones: forest recreation and outdoor leisure	Annual Plans of the Forestry Department National Master Plan 22	JNF: reclamation and afforestation Ministry of Agriculture: afforestation research but division still unclear
Lebanon	77	23				✓ Protection and conservation of forests	1949 Forest Code	Ministry of Agriculture Ministry of Environment Department of Antiquites

Country						Objectives	Laws/Policies	Institution
Syria	99	1				✓ Conservation and management of forest resources ✓ Conservation of biodiversity ✓ Establishing environmentally protected areas	1953 Decree, strictly regulatory, prohibiting customary rights of local users 1994 Forest Law but no Forest Plan 1994 Law on environmental protection	Ministry of Agriculture and Agrarian Reform
Turkey	99.9	0.1			83,000	✓ Protection of forests ✓ Improving productivity and supply to meet domestic timber demand ✓ Increasing provision of NWFPs ✓ Recovering degraded forestland ✓ Supporting local population		
Cyprus	40	60			2.1	Multiple use of forest resources: ✓ Improving forest conditions ✓ Conserving soils and watersheds ✓ Protecting biodiversity and heritage sites ✓ Promoting ecotourism ✓ Promoting sustainable production of wood and non-wood forest products	National Forest Programme National Forest Policy Statement	Ministry of Agriculture, Natural Resources and Environment
Northern Greece	74	6	20	> 100	< 2	✓ Multiple use of forest based on the sustained yield principle	1975 Constitution and a set of Laws	Ministry of Agriculture General Secretariat of Forest and the Natural Environment
Albania	96.5	0.5	3	1000	20	✓ Sustainable forest management ✓ Market-oriented aspects ✓ Environmental issues	1992 Law on Forestry 1993 Law on Environmental Protection 1995 Law on pastures and meadows	Ministry of Agriculture and Food General Directorate for Forest and Pastures
Croatia Slovenia[c]	80 30	20 70			2.3 > 2	✓ Close-to-nature and multi-purpose management: conservation and development of all forests and their functions	1993 Forest Act 1994 National Forest Development Programme	Ministry of Agriculture, Forestry and Food, Forest Department

continued

Table 23.1. *Continued.*

| Countries | Structural aspects | | | | | Policy objectives | Institutional framework | |
| | Forestland ownership % | | | Size of properties (ha) | | | Policy tools | Forest Institutions |
	Public	Private	Other[a]	Public	Private			
Italy	44	66		400	3–4	✓ Watershed protection ✓ Sustainable forest conservation	1923 Forest law 1988 National Forest Plan (not successful)	Ministry of Agricultural and Forest Policies Regional Forest Administrations
France	25	75			3–3.6	✓ Sustainable forest management in a European context ✓ Social dimension of forests: improvement of living conditions, promotion of recreation, preservation of the environment, air and water protection, production and utilization of wood-based ecomaterials ✓ Economic valorization of the increasing wood production potential	1827 French Forest Code 2001 Forest Orientation	French Forest Angency
Spain	34	63	3		Average 7–8 ha with very large regional differences	✓ Multiple use of forest based on the sustained yield principle	1957 Forest Law + Regional Law	Ministry of Environment Nature Conservation Agency
Portugal	6	94			North: < 10 ha; South > 10 ha	✓ Afforestation of uncultivated land to prevent erosion		Ministry for Agriculture

[a]Mainly collective

[b]Algeria: other types of ownership include 2.3% of collective property, while the remaining 5.7% is undefined.

[c]Slovenia: a process of privatization actually in place will bring it to a final situation of 80% private and 20% public.

Source: Chapters 5–22.

idea of the magnitude of the problem and of its perception by the forest managers in the country. In this context, the recent actions started in Algeria, aimed at assigning property rights to local people in exchange for rent agreements with specific engagements for maintenance and conservation, are remarkable.

The size of properties is closely connected with the distribution of land ownership. The possibility of carrying out sound and efficient forest management through economies of scale is of course very sensitive to this aspect. The data of Table 23.1 show clearly that: (i) a complete picture on the structure of the public and private forest estates is not available; and (ii) a dichotomy exists between public and private land ownership. The size of public forest enterprises – or management units – is such as to allow uniformity of management on a rather large plot of land and efficient use of production factors. Conversely, the very small average size of the private enterprises seems to be a common problem affecting the whole Mediterranean basin, from north to south. In this case, the result can be – and often is – lack of active management, meaning abandonment and a threat to the provision of public goods and services – in other words, the occurrence of negative externalities.

From a wider viewpoint, policy objectives, policy tools and forest institutions represent the governance framework in place in the 18 countries.

The identification of policy objectives is one of the most important steps of the policy process, generally included in the forest policy formation phase (Merlo and Paveri, 1997). Clear objectives greatly help policy implementation; while, conversely, ambiguous definitions of objectives may represent one of the main shortcomings of forest/environmental policies for rural areas.

As anticipated, the survey carried out in the 18 countries has shown that policies in place in the Mediterranean basin have – at least in terms of stated official objectives – a multi-functional nature. Several objectives are pursued simultaneously: primary and secondary, stated and unstated, short and long term. The approach of the 1992 Rio Conference for a Sustainable Forest Management seems to have been acknowledged in all forest policies, although different emphasis has been given to the diverse objectives in the individual countries. Protection and

conservation of the forest resources for present and future generations is the principle permeating all forest policies; the role of forests in relation to soil protection from erosion and desertification is the other paramount feature common to the whole Mediterranean forest scenario. Also socio-economic aspects – mainly job creation and subsistence of the local population – have received a great deal of attention, while recreation and leisure are more recent objectives, and, having high income elasticities, have received more attention in the northern group.

The problem here, argued also in Merlo and Paveri (1997), is the frequent discrepancy between the stated objectives and the actual achievements: 'the striking contrast between wishful policies and the poor results in practice' (p. 235). It is also argued that the more sophisticated a policy is, the more difficult is its implementation and, therefore, the poorer are the results in terms of implementation. The difficulties of achieving different objectives at the same time emerge as the other 'side of the coin' of multi-purpose policies, giving space to several contradictions and conflicts. The problem of identifying socially acceptable reference points or boundaries between goods and bads is dramatically important in this context.

The analysis of existing policy tools (Table 23.1) sheds further light on these issues. The data collected from the country chapters show that all countries have a well-established institutional framework in terms of juridical instruments, such as Forest Laws or Codes, which form the irremissible support for implementing any forest policy – including those based on economic and market approaches. According to specific devolution situations, the enforcement level has occurred either at a national level (southern, eastern and some northern countries) or at a regional one (Italy, France and Spain).[4] Top-down approaches seem to predominate, and the risk here is of overlooking local people's needs, thus lacking that social consensus crucial for the success of any policy action. A meaningful example is what occurred in Syria, where the 1953 Forest Law did not seem to acknowledge the preservation of existing use ownership rights in public forests. This issue was subsequently dealt with in the 1994 Forest Law, when forest villagers were assigned the right to use the neighbouring forests to the extent necessary to meet their own needs. Other interesting instances of the involvement of the local population have been registered in Morocco: rights of collection of wood and fruit are granted to local people, whereas the law establishes that at least 20% of the communal forest income is used for forest improvement or planting of fruit trees and management of communal shelters or roads. In Italy, the role of supporting the social communities was played in the past by the Village Communities (*Regole*), which – by means of the forest revenues – provided for the needs of the whole rural communities in terms of police, justice, defence, roads, education, health and assistance (Merlo, 1989). Nowadays, the low timber prices together with high management costs have worn away forest landowners' margins, so that their contribution to the life of the rural communities is no longer feasible.

Forest legislation has been accompanied, in some countries (such as Egypt, Palestine, Cyprus, Slovenia and Italy), by National Forest Plans, the operative tools through which the forest principles and objectives should be put into practice. These seem to be the real bottleneck of the process: the generalized lack of funds is such that these plans 'have often remained rather theoretical documents [. . .] out of touch with social and political realities' (Merlo and Paveri, 1997, p. 235, quoting the Food and Agriculture Organization's National Forestry Action Programmes-Operational Principles, 1994).

The next link in the policy process chain is represented by the forest institutions in place, which are responsible for forest policy enforcement. As already stressed, enforcement has proved illusory, even with social consensus, when adequate forest services were not available. It must be accepted, therefore, that the adoption of a 'tough line', the stick, is part of the game, and has proved to be an essential part of the policy package aimed at forest resource conservation. Even if coercive policy measures are no longer palatable, a certain level of regulation and institutional control remains essential for implementing any forest policy – including those based on economic and market approaches. Apart from the northern Mediterranean countries, a positive experience in this direction seems to be that of Algeria, where a very diffuse and minute Forest Administration Service exists, with the smallest unit possible being present in each

Municipality. The high costs of running such a structure must not, however, be overlooked, together with a certain animosity towards Forest Services; this was the case in Italy some decades ago and it is now the case in some Maghreb countries such as Morocco.

Finally, looking at the different components of the TEV and at how they have been achieved through policy tools in the 18 countries, the research has revealed how the water-related and soil conservation functions have traditionally been captured within institutional–legal tools, guiding Mediterranean forest management through several 'bonds' or 'obligations' aimed at protecting the welfare of society overall. Moreover, some forest externalities have been partially internalized by markets in the western European countries, whereas in the southern and eastern Mediterranean they are mainly enjoyed for free. The use of market-based tools has increasingly gained attention at a worldwide level (Pagiola *et al.*, 2002). Recreation is an example of positive externality that in the past was a pure non-market good all over the Mediterranean. Nowadays, in the Western European countries, various market-based mechanisms for its promotion have been established, such as entrance fees to parks or other recreational areas and other payments for connected goods (such as ski equipment, tourist facilities and hotels). In the southern Mediterranean basin, internalization by the market occurs very rarely: one outstanding example is the case of cedar forest reserves in Lebanon, now almost a relic, but highly valued by the Lebanese people as the image/identity of their country.

Conclusions: Policy Implications and Recommendations

In order to draw some conclusions regarding the improvement of current forest policies, the institutional analysis carried out in the previous sections should be cross-compared with the main findings of the TEV estimates carried out in Chapter 4.

From a geographical viewpoint, two main groups of countries – having similar characteristics in terms of institutional framework – can be thus identified in the Mediterranean basin:

1. The southern and eastern countries, except for Cyprus, characterized by the absolute dominance of public properties accompanied by a not always definite allocation of property rights. Here, a large share of the rural population is poor and depends, for its daily life, on the collection of forest products: fuelwood, timber, litter, fruits and wild animals. The forest is also a crucial source of fodder for grazing animals. Forest use values are therefore essential to satisfy vital requirements of rural inhabitants, while indirect use values and non-use values are not perceived as such, their importance being overshadowed by the needs of daily subsistence. This often results in 'tragedy of the commons' behaviour, resulting in an unsustainable use of forests: non-authorized timber and fuelwood collection, deforestation for agricultural use and overgrazing. These are all negative externalities which, under the wide TEV umbrella, can reduce the total estimated forest value by as much as 20%. Control authorities are present, but their action is not always effective or based on widespread social consensus.

2. The northern group of countries and Cyprus, where, with some exceptions, dominance of private properties – often associated with abandonment -- and efficient public forest management are the main features. Here, higher gross domestic product (GDP) *per capita* and greater availability of spare time have sharpened society's awareness of – and demand for – indirect, option and non-use values. The existence of attempts to estimate these values (see the results of the country chapters) is an acknowledgment in itself of their perception by the scientific community, which in turn is stimulated by a widespread social and political will. Here the main concern is represented by the abandonment of agricultural and forest activities due to the very low income they provide if compared with other economic activities (industry and services). This brings about negative externalities such as erosion, increased risk of forest fires, loss of landscape quality and fewer recreational opportunities.

In this context, it is clear that the identification of the right policy tool for increasing the provision of positive externalities and an improved allocation of the forest benefits and costs between forest owners, communities and society is a very complex issue, especially when referred to the Mediterranean region: a space embracing different economic, social, cultural and environmental conditions. It is very likely that a sustainable development could be helped through the adoption of a mix of policy tools, allowing the use of regulatory and voluntary approaches. The 'right recipe' should balance sustainability, distribution and equity issues and participation. The final aim should be the creation of 'multiple dividends' in terms of environmental quality, landscape, biodiversity conservation, employment and rural development, with all their associated cultural and social values. Given the starting situation, two main approaches can be considered.

The first one is targeted mainly at the less developed countries of the southern and eastern Mediterranean. Since the local population derives many benefits from forests, they would have a strong incentive to conserve them if the tenure status was more appropriate, resulting in a direct improvement of their living conditions. Negotiated agreements for community-based forest management could therefore be powerful tools to improve both rural income and environmental sustainability. Given that values other than those related to use are perceived mainly at a global level rather than at a local one, the negotiated international agreements and conventions could provide the general umbrella to channel funds. Bottom-up approaches seem essential to achieve acceptance by concerned parties and stakeholders in general, an issue that is dealt with in depth in Chapter 24. The experience of many Food and Agriculture Organization projects (Merlo and Paveri, 1997) and the initiatives under the Clean Development Mechanism and the joint implementation (see Box 23.1) give clear and sound indications in this direction.

The second one deals mainly with the northern Mediterranean industrialized countries. Here, water-related functions have proved to be the most important public goods produced by forests. Loss of this positive externality due to poor management or forest abandonment would result in high costs for society. Irreversibility, linked to soil erosion and fertility loss, also plays an essential role in this context, such as to justify mandatory imposition of hydrogeological bonds and controls over forest management operations. There are, however, other forest benefits – mainly recreation and landscape – for which *pro-capita* income and perceptions are such that direct selling through market instruments can represent a viable solution. Considerable work on this subject, fruits of a EU research project, has been published recently by Mantau *et al.* (2001). The rationale here is to transform public goods such as recreation and landscape into private goods by enforcing excludability and rivalry.

Marketability would be the result of introducing market-based mechanisms for selling environmental services supported by previous institutional changes related to clear definition, assignment and enforcement of property rights (Merlo, 2002). Various case studies have shown that dramatic changes of property rights are not feasible as they are unpalatable to the general public, and in any case they are not so necessary, as management and marketing approaches aimed at achieving 'exclusiveness' through additional services (such as ski lifts, parking places and hotels) are often enough. Italian case studies have shown that one-third of the path along the 'marketability arrow' is the result of institutional changes and two-thirds are due to management and marketing (Merlo, 2002).

One of the issues to be resolved here has mainly a distributional nature, being linked to ill-defined allocation of property rights. The forest benefits associated with recreation, landscape and biodiversity are usually perceived and enjoyed by segments of society, such as tourism operators and visitors, different from those supporting the costs of their production, i.e. the rural community. Forms of compensation need to be found, so as to reallocate the forest costs and benefits in a socially equitable way.

Finally, the essential role of information in the greatly advocated participation process must not be forgotten, especially given the high option and non-use values at stake, whose role this book has tried to bring to light. Therefore, each proposed specific package of policy measures should be based on transparency of information and accompanied by measures of persuasion and communication.

Notes

[1] This chapter, which was originally intended to be written by Professor Merlo, mainly draws together concepts and ideas on his recent work on the subject of Forest Policy Tools, part of which was carried out in the framework of some EU research. The following articles and papers in particular have provided most of the material used in this review: Gatto and Merlo (1997), Merlo and Paveri (1997), Gatto and Merlo (1999), Merlo and Rojas (2000). I am extremely grateful to Professor Merlo for all his precious and fruitful teachings which have helped me to develop this chapter.

[2] This section is taken entirely from Merlo and Paveri (1997) and Gatto and Merlo (1997).

[3] For a full updated description of forest policy in Spain, see Rojas (2000).

[4] The aspects linked to devolution and participation are examined in depth in Chapter 24.

References

Amano, M. and Sedjo, R. (2003) *Forest Carbon Sinks: European Union, Japanese, and Canadian Approaches.* RFF Discussion Paper 03-41, October. http://www.rff.org/rff/Documents/RFF-DP-03-41.pdf

Bishop, K.D. and Phillips, A.C. (1993) Seven steps to market: the development of the market-led approach to countryside conservation and recreation. *Journal of Rural Studies* 9, 315–338.

Bowes, M.D. and Krutilla, J.V. (1989) *Multiple Use Management: the Economics of Public Forestlands.* Resources for the Future, Washington, DC.

Bromley, D. (1991) *Environment and Economy.* Blackwells, Oxford, UK.

Bromley, D. and Hodge, I. (1990) Private property rights and presumptive policy entitlements: reconsidering the premises of rural policy *European Review of Agricultural Economic* 17, 197–214.

Coase, R. (1960) The problem of social cost. *Journal of Law and Economics* 3, 144–171.

Colman, D. (1994) Ethics and externalities: agricultural stewardship and other behaviour: Presidential Address. *Journal of Agricultural Economics* 45, No. 3.

Colman, D., Crabtree, B., Froud, J. and O'Carrol, L. (1992) *Comparative Effectiveness of Conservation Mechanisms.* Department of Agricultural Economics, University of Manchester, UK.

Corradini, E., Gennari, A. and Merlo, M. (1993) *Análisis Económico y Político del Sistema Forestal Argentino.* FAO TFAP, Buenos Aires.

Favaretti, G. (1976) Aspetti economici dell'-arboricoltura da legno nelle Venezie. *Agricoltura delle Venezie* 7/8.

Ferro, O., Merlo, M. and Povellato, A. (1995). Valuation and remuneration of countryside stewardship performed by agriculture and forestry. In: Peters, G.H. and Hedley, D.D. (eds) *Agricultural Competitiveness: Market Forces and Policy Choice*. Dartmouth, Oxford, UK, pp. 415–435.

Food and Agriculture Organization (1994) *National Forestry Action Programmes – Operational Principles* (draft paper). Forestry Department, Rome.

Gatto, P. and Merlo, M. (1997) Issues and Implications for agriculture and forestry: a focus on policy instruments. In: Adger, N., Pettenella, D. and Whitby, M. (eds) *Climate Change Mitigation and European Land Use Policies*. CAB International, Wallingford, UK, pp. 295–312.

Gatto, P. and Merlo, M (1999) The economic nature of stewardship: complementarity and trade-offs with food and fibre production. In: Van Huylenbroeck, G. and Whitby, M. (eds) *Countryside Stewardship: Policies, Farmers and Markets*. Pergamon, Oxford, UK, pp. 21–46.

Hamsvoort, C.P. and Latacz-Lohman, U. (1996) *Auctions as a Mechanism for Allocating Conservation Contracts Among Farmers*. Agricultural Economics Research Institute LEI-DLO, The Hague, Netherlands.

Hartwig Carte, F. (1994) *La Tierra que Recuperamos*. Edit Los Andes, Santiago, Chile.

Kengen, S. (1997) Linking forest valuation and financing. *Unasylva* 48, 44–49.

Mantau, U. (1994) Produktstrategien für kollektive Umweltgüter. *Zeitschrift für Umweltpolitik und Umweltrecht* 3.

Mantau, U., Merlo, M., Sekot, W. and Welcker, B. (2001) *Recreational and Environmental Markets for Forest Enterprises*. CAB International, Wallingford, UK.

Merlo, M. (1989) *Collective Forest-land Tenure and Rural Development: the Experience of the Village Communities in the North-Eastern Italian Alps*. FAO, Rome.

Merlo, M. (2002) Marketing public goods and externalities provided by agriculture and forestry In: Brouwer, F. and van der Straaten, J. (eds) *Nature and Agriculture in the European Union*. Edward Elgar, UK, pp. 207–232.

Merlo, M. and Paveri, M. (1997) Formation and implementation of forest policies: a focus on the policy tools mix. In: *Social Dimension of Forestry's Contribution to Sustainable Development – Policies, Institutions and Means Forestry*

Development. Proceedings of the XI World Forestry Congress, Antalya, Turkey, 13–22 October 1997, Vol. 5, pp. 233–254.

Merlo, M. and Rojas, E. (2000) Public goods and externalities linked to Mediterranean forests: economic nature and policy. *Land Use Policy* 17, 197–208.

Organization for Economic Cooperation and Development (1996) *Amenities for Rural Development*. OECD, Paris.

Pagiola, S., Bishop, J. and Lindell-Mills, N. (2002) *Selling Forest Environment Services. Market-based Mechanisms for Conservation and Development*, Earthscan, London.

Petroula, T. (2002) *Sinks as an Option to Meet CO_2 Emission Reduction Targets in Europe*. RIVM report 500005001/2002.

Pigou, A.C. (1920) *Economia del Benessere*. UTET, Turin, Italy.

Prevosto, M. (1986) Il mercato del pioppo: problemi e prospettive. *Cellulosa e Carta* 6.

Repetto, R. (1988) Overview. In: Repetto, R. and Gillis, M. (eds) *Public Policies and the Misuse of Forest Resources*. Cambridge University Press, Cambridge, UK.

Rojas, E. (1995) *Una Política Forestal para el Estado de las Autonomías*. Editorial Aedos, Madrid.

Rojas, E. (2000) Perspective for a new federal forest legislation in Spain. In: *Forging a New Framework for Sustainable Forestry: Recent Development in European Forest Law. IUFRO World Series* 10, 281–292.

Slangen, L.H. (1992) Policies for nature and landscape conservation in Dutch agriculture: an evaluation of objectives, means, effects and programme costs. *European Review of Agricultural Economics* 19, 331–350.

Steinlin, H. (1997) The main forest challenges worldwide. In: *II FORUM de Politica Forestal, Centre Tecnologic Forestal del Solsones*. 11–15 March 1997.

Whitby, M. (1995) Transaction costs and property rights: the omitted variables? In: Albisu, L.M. and Romero, C. (eds) *Environmental and Land Use Issues: an Economic Perspective*. Wissenschaftverlag Vauk, Kiel, Germany.

Whitby, M. and Saunders, C. (1996) Estimating conservation goods in Britain. *Land Economics* 72, 313–325.

Zamagni, S. (1994) Global environmental change, rationality and ethics. In: Campiglio, L. *et al*. (eds) *The Environment after Rio: International Law and Economics*. Graham and Trotman, UK.

24 Decentralization and Participation: Key Challenges for Mediterranean Public Forest Policy

Eduardo Rojas-Briales

Agriculture and Forest Faculty, Polytechnic University of Valencia, Campus de Vera s/n E-46022 Valencia, Spain

Introduction

Increased attention has been paid to forestry by the international arena through the means of, for example, the Rio Summit (Chapter 11 of Agenda 21, Forest principles), the International Panel for Forest/International Forum on Forests (IPF/IFF) process, the creation of the United Nations Forum on Forests, and the Paneuropean Ministerial Conferences on Protection of Forests (1993, 1998). This has overlapped with important developments in public governance, particularly linked with environmental and economic development issues. From these processes, two levels for action have been identified: operational and political. The operational level is linked to applied management, focusing on silviculture, best management practices, management planning, control and audit systems, and certification. The performance of this level is followed up by different criteria and indicators. The political level deals with the legal, institutional and economic framework of forestry, the accomplishment of which is manifested through national and subnational forest programmes.

National forest programmes are increasingly identified in the literature as processes in which the key elements, principles and requirements are tested to see if they are consistent with the measures and actions proposed (see Cost Action E19). From the different processes, the identified political principles can be grouped into two categories – good governance and planning, and policy design – as shown in Table 24.1. As the borders among these principles (objectives) are not clearly defined, there are links and frequent interactions between them.

Analysis of the Application of Political Principles for Forest Policy in the Mediterranean Region

The identified principles are analysed individually, considering their strengths and weaknesses, and are illustrated through encouraging examples and opportunities for development.

National sovereignty

Increasing environmental awareness has led to a rapid development of international environmental legislation, especially after the Rio Summit. However, this should not cause significant erosion of the country's primary responsibility. The different social, economic, cultural and natural conditions must be integrated into progressive and country-adapted strategies, for which the main duty lies with the country. This

Table 24.1.   Forest principles identified in the international processes.

Good governance	Planning and policy design
National sovereignty	Participation and partnership
Decentralization	Transparency
Governance by contract	Iterative, audit and monitored
Enforcement	Long-term commitment
Capacity building	Holistic
Sustainable finance	Consistency with international commitments
Minority protection (indigenous and local people)	Intersectorial coordination, consistency and synergies with other policies
Secure ownership rights (land tenure)	Integration with sustainable development strategies
Equity	Research and knowledge based
Separation of powers	Conflict resolution
Accountability	

principle was accepted after pressure from the developing countries to link environmental issues with development. The issue is more important for countries with a small population.

The practical application of national sovereignty in centralized countries is an issue that is reasonably clear. However, the decentralization and the development of supra-state conglomerates, such as the European Union (EU), poses complex problems, given that environmental policy is frequently shared between the different levels of government. The legal instrument of EU Directives presents an interesting example for respecting national or regional sovereignty, in which a higher level of government sets the objectives and standards, but does not enter into the means and strategies adopted for their fulfilment. In general, bottom-up processes are consistent with the respect for national sovereignty, whereas top-down processes generally are not.

Decentralization[1]

By bringing governmental decision-making closer to the citizenry, decentralization is widely believed to increase public sector accountability and, therefore, effectiveness. (Fox and Aranda, 1996)

At its most basic, decentralization aims to achieve one of the central aspirations of just political governance – democratization, or the desire that humans should have a say in their own affairs. In this sense, decentralization is a strategy of governance prompted by external or domestic pressures to facilitate transfers of

power closer to those who are most affected by the exercise of power. (Agrawal and Ostrom, 1999)

In the traditional approach to biodiversity conservation, local people and their economic activities were viewed as threats to the undisturbed functioning of natural ecosystems and were to be excluded from protected areas. However, it became evident that the social costs of exclusionary conservation projects were sometimes high, and that their success rate, even in biological terms, was disappointing. (Wells and Brandon, 1993)

Empirical political processes, together with progress in public sector administration, have pushed decentralization high up in the agenda during the past two decades. Globalization, although apparently contradictory, encourages decentralization by calling for more flexible, adaptive structures. Additionally, from the social perspective, the need for identity strengthens local and regional feelings as national identity loses its former strong position. Polycentrism is the new term for these emerging political structures (Ostrom and Ostrom, 1977; Ostrom et al., 1994).

Decentralization comes in different forms (political, administrative, fiscal or market) and intensities. Political decentralization may include regulatory faculties. Administrative decentralization may range from deconcentration and delegation up to devolution. Market decentralization and devolution are frequently linked with deregulation and privatization, whereas all decentralization processes are linked with participation, to the extent that decentralization requires a well-organized society including participatory culture. An interesting case is the integration of major stakeholders through new flexible bodies, such as boards, public entreprises, foundations and centres, not only in the consultative phase but also in decision-making and implementation processes.

A specific case of devolution is related to the acknowledgement of ownership rights of local communities and indigenous people. During the 1950s and 1960s, the ideal situation was considered to be State ownership of forests. More flexible approaches of recent times together with positive experiences of devolution have helped the recognition of these rights in the form of ownership or prevalent user rights.

Forestry today lacks the political weight to justify either specific decentralization processes or, the opposite, their exclusion. Important Mediterranean countries have recently gone through decentralization processes. The most significant cases are Italy since the 1970s, and Spain and France in the early 1980s. Smaller countries such as Greece or Portugal have yet to start significant decentralization that goes beyond deconcentration of the forest service. In fact, Portugal refused proposed decentralization in a referendum held in the late 1990s, despite the autonomous statutes for Madeira and the Azores. The countries on the southern and eastern rim of the Mediterranean are generally centralist, despite some slight signs of a decentralization in Turkey, Morocco and Tunisia.

In Italy, the main administrative and legislative powers for forestry have been assumed progressively by the regions. Despite this, the central government still controls the forest guards (Corpo Forestale dello Stato, CFS) and thus policy power in terms of national forest legislation, international and European processes. According to various authors, the decentralization process is, however, very dispersed and ineffective due to high transaction costs. Three of the five regions with a special autonomous regime and eight of 15 regions with normal autonomy have approved regional forest laws. However, the first 1923 Forest Law is still formally in force as the framework law, despite its lack of adaptation to modern demands and institutional changes during the last 80 years (Corrado and Merlo, 1999). In Spain, the main forest powers (civil servants, public forests, police, legislative and others) were

devolved to the 17 regions, whereas the central government retained the framework forest law, international regulation, inventory and national parks (Rojas 1992, 1995, 2000a). Some 20 years after devolution, the State has finally approved a new national law (Law No. 43/2003 of forests), whereas seven regions have approved their own regional law. Similar devolution has been applied to related fields such as nature protection, land planning or agriculture, but not to water management. France underwent a much weaker decentralization process in the early 1980s that had no direct influence on the forest sector. In a previous reform of the forest administration of the late 1960s, private forests were integrated into the 17 Regional Forest Centres (Centre Régional de la Propriété Forestière, CRPF), whereas for the public forests a specific public enterprise, the National Forest Office (Office National des Forêts, ONF), was established. The ONF had from the very beginning a deconcentrated regional structure. The new French regions gained no legislative power or further competence in forestry. In Finland, the structure of the public forest services is quite similar to the French model.

Spain and Italy show how shortcomings of devolution can generate legal insecurity and inefficient overlapping, if the institutional changes are not followed by a legal reform. In the UK, the devolution of Scotland and Wales has formally affected neither the Forestry Commission as public forest service nor the State forest ownership. This fact has caused much confusion concerning the real effect of devolution in forestry (Inglis, 1999). From experiences in developing countries, despite the differences in socio-economic frame conditions, the following conclusions have been drawn:

- Decentralization policies have positive social effects when they seek to empower local people and when those receiving powers are accountable to local people (Ribot, 1997; Larson, 2004).
- Decentralized structures are most appropriate to multi-functional landscapes, whereas strict protected wilderness areas are more likely to benefit under centralized structures (Sayer *et al.*, 2004).
- Devolution cannot be improvised. Hasty processes may leave political vacuums and

absence of accountability (Sayer *et al.*, 2004).

- Effective control of natural resources rests more with the forest users and the local population rather than in the formal control of governmental offices (Enters and Anderson, 1999).

The positive examples show that decentralization allows a more flexible answer to societal, technological and economic changes, as well as a higher degree of participation. Encouraging examples are found in new structures that allow the main stakeholders to be included in the decision-making and implementation process. The dysfunctions observed in management by centralized structures may be corrected by well-designed independent, mainly public, bodies. Contractual solutions, such as those explained in the next section, may help to capture the social objectives into management.

Despite this, some shortcuts are frequently taken in response to the shortcomings of a centralized public forest service. For example, in Spain, many regions as well as the central government have created their own public enterprises as mere operative tools with a high degree of flexibility (contracting, time), but suffer from high costs and non-compliance with the competition rules. Another frequent option has been the outsourcing for each single operation (such as management and felling), which usually is associated with high transaction costs.

Governance by contract[2]

Proactive public policies demand that autonomous agencies are able to achieve their targets efficiently. However, care needs to be taken not to allow any inside interests. As these agencies are normally not completely self-funding, a specific accountability scheme is increasingly established. It allows the achievement of operative targets and indicators as a basis for compensation for the public support, following the principle of cross-compliance by a frame contract. The degree of performance is also used as a key modulating factor. This kind of frame contract is increasingly used in many fields of public interest where a centralized provision has been shown to be ineffective and an autonomous

provision unit is expected to have higher performance rates.

Administrative law, designed for the State core tasks (*imperium function*: police, courts and defence), shows high rigidities in the provision of services (*service function*: education and health). There are few doubts that they can be managed more efficiently under civil law, introducing modern enterprise management rules, than under the traditional administrative statutes.

Examples of this are the railway companies, universities and research institutions, hospitals and public forest management agencies, such as the ONF (Office Nationale des Forêts, 1995) or Österreichische Bundesforste (ÖBF). A specific situation is identified in the case of a natural resource, such as agriculture, fisheries, forestry or protected areas. It should be remembered that, at least under Mediterranean and European conditions, in most cases the ownership and/or the direct management corresponds to a large number of local players (farmers, forest owners, hunters, fishermen and others). Comparative analysis shows that an efficient method to fulfil public objectives is deciding and implementing policies in a participatory manner within boards grouping representatives of both public institutions and main stakeholders. Public objectives may easily be ensured by well-negotiated and conditioned contracts supported by public financing.

On a more operative scale, contracts are being used increasingly for internalizing environmental objectives into forest management, for example for ensuring a determined management system, setting aside some parts out of management and taking protection into account. Whereas in agriculture these contractual agreements are adopted in the frame of agroenvironmental support systems (see French (1999) or Catalan (2001) laws), in forestry, management plans seem to be the most suitable frame for an efficient meeting of the increasing demand for non-market goods and services.

The implementation of forestry contracts in the Mediterranean countries is still at a low level of development, with the exception of France. Management of public forests is still conducted by the public administration, though this will probably change in the near future. Contracts on the operational scale are also scarce, as protected areas are generally located in public forests, especially in the eastern and southern Mediterranean areas. The increasing protection of private forests in the southern EU countries, especially in connection with the EU network Natura 2000 (Directive 92/43), is generalizing the demand for contractual solutions (see Estrategia Forestal Española, 2000).

Even if in the east and south of the Mediterranean region State ownership prevails, the oncoming decline in marginal agriculture, the recognition of traditional user rights and decentralization processes will increase the relevance of private and communal forestry in this area and by this the need to find specific contractual solutions for demands not covered by the market.

Enforcement

Once a policy has been agreed on, its enforcement becomes the key issue. However, enforcement shortcomings have been very frequent in developing countries, and also to some degree in developed countries. The two main shortcomings observed in forestry lie in finance and in political weight.

In contrast to other policy planning processes, in forestry and nature protection, policy setting is uncoupled from the financial debate. Many forest strategies and programmes do not include financial estimates; in other words, the gap between the planned goals and the resources available is more than evident, as is the case with the EU forest strategy (1998), Italian Forest Plan (1987), Spanish Forest Plan (Ministerio de Medio Ambiente, 2002), French Forest Strategy (2000) and others.

Increasingly, the main challenges for forests and nature protection lie in cross-sectorial issues. A weak political presence is frequently the cause of the shortcomings in the enforcement of forest policies in cross-sectorial issues. As a consequence of an extreme interpretation of the federal principle of subsidiarity, forest political presence in federally structured countries and in the EU is disperse and extremely weak. This shows an inability to guarantee that forest interests are adequately represented, where they overlap with others, something

characteristic for cross-sectorial fields such as forestry.

Enforcement failures may also be related to the so-called 'virtual reality political vicious circle' in which the undisturbed – mainly due to electoral periods – time available for an exercising politician is used for previously non-implemented measures (white papers, strategies, laws, programmes, plans and others). When the circle is more or less closed, a change in government may help to restart the process.

Great political changes in the constitutional, economic or institutional framework place very significant restrictions on enforcement, especially of long-term policies as in the case of forestry. Democratization in Greece, Portugal and Spain provoked a large increase in forest fires, which would come under control only years after the consolidation of the democratic regime.

Enforcement may be increased by institutionalized participation capable of maintaining the overall perspective and avoiding *ad hoc* processes feeding back on themselves. Frequent auditing and revision systems may help the respect, enforcement and improvement of the approved plans.

Capacity building

Generally speaking, public forest services in Mediterranean countries do not lend themselves well to capacity building. On the one hand, they are too understaffed and lack other resources to take over the general management of the forest resources. On the other, they are too big and inflexible as a result of administrative rules and their tasks are too contradictory to perform efficiently. Problems arising may include:

- The uncoupling of the forest rangers, due to a prevailing police orientation (environmental, rural),[3] from the rest of the forest service.
- Centralized organizations with a limited presence of forestry graduates in the countryside.
- Unmotivated and, in many countries, underpaid staff.
- Lack of knowledge updating and specialization.

- Structures that are too hierarchical for cost-efficient decision making and management.
- Contradictory situations (management and control in one hand).
- Inflexible administrative rules (such as lifetime commitment, purchases, marketing, responsibilities).
- Contradictory laws and institutional overlap.

A requisite for promising capacity building is the establishment of a modern organization of the forest sector in accordance with the development of the general economy. Policing, administrative, managerial and service functions require different bodies that avoid internal conflicts of interest as well as different degrees of participation by the main stakeholders and, last but not least, the use of market competition between the operators in all suitable fields. Decentralization and knowledge through research and its transfer become decisive elements for a qualitative capacity building.

Sustainable finance

Centuries of human interaction with Mediterranean forests have resulted in deforestation and degradation. These forests probably have the highest world rate in the quotient:

FoEx+/CoFoOp

(where FoEx+ = positive forest externalities and public goods, and CoFoOp = commercial forest output) as they are either mountain or coastal and in all cases host high biodiversity values. Therefore, public finance is one of the most relevant issues and presents the most common shortcoming, even in the developed EU countries. As an example, only 1.25% of the Common Agricultural Policy (CAP) is devoted to forests (an annual average of 4€/ha; Morcillo, 2001) and there is no specific treatment of Mediterranean forests. In fact, the execution rate in financial terms of the existing forest plans is low; in Spain it is around 20% (Corrado and Merlo, 1999; Alcanda, 2001). One notes that forests have traditionally been a source of income and not expenditure. Exceptional situations were

generally linked with Keynes' policies in order to ensure work and income in times of depression, especially under dictatorial regimes: for example in Spain, the forest investments during the Franco Regime (Rojas, 1992, 1995), in Italy under Mussolini, or in Portugal under the Salazar regime. Inefficiencies are linked not only to the general funding insufficiency, but also to the wide oscillations in the available budgets. The short-term peaks of investment prove unsustainable in the long run, whereas long-term investment and employment is more favourable. This can be achieved by strengthening the link of forest resources to rural development and the local economy, one of the main driving forces for forest decentralization (Inglis, 1999).

Investment distribution and execution is also quite important and may have a strongly negative or positive stabilization effect on local employment. For example, forest funds could be considered as more favourable than annual budgets, as the rigidities and oscillations between years are overcome. They can also mobilize forest-related incomes that otherwise would not be available for the general public budget, such as fines, taxes, State forest revenues, public forests used for other public purposes, earmarked environmental taxes on carbon dioxide (CO_2) emissions, water consumption or land occupation.[4]

Finally, even if in certain cases there seems to be a significant budget available, the degree of freedom in deciding its allocation is a premise for its efficient application. Even if these budgets might help to compensate for the structural lack of financial resources, they are frequently externally designed and follow different logic and objectives. In other words, the problem lies in their tangential perspective towards forests. Examples include the EU afforestation policy of agricultural land (see Regulation 2080/1992) or investments for job creation.

Together with international processes, funding is the driving force for the engagement of the higher levels of governance, despite their formal lack of direct competence. Some recently decentralized states (Italy, Spain and Belgium) seem not to have understood this potential instrument in comparison with the even weaker juridical basement for the EU's subsidies policy.

Minority protection (indigenous and local people)

Many national laws continue to reflect a state-centred approach to resource management and a restricted philosophy of property rights which has tended to undermine existing community-based systems and has seriously constrained local people and progressive government officials in the search for new community-based solutions. (Bruce, 1999)

Due to the characteristics of local peoples in the Mediterranean area, historically made up of many cultures, customs and traditions, there is a real need for local involvement in the governance and management of the natural resources. This issue is often reflected by the land tenure system in terms of either formal or factual traditional rights over the natural resources (forests, hunting, fisheries, grazing and others). In Spain, Portugal, France and Italy, most of the mountain forests are communally owned and worked by local people, whereas in the southern and eastern Mediterranean countries, forests are generally State owned, with user rights acknowledged for local people. Many of the problems affecting sustainability of Mediterranean forests (such as fires or grazing) arise when these traditional rights clash with the formal State ownership.

These rights should be formalized and updated to ensure the future preservation of their content under changing conditions (Lindsay, 1999). Less appropriate are the extreme options of either their abolishment or their rigid preservation. Conflicting rights can also be resolved by dividing the estates proportionally. Nevertheless, both owners and users should be integrated into forest management planning.

Together with the ownership issues, governance of natural resources should include local representatives as the most important stakeholders, holding the majority on ruling bodies, such as the boards for protected areas. Top-down representations, such as forest owners and non-governmental organizations (NGOs), should be avoided.

Secure ownership rights

For centuries, many, generally legitimate, rights have clashed in rural and forest areas. Changes in legislation and insufficient enforcement have favoured conflicts; for example, problems arising from the lack of clear cartographic and physical delimitation of the estate boundaries.

Forest management is based on ownership rights. Both rights and duties are very relevant in forestry and demand a clear ownership situation. For example, Lindsay (1999) has listed key elements required for an efficient and legally secure local forest management.

Conflict-resolving mechanisms and pragmatism in designing and applying legislation, together with the formal and cartographic delimitation of the estates, may help to solve these conflicts. Traditional user rights should either be formalized – including area, intensity and responsibilities – or compensated for in ownership or money. Rigidities in forest ownership may impede adequate solutions.

Growing environmental awareness in many Mediterranean countries has often increased the duties imposed on forest owners and managers without mobilizing the appropriate public funding. The situation is similar to the 'tragedy of the commons', where forest management costs are supported by the owners, while forest benefits are enjoyed by the general public, thus increasing land abandonment. Duties must be objectively based and balanced with rights; otherwise ownership rights would be unequal and insecure. In this sense, the Spanish Forest Strategy (Ministerio de Medio Ambiente, 2000) recognized the need to set clear limits to the restriction faculties so as to ensure the respect for ownership rights in the so-called 'Statute of Forest Ownership' (International Conference on Private Forest Statute; Conferencia sobre el Estatuto de la Propiedad Forestal, 2004).

In this regard, it should be remembered that the undefined legal situation of most volatile forest products (mushrooms, flowers, berries, medicinal plants, cones or even grazing) and services runs contrary to efficient output and sustainable use of forest resources.[5] This situation is to a certain extent comparable with the question of the copyright of books, CDs or videos. Changes in legislation will have to adapt to the present nature of forests as service-providing assets and no longer as primary production-oriented assets.

Equity

Public policies should be based on the principle of equity – a complex and delicate issue in the governance of natural resources. Conflicts between rural and urban populations, long-term commitment and short-term benefits, social externalities and public goods and private costs may undermine the policy output. The consequences of public policies should be analysed on equity grounds. If the burden and benefits are distributed unequally, balancing actions must become an integral part if the policy is not to fail. Frequently, forest and nature protection policies have been seen from a mere technical perspective, neglecting any social analysis, and especially their conse-quences on equity. The Natura 2000 network in the EU is a paradigmatic example of this (Rojas, 2000b; Centre Tecnològic Forestal de Catalunya, 2001).

Separation of powers

Control and balance systems established since Montesquieu have been progressively enlarged with other elements (Fernández-González and Alyward, 1999), such as multi-layer gover-nance, public participation and transparency, by using modern techniques. However, contra-dictory or conflicting interests must be avoided; for example, the regulator must be a different body to that of the operator.

Accountability

Several authors insist on the importance of accountability in the context of modern gover-nance. In the specific case of decentralization and devolution, an efficient control system based on permanent reliable monitoring and revision processes requires locally account-able representation. Empowerment of local authorities is also seen as a prerequisite for efficient accountability. Accountability requires elective processes that should be comple-mented by other more informal and iterative means suitable to the cultural background (Ribot, 1999).

Participation and partnership[6]

The UNCED Conference reinforced the impor-tance of a strong social component to environ-mental policies, where participation and partnership become the clearest elements. Public participation conducted only by the political parties and regular elections (4–5 years) is a very limited and restricted way of channelling the views and demands of social groups and individuals. Much better adjusted governance ensures that the different social interests – both economic and non-economic – are gathered through efficient entities that defend and represent their interests.

Participatory processes in environmental issues have grown enormously since Rio. Normally they function on previously set com-mon or political general objectives by develop-ing agreed targets and measures to achieve them as well as on permanent auditing and adaptation.

There are many forms of participatory processes, including formal or _ad hoc_, technical or political, wide or sectorial, informative or executive, private or public. _Ad hoc_ processes generally approach specific temporary tasks or problems; also, frequently, they represent the first step towards a formal institutionalized body, such as the Forest Advisory Council of the DG Agriculture of the European Commission. Technical bodies[7] analyse highly complex oper-ational questions, normally under the umbrella of political boards that control and guide their work. Participatory processes normally do not include voting rights, as they formally advise governments but do not themselves include decision making. Despite this, in some cases, the frequency of meeting, number and scope of the board only allow an information flow, some-times uni-directional, sometimes bi-directional. Other bodies meet with sufficient frequency, are small enough and have a level of commitment that allows them to pass from the simple information stage to debate and decision, even if this is not their formal task. Finally, participatory processes, especially partnership developing, may take the form of *ad hoc* NGO-driven processes where the public authorities are either absent or are a non-determining part of the process, e.g. certification is an example of a privately driven participatory process, in both the Pan-European Forest Certification (PEFC) and the Forest Stewardship Council (FSC), or the recent accord on Natura 2000 by forest owners and NGOs in Spain (2002).

A clear indicator of the degree of political dominance or participatory weight is the presidency of the Board and the proportion and designation procedure of the representatives from the public and private sectors, including NGOs. Undesirable power struggles between entities representing specific sectors are well known and should be avoided, for example by previous *ad hoc* solutions.

The independent professional moderator may be a very useful tool in drafting processes such as forest programmes to ensure equal participation. Participatory bodies allow the mobilization of a wide expertise (panel of experts) with very low costs in the process, although the methodology – moderator, openness and a wide range of representatives – has to be sound.

Selective participation, harming the overall output, is a frequent consequence of disproportionate strength of the different stakeholders (international NGOs or strong economic interest groups versus local players) or a lack of interest by some stakeholders that have been free-riding forest services – such as tourism, water, energy – and would evidently be asked to contribute as a result of the process.

If participatory processes have achieved a partnership of the participating entities, previous conflicts will decrease significantly as official and parallel channels for conflict resolution will be open, and no participating entity will benefit from an open conflict.

Most of the modern national and subnational forest programmes – Spanish Forest Strategy, French Forest Strategy, Spanish Forest Plan – have passed through or are even the result of a participatory process, e.g. Navarra in Spain and Baden Württemberg in Germany. Many countries have forest and/or environmental or agricultural advisory bodies in which forest matters are dealt with.

Transparency

Participation and partnership require transparency not only from the participating public bodies, but also from private entities. Proposals, drafts and positions should be clearly stated and known by all participants in time to allow feedback. The Natura 2000 network of the EU is a paradigmatic example of non-transparent procedures and selective participation. Whereas the affected population and local NGOs were not informed, the leading international NGOs had important inside knowledge that enabled them to adapt their lobbying and public relations campaigns.

Multi-faceted interests and debate as a sign of social diversity is an enriching and sound process in the long run and should be favoured – an issue greatly assisted by modern communications. It is a sound practice for stakeholders to set their goals and plan in an internal participatory and transparent way.[8]

Iterative, audit and monitored

Governance of natural resources, including forestry, is a long-lasting never-ending process. As the social, economic and even natural – due mainly to climate change – conditions change very quickly, even well-designed forest programmes need to adapt interactively over relatively short periods. Flexible planning is needed to substitute former rigid planning structures.

As defined by Worrel (1970), a policy is formed by a chain starting with the analysis of a given problem, followed by the setting of goals and objectives from which the adequate policy instruments are derived. Once this basic framework is defined, its only sense is its implementation. Periodic evaluation based on the report of the monitoring process and, if needed, revision forms a complete and harmonic chain able to adapt to a changing environment. Efficient feedback mechanisms are required, mainly through participatory processes, statistics, inventories, regular publications regarding the state of the forests[9] and execution of the programme, and mid-term reports. External audits may help to identify new developments, risks, deviations or emerging demands. An interesting example of external revision occurred in the late 1990s in Switzerland. A team of prestigious foreign specialists with a broad background was appointed for the revision team, giving useful analysis and proposals for the further development of Swiss forest policy.

Frequently after a strong social and political commitment in the programme elaboration phase, the implementation phase is undervalued, ignoring the fact that the feedback during this stage may be of high relevance in order to adapt it and ensure a high degree of achievement of the stated goals.

Long-term commitment

Despite the need for flexible iterative planning, forest plans have to keep a long-term perspective and commitment. Sustainability in the long term as a broad concept has to be ensured and demands non-economic long-term investments. This fact can be integrated by separating the strategic objectives from the operational ones, measurable by specific indicators. The first would be determined by law and should undergo very limited and progressive changes, whereas the second should be reviewed in the framework of the cited interactive processes.

Holistic

Forest resources have changed from a clear vertical 'sectorial' structure to a field of converging social interests and demands with a clear cross-sectorial character (Fig. 24.1). Any political planning of resources must analyse all these complex linkages and propose action in order to ensure sustainable integration, including new definitions of rights and duties. Whereas long-term commitment is oriented towards the vertical perspective of sustainability, a holistic perspective is orientated horizontally to the present demands of different social groups and territories on, and threats to, forests from society.

Consistency with international commitments

The intensive international environmental processes since Rio and the strong linkages of the main environmental issues with forests require appropriate consideration both in the setting of goals and in the specific measures agreed. The protection of specific species and habitats according to the Biodiversity Convention and others (such as Ramsar, Bern), the role of forests in climate change in the light of the Kyoto protocol, desertification and soil protection of forests in the framework of the Desertification Convention are the most relevant examples (United Nations, 1995).

However, there are also an important number of worldwide forest commitments: Chapter 11 of Agenda 21, non-legally binding forest principles, proposals for action of the IPF/IFF (United Nations, 2002). Other commitments are on a regional level and include the four Pan-European Ministerial Conferences on Protection of Forests (especially II 1993 and IV 1998) regarding sustainable forest management in Europe. An important role in the Mediterranean area is recognized with the *Silva Mediterranea* network coordinated by the Food and Agriculture Organization.

Although these commitments are not legally binding, they are being introduced progressively into the national law and are acquiring legal force, for example national forest programmes. The EU has strengthened their statutes as advised measures to be implemented (Proposals for action of the IPF/IFF process) by requiring that the regions who demand rural development funding from the EU present a regional forest plan. Another example is the criteria and indicators approved in Lisbon (1998), having profound implications for forest policy in all European countries.

Intersectorial coordination, consistency and synergies with other policies

As shown in Fig. 24.1, the strong linkages of Mediterranean forests with other sectors require the integration of all overlapping policies. This is quite complicated due to the weak political position of forestry and the strong overlapping

Fig. 24.1. Forest resource linkages with other activities and fields of social interest.

interests. Many measures generally included in forest plans have not been negotiated with the affected bodies and therefore may not be implemented.

Forestry should have a proactive attitude in other policies, such as coordination and consistency, regarding, for example, land use and watershed management plans, rural development programmes and energy policies. Sound research-based information, together with a communication policy that avoids a negative and catastrophic image of forestry, are very relevant supporting factors for intersectorial coordination. Negotiation with other sectors must search for progressive solutions.

Afforestation is a useful tool for rural development, erosion control, water quality improvement or CO_2 sequestration. However, afforestation must be integrated into forest planning, taking fire prevention, landscape values or biodiversity maintenance into account, and should not be designed in isolation as has been the case in the EU set-aside programmes (*see* Regulation 2080/1992). In addition, other important political issues such as climate change, renewable energy, land use policies, regional policies and water policies show important overlaps with forest policies requiring an efficient and proactive intersectorial coordination.

Integration with sustainable development policies

In the framework of the international environmental processes, sustainable development plans should be drafted, discussed and enforced. As cited above, forestry is a key element for ensuring the inclusion of environmental services in any sustainable development policy. The risk is quite evident that the free services of forests acquire a high relevance in sustainable development strategies. However, the only consequence of this is an increasing regulatory pressure on forestry, whereas the bottlenecks such as competition with non-renewable resources, environmental taxation, earmarking of environmental taxes for positive environmental policies, distribution between agriculture and forestry of the rural policy funds among others – in forestry are not faced (even if doing so would have important environmental benefits) because they affect stronger economic sectors.

Research and technology based

Forestry today is evidently a tertiary (services) resource but increasingly has elements of the quaternary society (information). Forestry's wide range of interactions and its relatively weak economic performance explain the low level of knowledge and the frequent decisions based on dilettante and poor information sources. Modern governance is based on a sound information base that ensures high qualitative information, an issue that has been recognized in the different international forest processes such as the Ministerial Conference on Protection of Forests or the United Nations Forum on Forests. The question arising is no longer whether forests are 'good or bad', but is more complex (where and which kind of forests and how are they managed), and the answer must be based on knowledge. It is also very important to integrate research and technological progress from other sectors into forestry in order to reduce costs and ensure the best information base possible.

Conflict resolution

Due to the wide range of overlapping interests and the spatial dimension of forests, conflicts are unavoidable. It has proved much more efficient to channel conflicts and establish conflict resolution procedures than to avoid the evidence. Conflict resolution via the traditional way (law and order) is expensive and not very efficient, due to high costs and long delays.

While breaches of the law must be controlled legally, conflicting interests need specific procedures. Participatory processes and the development of partnerships are the basis for approaching this question appropriately. The maturity of the stakeholders is of key relevance for avoiding a single case perspective integrating into a wider perspective. On the other hand, mixed bodies may be of great utility in terms of cost and time in order to evaluate compensation in forestry. Arbitral solutions, as used in other similar cases such as in consumer legislation, may be an efficient, cheap and conflict-avoiding option.

Comparative Analysis of Key Forest Policy Elements in Different Mediterranean Countries

In the frame of this book, a survey between different Mediterranean countries was realized. The survey asked for information regarding decentralization processes, allocation and reforms of the public forest services, National Forest Programmes, and participatory boards. Its main results are shown in Table 24.2.

From the comparative analysis, the strong tradition of the French centralistic model in the entire region becomes evident. Despite this background, in recent times and for the near future, decentralization tendencies are evident in practically all the countries despite significant differences in their scope.

The Forest Service is structured regionally in many countries so that the forthcoming tendencies will not affect their internal structure greatly, but probably its political command. Only Turkey and Croatia may have to adapt to decentralization processes. In the Italian case, the centralized structure of the Corpo Forestale dello Stato may collide with the strong degree of decentralization achieved in Italy.

Concerning the Ministerial allocation, a clear shift from Agriculture to Environment is to be observed except in the EU countries where the dependence on EU funding based on the

Table 24.2. Main findings of the comparative survey on key forest policy elements in Mediterranean countries.

1. Present situation	A. Federal including legislative power	B. Decentralized regions	C. Some decentralized regions	D. Fully centralized
Countries	E, I	AL, GR, F	MA, P	PAL, CHI, SLO, TR, CRO

2. Tendencies in case 1 C) or D)	No decentralization planned	Yes, but open output	Yes, but only for specific region	Yes, for the whole country
Countries	CHI, P		CRO	PAL, SLO, TR, MA

3. The Forest Service is	Strongly centralized	Mixed	Regionally organized with some national coordination	
Countries	CHI, TR, CRO	SLO, F, I, P	AL, GR, MA, E	

4. Forestry is allocated in the Ministry of	Internal Affairs	Agriculture	Environment	Others
Countries	–	PAL, SLO, GR, CRO, I, P	E, TR[a]	CHI[b], AL[b], F[c], MA[d]

5. NFP[e]	Ongoing		Existing, not applied	In discussion	Plans to start	No discussion at all	NFP regionally	Both regional and national	Integrated in other plans
Countries	AL, CHI, SLO, TR, CRO, F, E		I	P			GR	AL, I, F, E	AL, PAL

6. Reform plans for the Forest Service	No			Yes	
Countries	PAL, CHI, F, GR, MA, CRO, E			AL, SLO, TR, P, I	

7. Participatory boards	Existing and relevant	Existing but without significant relevance	Exist but exclude a significant part of stakeholders	Not existing
Countries	PAL, MA, P, E	AL, F, I	CHI	SLO, TR, GR, CRO

AL = Algeria; CHI = Cyprus; CRO = Croatia; E = Spain; F = France; GR = Greece; I = Italy; MA = Morocco; PAL = Palestine; P = Portugal; SLO = Slovenia; TR = Turkey.
[a]Environment and Forestry. [b]Agriculture and Environment. [c]Partly depending on both Agriculture and Environment. [d]Water, forests and desertification. [e]National Forest Programmes. Several options possible.

CAP probably breaks this tendency (Italy, France and Greece). Only Palestine, Slovenia and Croatia maintain forestry in Agriculture. An interesting example is the case of Morocco where forestry and water have their own specific Ministry.

Practically all Mediterranean countries have approved and are implementing national or regional forest programmes, or even both. About half of the countries apply these plans either exclusively at the regional scale or at both scales (Algeria, Greece, Italy, Spain and France).

Major changes in the Forest Services are under discussion in practically all countries (Turkey, Algeria, Slovenia, Portugal and Italy). Decentralization, economic and social demands and administrative reforms are the major driving forces.

Concerning public participation, one-third have active participation fora (Spain, Morocco, Portugal and Palestine), one-third have restricted fora (Algeria, France, Italy and Cyprus) and the rest have no participation board at all.

Some Encouraging Examples

The cited principles are normally not applied in isolation but instead are often integrated, as has been achieved to some degree in certain countries. The examples cited fall into two categories. The first category is formed by mixed or shared bodies in which specific public tasks are delegated to agencies run by representatives of the public and private sectors. In these cases, the participation is not of an advisory nature but of shared governance. In this sense, it is important that the policies are set previously, if possible in a participatory way, and that the shared bodies are limited to enforcement. With regard to voting rights, consensus building should be structurally enhanced. If the tasks affect key stakeholders to a high degree, it might be appropriate for a slight majority of votes to be kept for them, perhaps even including the presidency. In the second category, different approaches to introducing entrepreneurship in the management of state forests are identified. Finally, the relevance of different contractual options is analysed.

Shared or mixed bodies

Regional forestry centres

Regional forestry centres have been established in the past in countries with a high proportion of private forests. In the Mediterranean area, France was the groundbreaker in the framework of the late 1960s' forest service reform. Later on, some German Länder, Finland and Catalonia (Spain) established a similar situation in the early 1990s (Rojas, 2002), and some are soon to be established in other Spanish regions.

Regional forestry centres are normally responsible for the majority of the public tasks in private forestry: mainly approval of management plans and felling licences, incentive systems, representation of private forestry on advisory boards, information regarding plans that affect private forests and extension. The board is mixed, with the majority, including the president, being elected by the forest owners in possession of a current management plan, while the government designates the rest of the representatives. The police function is not normally delegated to the boards. The director is a civil servant appointed by agreement between the Ministry and the board.

Whereas in France, from the very beginning, the centres had a high degree of autonomy that has not changed significantly, the Catalonian case has shown a clear progression

in competence and political weight. It started as a deconcentrated organ of the forest service without its own budget and with limited powers in private forests with management plans (1990–1999). In 1999, a specific Forest Centre Act was passed by the Parliament that gave the centre very relevant powers, such as financial autonomy, competence over all authorizations in private forests and incentives, as well as the establishment of a Private Forest Fund.

The output of the forest centres has clearly been positive. Forest management plans have become generalized in the areas with forest centres in comparison with other Mediterranean regions. For example, in Catalonia, 12 years after the start of forest management plans in private forests, more than 2000 management plans have been approved, covering more than 300,000 ha.

All elements of performance in forest policy (such as fire prevention, investments, felling and conflict resolution) show encouraging trends when there are active forest centres.

It might be interesting to point out here a related situation outside the Mediterranean area. In its Forest Act (1976), Baden-Württemberg (Germany) established a mixed body for all administrative decisions concerning communal forests in which the forest service and representatives of the communes each hold 50% of the voting rights, something that forces structural consensus building.

Boards of protected areas[10]

In many countries, protected areas have boards that either advise or control the park service. These boards are normally participatory and include both local and regional major stakeholders. The chairman of the board is generally appointed politically, but sometimes also elected, whereas the director is a civil servant appointed by the Ministry. There is frequent friction due to unclear distribution of tasks and chains of command.

Performance in protected areas improves with higher degrees of participation, especially by local stakeholders. Their importance should be recognized by a majority representation. However, the distribution of representatives should structurally favour consensus building between the public and private sectors or between the local and regional players as well as between different institutions. The political profile of the chairman should be strengthened and, if possible, he/she should be appointed by the board, whereas the director should maintain a professional profile directly dependent on the board and its chairman.

There are interesting parallel figures that promote sustainable development in protected areas run by local stakeholders based on the publicity of the protected area (see the Foundation Doñana 21). The registration of the name of the area by these agencies can help to capture and internalize part of the positive externalities of these areas.

Enterprises for state forests[11]

Management of public forests by a conventional administration has been shown to be inefficient in present times. Countries with important public, especially State-owned, forests have designed specific solutions for the management of these State forests. These range from mixed figures between administration and public enterprise to the privatization of the State forests. In the case of communal forests, a clear tendency is observed towards devolution of the management to the communes or intermediate bodies.

In France, with the reform of the Forest service in the late 1960s, the ONF was created as a mixed body between the administration and public enterprise. The technical staff are still civil servants but the formal status is that of a commercial public enterprise. The board of the ONF is comprised of representatives of the State and of the forest-owning municipalities, as these are managed on a compulsory basis by the ONF. The financial relationship between the state and the ONF is governed by a 5-year-frame contract.

In Austria, the ÖBF (federal forests) passed from an administrative to a public enterprise status with full commercial freedom (Österreichishe Budesforste, 1997). New personnel are contracted by civil law. Specific public outputs are included in the frame contract established between the government and the ÖBF. In Germany, the Saar region broke the ice on this front, creating a specific enterprise for managing the State forests (Saar Forst), closely following the Austrian example. Finally, in Sweden, the State forests were first transformed into a public company (Assi Domän), and later on partly privatized (65%) in the form of a stock market company. Presently there is a lively ongoing debate concerning the future of the state shares in the company. The state retained control of those forests with major public functions (protected or recreation areas).

Contractual solutions

Contractual options are gaining relevance in forestry. On the one hand, the frame contracts are interesting instruments for ensuring the internalization of public interests by major agencies while keeping their autonomy and primary responsibility. On the other, land contracts have gained considerable importance, especially in the case of France, by Forest Law (2001), Agricultural Orientation Law (1999) and *Circulaire des Chartes de Territoire Forestier* (Ministère de l'Agriculture et de la Pêche, 2001). These contracts are established parallel to the drafting of planning instruments (such as management and protection area plans) and transform the content of the plans into contractual obligations, including the distribution of responsibilities and funding.

Conclusions

Due to the relevant environmental output of forests, international environmental processes have been strongly linked to forest resources. As a consequence of these processes, key guiding principles for the orientation of forest policy have been identified and generally accepted. The state or art of their application in the Mediterranean basin and the potential for future implementation have been analysed, one by one, for each of the 21 principles identified, the most relevant being decentralization and participation.

One of the main conclusions is that, in practice, most of these principles are applied in a combined way. Encouraging examples of combined application are described as possible guidance for future developments in this area.

Another important conclusion is related to National Forest Programmes. Together with sound participatory processes, they give coherence and stability to forest policy. They allow the forest agenda to be determined proactively and not as a reaction to external factors. New situations such as those cited may help to combine efficiency with adequate participation and partnership of the stakeholders. Public policies may achieve their objectives efficiently in social and economic terms when the distribution of tasks between the public administration, society and the market is: rational, according to modern principles of public administration; and coherent, with progress in the rest of societal and economic development.

Notes

[1] See Clément and Kabamdana (1995), Agrawal and Ostrom (1999), Enters and Andersen (1999), Inglis (1999), Lindsay (1999), Ribot (1999), Neven (2002), Larson (2004) and Sayer (2004).
[2] See Davies (2001).
[3] Italy (Corpo Forestale del Stato), many regional forest guard corps in Spain and others (Corrado and Merlo, 1999).
[4] See Costa Rica's Biodiversity (1998) and Forest Act (1996).
[5] See the effects of microdecoupling in Mediterranean forests in Mendes (1999). See also Mantau et al. (2001).
[6] See Buttoud (2001).
[7] These kinds of bodies are very frequent in relation to the complex EU processes and legislation, at both EU and national levels.
[8] The Finnish farmers and forest owners' union MTK has recently carried out exemplary work in this field (2001).
[9] Annual or biannual. See FAO biannual World Forest Report (Food and Agriculture Organization, 2002).
[10] See Rojas (2000b) and Corraliza et al.
[11] See Borchers (1996, 1998).

References

Agrawal, A. and Ostrom, E. (1999) *Collective Action, Property Rights and Decentralization: Comparing Forest and Protected Area Management in India and Nepal*. (Cited by Ribot, 1999).

Alcanda, A. (2001) España: 10 años de experiencia en Planes Forestales Autonómicos. In: *Actas del Preseminario sobre Planes Forestales Nacionales. Contexto Social y Político*. Ministerio de Medio Ambiente, Madrid, CD, 9–24.

Borchers, J. (1996) *Privatisierung Statlicher Forstbetriebe*. Sauerläder's Verlag, Frankfurt, Germany.

Borchers, J. (1998) Alternativas para mejorar la Organización del Sector Forestal. Qué hacer con los bosques estatales? In: *Actas del II Forum Internacional de Política Forestal*. CTFC vol. 4, pp. 191–200.

Bruce, J.W. (1999) *Legal Bases for the Management of Forest Resources as Common Property*. Common Forestry Note No. 14, FAO, Rome.

Buttoud, G. (2001) *Gérer les Forêts du Sud*. L'Harmattan, Paris.

Centre Tecnològic Forestal de Catalunya (2001) *IV Forum de Política Forestal*. Actas. CTFC, Vol. 9. Solsona, Spain.

Clément, F. and Kabamdana, G. (1995) *La Participación Descentralizada. Nota de Orientación para los Programas Forestales Nacionales*. FAO, Rome.

Conferencia sobre el Estatuto de la Propiedad Forestal (COSE) (2004) *General Declaration, Conferencia sobre el Estatuto de la Propiedad Forestal*. Barcelona, Spain (in press).

Corrado, G. and Merlo, M. (1999) The state of art of National Forest Programmes in Italy. In: *Formulation and Implementation of National Forest Programmes*. EFI Proceedings No. 30 Vol. II, Joensuu, Finland, pp. 157–175.

Corraliza, J.A., García Navarro, J. and Gutiérrez del Olmo, E.V. (2002) *Los Parques Naturales en España: Conservación y Disfrute*. Fundación Alfonso Martín Escudero, Madrid.

Davies, A.C.L. (2001) *Accountability Oxford Socio-Legal Studies*. Oxford University Press, Oxford, UK.

Enters, T. and Anderson, J. (1999) Rethinking devolution of biodiversity conservation. *Unasylva*, FAO.

European Union (1998) *Resolución del Consejo sobre una Estrategia Forestal para la Unión Europea*. Documento 14244/98. 16.12.

Fernández-González, A. and Alyward, B. (1999) Participation, pluralism and polycentrism: reflections on watershed management in Costa Rica. *Unasylva*, FAO.

Food and Agriculture Organization (2002) *The State of the World Forests*. FAO, Rome.

Fox, J. and Aranda, J. (1996) *Decentralization and Rural Development in Mexico: Community Participation in Oaxaca's Municipal Funds Program*. Center for US–Mexican Studies, University of California, San Diego, California.

Inglis, A.S. (1999) Implications of devolution for participatory forestry in Scotland. *Unasylva*, FAO.

Larson, A.M. (2004) Democratic decentralization in the forest sector: lessons learned from Africa, Asia and Latin America. In: *UNFF Inter-sessional Workshop on Decentralization*. United Nations.

Lindsay, J.M. (1999) Creating a legal framework for community-based management: principles and dilemmas. *Unasylva*, FAO.

Mantau, U., Merlo, M., Sekot, W. and Welcker, B. (2001) *Recreational and Environmental Markets for Forest Enterprises*. CAB International, Wallingford, UK.

Mendes, A.M.S.C. (1999) Portugal. In: *Forestry in Changing Societies. Information for Teaching Module. Part II: Country Reports*. Silva-network, ICA and University of Joensuu, Joensuu, Finland, pp. 295–322.

Ministère de l'Agriculture et de la Pêche (2001) *Circulaire Chartes de Territoire Forestier*. DERF/SDF/SDIB/C2001-3004, 15.2.

Ministerial Conference on the Protection of Forests in Europe (1993) *General Declaration, Resolutions and Annexes of the Second Conference*. Helsinki.

Ministerial Conference on the Protection of Forests in Europe (1998) *General Declaration, Resolutions and Annexes of the Third Conference*. Lisbon.

Ministerio de Medio Ambiente (2000) *Estrategia Forestal Española*. DGCN, Madrid. 3 Vol. 111.

Ministerio de Medio Ambiente (2002) *Plan Forestal Español*. Madrid.

Morcillo, A. (2001) *El Sector Forestal y la Unión Europea*. Mundi Prensa, Madrid.

Neven, I. (2002) *Background Paper on Decentralization. Seminar National Forest Programmes*.; Savonlinna, Finland. COST E 19. EFI proceedings.

Office Nationale des Forêts (1995) *Contrat Etat/ONF*. ONF, France.

Österreichishe Budesforste (1997) *Unternehmenskonzept*. OBF, Austria.

Ostrom, V. and Ostrom, E. (1977) Public goods and public choices. In: *Alternatives for Delivering Public Services: Toward Improved Performance*. Westview Press, Boulder, Colorado, pp. 167–197.

Ostrom, V. *et al.* (1984) *Rules, Games and Common-pool Resources*. University of Michigan Press, Ann Arbor, Michigan.

Ribot, J.C. (1997) *Decentralization without Representation: Rural Authority and Popular Participation in the Sahelian Forestry*. Paper presented at FAO (cited by Lindsay, 1999).

Ribot, J.C. (1999) Accountable representation and power in participatory and decentralized environmental management. *Unasylva*, FAO.

Rojas, E. (1992) *Evolución de la Legislación Forestal en España. Desarrollo, Situación Actual y Perspectiva*. Report of the IUFRO WP S6.13-00, Forstwissenschaftliche Beiträge No. 11, Professur Forstpolitik und Forstökonomie, ETH-Zürich, pp. 232–258.

Rojas, E. (1995) *Una Política Forestal para el Estado de las Autonomías*. AEDOS/Mundi Prensa, Madrid–Barcelona–México.

Rojas, E. (2000a) Perspective for a new federal forest legislation in Spain. In: *Forging for a New Framework for Sustainable Forestry: Recent Development in European Forest Law*. IUFRO World Series Vol. 10, pp. 281–292.

Rojas, E. (2000b) Socio-economics of nature protection policies in the perspective of the implementation of Natura 2000 Network: the Spanish case. *Forestry*, 199–207.

Rojas, E. (2002) *A Spanish Experience: the Forest Plan of Catalonia and its Consequences for Private Forest Management. National Forest Programmes in a European Context*. EFI Proceedings No. 44, Joensuu, pp. 93–98.

Sayer, J., Elliott, C., Barrow, E., Gretzinger, S., Maginnis, S., McShane, T. and Shepherd, G. (2004) The implications for biodiversity conservation of decentralized forest resources management. *UNFF Inter-sessional Workshop on Decentralization*.

United Nations (1995) *Declaració de Rio (et al.) y Guia de l'Agenda 21*. Departament de Medi Ambient, Barcelona, Spain.

United Nations (2002) *The IPF/IFF Proposals for Action*. Secretariat of the UN, New York.

Wells, M.P. and Brandon, K.E. (1993) The principles and practices of buffer zones and local participation in biodiversity conservation. *Ambio* 22, 157–162.

Worrel, A.C. (1970) *Principles of Forest Policy*. McGraw Hill, New York.

Legislation

Waldgesetz (Baden-Württemberg) (1976), 10.02. GBl.: 99 + 524.

Forest Law (Catalonia) 6/1988, 30.3. DGCC No. 978, 15.4.

Directive EEC 92/43 21.5 on conservation of natural habitats and wild fauna and flora. OJEC L 206/7, 22.7.

Regulation EEC 2080/1992 on forest measures in agriculture, 30.6. DOCE: 215/96, 30.7.

Ley Forestal (Costa Rica) No. 7575, 13.2.1996. Asamblea Legislativa.

Ley de Biodiversidad (Costa Rica) No. 7788, 23.4.1998. Asamblea Legislativa.

Law of the Forest Center (Catalonia) 7/1999, 30.7. DOGC 9.8 No. 2948, 10701–10703.

Law 99–574 of Agricultural Orientation (France), 9.7. JO No. 158, 10.7: 10231.

Act of Agricultural Orientation (Catalonia) 18/2001, 31.12. DOGC No. 3549, 9.1.2002.

Forest Orientation Law (France), No. 2001–602, 9.7. JO No. 159, 11.7.

Decree 686/2002 regulating the Environmental Advisory Council (Spain), 12.7. BOE 17.7: 26178–26180.

Forest Law (Spain) 43/2003. 21.12. BOE: No. 280, 22.11: 41422–41442.

Websites

www.donana.es (Fundation Doñana)

www.fudacionentorno.org/notis (Agreement of forest owners and NGOs on Natura 2000)

www.mcpfe.org (Ministerial Conference on Protection of Forests in Europe)

www.metla.fi/eu/cost/e19/ (Action COST E 19)

25 The Need for an International Agreement on Mediterranean Forests

Américo M.S. Carvalho Mendes

Faculty of Economics and Management of the Portuguese Catholic University, Rua Diogo Botelho, 1327, 4169-005 Porto, Portugal

Mediterranean Forests: Important Ecosystems at Risk

Economic importance and economic risks

The country chapters of this book provide estimates of the economic values of the Mediterranean forests and of their contribution to the economies of the countries in this part of the world. These values and contributions vary across countries and, in general, they are not negligible. However, the problem is that most of these values correspond to non-market outputs and/or are not internalized by the forest producers in a suitable manner that provides them with the necessary financial means to maintain and develop the forestry activity. This situation seriously undermines the possibility of maintaining and developing the current forest cover in Mediterranean countries.

The fact that an important part of Mediterranean forest products are non-market goods and services does not necessarily mean that they do not have an economic significance. In fact, an important part of these goods and services provides intermediate consumption for other sectors that often use them free of charge, while the corresponding benefits are not internalized by forest producers. It is the case for grazing resources, water resources protected by forests, forest landscape quality consumed by tourist activities, and others. Therefore, some of the economic values of these forest outputs are included in the economic value of non-forest activities, such as livestock production or tourism.

Other economic factors put Mediterranean forests at risk, again because of the lack of internalization of some non-market benefits of forests. In some countries, especially in the southern part of the region, expansion of agricultural and pastoral land uses at the cost of forestland is an important factor. In these areas, the population is generally poor and no mechanism exists to internalize forest benefits, such as soil conservation and water protection. Consequently, people use land for what they need most in daily life, which is food. These negative effects on soil and water are fundamental issues, considering the following two facts: (i) in many Mediterranean areas, water resources are scarce; and (ii) the Mediterranean climate favours soil erosion. Forests are the main and sometimes the only land use that can cope with these problems.

In other cases where farming and forestry are coexisting in agroforestry systems, the harmony between the two kinds of land use is not always easy to maintain:

- Mechanization of agriculture may have negative effects on the density and health of trees.
- Increase in livestock production may also be detrimental to forestry.

- Decline in the economic value of some livestock products (e.g. the Iberian pig when affected by swine fever) using forest products as intermediate consumption (acorns, for example) may also lower the economic value of the trees (such as holm oak).

In other countries, especially in the coastal areas of the northern part of the region, forestland uses sometimes have to compete with urban and touristic uses that are not always respectful of sustainable forest management. In the areas where this tourist and urban pressure is lower, there are still problems for forestry, but of a different nature. These problems arise from the fact that, due to farm outmigration, agricultural land uses are declining, and forest and shrub lands are expanding, but often without appropriate management.

This urban pressure, and rural abandonment, combined with the characteristics of Mediterranean climates (wet winters and dry summers) contribute to what is nowadays the main threat to forestry in many Mediterranean areas, especially in the northern rim. This risk is the destruction by fire. Estimates of the costs of this phenomenon provided in the country chapters of this book show that there are countries (Portugal, for example) where every year they consume a substantial part of the benefits generated by forest production. For obvious reasons, these fires are a very important factor impeding private forest investment.

Ecological importance and ecological risks

Mediterranean forests, together with tropical forests, have the richest biodiversity in the world. At the same time, they are also among the most threatened by ecological risks: (i) climate favouring forest fires, drought and erosion; and (ii) climate changes.

This second factor is worthy of being stressed because the geographical location of Mediterranean areas puts them in a situation where they are among the regions in the world where the effects of climate changes are likely to be more intense.

Cultural importance and cultural risks

For historical reasons related to the geographical patterns of emergence and diffusion of human civilizations, Mediterranean forests are among those where the presence and effects of human actions are more important. Therefore, their characteristics are a result of that human intervention and those characteristics are also part of the cultural identity of Mediterranean regions. In spite of this cultural importance, probably because of the economic factors mentioned above, forests are very rarely part of the political agenda and social debate. When they are considered by politicians or by the (urban) population, it is not always for good reasons (e.g. forest fires). So what is found most of the time is ignorance, neglect or indifference.

Regional and Global Mediterranean Forest Problems

The previous section highlighted the following global public goods involved in the management of Mediterranean forests: (i) biodiversity; and (ii) cultural value.

There was also mention of the following public goods that may be considered to have regional scope, at least:

- Soil protection.
- Water protection.
- Landscape quality.
- Knowledge about the specificities of Mediterranean forest and agroforestry systems.

With these global and regional public goods provided by Mediterranean forests, there is already plenty of room for inter-regional cooperation. Other areas for this kind of cooperation can be added to that list. An important one – already mentioned in the first section – has to do with forest fires. This is a problem common to many Mediterranean countries. Therefore, there is need for an exchange of experiences about good practices in fire prevention and firefighting and for assistance to countries or regions in situations of emergency.

Another area for international cooperation is integrated rural development. As mentioned in the first section, the human pressure on existing or potential forestland in the southern Mediterranean countries is still strong in many places and the standards of living of the rural population are low. Without successful integrated rural development projects, the pressure on forests will continue to be strong, the living standards of the population will not rise and, in the long run, there will be risks of desertification and emigration to the urban areas of the country, or abroad, mainly to the northern Mediterranean rim and to other European countries.

Current Networks of Cooperation Among Mediterranean Countries

In spite of the need for international cooperation on Mediterranean forests, in the areas mentioned previously, there is no forum where they can, or have been, addressed in a specific, comprehensive and ongoing way.

Silva Mediterranea

Silva Mediterranea is the oldest network dealing specifically with Mediterranean forests. It is an international network of technical and scientific cooperation with an intergovernmental nature, since it is part of the Food and Agriculture Organization. It started its activities within the framework of this organization in 1948, more than 30 years after the idea was proposed by the French forester Robert Hickel, who succeeded in creating a network with the same name that functioned from 1922 until the end of the 1930s (Morandini, 1999). Since then, the activities of Silva Mediterranea have been very much dependent on the will of governments and the initiative of groups of researchers who want to push forward research and international cooperation on specific technical issues of Mediterranean silviculture. Socio-economic issues were on the agenda in the early days of Silva Mediterranea, but they lost ground later on. In more recent years, there have been active networks dealing with forest fires, cork and cork oak, stone pine and cedar.

The weak commitment of many countries to support the activities of this network reached a point where the extinction of Silva Mediterranea was considered as a possible option some years ago. This drastic decision was not taken, perhaps because this would have put an end to the only intergovernmental body specifically concerned with Mediterranean forests. However, the fact remains that, after that crisis, Silva Mediterranea was not reformed up to the point of getting a strong political commitment from the participating countries, translated into effective support for an international cooperation comprehensive enough for covering the main areas where this kind of action is most needed.

A positive characteristic of Silva Mediterranea is that it includes all countries in Europe, northern Africa and the Middle East that have Mediterranean forests, and not only the countries bordering the Mediterranean Sea.

Mediterranean Action Plan

In 1975, the United Nations Environment Programme (UNEP) and the European Commission adopted a convention supplemented by an action plan (Mediterranean Action Plan) whose main objective was the protection of the Mediterranean Sea and of its coastal environments.

The pillars of this initiative involving the 21 bordering countries were as follows:

- A set of legally binding agreements for the contracting parties, more precisely the Barcelona Convention, supplemented by specific protocols (dumping protocol, emergency protocol and new emergency protocol) related to marine pollution.
- A set of 'regional activity centres' located in some of the participating countries to carry out research and technical work on specific problem areas fitting within the general framework of the Barcelona Convention (monitoring and research of maritime pollution, prevention and emergency intervention for marine pollution accidents, prospective studies about sustainable development, planning of integrated coastal development, protection of the coastal environment and endangered marine species, protection of historical heritage and environmental remote sensing).

In the follow-up to the 1992 Rio United Nations Conference on Environment and Development (UNCED), this initiative was relaunched with the adoption of an Agenda 21 MED in Tunis in 1994, the revision of the Mediterranean Action Plan (MAP II) in Barcelona in 1995, and the creation of a Mediterranean Commission on Sustainable Development (MCSD) in Montpellier, in 1996. MCSD works as an advisory body of MAP, meeting every year and including representatives of the contracting parties and non-governmental organizations (NGOs) concerned with the issues covered by the MAP.

With this relaunching of MAP, the initial focus on marine pollution was expanded to the sustainable development of the coastal regions. This shift is important for the issues dealt with in this present study because it facilitates the inclusion of forest ecosystems in the work programme of MAP II. The regional activity centres of MAP, and especially the Plan Bleu, produced very interesting prospective work and organized useful symposia about Mediterranean forests and about the interactions between forests, soils, water resources, tourism, urbanization and population (Marchand, 1990; Lanquar, 1995; Boisvert et al., 1997; Ramade, 1997; Villevieille, 1997; Margat and Vallée, 2000; Attane and Courbage, 2001; Moriconi-Ebrard, 2001; de Montgolfier, 2002; Margat, 2002; de Franchis, 2003; Margat and Treyer, 2004).

Compared with Silva Mediterranea as far as forest issues are concerned, MAP loses in the specificity of its focus, but gains in putting those issues in the broader context of sustainable development. Keeping a predominantly technical nature, as did Silva Mediterranea, it relies on more transdisciplinary capacities and on more political commitment by the contracting parties. It also has the advantage of providing a forum for multiple stakeholder dialogue (MCSD). The problem for forest development is that it has not been raised to a sufficiently high profile in the agenda of MAP to lead to a strong work programme in this area. The main focus is still the protection of the Mediterranean Sea and of its coastal regions. Other Mediterranean regions, non-bordering countries and terrestrial Mediterranean ecosystems are not a priority.

Euro–Mediterranean Partnership

The Euro–Mediterranean Partnership was launched in 1995, with the Barcelona Declaration signed by all the 15 EU countries and by 12 non-EU Mediterranean countries. This declaration shows the political will of the EU to contribute to peace and stability in the region, not only by promoting cultural exchanges and political dialogue, especially in the Middle East, but also by contributing to building up the economic basis on which such peace and stability can be founded in a sustainable way. More precisely, this partnership has the following aims:

- Progressive establishment of a free trade area.
- Economic and financial cooperation and concerted action namely in industry, agriculture, transport, energy, telecommunications and information technology, regional planning, tourism, environment, science and technology, water and fisheries.
- Development of human resources and cultural exchanges.

It is interesting to note that even though forest ecosystems are not excluded from the Barcelona Declaration, they do not deserve a special mention among the areas for cooperation and concerted action.

There are four important points to note about this partnership. One is that it involves all the current and possibly future EU countries and is open to all the non-EU Mediterranean countries. Therefore, it is not limited to the countries bordering the Mediterranean Sea. Another point to note is that this partnership embraces the MAP, but goes beyond, not only in terms of scope, but also in terms of Contracting Parties, political commitment, and resources allocated to the work programme. This leads to the third important point to note about this initiative, which has to do with the fact that the EU has funds specially allocated to programmes of cooperation (MEDA, for example) and concerted action fitting in with the framework of this partnership. Finally, it is stated within the Barcelona Declaration that periodic ministerial conferences can be organized on specific areas, which has been the case in Foreign Affairs and Environment. Therefore, if there is enough political will and pressure from society for that, the door could be open for the organization of a series of Ministerial Conferences for the Protection of Mediterranean Forests.

Measures Required and the Nature and Role of an International Agreement on Mediterranean Forests

These Ministerial Conferences for the Protection of Mediterranean Forests (MCPMF) should take advantage of the experience of the Ministerial Conferences for the Protection of Forests in Europe and should work in close cooperation with this pan-European process. It could also take advantage of the scientific and technical basis provided by Silva Mediterranea and by the regional activity centres of MAP. It could also lead to relevant development projects using this knowledge basis and the financial instruments put forward by the EU for the Euro-Mediterranean Partnership.

The major step required to get this kind of initiative started is to build up sufficient awareness in society and among policy makers about the importance and risks concerning Mediterranean forests. The research community and the NGOs concerned with these issues have an important civic role to play here, and some are actively working in this direction. Good examples are the MEDFOREX Regional Centre of the European Forest Institute and the International Association of Mediterranean Forests (Bonnier and Poulet, 2002).

The MCPMF cannot and should not be based on some kind of legally binding agreement (Glück *et al.*, 1997). The socio-economic and political conditions relevant for forest development vary greatly from country to country, around the Mediterranean basin. Imposing common rules on everybody would be bound to result in failure and is unnecessary. Much more important would be the setting up of a regular forum of effective political commitment to develop and share knowledge, as well as technical and financial resources, and to debate, design and implement concerted policies in those areas more concerned by the regional and global public goods provided by the Mediterranean forests.

References

Attane, I. and Courbage Y. (2001) *La Démographie en Méditerranée. Situations et Projections*. Les Fascicules du Plan Bleu No. 11. Economica, Paris.

Boisvert, V., de Montgolfier, J., Vallée, D. and Glass, B. (1997) *Vers des Indicateurs de Suivi des Espaces Boisés en Méditerranée*. Plan Bleu-Centre d'Activités Régionales, Sophia-Antipolis, France.

Bonnier, J. and Poulet, D. (eds) (2002) Problématique de la forêt méditerranéenne. *Forêt Méditerranéenne* Hors Série No. 1.

de Franchis, L. (2003) *Threats to Soils in Mediterranean Countries. Document Review*. Plan Bleu d'Activités Régionales, Sophia-Antipolis, France.

de Montgolfier, J. (2002) *Les Espaces Boisés Méditerranéens. Situation et Perspectives*. Les Fascicules du Plan Bleu No. 12. Economica, Paris.

Glück, P., Tarasofsky, Byron, N. and Tikkanen, I. (1997) *Options for Strengthening the International Legal Regime for Forests*. European Forest Institute, Joensuu, Finland.

Lanquar, R. (1995) *Tourisme et Environnement en Méditerranée. Enjeux et Prospective*. Les Fascicules du Plan Bleu No. 8. Economica, Paris.

Marchand, H. (1990) *Les Forêts Méditerranéennes. Enjeux et Perspectives*. Les Fascicules du Plan Bleu No. 2. Economica, Paris.

Margat, J. (2002) *Are Water Shortages a Long-range Outlook in Mediterranean Europe?* Plan Bleu-Centre d'Activités Régionales, Sophia-Antipolis, France.

Margat, J. and Treyer, S. (2004) *L'Eau des Méditerranéens: Situation et Perspectives*. Plan Bleu-Centre d'Activités Régionales, Sophia-Antipolis, France.

Margat, J. and Vallée, D. (2000) *Mediterranean Vision on Water, Population and the Environment for the 21st Century*. Plan Bleu-Centre d'Activités Régionales, Sophia-Antipolis, France.

Morandini, R. (1999) Silva Mediterranea – 50 ans de Cooperation dans le Domaine Forestier Méditerranéen. *Unasylva* No. 197.

Moriconi-Ebrard, F. (2001) *Urbanisation in the Mediterranean Region from 1950 to 1995*. Plan Bleu-Centre d'Activités Régionales, Sophia-Antipolis, France.

Ramade, F. (1997) *Conservation des Ecosystèmes Méditerranéens. Enjeux et Perspectives*. Les Fascicules du Plan Bleu No. 3. Economica, Paris.

Villevieille, A. (1997) *Les Risques Naturels and Méditerranée. Situation et Perspectives*. Les Fascicules du Plan Bleu No. 10. Economica, Paris.

Index

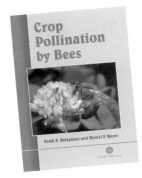
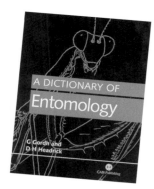
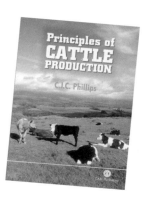